Deep Space Communications

DEEP SPACE COMMUNICATIONS AND NAVIGATION SERIES

Issued by the Deep Space Communications and Navigation Systems
Center of Excellence, Jet Propulsion Laboratory
California Institute of Technology

Joseph H. Yuen, Editor-in-Chief

Published Titles in this Series

Radiometric Tracking Techniques for Deep-Space Navigation
Catherine L. Thornton and James S. Border

Formulation for Observed and Computed Values of Deep Space Network Data Types for Navigation
Theodore D. Moyer

Bandwidth-Efficient Digital Modulation with Application to Deep-Space Communications
Marvin K. Simon

Large Antennas of the Deep Space Network
William A. Imbriale

Antenna Arraying Techniques in the Deep Space Network
David H. Rogstad, Alexander Mileant, and Timothy T. Pham

Radio Occultations Using Earth Satellites: A Wave Theory Treatment
William G. Melbourne

Deep Space Optical Communications
Hamid Hemmati, Editor

Spaceborne Antennas for Planetary Exploration
William A. Imbriale, Editor

Autonomous Software-Defined Radio Receivers for Deep Space Applications
Jon Hamkins and Marvin K. Simon, Editors

Low-Noise Systems in the Deep Space Network
Macgregor S. Reid, Editor

Coupled-Oscillator Based Active-Array Antennas
Ronald J. Pogorzelski and Apostolos Georgiadis

Low-Energy Lunar Trajectory Design
Jeffrey S. Parker and Rodney L. Anderson

Deep Space Communications
Jim Taylor, Editor

Deep Space Communications

Edited by Jim Taylor

Jet Propulsion Laboratory
California Institute of Technology

DEEP SPACE COMMUNICATIONS AND NAVIGATION SERIES

WILEY

Published by John Wiley & Sons, Inc., Hoboken, New Jersey
Published simultaneously in Canada

Library of Congress Cataloging-in-Publication Data:

Deep space communications / Jim Taylor, editor.
1 online resource.
Includes bibliographical references and index.
Description based on print version record and CIP data provided by publisher; resource not viewed.
ISBN 978-1-119-16905-5 (Adobe PDF) – ISBN 978-1-119-16906-2 (ePub) – ISBN 978-1-119-16902-4 (cloth)
1. Deep Space Network. 2. Astronautics–Communication systems.
3. Space probes. I. Taylor, Jim, 1939– editor.
TL3026
621.382′38–dc23

2015033183

Set in 11/12.5pt Times New Roman by SPi Global, Pondicherry, India

Printed in the United States of America

10 9 8 7 6 5 4 3 2 1

Deep Space Communications

May 2016

The research described in this publication was carried out at the Jet Propulsion Laboratory, California Institute of Technology, under a contract with the National Aeronautics and Space Administration.

Contents

Foreword

The Deep Space Communications and Navigation Systems Center of Excellence (DESCANSO) was established in 1998 by the National Aeronautics and Space Administration (NASA) at the California Institute of Technology's Jet Propulsion Laboratory (JPL). DESCANSO is chartered to harness and promote excellence and innovation to meet the communications and navigation needs of future deep-space exploration.

DESCANSO's vision is to achieve continuous communications and precise navigation—any time, anywhere. In support of that vision, DESCANSO aims to seek out and advocate new concepts, systems, and technologies; foster key technical talents; and sponsor seminars, workshops, and symposia to facilitate interaction and idea exchange.

The Deep Space Communications and Navigation Series, authored by scientists and engineers with many years of experience in their respective fields, lays a foundation for innovation by communicating state-of-the-art knowledge in key technologies. The series also captures fundamental principles and practices developed during decades of deep-space exploration at JPL. In addition, it celebrates successes and imparts lessons learned. Finally, the series will serve to guide a new generation of scientists and engineers.

Joseph H. Yuen
DESCANSO Leader

Preface

This book presents the planetary communications design as developed by the Deep Space Network (DSN) and Jet Propulsion Laboratory (JPL) flight projects. It uses a case study approach and shows the communications link performance resulting from the design. This is accomplished through a description of the design and performance of six representative planetary missions. These six cases illustrate progression through time of the DSN and onboard hardware and software techniques, capabilities, and performance from 1970s technology to the most recent missions.

The first chapter presents an overview of deep space communication capability over the last five decades. It also describes the design process for these links. The second chapter gives an overview and the current capabilities of the Deep Space Network (DSN).

In Chapters 3 through 8, from Voyager in the 1970s to the Mars Science Laboratory in the 2010s, the six missions represent all those that have communicated directly with the Deep Space Network. Two of six also communicated from the surface of Mars to orbiting spacecraft that in turn communicated with the Earth.

The Voyager mission was intended as a flyby mission to Jupiter and Saturn, and its X-band communications system was sized for science return from those planets. With improvements in the Deep Space Network, the two Voyager spacecraft continue to return data from beyond the distances of additional flybys at Neptune and Uranus.

The Galileo mission was intended to transmit the majority of its science data from orbit around Jupiter via a high-gain antenna at X-band. When that antenna failed to unfurl properly early in the cruise to Jupiter, JPL mounted a major effort to change the spacecraft's software and the ground stations' telemetry system to achieve the return of a high amount of science information via a low-gain antenna at S-band.

Deep Space 1 was intended as an in-flight test bed of new capabilities, with an incidental science mission. This spacecraft carried the first Small Deep Space Transponder (SDST) that has been the mainstay of many missions communicating with the DSN since it worked in Deep Space 1. It also carried a solid state Ka-band amplifier to provide the means to test deep space communications at this frequency. Notable also was the development—after a failure in the spacecraft's attitude control system—of an operational workaround using the spacecraft signal and the station to achieve control of the spacecraft antenna pointing during data return.

The Mars Reconnaissance Orbiter, besides having an ambitious science mission that continues today, was intended to serve as a communications relay terminal for surface landers and rovers. For direct communications with Earth, in addition to using the SDST, it has a 100-watt traveling wave tube amplifier at X-band and a large fixed high-gain antenna. For UHF relay communications with the surface, it uses an Electra transceiver and the Proximity-1 communications protocol.

The two Mars Exploration Rovers, Spirit and Opportunity, for surface operations were each equipped with the SDST and two redundant solid state X-band amplifiers for direct communications and UHF transceivers for relay operations with Mars orbiters. Intended for a 90-day mission after landing on Mars, both rovers outdid themselves, and Opportunity is still continuing its exploration.

The Mars Science Laboratory reflects the maturity of the communications systems first operated on Deep Space 1, Mars Reconnaissance Orbiter, and the Mars Exploration Rovers. After landing on Mars in August 2012, this large rover has been communicating via X-band equipment similar to the Mars Exploration Rover, and a version of the Electra transceiver in the rover as well as in the Mars Reconnaissance Orbiter at the other end of the relay link.

Jim Taylor
Pasadena, California
January 2016

Acknowledgments

This book presents results of research carried out at the Jet Propulsion Laboratory, California Institute of Technology, under a contract with the National Aeronautics and Space Administration.

As editor and authors, we acknowledge the constant advice, help, comments, and encouragement from Joseph H. Yuen, the editor-in-chief of the Deep Space Communications and Navigations Systems (DESCANSO) series. David J. Bell's thorough review resulted in much improved clarity and completeness of the spacecraft chapters. In addition, they acknowledge the always helpful inputs and advice of Roger V. Carlson, the JPL editor.

Chapter 1, *Telecommunications Link Analysis*, lays out the principles and the statistical mathematical models for the design control table, or "link budget" in much the original form created by its author Joseph H. Yuen in the 1970s. We have retained this form in sections of Chapter 1 basically unchanged from the version as written by Dr. Yuen in the 1982 edition of *Deep Space Telecommunications Systems Engineering*.

Chapter 2, *the Deep Space Network*, draws extensively from the Deep Space Network (DSN) Telecommunications Link Design Handbook (810-005) as do sections of the other chapters. As the DSN evolved, many people have created and updated the material in the modules of this handbook. For this chapter we are most indebted to Robert Sniffin, the principal author and editor of the handbook for many decades. Some of the modules in 810-005 carry the names of the principal authors of those modules. In addition, numerous individuals, far too many to name, have contributed to the technology that is documented in their papers in the *Interplanetary Network (IPN) Progress Reports*. This technology has gone into the stations and the control centers, the hardware and software development, the thorough testing involved for each new system or software update, and the continued operation of the network in support of the missions represented in the following chapters.

Each of the remaining chapters describes the telecommunications system aboard a particular spacecraft. Each such subsystem operates in one to three frequency bands and consists generally of multiple antennas and their routing elements, one or more transponders or individual receivers and transmitters, and the command-receiving and telemetry-generating elements. Each subsystem involved the efforts and time of dozens of individuals or teams that developed the technologies, the functional design, the hardware, and the software. Many others performed the integration of the subsystem into the whole spacecraft and the testing before and after integration. The references in each chapter are those project documents, papers, conference proceedings, and websites that the coauthors and editor of this book drew on directly.

In Chapter 3, *Voyager*, much of the telecom design information was obtained from original Voyager prime mission design documentation: the design control document for the telecommunications links, the functional description of the telecommunications system, and the hardware design requirement for the modulation demodulation subsystem. Much of the mission and operational information was obtained from the Voyager Operational Handbook and the Voyager Neptune Travel Guide. The authors are grateful to Dave Bell, Kar-Ming Cheung, and Ed Massey for their Voyager background and development information.

For Chapter 4, *Galileo*, the authors express their appreciation to many individuals in the Interplanetary Network Directorate and the Telecommunications Science and Engineering Division (33) at JPL and other organizations who contributed directly to the success of the Galileo mission. In addition to the papers cited in the references, Richard Brace, Vic Albrecht, Rick Nybakken, and Gordon Wood, as the spacecraft telecom hardware cognizant engineers, oversaw the development and testing of the transponder, the modulation-demodulation subsystem, the antennas, and the ultra-stable oscillator. Prof. Wai-Kuen Cham of the Chinese University of Hong Kong invented the integer cosine transform (ICT) and helped to refine the ICT algorithm that was used as a form of lossy compression mostly for image data after the high-gain antenna (HGA) failed to deploy. The compression reduced the data volume to fit through the smaller bandwidth provided by the low-gain antenna link to Earth. Dr. Andrew Watson and Sherry Chuang at NASA Ames designed the ICT quantization matrix and performed subject tests with Galileo scientists to determine acceptable fidelity with ICT lossy compression. The authors are especially grateful to Eilene Theilig for her project systems perspective and to Steve Townes for his organizational recommendations that carried over to subsequent chapters.

For Chapter 5, *Deep Space 1*, the on-board telecom hardware development and testing was overseen by cognizant engineer Sam Valas and the antennas by Joe Vacchione. DS1 carried the first Small Deep Space Transponder, developed by Motorola with Keith Siemsen as the technical lead. Marc Rayman, as DS1 system engineer, added much to the clarity of the DESCANSO article the chapter is based on, as well as the success of the mission. The DS1 software team led by Dan Eldred helped the AutoNav team as it ventured for the first time into the realm of deep-space flight software. Finally, and perhaps most importantly, thanks go to the attitude control system (ACS) team, led by Sima Lisman, and including Tony Vanelli and Steve Collins. Besides their function of keeping the spacecraft pointed correctly, ACS was key to working with Jim Taylor of telecom to develop the HGA pointing workaround after the in-flight failure of the stellar reference unit.

For Chapter 6, *Mars Reconnaissance Orbiter (MRO)*, as well as for the two Mars rover missions (Mars Exploration Rover [MER] and Mars Science Laboratory [MSL]) that conclude the book, Chad Edwards was the Mars Program office sponsor, and Edgar Satorius was the relay link signal processing lead. Key to the technical and managerial success of the Electra relay radio program development was Tom Jedrey. Eric Schwartzbaum was the Electra program manager, Ann Devereaux was responsible for the baseband processor, Ken Peters was responsible for software, and Todd Ely was responsible for radiometrics. The Electra first flew in the MRO and first operated on the surface in MSL. As MRO telecom lead, Stan Butman provided several suggestions and graphics for the chapter. Ricardo Mendoza conceived and developed the data volume capability file that is fundamental for science data return planning for all the Mars relay missions and inter-project coordination. The authors are grateful to David Bell, Tom Jedrey, and Ramona Tung for the information they contributed to the descriptions of the Electra transceiver and its use in relaying information with landers on the surface. We thank Charles Lee for the surface communications opportunities simulation, James Border for the information on delta differential one-way ranging (delta-DOR), and David Morabito for the discussion of solar conjunction effects on communications and the experiment plans during the MRO solar conjunctions.

For Chapter 7, *Mars Exploration Rover*, the authors express their appreciation to Brian Cook and Peter Ilott for the wealth of information on the heritage, performance, and testing of the X-band and UHF subsystems, respectively. Thanks also to Bill Adams, the Odyssey and MGS Flight Team Telecom lead at Lockheed Martin Aerospace. We also, thank Jan Ludwinski, whose excellent mission plan became an integral part of this chapter. The authors are grateful to Monika Danos, for scripts that made summaries of data from years of MER prime and extended mission flight operations consistent. Ramona Tung and Ricardo Mendoza were key to developing seamless relay link prediction capabilities involving the geometric intricacies of a roving surface vehicle working with several relay orbiters. Finally, the authors are indebted to Ed Satorius, Sue Finley, Christine Chang, Doug Johnston, Dave Fort, and Sami Asmar for their contributions to the analysis, development, testing, and operation of the Entry, Descent, and Landing (EDL) Data Analysis (EDA) system that enabled the return of the intricate series of signals with their rapidly varying signal levels and frequencies.

For Chapter 8, *Mars Science Laboratory*, the authors appreciate the access to information and documents provided by members of the MSL spacecraft development, test and flight operations teams. Many of those acknowledged for MRO and MER also participated in MSL. Peter Ilott was the telecom cognizant engineer and provided "better art" for many of the graphics in this chapter. Melissa Soriano, Sue Finley, and Polly Estabrook developed the EDA configuration to receive the X-band signal during EDL. Melissa Soriano, Sue Finley, Kamal Oudrhiri, and Daniel Kahan tested and operated the EDA and Radio Science Receiver during EDL. Mazen Shihabi and David Bell adapted telemetry analysis tools, first developed by Brad Arnold and Tom Jedrey, to rapidly process analyses of relay link performance the first month post-landing. This led to the successful "tuning" of the MRO Electra radio to overcome electromagnetic interference from MRO science instruments that degraded the MSL relay link performance. The new adaptive data rate mode, first used on the MSL/MRO return link was also tuned to maximize data return volume. Much of the information on spacecraft configuration, the science payload, and subsystems other than telecommunications came from the excellent project review information in the project's DocuShare library, maintained by Marie-Ann Carroll. We made particular use of material from the Mission Plan, as prepared for the 2009 mission by Bobak Ferdowsi and John Gilbert and as extensively updated for the 2011 mission by Charles Baker. Brian Schratz, the lead of the Entry, Descent, and Landing (EDL) mission phase provided the description of the data analysis of X-band and UHF signals received during the "seven seconds of terror" culminating in touchdown on the surface.

January 2016

Contributors

Ana I. Bolea-Alamañac received her undergraduate degree in telecommunications engineering jointly from the Centro Politécnico Superior (Zaragoza, Spain) and the Ecole Centrale de Lille (Lille, France) in 1995, her master's in space studies from the International Space University (Strasbourg, France) in 2001, and her Ph.D. from École nationale supérieure de l'aéronautique et de l'espace (SUPAERO), in Toulouse, France in 2004, with her work focused on the implementation of fade mitigation techniques applied to advanced satellite communication systems.

From 1996 to 2000, Dr. Bolea-Alamañac worked as an engineer and manager engineer at Telefónica de España, where she worked in the international network planning and operation. In 2001, to fulfill a master's degree requirement, she did an internship in deep-space communications at the Jet Propulsion Laboratory. Her work focused on the Deep Space 1 mission, especially the Beacon Monitor Operations Experiment (BMOX) and its implications for a Deep Space Network operational beacon monitoring system. After her Ph.D., she continued her work in the area of advanced digital communication techniques with the Research and Development Department of Thales Alenia Space in Toulouse, France. In 2005, she joined the European Space Agency (ESA) Research and Technology Centre (ESTEC), Noordwijk, The Netherlands, as a communication system engineer.

Andrea Barbieri received a degree ("laurea") in Electrical Engineering in 1995 from the University of Padova, Italy.

Mr. Barbieri joined JPL in 1997. He served as lead system engineer for the UHF subsystem on the Mars Odyssey 2001 orbiter and the 2003 Mars Exploration rovers, and was the telecom system engineer for various deep space science projects. As a senior engineer in the JPL Communications Systems and Operations Group of the Flight Communications Systems Section, he became the telecommunications system engineer for the Mars Science Laboratory project in 2004.

Mr. Barbieri has been with Airbus Defence and Space in the United Kingdom since 2007. He was team leader for the payload validation of the first four Galileo navigation satellites, and he is currently test manager for telecommunications payloads.

Kar-Ming Cheung received his B.S. EE from the University of Michigan, Ann Arbor in 1984; and his M.S. and Ph.D. from California Institute of Technology (Pasadena) in 1985 and 1987, respectively.

In 1987, he joined the Communications Systems and Research Section at the Jet Propulsion Laboratory (JPL). In the earlier part of his career at JPL, he worked on the research and development of error-correction coding, data and image compression, synchronization, and system analysis of the NASA's Deep Space Network (DSN) communications system.

In the mid-1990s, when the Galileo spacecraft lost the use of its high-gain antenna, Dr. Cheung implemented a low-complexity data compression software scheme that increased Galileo's data return by a factor of 10. He also led development of the error-correction coding scheme, which was part of the ground-enhancement effort to further increase Galileo's data return. He received the NASA's Exceptional Achievement Medal for his contribution to the Galileo Project.

Dr. Cheung is currently a principal engineer and technical group supervisor of the Communications Systems and Research Section's Telecom Architecture Group. He continues to support the multi-mission telecom link analysis tools development and to lead a multi-center team on architecture and system engineering trade studies to support the Space Communications and Navigation (SCaN) Office of NASA.

Polly Estabrook received a B.S. in engineering physics and a B.A. in economics in 1975 from the University of California, Berkeley. She received her M.S. and Ph.D. in electrical engineering in 1981 and 1989, respectively, from Stanford University (Palo Alto, California).

Dr. Estabrook is Deputy Section Manager of the Communications Architectures and Research section. Her research interests lie in the areas of signal detection during critical events, deep-space telecom system design, and the application of new communication technologies to space exploration.

Dr. Estabrook joined JPL in 1987 as a member of the technical staff, working on satellite communications. She was the lead telecommunications system engineer for the Mars Exploration Rover (MER) project, responsible for the performance of the entry, descent, and landing telecommunications system and for the overall design and performance of the direct-to-Earth and relay communications system. After the landing of the MER rovers, she worked on the design of the communication system for NASA's Vision for Space Exploration Program supporting human exploration of the Moon and Mars. From 1992 to 2004, she was supervisor of the Advanced Communications Systems Concepts Group in the JPL Communication Systems and

Research Section. Her group provided the telecommunication system engineers for Cassini, Mars Pathfinder, Deep Space 1, Deep Space 2, Space Technology 3, ST-4, 2001 Mars Odyssey, Mars Sample Return, and MER.

Peter Ilott received his B.Sc. in physics and Ph.D. in electrical engineering at McGill University (Quebec) in 1980 and 1988, and he received his MSc in plasma physics at the Université de Montréal in 1982.

Dr. Ilott came to JPL in 2000. He is the telecommunications lead at JPL for Mars Science Laboratory (launched in 2011) which includes the mission design for entry, descent, and landing (EDL). For the Phoenix Mars Lander, he was a telecom design consultant and participated in EDL operations. For the MER mission, he was the UHF relay engineer and worked EDL and surface operations.

At JPL, Dr. Ilott also worked on Odyssey, Mars Reconnaissance Orbiter (MRO), Deep Impact, and CloudSat. Prior to JPL he worked for 12 years at SPAR Aerospace (now part of MDA) and Hughes Space and Communications (now Boeing Satellite Development Center) on telecommunications systems for commercial mobile communications spacecraft such as MSAT, ICO, and Thuraya, and on remote sensing spacecraft such as RadarSat.

Dennis K. Lee earned his B.S. from Case Western Reserve University (Cleveland, Ohio) in 1997 and his MS from Rensselaer Polytechnic Institute (Troy, New York) in 1998, both in electrical engineering.

He joined JPL in 1999 as a member of technical staff in the Digital Signal Processing Research group. He has worked on telecommunications systems development for several JPL missions including the Gravity Recovery and Climate Experiment (GRACE), Dawn, Mars Reconnaissance Orbiter (MRO), Kepler, Phoenix, and the Soil Moisture Active Passive mission (SMAP). He is currently serving as the NASA Rapporteur for the Consultative Committee for Space Data Systems (CCSDS) Radio Frequency and Modulation Working Group, and is also the principal investigator on a research and development task to develop a wideband integrated-array combiner and telemetry receiver for the Deep Space Network. His research interests include bandwidth-efficient and multi-carrier modulations, array signal processing, and ultra-high rate communication systems. Mr. Lee is currently a senior engineer in the Communications Architectures and Research section.

Roger Ludwig received his B.S. in management from Oakland University (Rochester, Michigan) in 1981 and his MBA in finance from California State Polytechnic University, Pomona in 1996. He began his aerospace career in 1981 at Altair Radar, Kwajalein Missile Range, the Marshall Islands where, as operations director, he conducted radar "skin" tracks of Earth-orbiting objects for the North American Aerospace Defense Command Space Detection and Tracking System.

In 1985, Mr. Ludwig joined the Jet Propulsion Laboratory as a spacecraft mission controller on the Active Magnetospheric Particle Tracer Explorers

(Earth-orbiter) mission. He first performed spacecraft telecommunications analysis supporting Voyager Neptune encounter in 1989, followed by Magellan (Venus), Galileo (Jupiter), Mars Observer, the Defense Support Program, and Cassini (Saturn). He returned to Voyager in 2000, assuming system-level responsibilities, and in 2004, he added command sequence integration engineer to his portfolio of Voyager specialties.

Andre Makovsky received his B.S. (1984) and M.S. (1985) in electrical engineering from the Rensselaer Polytechnic Institute (Troy, New York). He joined JPL in January 1986 and has worked in telecommunications system development on Galileo, Cassini, Deep Space 1 (DS1), 2001 Mars Odyssey, Mars Exploration Rover (MER), Deep Impact, and Mars Science Laboratory (MSL). For each project, he developed the telecommunications link performance tools used during preproject, prelaunch, and in-flight. Prior to his retirement in 2014, he was a senior engineer in the Communications Systems and Operations Group of the Flight Communications Systems Section.

Mr. Makovsky did extensive prelaunch system performance tests in the JPL Telecommunications Development Laboratory and with the Deep Space Network's Compatibility Test facilities at JPL, in a mobile Compatibility Test Trailer at spacecraft contractor sites, and at the Eastern Test Range at Cape Kennedy, Florida. He participated in extensive testing of Small Deep Space Transponders both at the contractor (General Dynamics, formerly Motorola) in Phoenix, Arizona and at JPL for DS1, Mars Odyssey, Space Infrared Telescope Facility, MER, and MSL.

Michela Muñoz Fernández received her M.S. and Ph.D. degrees in electrical engineering from the California Institute of Technology (Pasadena, California) in 2001 and 2005, respectively. She previously earned an M.S. in Space Studies from the International Space University (France), an M.S.E.E., and two B.Sc. degrees summa cum laude in electrical engineering and telecommunications engineering in Madrid (Spain). Her Ph.D. thesis was based on a coherent-optical-array receiver for pulse-position modulated signals under atmospheric turbulence for deep-space communications.

Dr. Muñoz Fernández has been working on flight, payload, instrument operations, telecommunications, and ground systems engineering for NASA and ESA missions (Orion, Juno, SIM, Deep Space 1, ROSETTA) since 1998. She is the principal investigator of the Juno mission-modeling task for NASA HQ, a task manager in the information architecture standards area at JPL, and also works on DSN related research applied to link complexity and RF/optical communications.

She is a Society of Photo-Optical Instrumentation Engineers (SPIE) Free Space Laser Communications Program Committee member, and she has been a chair of sessions on Free-Space Link Performance. She was selected in 2008 to the first rounds of the European Space Agency's astronaut program. She won two Amelia Earhart Awards in 2002 and 2004 respectively, plus other ESA and NASA awards.

Dongae Seo received a B.S. degree in physics from Chung-Nam National University, Deajun, Korea, in 1991 and a master's in space studies from the International Space University (Strasbourg, France) in 2002.

In 1992, she joined Korea Telecom as an engineer and worked on operations with fixed telephone lines including fiber cable. She transferred to the Satellite Business Center in 1994 and worked on interference analysis and technical coordination between satellite networks, designing communication satellite networks (including meteorological satellites) and the management of satellite and ground control systems (Koreasat). In 2002, for the master's degree requirement, she completed an internship at the Communications Systems and Research Section's Flight Communications Systems Group in the Jet Propulsion Laboratory. The internship, during which she co-authored the DESCANSO article Galileo Telecommunications, was mainly about the Galileo mission and Galileo telecommunication link analysis. She also studied the propagation effects of a cometary environment on deep space communication, with application to the Deep Space 1 mission.

Shervin Shambayati received his B.S. in applied mathematics and engineering from California State University, Northridge, in 1989, and his engineering degree, M.S.E.E., and Ph.D. in electrical engineering from the University of California, Los Angeles, in 1991, 1993, and 2002, respectively. Dr. Shambayati joined the JPL Deep Space Telecommunications Group in 1993. With that group, he participated in the Deep Space Network Galileo Telemetry Receiver (DGT) and the 34-m Arraying Task.

In 1997, Dr. Shambayati transferred to the Information Processing Group at JPL. With that group, he participated in the Mars Global Surveyor Ka-band Link Experiment II (MGS/KaBLE II) and Deep Space 1 in-flight Ka-band tests. Dr. Shambayati was also the principal investigator for the Mars Reconnaissance Orbiter Ka-Band Demonstration. Later on, Dr. Shambayati became a member of Telecommunications Architecture Group at JPL, where his research areas included evaluation of weather effects on the end-to-end performance of the deep-space link; end-to-end link design and spacecraft downlink operations over Ka-band; and the end-to-end performance of optical links.

In 2012, Dr. Shambayati joined SSL in Palo Alto, California. Currently he is a senior systems engineering specialist at SSL with the Digital Communications Engineering Section. His current responsibilities include design and analysis of telemetry, command, and ranging systems for commercial and non-commercial spacecraft as well as design and analysis of communications systems for hosted payloads onboard commercial satellites.

Jim Taylor received a B.S., magna cum laude, in 1961 and an M.S. in 1962, both in electrical engineering from Stanford University (Palo Alto, California). His career has been in spacecraft communications. At JPL since 1970, he has been centrally involved in planning and assessing communications links between interplanetary spacecraft and the Deep Space Network (DSN).

He has pursued a special interest in the effects of solar interference on radio communications between interplanetary spacecraft and their supporting ground stations, starting with Mars Mariner in 1971, and most recently the Mars Science Laboratory (MSL). In 2003 for the Mars Exploration Rover project, Mr. Taylor brought the concept of communications windows to maturity. (Now also on MSL, communications windows are intervals in which the flight receiver, transmitter, and antenna are configured by the flight software from on-board tables rather than by sequenced commands.)

Mr. Taylor is a principal engineer at JPL, working on telecommunications analysis, ground-system implementation, and flight operations for deep-space and Earth-orbiting projects. He was the founding telecommunications member of JPL's Spaceflight Significant Events Group, now called Lessons Learned. He received the NASA Exceptional Achievement Medal in 2000 for innovative use of the DS1 communications systems and the NASA Exceptional Service Medal in 2006 for operational development and support on Deep Impact.

A. Gail Thomas received a B.A. from the University of California, Los Angeles (California) in 1981 with a major in music. She began her career at JPL in 1984 in the Parts Reliability Section but soon moved to the Spacecraft Telecommunications Equipment Section, now the Flight Communications Systems Section.

She worked on ground support equipment (GSE) hardware and software for the Cassini telemetry control unit. She more recently worked on the GSE software for the Mars Exploration Rover (MER) and the Mars Science Laboratory (MSL) telecommunications subsystems.

Ms. Thomas began working in MER flight operations as a "data miner" during the MER prime mission, then transitioned to become a telecom analyst during the first MER extended mission in 2004, and most recently became a telecom analyst for Cassini. She has also served as the telecom lead in subsequent MER extended missions through 2007.

Ramona Tung received her B.S. (1992) and M.S. (1994) in electrical engineering from the Massachusetts Institute of Technology (MIT, Cambridge). At JPL in the Communications Ground Systems Section, she did simulations and analysis to support the development of the Block V Receiver and the Block III Maximum-Likelihood Convolutional Decoder (B3MCD), both of which are standard equipment at the three Deep Space Network sites. In the Communications Systems and Research Section, she did high-level telecommunications systems analysis and led the development of the Telecom Forecaster Predictor (TFP) multi-mission telecom analysis tool. TFP has supported many JPL deep-space projects, including Mars Reconnaissance Orbiter (MRO), MER, MSL, and Deep Space 1.

On the MER telecom operations team, Ms. Tung was one of the two telecom analysts providing the majority of telecom support for Spirit during the prime mission, including the resolution of the Sol 18 anomaly. She has worked closely with telecom system engineers at JPL and Lockheed Martin. On MRO, she was the Mission Operations System/Ground Data System lead for Electra, the next-generation UHF proximity radio for communicating between MRO and the MER and MSL rovers. Now in the Mission Systems and Operations section, she is the Flight Operations Team lead for the Soil Moisture Active Passive (SMAP) Earth-orbiter launched in 2015.

Joseph H. Yuen received his Ph.D. in electrical engineering from the University of Southern California (Los Angeles). He is the Chief Technologist of the Communications, Tracking, and Radar Division, and the Leader of the Deep Space Communications and Navigation Systems Center of Excellence, Jet Propulsion Laboratory (JPL), California Institute of Technology, Pasadena, California. He is the Series Editor of two active book series: JPL Deep Space Communications and Navigation, and JPL Space Science and Technology; both by John Wiley and Sons, Hoboken, New Jersey. He published a book, *Deep Space Telecommunications Systems Engineering* (ed.), by NASA, Washington, D.C., 1982, and later commercially by Plenum, New York, 1983.

Chapter 1
Deep Space Communications: An Introduction

Joseph H. Yuen

1.1 Introduction and Overview

Communications are required and critical to the success of space missions. From the moment of launch, the only connection between a spacecraft and the Earth is the communications system. This system enables return of data from spacecraft to Earth, the tracking of the spacecraft, and commanding the spacecraft to perform any actions that it cannot perform automatically.

Since the beginning, with Sputnik in 1957 and Explorer in 1958, space missions have gone farther and have become more and more demanding in data return to enable far more ambitious science goals. This is particularly so for probes in deep space—at the Moon and farther. In the 1960s and 1970s, the missions were planet flybys, which typically have short encounter periods. Then the missions progressed in the 1980s and 1990s to plant orbiters, which have long and sustained scientific observations—often years of continuous operation. In the 2000s, missions involved landing rovers that moved around on the surface of planets to engage in scientific investigations. In 2012, the latest of these, Mars Science Laboratory (MSL) rover was landed on Mars for years of continuous active scientific investigations.

Deep Space Communications, First Edition. Edited by Jim Taylor.

To overcome the enormous communication distance and the limited spacecraft mass and power available in space, the Jet Propulsion Laboratory's (JPL's) deep space communications technologies developed for spacecraft of the National Aeronautics and Space Administration (NASA) and NASA's Deep Space Network (DSN)[1] have enabled every JPL deep space mission ever flown, and contributed to the development of exciting new mission concepts. Figure 1-1 summarizes the evolution of deep space communications capabilities and performance of our spacecraft since the first NASA spacecraft in 1958, and it projects to the future capabilities. One can see the tremendous improvements over the years. To continue meeting the increasing demand on deep space communications systems, JPL will need to increase its capability by a factor of ten during each of the coming decades.

This book is a collection of some JPL space missions selected to represent typical designs for various types of missions; namely, Voyager for fly-bys in the 1970s, Galileo for orbiters in the 1980s, Deep Space 1 for the 1990s, Mars Reconnaissance Orbiter (MRO) for planetary orbiters, Mars Exploration Rover (MER) for planetary rovers in the 2000s, and the MSL rover in the 2010s. The cases we have selected were chosen from the JPL Design and Performance Summary series, issued by the Deep Space Communications and Navigation Systems Center of Excellence (DESCANSO) [1]. The case studies of this book illustrate the progression of system design and performance from mission to mission; as stated in the foreword of the series by DESCANSO leader Joseph H. Yuen when the series was launched. The case studies provide the reader with a broad overview of the missions systems described. Besides the systems designs, the case studies provide actual flight mission performance details of each system.

We have provided only the necessary editing to fit within the book and some updates as missions have progressed. As much as possible, we have preserved the original authors' content largely unchanged.

---

[1] NASA missions in low Earth orbit communicate through either the Near Earth Network (NEN) or the SN (space network), both operated by the NASA Goddard Space Flight Center (GSFC). The SN has of a number of Tracking and Data Relay Satellites (TDRS) in geosynchronous orbits. In addition, the European Space Agency (ESA) operates a number of ground stations that may be used to track NASA deep space missions during the hours after launch. Also, commercial companies operate ground stations that can communicate with NASA missions. The remainder of this book primarily describes communication performed by the Deep Space Network operated for NASA by JPL. (Because of their specific locations, stations of these other networks sometimes are planned to provide time-critical tracking assistance during launch and early mission phases of NASA's deep space missions.)

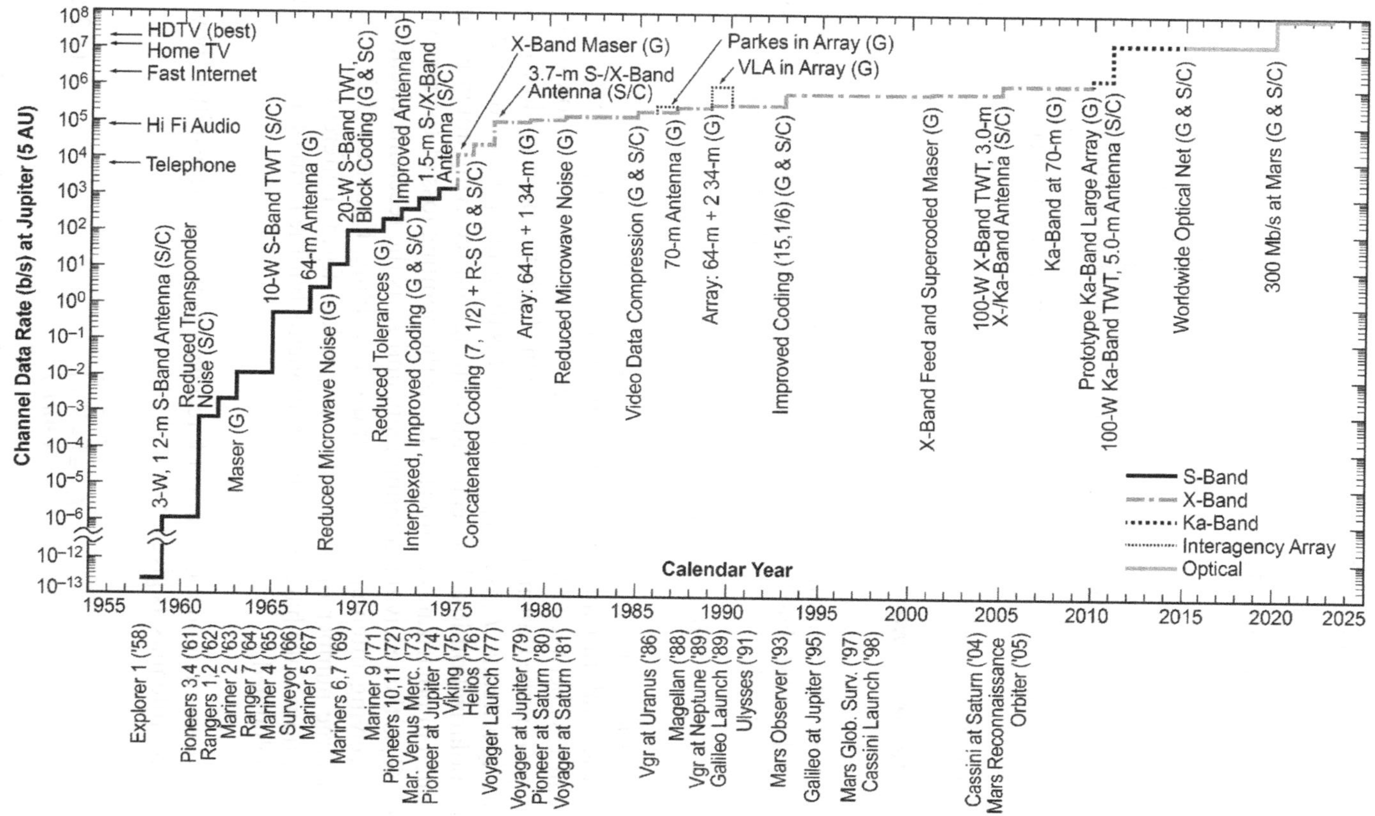

Fig. 1-1. Deep space communications downlink data rate evolution.

This chapter summarizes the theoretical background for telecommunications link analysis and telecommunications design control, respectively. The chapter has been adopted from Yuen, 1982 [2], and refers to chapters of that monograph for greater detail.

1.2 Telecommunications Link Analysis

The performance of a telecommunications system depends on numerous link parameters. Advanced modulation techniques, coding schemes, modern antennas, transmitters, and other advances all improve communications efficiency in their own ways. For designing an entire communications system, communications engineers put all the components or subsystems together and determine performance capability. Signal performance metrics, such as signal-to-noise-spectral-density ratios, are defined in this section. In addition, the component link parameters that enhance or impair the performance are defined.

1.2.1 Received Power

General questions used for performance computation are derived from the basic equations of communications in the medium between transmitting and receiving systems [3]. The first step in link analysis is to calculate the received signal power. Received power P_R is computed by the following equation:

$$P_R = P_T L_T G_T L_{TP} L_S L_A L_P L_{RP} G_R L_R \tag{1.2-1}$$

Where P_R is the received signal power at the input to the receiver or preamplifier, P_T is the total transmitted power at antenna terminals, L_T is the transmitting circuit loss between transmitting antenna terminals and radio case due to cabling, G_T is the transmitting antenna gain, L_{TP} is the pointing loss of the transmitting antenna, L_S is the space loss, LA is the atmospheric attenuation, L_P is the polarization loss between transmitting and receiving antennas due to mismatch in polarization patterns, L_{RP} is the pointing loss of the receiving antenna, G_R is the receiving antenna gain, and L_R is the receiving circuit loss between receiving antenna and receiver due to cabling. Equation (1.2-1) consists of a large number of parameters in product form. Different types of communications links have different components but the form of Eq. (1.2-1) remains unchanged.

The space loss, or numerical ratio of received power to transmitted power between two antennas, is given by

$$L_S = \left(\frac{\lambda}{4\pi r} \right)^2 \tag{1.2-2}$$

where λ is the wavelength of the radio signal and r is the distance between spacecraft and ground antennas.

The transmitting antenna gain G_T can be related to the effective antenna aperture A_T as

$$G_T = \frac{4\pi A_T}{\lambda^2} \tag{1.2-3}$$

where λ is the wavelength of radio signal. The effective antenna aperture A_T is related to the actual antenna aperture A_t by the relation

$$A_T = \mu A_t \tag{1.2-4}$$

where μ is the antenna efficiency factor. The receiving antenna gain is similarly defined (see Chapter 8 of Ref. [2] for more detailed discussions).

Some of the parameters in Eq. (1.2-1) are not defined in exactly the same way on all projects. For example, the transmitting circuit loss L_T is sometimes accounted for by decreasing the effective transmit antenna gain and/or by decreasing the effective transmitted power, obviating L_T. Also, the atmospheric attenuation (for clear, dry weather) is ordinarily accounted for in the ground antenna gain. No matter what the precise definitions are, the parameters must account for the entire telecommunications link.

The received power is referenced to some point in the receiving circuit. Of course, the choice of reference point affects L_R. On the uplink (from ground to spacecraft), the point of reference is usually the input port of the spacecraft transponder. On the downlink (from spacecraft to ground), the point of reference is the input to the maser amplifier. Whatever the reference, the noise equivalent temperature of the receiving system must be referenced to that same point if signal-to-noise ratios (SNRs) are to be computed correctly.

1.2.2 Noise Spectral Density

The noise for an uplink is dominantly thermal, internal to the amplifier in the front end of the spacecraft receiver. For the downlink, thermal noise in the station's low noise amplifier (LNA) is minimized by using helium-based cooling of the LNA. A large portion of the total receive system noise comes from outside the LNA, in particular from the atmosphere, hot bodies in the field-of-view of the antenna, the 2.7-kelvin (K) cosmic background, and that portion of the ground seen by antenna sidelobes.

It is assumed that the receiving system noise has uniform spectral density in the frequency band containing the signal. The one-sided noise spectral density N_0 (in units of watts/hertz, W/Hz) is defined as

$$N_0 = kT \tag{1.2-5}$$

where k is Boltzmann's constant = 1.380 × 10^{-23} J/K, $\tilde{k}$ = 10 log k = –198.6 decibels referenced to milliwatts (dBm)/(Hz K), and T is the system equivalent noise temperature. The uniform spectral density assumption and Eq. (1.2-5) are valid for the microwave frequency signals that are currently being used for deep space telecommunications. For signals in other frequency regions, such as in optical frequencies, different expressions should be used [3].

1.2.3 Carrier Performance Margin

Carrier phase tracking performance is dependent on the signal-to-noise ratio (SNR) in the carrier tracking loop. For either an uplink or a downlink, the carrier SNR in a bandwidth $2B_{\mathrm{LO}}$ is defined as M_c where

$$M_c = \frac{P_c}{2B_{\mathrm{LO}}N_0} \tag{1.2-6}$$

where P_c = portion of received power in the residual carrier, and B_{LO} = one-sided threshold loop noise bandwidth. Here, P_c is calculated from P_R using the modulation indices of the link and depends on the type of modulation used (see Chapter 5 of Ref. 2).

The above definition of carrier margin was chosen because a phase-locked loop receiver loses lock when P_c drops below $2B_{\mathrm{LO}}N_0$ watts (W) (see Chapter 3 of Ref. 2). Thus, $P_c = 2B_{\mathrm{LO}}N_0$ defines carrier threshold. $\widetilde{M}_c$ is calculated as

$$\widetilde{M}_c = \widetilde{P}_c - \widetilde{2B}_{\mathrm{LO}} - \widetilde{N}_0 \tag{1.2-7}$$

and represents the number of decibels the received residual carrier is above carrier threshold. Another popular name for M_c is carrier SNR in $2B_{\mathrm{LO}}$. However, this is a misnomer since $B_{\mathrm{LO}}N_0$, not $2B_{\mathrm{LO}}N_0$, is the noise power in a thresholding loop. So carrier SNR in $2B_{\mathrm{LO}}$ equals one-half the carrier SNR in a thresholding loop.

The minimum acceptable carrier margin, in general, is not 0 dB. For swept acquisition of the uplink, the minimum useful P_c is in the neighborhood of $2B_{\mathrm{LO}}N_0$ watts (W). That is, the minimum useful carrier margin is about 10 dB. For the downlink, the DSN recommends that carrier margin be at least 10 dB. Furthermore, carrier margins for two-way Doppler may need to be larger than 10 dB, depending on required radiometric accuracies.

1.2.4 Telemetry and Command Performance Margins

For both telemetry and command,

$$ST/N_0 \text{ to receiver} = \frac{S}{RN_0} \tag{1.2-8}$$

Where S is the portion of received power in the data modulation sidebands, and R is the data bit rate. Here S is calculated from P_R using the modulation indices of the link. The parameter ST/N_0 to receiver is sometimes denoted by E_b/N_0, which is the signal energy per bit to noise spectral density ratio. And

$$ST/N_0\,\text{output} = \left[ST/N_0\,\text{to receiver}\right]L_{\text{system}} \tag{1.2-9}$$

where L_{system} is the system losses. Threshold ST/N_0 is defined by the bit error probability required of a link. The bottom line of a telemetry or command link analysis is the performance margin. In decibels (dB)

$$\text{performance margin} = ST/N_0\,\text{output} - \text{threshold}\,ST\;ST/N_0 \tag{1.2-10}$$

1.2.5 Ranging Performance Margin

The ranging channel involves transmitting a ranging modulation or code from the DSN to the spacecraft, where it is modulated and then, together with receiver noise, is used to modulate the downlink from the spacecraft to the DSN (see Chapter 4 of Ref. 2). The ranging SNR at the spacecraft is

$$\text{ranging input SNR} = \frac{P_{R(u/l)}}{B_R N_{0(u/l)}} \tag{1.2-11}$$

where $P_{R(u/l)}$ is the portion of received uplink power in the ranging modulation sidebands, $N_{0(u/l)}$ is the uplink (that is, one-sided noise spectral density of the spacecraft receiver), and B_R is the one-sided noise bandwidth of the transponder ranging channel. Here $P_{R(u/l)}$ is calculated from the uplink P_R using the modulation indices of the uplink. The ranging signal-to-noise-spectral-density ratio returned to the DSN is

$$\text{received SNR} = \frac{P_{R(u/l)}}{N_{0(d/l)}} \tag{1.2-12}$$

where $P_{R(d/l)}$ is the portion of received downlink power in the ranging modulation sidebands, and $N_{0(d/l)}$ is the downlink one-sided noise spectral density. Here, $P_{R(d/l)}$ is a function not only of the downlink P_R and the downlink modulation indices but also of ranging input SNR. This is because ranging is a turnaround channel. Some of the modulation sidebands on the downlink are turnaround noise sidebands.

$$\text{output SNR} = \left[\text{received SNR}\right]L_{\text{radio}} \tag{1.2-13}$$

where L_{radio} is the radio loss of the ranging system. The value of the required SNR is specified by required radiometric accuracies and desired integration time (see Chapter 4 of Ref. 2).

The bottom line of a ranging link analysis is the performance margin, in dB,

$$\text{ranging performance margin} = \text{output SNR} - \text{required SNR} \qquad (1.2\text{-}14)$$

1.3 Communications Design Control

A small number of decibels is usually all that separates an inadequate link design from a costly overdesign. For this reason, extreme attention must be paid to performance prediction for deep space telecommunications systems.

If all link parameters were constant and precisely known to the telecommunications engineer, a simple accounting of the link parameters could predict performance. The real world is not so accommodating, however. Some link parameters vary with spacecraft environment, others with ground station parameters and the communications channel conditions. Some are associated with link components that have manufacturing tolerances.

In the early days of space exploration, engineers had little data and were relatively inexperienced in designing deep space telecommunications systems. Hence, they tended to be very conservative; they used a deterministic worst-case criterion [4–6] to assure sufficient link margins in guarding against uncertainties. Experience over many lunar and planetary flight projects has demonstrated that this approach is practical from the point of view of engineering and management [6–8]. The major disadvantages of this deterministic worst-case criterion are that it provides no information about the likelihood of achieving a particular design value. Hence, cost tradeoff and risk assessment cannot be done quantitatively.

Over the years, more experience was gained in deep space telecommunications systems design. Designers evolved a technique for treating telecommunications performance statistically [8,9], removing the major disadvantages of the deterministic approach while preserving its advantages. Since 1975 this statistical technique has been used in the design of deep space telecommunications systems. It is described in this section.

1.3.1 Design Control Tables

The communication link margin is computed using an equation of the following form:

$$y = y_1 y_2 \cdots y_k \qquad (1.3\text{-}1)$$

where y_i, $i = 1, 2, \ldots, K$ are parameters of the communication link such as in Eqs. (1.2-1) and (1.2-6). This equation is presented in its general form, without its detail components. Different types of communications links have different components, but the form of this equation remains unchanged.

The overall telecommunications system consists of a large number of parameters in product form. Hence, expressed in the dB domain, it becomes a sum of these parameters; that is,

$$x = x_1 + x_2 \ldots x_K \tag{1.3-2}$$

where

$$x = 10 \log_{10} y \tag{1.3-3}$$

and

$$x_i = 10 \log_{10} y_i \, , \; i = 1,2,\ldots,K \tag{1.3-4}$$

In managing the system design, it is most convenient to put this in tabular form with these parameters and entries. This table is referred to as a design control table, or DCT. All of the factors that contribute to system performance are listed in the order that one would find in tracing a signal through the system. Sample DCTs of the telemetry, command and ranging links can be found in Chapters 2 through 7 for six case study missions.

To every parameter in a DCT a design value, along with its favorable and adverse tolerances, is assigned by designers. These tolerances are used not as a hidden safety margin of each parameter; rather, they reflect probable uncertainties, including measurement tolerance, manufacturing tolerance, environmental tolerance, drift of elements, aging of elements, parameter modeling errors, and others. The table readily indicates the parameters with the largest tolerances—hence, the areas where more knowledge and hardware improvement might be most profitable.

The design procedure and performance criterion selection for deep space communications links are described in the following subsection.

1.3.2 Design Procedure and Performance Criterion Selection

The design procedure for deep space telecommunications systems design and the selection of a particular criterion for conservatism are both driven by weather conditions in the signal path between the ground station antenna and the spacecraft on telecommunications performance. “Weather” refers both to conditions in the Earth’s atmosphere (humidity, precipitation, wind) and to the presence of charged particles in the path through space from the top of the atmosphere to the spacecraft.

1.3.2.1 Weather Effects. Weather requires special consideration. For carrier frequencies at or above X-band, the randomness that weather introduces to the link dominates all other sources of randomness. There are two techniques for incorporating weather into telecommunications design control. The simpler one, the percentile weather technique, is described in this section. It is a reasonable estimate of the weather effects on link performance. Often a reasonable estimate suffices for preliminary system design and performance assessment purposes. The percentile weather technique is attractive for its simplicity. Conversely, for detailed design and link performance monitoring purposes, a more accurate estimate is required.

The percentile technique for incorporating weather into telecommunications design control requires the preparations of two design control tables. In the first design control table, a dry atmosphere and clear sky over the Deep Space Station (DSS) is assumed. In the second design control table, x-percentile inclement weather is assumed. What is meant by "x-percentile" weather is that with x percent probability a pessimistic assumption is being made about weather effects; moreover, with $(100 - x)$ percent probability an optimistic assumption is being made. As an example, 95-percentile means that 95 percent of the time the degradation due to weather is less than predicted, while 5 percent of the time the weather degradation is worse.

1.3.2.2 Design Procedure. The design procedure is described here. The procedure unfolds as a sequence of six steps during which the philosophy of telecommunications design control reveals itself. The discussion below follows Refs. [8] and [9].

Step 1. Three values are assigned to most link parameters: design, favorable tolerance, and adverse tolerance. All three values are to be in decibel representation. Those parameters that are not assigned three values should receive only a design value (in decibels). Data bit rate, space loss, and threshold (or required) SNR ratios are regarded as deterministic and only have design values. The weather-dependent parameters—atmospheric attenuation and, on the downlink, incremental noise temperature due to clouds—should be assigned only design values (noise temperature is in units of kelvin, not decibels). In fact, the design values of the weather-dependent parameters should be based on the assumption of clear, dry weather. Later, as explained in the previous paragraph, the design procedure is to be repeated with weather-dependent design values assigned on the basis of x-percentile inclement weather. The following definitions serve as a guide in the assignment of values to a link parameter:

- Design value = the a priori estimate of a parameter,
- Favorable tolerance = the best case of a parameter minus the design value,
- Adverse tolerance = the worst case of a parameter, short of failure, minus the design value.

Noise temperatures, noise spectral densities, and noise bandwidths have favorable tolerances with negative values and adverse tolerances with positive values. The opposite is true of all other link parameters that get assigned tolerances. Tolerances reflect one of more of the following: a limit cycle, a manufacturing tolerance associated with a link hardware component, a dependence on spacecraft environment, and other uncertainties.

Step 2. Arrange the link parameters in a vertical listing—a design control table—and identify independent groups among them.

Step 3. Within each of the independent groups, add the design values and the favorable and adverse tolerances so that there is only one design value with its associated favorable and adverse tolerances for each group.

Step 4. Assign a probability density function (pdf) to each independent group. Typically, on uniform, triangular, Gaussian, and Dirac-delta (for those groups without tolerance) pdf's are used. The assignment made by the Telecommunications Prediction and Analysis Program (TPAP) are tabulated in Chapter 10 of Ref. 2. (Note: TPAP was replaced by the Telecom Forecaster Predictor (TFP) in the late 1990s.) In case a probability density function is nonzero over the entire real line such as the Gaussian density function, use the absolute sum of its favorable and adverse tolerances as its 6-sigma (6 standard deviations) measure.

Step 5. Compute for each independent group (random variable) its mean and variance. Having been computed from a design value and tolerances all expressed in decibels, the mean will, of course, be in decibels (and the variance in decibels squared).

Step 6. Compute the mean and variance of the desired performance or carrier margin by algebraically summing the means and adding the variances obtained in step 5. What is meant by "algebraically summing" is that some means—those corresponding to noise spectral density, noise bandwidth, data bit rate, and threshold (or required) SNRs—are subtracted rather than added.

It is certainly true that a precise pdf of the overall link margin can be obtained by convolving the pdf's of the K independent random variables. However, the link margin tolerance distribution is well approximated by a Gaussian distribution by invoking the central limit theorem, since the overall link consists of K independent random variables formed in step 2 above. This simplifies the computational complexity to the point that hand calculation is indeed practical. Moreover, the pdf's of the K independent random variables were only estimated. It seems difficult to justify using tedious convolution to achieve a precise solution based on imprecise information, if an approximation is indeed satisfactory. A more worthwhile effort would be making a more accurate estimate of the pdf's of the K independent random variables.

The above procedure is repeated with the weather-dependent parameters being assigned design values based on *x*-percentile inclement weather. The weather percentile not only differs from site to site but may also may be defined on a monthly or seasonal basis. The performance or carrier margin is finally considered predicted with the specification of four members:

1) Mean margin with clear, dry weather

2) Mean margin with *x*-percentile weather

3) *n*-sigma margin with clear, dry weather

4) *n*-sigma margin with *x*-percentile weather

where "*n*-sigma margin" equals mean margin minus *n* standard deviations. The value *n* is typically 3 for command links and 2 for links carrying telemetry or providing radiometric data.

1.3.2.3 Performance Criterion Selection. In order to assure successful operation and guard against adverse situations, we must provide sufficient link margins. Based on the design procedure described previously, one can choose link performance that will not deviate from its mean margin by more than 3-sigma (three standard deviations) with probability 0.99. This 3-sigma value is used as an uncertainty measure for the link. Depending on how much risk is acceptable, we can choose any number of sigma values. Hence, a useful design criterion is: *the mean value of the link SNR must exceed the required SNR by an amount equal to or larger than the n-sigma.* Alternatively, one can choose a level of probability of success, then use the corresponding required SNR.

References

[1] DESCANSO: *Deep Space Communications and Navigation Systems*, *http://descanso.jpl.nasa.gov/* (website), Jet Propulsion Laboratory, California Institute of Technology, Pasadena California, includes *Design and Performance Summary Series*. http://descanso.jpl.nasa.gov/DPSummary/summary.html (accessed October 30, 2014)

[2] J. H. Yuen, editor, *Deep Space Telecommunications System Engineering*, National Aeronautics and Space Administration, Washington, District of Columbia, July 1982. (Also reprinted by Plenum Press, New York, 1983.)

[3] J. R. Pierce and E. C. Posner, *Introduction to Communication Science and Systems*, Plenum Press, New York, 1980.

[4] R. E. Edelson, B. D. Madsen, E. K. Davis, and G. W. Garrison, "Voyager Telecommunications: The Broadcast from Jupiter," *Science*, vol. 204, no. 4396, pp. 913–921, June 1, 1979.

[5] R. E. Edelson, editor, *Telecommunications Systems Design Techniques Handbook*, Technical Memorandum 33-571, Jet Propulsion Laboratory, Pasadena, California, July 15, 1972.

[6] M. F. Easterling, "From 8-1/3 Bits per Second to 100,000 Bits per Second in Ten Years," *Conference Proceedings, ITC'74*, Los Angeles, California, 1974.

[7] V. L. Evanchuk, "117.6 kbps Telemetry from Mercury In-Flight System Analysis," *Conference Proceedings, ITC'74*, Los Angeles, California, 1974.

[8] J. H. Yuen, *A Practical Statistical Model for Telecommunications Performance Uncertainty*, Technical Memorandum 33-732, Jet Propulsion Laboratory, California Institute of Technology, Pasadena, California, June 15, 1975.

[9] J. H. Yuen, "A Statistical Model for Telecommunication Link Design," *Proceedings, National Telecommunications Conference*, New Orleans, Louisiana, A77-15115 04-32, Institute of Electrical and Electronics Engineers, New York, pp. 25–17 to 25–21, December 1976.

Chapter 2
The Deep Space Network: A Functional Description

Jim Taylor

All deep-space missions—defined as those operating at or beyond the orbit of the Earth's Moon—require some form of telecommunications network with a ground system to transmit to and receive data from the spacecraft. The Deep Space Network or DSN is one of the largest and most sophisticated of such networks.

NASA missions in low Earth orbit communicate through either the Near Earth Network (NEN) or the SN (Space Network), with the SN, both operated by the NASA Goddard Space Flight Center (GSFC). The SN has of a number of Tracking and Data Relay Satellites (TDRS) in geosynchronous orbits. In addition, the European Space Agency operates a number of ground stations that may be used to track NASA deep space missions during the hours after launch. In addition, commercial companies operate ground stations that can communicate with NASA missions. The remainder of this book describes only the Deep Space Network operated for NASA by JPL.

The lessons and techniques of the DSN replicate many comparable issues of the other networks. The lessons from the missions described in the following chapters are widely applicable to all deep space telecommunications systems. This includes post-launch support that was negotiated and planned using stations belonging to networks other than the DSN.

Some of the grey scale figures in this book have color versions in the companion website www.wiley.com/go/deepspacecommunications. Please refer to this website where the text refers to colors.

Deep Space Communications, First Edition. Edited by Jim Taylor.

The description and performance summary of the DSN in this chapter come from the *DSN Telecommunications Link Design Handbook*, widely known within NASA as the 810-5 document [1]. This modular handbook has been approved by the DSN Project Office, and its modules are updated to define current DSN capabilities. It is an online source of technical information for all flight projects using the DSN. The following description is taken from 810-5 modules that provide technical information applicable to the current DSN configurations that provide carrier tracking, radiometric data, command transmission and telemetry reception.

The DSN is an international network of ground stations (antennas, transmitters, receivers, and associated systems) that operated intensively at S-band in the 1960s and 1970s, moving into X-band in the 1980s and 1990s, and more into Ka-band in the 2000s. The DSN supports interplanetary-spacecraft missions and radio- and radar-astronomy observations for the exploration of the Solar System and beyond. The DSN consists of three Deep Space Communications Complexes (DSCCs) placed approximately 120 degrees (deg) apart around the world: at Goldstone, in California's Mojave Desert; near Madrid, Spain; and near Canberra, Australia.

The DSN's hardware and software systems and their interconnected facilities have evolved over the decades. This chapter describes the DSN as it is today. The subsequent chapters describe the spacecraft designs of individual missions. Each chapter includes a description of the unique aspects of the ground system that supported the mission at that time.

Because each mission is unique, the telecommunications system for the mission is also unique. While the subsequent chapters describe the spacecraft designs of individual missions, these chapters will describe only the historical or unique aspects of the ground system as it supported that mission at the time.

This chapter includes brief descriptions and functional block diagrams of DSN systems at the DSCCs that provide carrier tracking, radiometric data (Doppler and ranging) collection, command uplinking, and telemetry reception and decoding for deep space missions, those defined at lunar distances or greater.

Each antenna (or Deep Space Station, DSS) in the DSN is capable of sending commands to one spacecraft at a time. Each DSCC contains one 70-meter (m) and from two to five 34-m antennas. There are two types of 34-m antennas. The first is the so-called high efficiency (HEF) antennas that have their feed, low-noise amplifiers, and transmitter located on the tilting structure of the antenna. These antennas were named when a less-efficient 34-m antenna was in use by the DSN and the name has survived. The efficiency of all DSN 34-m antennas is now approximately the same. The second type of

34-m antenna is the beam waveguide (BWG) antenna where the feeds, low-noise amplifiers and transmitters are located in a room below the antenna structure and the radio frequency energy is transferred to and from the antenna surface by a series of mirrors encased in a protective tube.

The capabilities of the antennas differ slightly depending on the microwave, transmitting, and receiving equipment installed.

2.1 Uplink and Downlink Carrier Operation

DSN stations are grouped by antenna size (26 m, 34 m, and 70 m), and for the 34-m antennas by type—BWG or HEF. The *DSN Telecommunications Link Design Handbook* [1] includes functional capability descriptions of each antenna size and type for the purpose of modeling link capability between a spacecraft and that station type.

2.1.1 The 34-m BWG Stations

The 34-meter diameter BWG (beam waveguide) and HSB (high angular-tracking speed beam waveguide) antennas are the latest generation of antennas built for use in the DSN. The newest of these, Deep Space Station 35 (DSS-35) at Canberra, became operational in October 2014. This section describes, as representative of the 34-m stations, the system functions at Deep Space Station 25 (DSS-25), a 34-m BWG station currently in use at Goldstone.

In general, each antenna has one LNA for each supported frequency band. However, stations that can support simultaneous right circular polarization (RCP) and left circular polarization (LCP) in the same band have an LNA for each. In addition, the stations that support Ka-Band contain an additional LNA to enable monopulse tracking when using RCP polarization. Each antenna also has at least one transmitter. Antennas with more than one transmitter can operate only one of them at a time.

DSS 25 is an exception and has a Ka-band transmitter that can be operated at the same time as its X-band transmitter. In Fig. 2-1, the radio frequency (RF) output from the 20-kW X-band transmitter goes through the X-band diplexer, then through an orthomode junction and polarizer to the X-band feed. The X-band uplink continues to the subreflector via an X-band/Ka-band dichroic plate, if simultaneous Ka-band is required. From the subreflector, the X-band uplink is focused to the 34-m main reflector, which is oriented in the direction of the spacecraft during the active track.

The X-band downlink signal from the spacecraft is collected by the 34-m main reflector. Then it is focused by the subreflector to the X-band feed (again via the X-band/Ka-band dichroic when there is also a Ka-band

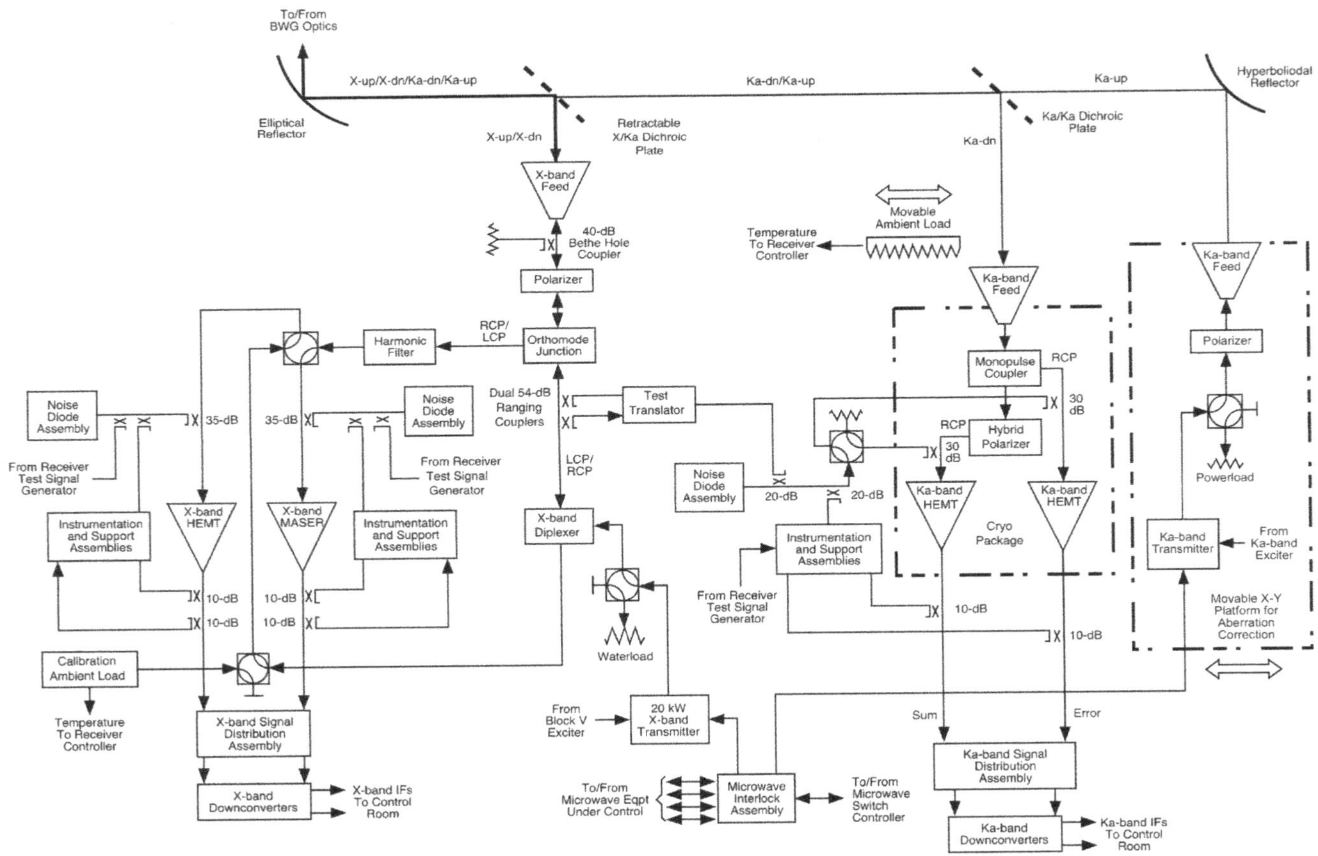

Fig. 2-1. Functional block diagram of the DSS-25 microwave and transmitter.

downlink from the spacecraft). The orthomode junction is the part of the antenna feed that combines or separates left-circularly polarized and right-circularly polarized (LCP and RCP) signals. From the feed the X-band RF signal goes to the X-band maser preamplifier.

When simultaneous X-Band uplink and downlink of the same polarization are required at stations with waveguide diplexers, reception must be through the diplexer, and the noise will be increased over that of the non-diplexed path.

After low-noise amplification, the downlink is frequency down-converted to a 300-megahertz (MHz) intermediate frequency (IF) for input to the Block V Receiver (BVR). All DSN antennas employ a receiver architecture where one or both circular polarizations of the received spectrum are amplified by a low-noise amplifier (LNA) and downconverted to IF. The antennas are designed to receive extremely weak signals and can be overloaded by signals in excess of –90 dBm. Antennas supporting 26 gigahertz (GHz) have a special low-gain mode that permits operation up to –50 dBm with degraded G/T.

The Ka-band downlink also is collected by the 34-m main reflector and focused by the subreflector. It passes through the dichroic plate to separate it from the X-band downlink signal path, on its way to the Ka-band feed. DSS-25 is equipped for RCP or LCP at Ka-band. The Ka-band preamplifier is a high-electron-mobility transistor (HEMT). Like the X-band downlink, after low-noise pre-amplification, Ka-band downlink is frequency down-converted for input to the BVR.

2.1.2 The 70-m (DSS-14 and DSS-43) Stations

Figure 2-2 shows the antenna, microwave and transmitter sections of the 70-m stations, DSS-14 and DSS-43.

The 20-kW X-band transmitter output goes through a polarizer and a diplexing junction to the X-band feed. From there, it passes through an S-band/X-band dichroic reflector on its way to the subreflector and the main 70-m reflector that sends the uplink on its way to the spacecraft.

The S-band uplink carrier, modulated with a command subcarrier when required, can be transmitted by a 20-kW transmitter or (at DSS-43 only) a 400-kW transmitter. The transmitter output goes through an S-band diplexer, orthomode junction and polarizer to the S-band feed. From there, as the block diagram shows, the S-band uplink path is via three smaller reflectors and the 70-m reflector before radiation to the spacecraft.

The X-band downlink from the main reflector is focused by the subreflector and passes through the dichroic reflector to separate it from the S-band signal

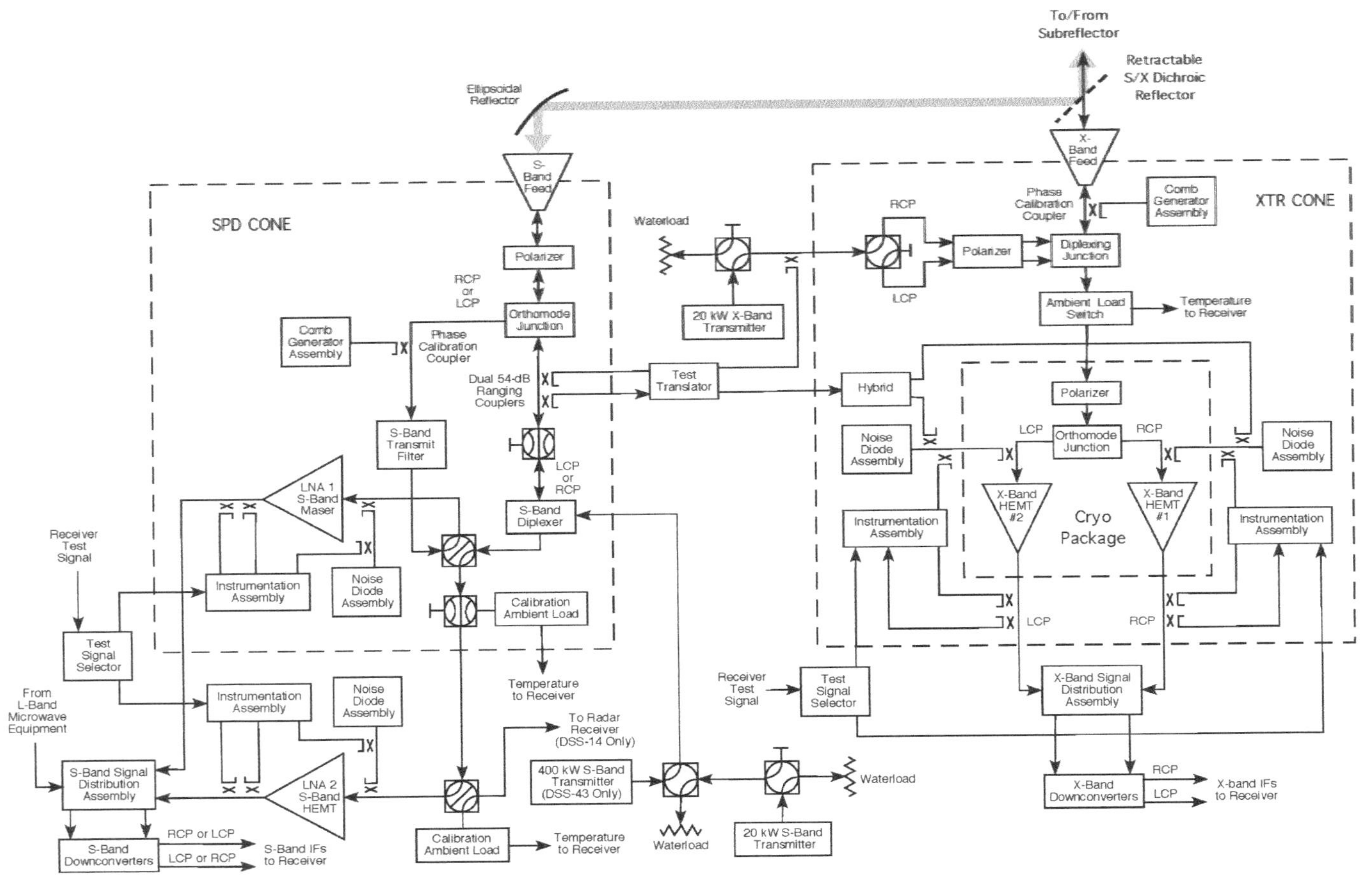

Fig. 2-2. Functional block diagram of 70-m microwave and transmitter.

path. From the diplexing junction, the X-band downlink goes to a polarizer to select (match) the left circular or right circular polarization of the downlink transmitted from the spacecraft. The X-band downlink from the X-band HEMT preamplifier is frequency-downconverted for input to the BVR.

From the 70-m reflector, the S-band downlink is directed by the subreflector to the S/X dichroic reflector. A dichroic surface is reflective at one frequency band and transparent at another, thus allowing the S-band frequencies to be separated from X-band frequencies for individual processing. The dichroic reflects the S-band downlink to the path shown by the thick line in the block diagram to the S-band feed. Reversing the path taken by the uplink, the downlink is directed by the diplexer to an S-band maser preamplifier, and its frequency is down-converted for input to the block V receiver (BVR).

2.2 Radiometric Data (Doppler and Ranging)

The relative motion of a transmitter and receiver causes the received frequency to differ from that of the transmitter. In deep space communications it is usual to define Doppler as the transmitted frequency (the uplink) minus the received frequency (the downlink) divided by the ratio that was used onboard the spacecraft (the transponding ratio) to generate the downlink frequency. Because the frequency of a carrier equals the rate-of-change of carrier phase, the Downlink Channel supports Doppler measurement by extracting the phase of the downlink carrier.

There are three types of Doppler measurement: one-way, two-way, and three-way. One-way refers to the radio-frequency (RF) carrier frequency being generated by an on-board oscillator in the spacecraft and received at the station. Two-way refers to the carrier being generated at the station, transmitted to a coherent transponder in the spacecraft, transmitted from the transponder and received back at the transmitting station. Three-way is the same as two-way, except that the downlink carrier is received at a second ground station, either in the same DSN complex or at another complex. In all of these cases, the accumulating downlink carrier phase is measured and recorded. Because ground-station oscillators have greater frequency stability than spacecraft oscillators, two-way or three-way Doppler measurements are used in deep space navigation.

At the station, the two-way downlink signal from the spacecraft is routed from the antenna feed/low-noise amplifier (LNA) to the downlink channel, as shown in Fig. 2-3. If the downlink is one-way, the uplink sections at the bottom of the figure do not play a part. Within the RF to intermediate-frequency downconverter (RID), which is located at the antenna, a local oscillator is generated by frequency multiplication of a highly stable frequency reference from the frequency and timing system (FTS) and the incoming

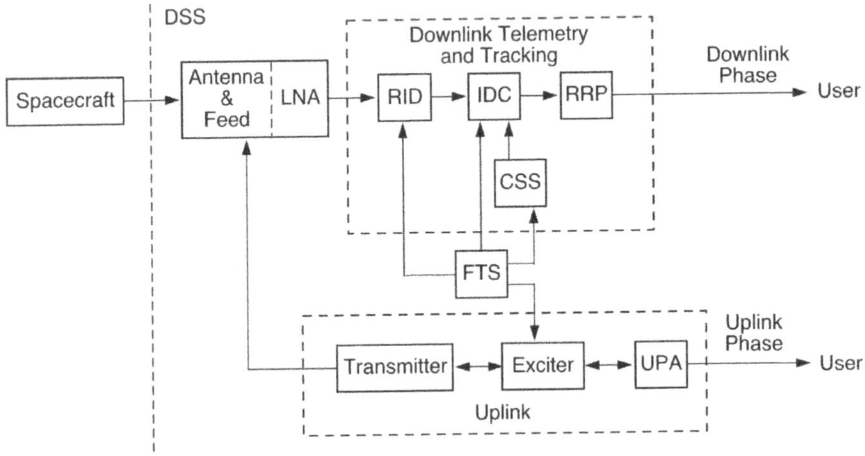

Fig. 2-3. Two-way or three-way Doppler measurement.

downlink signal is heterodyned with this local oscillator. The intermediate-frequency (IF) signal that results is sent to the signal processing center (SPC).

Two-way or three-way Doppler data points consist of uplink and downlink phase counts at sky frequency (only downlink phase counts in the case of a one-way measurement). The downlink phase counts are available at 0.1-second (s) intervals. The uplink phase counts are available from the uplink processor assembly (UPA) at 1.0-s intervals.

The Doppler measurements establish the spacecraft-station velocity as a function of time and can be compared with the expected or modeled velocity. This velocity measurement includes the motion of the station in inertial space due to the Earth's rotation.

The DSN ranging system measures the round-trip phase delay of a ranging signal sent from an uplink DSS to a spacecraft and back to a downlink DSS. In the most common configuration, known as two-way ranging, the uplink and downlink stations are the same, and the measured two-way phase delay permits the determination of the round-trip light time (RTLT) between the DSS and spacecraft.

A range measurement may be made with a constant uplink carrier frequency or when the transmitted uplink carrier frequency is time varying. For some missions, it is desirable to anticipate the uplink Doppler effect and to transmit an uplink carrier whose frequency varies in such a way that the uplink carrier arrives at the spacecraft with minimal offset from channel center. This is called uplink Doppler compensation and has the advantage of reducing the stress on the carrier tracking loop in the spacecraft receiver. The DSN

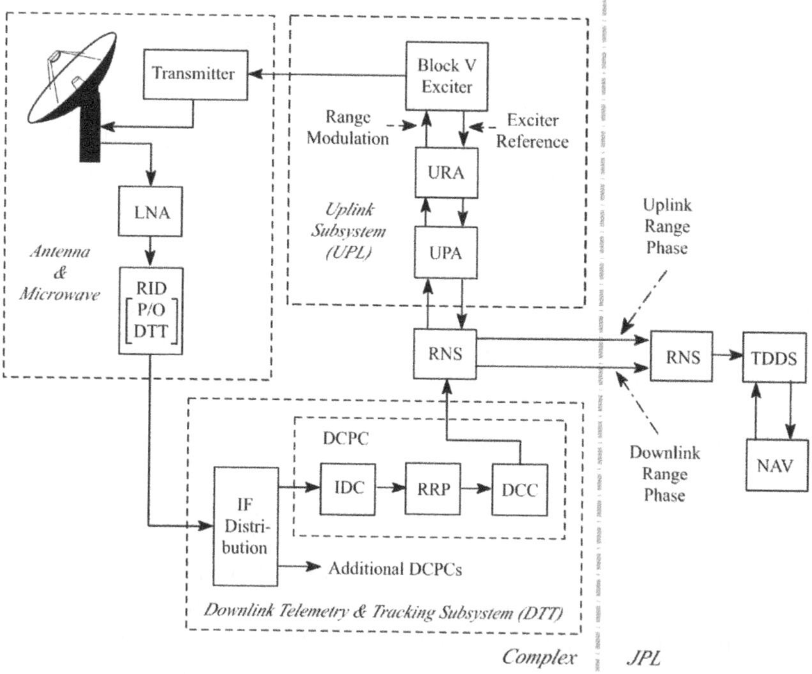

Fig. 2-4. The DSN ranging system architecture (P/O = part of).

ranging system accommodates either time-varying or constant transmitted uplink frequency.

The architecture for the DSN ranging system is shown in Fig. 2-4. The system consists of a front-end portion, an uplink subsystem (UPL), and a downlink telemetry and tracking subsystem (DTT). The front-end portion includes the microwave components, including a low-noise amplifier (LNA), the transmitter, and the antenna. The UPL includes the uplink ranging assembly (URA), the exciter, and their controller, referred to as the uplink processor assembly (UPA). The DTT includes a downconverter (the RID) located on the antenna, the IF-to-digital converter (IDC), the receiver and ranging processor (RRP) and the downlink channel controller (DCC).

The uplink ranging assembly (URA) uplink ranging function is controlled by the UPA. The RRP downlink ranging function is controlled by the DCC. Each measures the range phase and sends the measurements to the tracking data delivery subsystem (TDDS) via the reliable network service (RNS). The URA generates the uplink ranging signal and measures its phase before passing it to the exciter. The signal is phase-modulated onto the carrier by the exciter and, after being amplified to a suitable level by the transmitter, is forwarded to the spacecraft. The spacecraft coherently transponds the signal, sending it back to the DSN receiver.

The amplified downlink signal from the antenna is downconverted by the RID (located on the antenna) and fed to an intermediate frequency (IF) distribution assembly in the control room. The IF is fed to one or more downlink channel processing cabinets (DCPCs) as required. Each DCPC is equipped with a single channel, which includes a single IDC and RRP. For a spacecraft that requires ranging on multiple downlinks (for example, S-band and X-band), multiple DCPCs will be assigned to that antenna.

Three-way ranging is accomplished in essentially the same way as two-way ranging, except that there are two stations. The UPL, transmitter, and antenna of one DSS are used to transmit an uplink carrier modulated with a ranging signal. The uplink range phase is recorded at that station. At a second DSS, the antenna, LNA and DTT receive the downlink from the spacecraft and record the downlink range phase.

The sequential ranging signal is a sequence of periodic signals. These periodic signals are all coherently related to each other and to the uplink carrier. The basis for these periodic ranging signals is a table of well-defined range components. Each component is assigned a number. A larger number represents a component with a smaller frequency (but a larger period). The components that are used in ranging are assigned the numbers 4 through 24 and are ordered according to these component numbers. The frequency of component 4 is always approximately 1 MHz, and it is often called the "1 MHz component" and used as the "clock." The frequency of components 5 through 24 is exactly half of their immediate predecessor.

At JPL, the radiometric data conditioning group, part of the multimission navigation function, processes and delivers the Doppler and ranging data to project navigation. The radio-navigation data sets are also used to generate prediction files (P-files) for delivery back to the DSN, for use in creating the frequency and pointing predicts for subsequent tracking passes. Frequency predicts are input to the BVR to assist in locking the receiver to expected periods of one-way, two-way, or three-way data. Pointing predicts are used to drive the station antenna in elevation and azimuth angle during the pass. Pointing predicts are supplemented by several tables specific to the station type, location, and the general declination of the spacecraft. These supplementary tables include corrections for atmospheric refraction as a function of elevation angle and azimuth as well as for deformation of the antenna structures (and thus, changes in the beam direction) as a function of elevation angle.

2.3 Delta Differential One-Way Ranging

Delta-differential one-way ranging (delta-DOR) is a very long baseline interferometric (VLBI) radio-tracking technique using two deep-space stations located at different complexes for a single measurement. It uses the differential one-way range technique to provide information about the angular location of a target spacecraft relative to a reference direction where the reference direction is defined by the direction of arrival of radio waves from a distant known source, such as a quasar, whose direction is well known and catalogued.

Delta-DOR provides a direct geometric determination of spacecraft angular position and is therefore especially useful when line-of-sight measurements (Doppler and ranging) have weaknesses such as spacecraft traveling near zero declination and spacecraft with small, unmodeled dynamic forces affecting their motion. Another advantage of delta-DOR is that measurements are of relatively short duration (approximately 1 hour) as compared with the hours of tracking typically required for Doppler and ranging.

Measurements are conducted using either the Goldstone-Madrid baseline or the Goldstone-Canberra baseline. Two baselines with orthogonal components are needed to measure both the right ascension and declination coordinates of angular position. The Goldstone–Madrid baseline is oriented east–west and is most sensitive to right ascension for spacecraft near the ecliptic plane. The Goldstone–Canberra baseline is canted and has most sensitivity in the direction that splits the axes of right ascension and declination.

Differential one-way ranging is supported by the 34-m and 70-m antennas. Other equipment includes the VLBI science receivers (VSRs) in the signal processing center (SPC), the ground communications infrastructure, and the DOR correlator at JPL. Figure 2-5 shows the DSN equipment used.

Planning a measurement involves scheduling the stations and identifying the appropriate distant reference sources. The spacecraft's contribution to making a delta-DOR measurement is to provide a one-way downlink carrier modulated by a set of continuous-wave tones. The "delta DOR module" in the Small Deep-Space Transponder (SDST) generates a tone at approximately 19.1 MHz.

On many projects, Navigation also requests that a telemetry subcarrier (preferably without telemetry symbols) modulate the downlink carrier to provide a second delta-DOR "tone." Though subcarrier frequencies in the range of 375 kilohertz (kHz) are preferred for this purpose, some projects provide a subcarrier frequency in the range of 25 kHz. A subcarrier modulation index of about 30 deg provides a good balance of power between the carrier the delta-DOR tone and the subcarrier "tone."

2.4 Command Processing and Radiation

Uplink data are delivered to the DSN using one of three services, named stream mode radiation, file mode radiation, and command delivery.

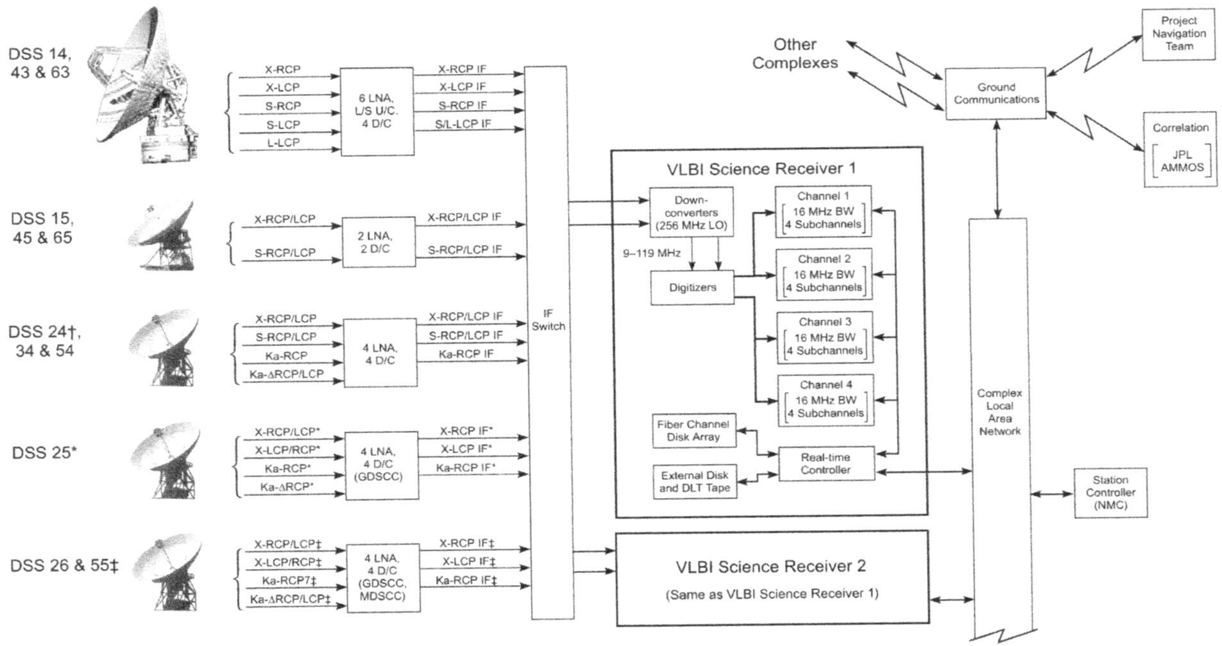

* Equipment or interface only exists at Goldstone DSCC

‡ Equipment or interface only exists at Goldstone and Madrid DSCCs

† DSS 24 is not recommended for ΔDOR measurements.

Fig. 2-5. DSN equipment for delta-DOR support.

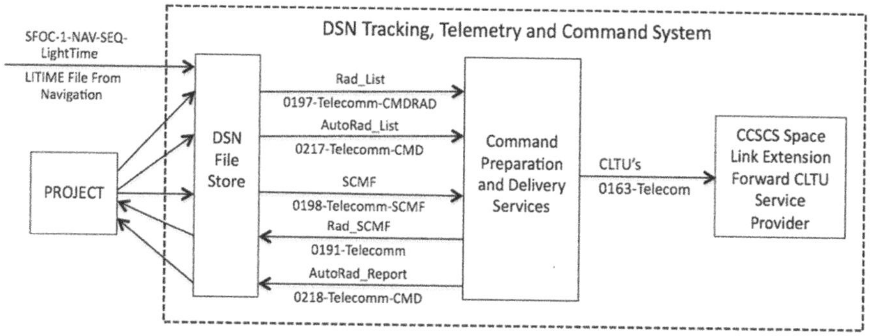

Fig. 2-6. Command radiation service data flow for "File Mode" (SCMF).

Stream mode command radiation uses the space link extension (SLE) forward service, an implementation of the Consultative Committee for Space Data Systems (CCSDS) recommendation 912.1, SLE command link transmission unit (CLTU) Service [2], and is described in DSN Document 820-013, module 0163-Telecomm [3], an internal JPL document. The SLE forward service is an online only service in which the service users (flight projects) provide command symbols to be transferred to the spacecraft and ancillary information such as routing, ensuring the integrity of the Earth segment of the communications link, and providing the project limited control of the command process.

File mode command radiation accesses a file of CLTUs from project's mission support area via DSN File Store where the individual CLTUs are extracted and passed on to the station for modulation onto the uplink carrier and radiation to the spacecraft. The file of CLTUs is referred to as a spacecraft command message file (SCMF). Refer to Fig. 2-6 for "file mode" data flow. This service is an online or offline store and forward service that allows management of multiple stored command files.

Command delivery service, uses the CCSDS file delivery protocol (CFDP) and is available for spacecraft that employ this protocol. It is described in Ref. [4]. The service is provided by accessing files from the MSA via DSN file store where the files are converted to CLTUs, which are then passed to the tracking station for modulation onto the uplink carrier and radiation to the spacecraft. As shown in Fig. 2-7, the only function performed at the stations is the mechanism whereby command data are extracted from the delivery format and converted to an RF signal suitable for reception by the spacecraft.

In the mission support area, the project ACE (call sign for project real-time mission controller) operates the multimission command system from a workstation. An ACE is able to activate command transmission within 2 s of the

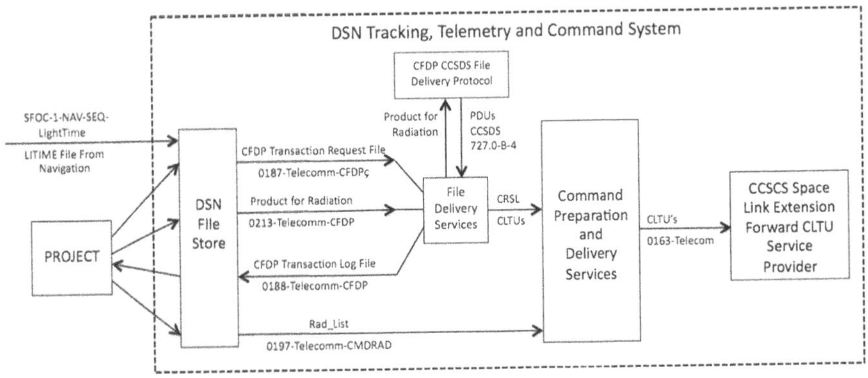

Fig. 2-7. Command delivery service (CFDP) data flow (see Ref. 13 regarding 727.0B-4).

nominal time. To begin or end a command session, the ACE requests the station to turn the command modulation on or off, respectively.

The RNS transfers the command files to the station in the staging process, as well as the ACE directives for radiation of the staged commands. At the station, the command processor assembly (also part of the "service provider") performs the digital processing to create the command-bit stream from the command files and the activation signal.

2.5 Telemetry Demodulation and Decoding

In general, telemetry service support requires one antenna, at least one receiver, and telemetry processing equipment for each spacecraft. Additional receivers and telemetry processing equipment can be added for spacecraft with multiple downlinks or for redundancy. In addition, the DSN is capable of tracking two spacecraft per antenna (multiple spacecraft per aperture, MSPA) if they both are within the scheduled antenna's beamwidth.

The telemetry system performs three main functions: data acquisition, data conditioning and transmission to projects, and telemetry-system validation. Data acquisition includes receiving and tracking the downlink carrier and subcarrier (if used), detecting and synchronizing the telemetry symbols, and decoding the symbol stream for input into telemetry frames. These functions are within the "downlink channel" block in Figs. 2-8 and 2-9.

As described for downlink carrier tracking, the arriving signal is routed from the antenna feed and LNA to the downlink channel. After the frequency-downconverted IF signal reaches the SPC, the intermediate-frequency to digital converter (IDC) alters the frequency of the IF signal by a combination of up-conversion and down-conversion to a final analog frequency of approximately 200 MHz and then performs analog-to-digital conversion. The final

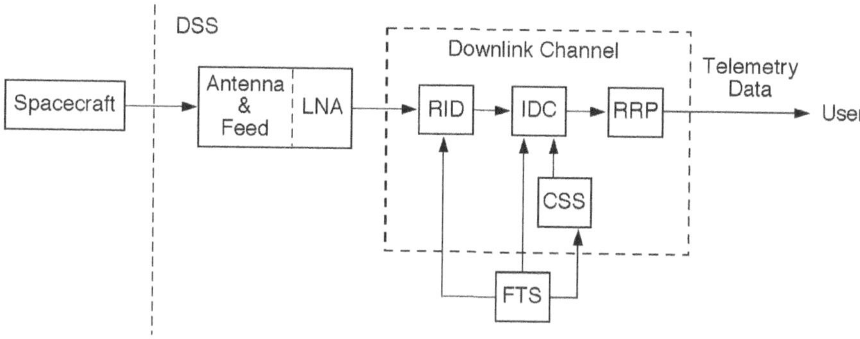

Fig. 2-8. Receiver architecture for downlink telemetry.

analog stage of down-conversion uses a local oscillator supplied by the channel-select synthesizer (CSS), which is also part of the downlink channel.

The channel select synthesizer (CSS) is adjusted before the beginning of a pass to a frequency appropriate for the channel of the incoming downlink signal; during the pass, the frequency of the CSS remains constant. The frequency of the CSS (and, indeed, of all local oscillators in the analog chain of downconversion) are synthesized within the downlink channel from highly stable frequency references provided by the frequency and timing system (FTS). The receiver and ranging processor (RRP) accepts the digital signal and performs carrier, subcarrier, and symbol synchronization, Doppler compensation, and data demodulation. For purposes of telemetry, the output of the RRP is a stream of soft-quantized symbols, suitable for input to a decoder.

Almost all spacecraft employ forward error-correcting (FEC) codes to make more efficient use of the communications channel. FEC codes add additional symbols to the transmitted data stream that the decoder can use to improve its estimate of the encoded bit stream. The exceptions to FEC use would likely be extremely high data rate transmissions where adequate signal power is available to make the gain achieved by coding unnecessary and any bandwidth needed for the symbols added by coding is unavailable.

The DSN supports two convolutional codes, the Consultative Committee for Space Data Systems (CCSDS) standard Reed-Solomon code and the CCSDS Turbo codes. Convolutional codes are used because they achieve significant coding gain with simple, highly reliable encoders and their decoders are of reasonable complexity. They also provide low latency and are useful when conditions prevent reception of a block of symbols. The Reed–Solomon code provides excellent performance with minimum bandwidth expansion in a high signal-to-noise environment. It is most often used as an outer code in combination with a convolutional inner code but may be used by itself under

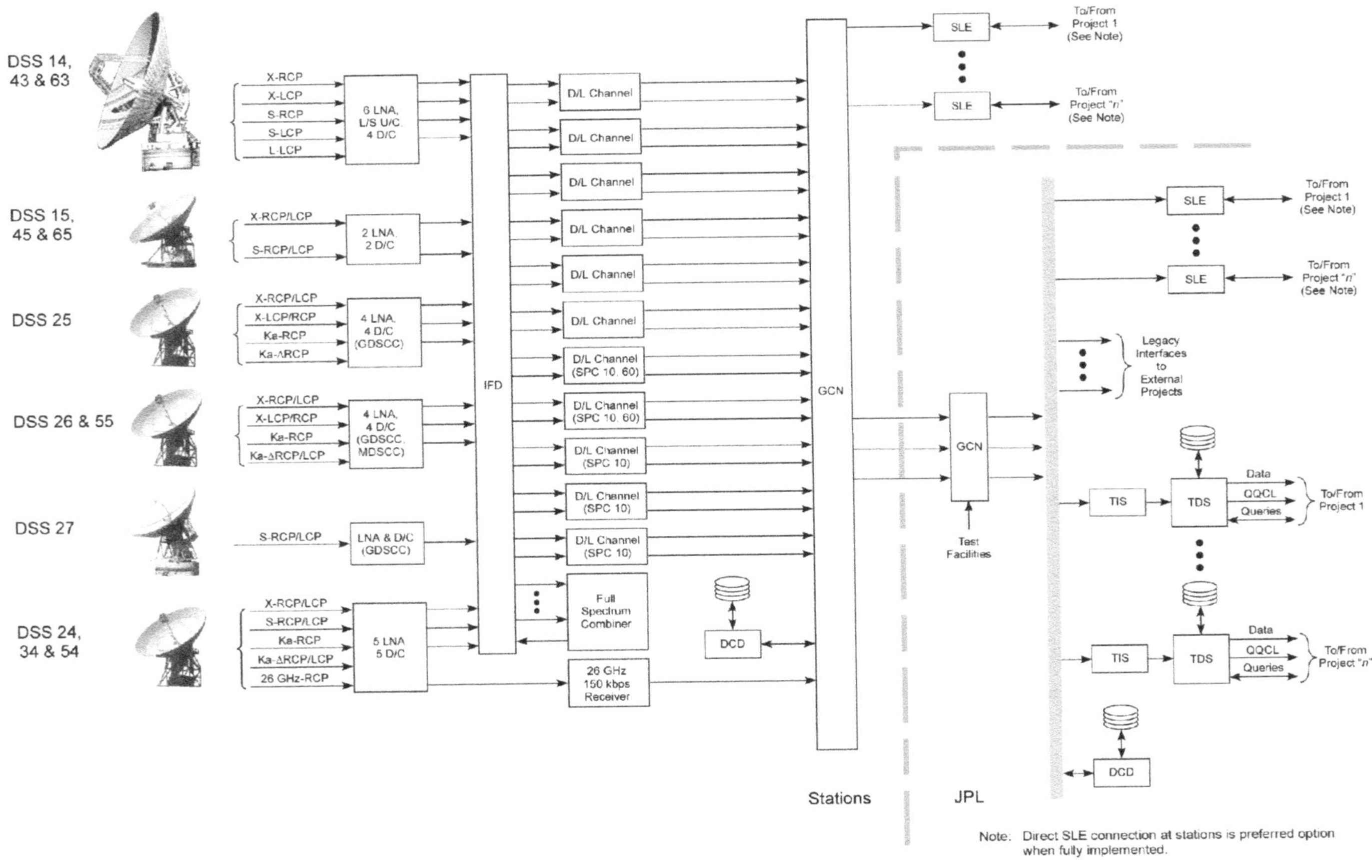

Fig. 2-9. DSN telemetry equipment for spacecraft support.

appropriate signal conditions. Turbo codes provide near-Shannon-limit error-correction performance with reasonable encoding and decoding complexity.

Frame synchronization must be established before processing any block code such as Reed–Solomon or Turbo codes or before formatting the data for delivery. Synchronization is accomplished by preceding each code block or transfer frame with a fixed-length attached synchronization marker (ASM). This known bit pattern can be recognized to determine the start of the code blocks or transfer frames. It also can be used to resolve the phase ambiguity associated with binary phase-shift keying (BPSK) or quadrature phase-shift keying (QPSK) staggered-quadrature phase-shift keying or offset-quadrature phase-shift keying (SQPSK or OQPSK) modulation. The DSN contains two frame synchronizers. The first of these operates in the bit domain and is used with convolutionally coded, Reed–Solomon coded or uncoded data. The second operates in the symbol domain and is used with Turbo coded data.

The ground communications network (GCN) uses communications circuits provided by the NASA Integrated Services Network (NISN) to connect the stations to JPL Central and users. The DSN provides CCSDS SLE data delivery directly from the station at which it is received or through the DSN central facility at JPL. Data storage, buffering against line outages, access, retrieval, and query are provided at all locations. Data delivery for additional telemetry functions such as packet extraction and CFDP file processing is from the DSN central facility where these functions are performed.

The Advanced Multimission Operations System (AMMOS) processes telemetry in both near-real time (delays as long as 1 minute) and in nonreal time (as complete a record as possible, but with a delivery time guaranteed within 2 hours of the end of track). The non-real time version includes retransmission of data lost between the station and JPL and replays from the central data recorder (CDR) as necessary.

Telemetry processing by AMMOS at JPL includes “channelizing” the data from the packets received, ordering the telemetry data that may have been transmitted in real time or from spacecraft storage, and time-tagging the data either by Earth-received time (ERT) or spacecraft-event time (SCET). Starting in 2011, the MSL project was the first to use the Mission-data Processing and Control Subsystem (MPCS) rather than AMMOS for telemetry data storage, display, and query.

2.6 DSN Performance

This section summarizes the major uplink and downlink characteristics of the stations when operating in the DSN frequency bands. Tabulated values are for the 34-m BWG stations DSS-24 (S-band) and DSS-25 (X-band and

Ka-band) at Goldstone and the 70-m station DSS-14 at Goldstone. Refer to the more detailed tables in 810-5 [1] for other stations and other parameters.

2.6.1 Antenna Gain

Acquisition (AC) aid antennas operate at X-band only and at downlink only (Table 2-1).

Table 2-1. 34-m and 70-m antenna gain and beamwidth.

Station	Parameter	Unit	Value	Remarks
34-m BWG	Uplink gain	dBi	56.3	S-band
			67.1	X-band
			79.5	Ka-band
70-m	Uplink gain	dBi	63.0	S-band
			73.2	X-band
34-m BWG	Uplink 3 dB beamwidth	deg	0.263	S-band, DSS-24
			0.077	X-band
			0.016	Ka-band, DSS-25
70-m	Uplink 3 dB beamwidth	deg	0.128	S-band
			0.038	X-band
34-m BWG	Downlink gain	dBi	56.84	S-band DSS-24
			68.2	X-band main
			38.0	Ac aid
			78.9	Ka-band, DSS-25
70-m	Downlink gain	dBi	63.6	S-band
			74.6	X-band
34-m BWG	Downlink 3 dB beamwidth	deg	0.242	S-band DSS-24
			0.066	X-band main
			2.1	Ac aid
			0.017	Ka-band, DSS-25
70-m	Downlink 3 dB beamwidth	deg	0.118	S-band

2.6.2 Transmitter Power

The 20-kW S-band and X-band transmitters (Table 2-2) can be operated at levels between 200 W and the full rated 20 kW.

Only DSS-43 has a 400-kW S-band transmitter. The Ka-band transmitter at DSS-25 can be operated at levels between 50 W and the 800 W maximum.

2.6.3 System Noise Temperature

These specific S-band values (Table 2-3 apply to DSS-24. The X-band and Ka-band values apply to DSS-25.

2.6.4 Thresholds and Limits

The downlink carrier acquisition and tracking threshold depends on the receiver bandwidth. Tracking bandwidths of less than 1 hertz (Hz) are not recommended, and this equates to a minimum downlink carrier power of about –172 dBm.

Table 2-2. 34-m and 70-m transmitter power, EIRP and frequency bands.

Station	Parameter	Unit	Value	Remarks
34-m BWG	Power output	kW	20.0	S-band
			20.0	X-band
			0.8	Ka-band
70-m	Power output	kW	20.0	S-band main
			400	High power
			20.0	X-band
34-m BWG	EIRP	dBm	128.7	S-band
			139.6	X-band
			138.2	Ka-band
70-m	EIRP	dBm	135.6	S-band main
			148.7	High power
			145.8	X-band
34-m BWG	Frequency band	MHz	2110–2118	S-band
				DSS-24
			7149–7188	X-band
			34315–34415	Ka-band
				DSS-25

Table 2-3. 34-m and 70-m downlink system noise temperature and polarization.

Station	Parameter	Unit	Value	Remarks
34-m BWG	System noise temp (LNA1)	K	26.1 nondiplex	S-band
			33.5 diplex	DSS-24
			20.2 nondiplex	X-band main
			29.2 diplex	
			280	Ac aid
			27.9 Ka-only	Ka-band
			31.4 X/Ka	DSS-25
70-m	System noise temp (LNA1)	K	10.5 nondiplex	S-band
			15.0 diplex	DSS-14
			20.2 nondiplex	X-band
			29.2 diplex	
34-m BWG	Receive polarization		RCP or LCP	S-band
			RCP or LCP	X-band main
			RCP	Ac Aid
			RCP or LCP	Ka-band
70-m	Receive polarization		RCP or LCP	S-band
			RCP or LCP	X-band
34-m BWG	Frequency band	MHz	2200–2300	S-band
				DSS-24
			8400–8500	X-band
			31800–32300	Ka-band
				DSS-25

Terms: LCP = left circularly polarized; LNA1 = low-noise amplifier 1; RCP = right circularly polarized

The recommended maximum downlink received total power at the station's low noise amplifier is –90 dBm.

The maximum uplink ranging carrier suppression for reliable operations is –6 decibels (dB). The minimum downlink ranging Pr/No for reliable operation is –8 decibel Hertz (dB-Hz).

The minimum recommended transmitter power for normal operations with a 20-kW transmitter is 2 kW. For the initial acquisition day, 200 W is often used.

References

[1] *DSN Telecommunications Link Design Handbook*, 810-005, Jet Propulsion Laboratory, Pasadena, California, periodically updated. http://deepspace.jpl.nasa.gov/dsndocs/810-005/ (accessed October 30, 2014)

[2] *Space Link Extension—Forward CLTU Service Specification, CCSDS*, 912.1-B-1, Blue Book, Consultative Committee for Space Data Systems, April 2002. http://public.ccsds.org/publications/archive/912x1b3.pdf (Accessed November 14, 2013)

[3] "Space Link Extension Forward Link Service and Return Link Service," Document, 820-013-Telecom Rev. E (internal document, also D-76496), Jet Propulsion Laboratory, California Institute of Technology, Pasadena, California, October 20, 2015. https://jaguar.jpl.nasa.gov/SW_detail.php?modid=2203 (accessed January 25, 2016)

[4] *CCSDS File Delivery Protocol (CFDP) Recommended Standard*, CCSDS 727.0-B-4, Blue Book, issue 4, January 2007. http://public.ccsds.org/publications/archive/727x0b4.pdf (accessed May 2, 2014)

Chapter 3
Voyager Telecommunications

Roger Ludwig and Jim Taylor

This chapter describes how the two Voyager spacecraft and the Deep Space Network (DSN) ground systems receive and transmit data. The primary purpose of this chapter is to provide a reasonably complete single source from which to look up specifics of the Voyager radio communications.

The description is at a functional level, intended to illuminate the unique Voyager mission requirements and constraints that led to the design of the Voyager spacecraft communications system in the 1970s and the upgrade of flight software and the ground communication system in the 1980s. The chapter emphasizes how the end-to-end communication system continues to serve the Voyager Interstellar Mission (VIM) that began in the 1990s and continues in the 2010s [1–3].

The Voyager spacecraft were designed and constructed at the Jet Propulsion Laboratory (JPL) in Pasadena, California. The flight team, very much reduced in size more than 30 years after the launches, is also located at JPL.

3.1 Voyager Interstellar Mission Description

The two Voyager spacecraft are continuing on long-term (1977–2025) exploratory mission. After exploring the outer planets—Jupiter, Saturn, Uranus, and Neptune—the Voyager spacecraft reached the edge of the Solar System and continue heading toward their final destination: interstellar

Some of the grey scale figures in this book have color versions in the companion website www.wiley.com/go/deepspacecommunications. Please refer to this website where the text refers to colors.

Deep Space Communications, First Edition. Edited by Jim Taylor.

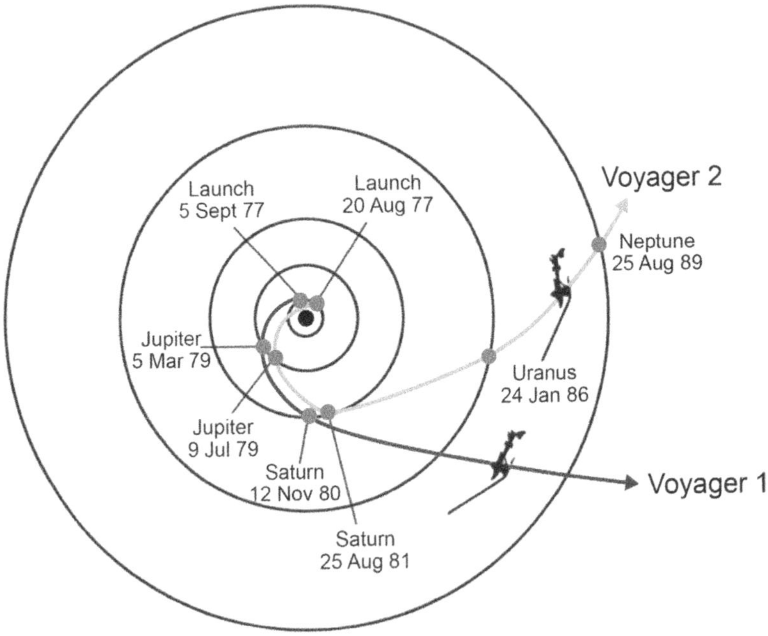

Fig. 3-1. Voyager flight paths.

space. They are now traversing regions of space never before encountered, building on the legacy of the National Aeronautics and Space Administration's (NASA's) most successful and productive interplanetary exploration endeavor [1].

Voyager 1 and Voyager 2 were launched in 1977, within the 3-year period that occurs once every 176 years when a unique alignment of Earth, Jupiter, Saturn, Uranus, and Neptune presents the opportunity for a "Grand Tour." Both spacecraft had close encounters with Jupiter and Saturn. Voyager 1 (launched second) arrived at Saturn first and successfully scanned the scientifically interesting and high-priority moon Titan, then passed somewhat "beneath" Saturn and was deflected "up," north of the ecliptic plane at an angle of approximately 35 deg. This freed the later-arriving Voyager 2 (launched first) from the Titan obligation, allowing it to be targeted on to Uranus and Neptune. Voyager 2 departed Neptune and the ecliptic heading approximately 48 deg south. Voyager flight paths are displayed in Fig. 3-1.

The remainder of this section focuses on the Voyager Interstellar Mission (VIM), the current mission phase,[1] which began in January 1990. The VIM is

---

[1] Earlier mission phases included launch and Earth-Jupiter cruise and the planetary mission (Jupiter, Saturn, Uranus, and Neptune encounters). These phases are archived in a section of Ref. 1, http://voyager.jpl.nasa.gov/news/index.html#

critical for meeting certain science objectives as defined in NASA's Space Science Enterprise 2000 Strategic Plan.[2] One objective in the plan that year was to "understand our changing Sun and its effects throughout the Solar System." A dozen years later, the Voyager mission was the only one continuing to explore the outer heliosphere. The Voyager spacecraft remain on trajectories that are ideally situated to contribute to our understanding of events occurring within and eventually beyond the farthest reaches of the immense region carved out of the interstellar medium by the Sun.

Other Strategic Plan objectives defined in 2000 were to "Learn how galaxies, stars, and planets form, interact, and evolve" and to "Use the exotic space environments within our Solar System as natural science laboratories and cross the outer boundary of the Solar System to explore the nearby environments of our galaxy." The Voyager spacecraft are the only ones in position to carry out the objective of exploring nearby environments of our Galaxy. The longevity of the Voyagers makes them ideal platforms for studying long-term solar wind variations. Their distance makes them ideal for studying the evolution of the solar wind, shocks, and cosmic rays. The interpretation of Voyager data is greatly enhanced by the ability to compare it with data from Earth-orbiting spacecraft (IMP 8, WIND, ACE, SAMPEX, and IBEX) and Ulysses traveling far south and north of the ecliptic.

The Voyagers and Pioneers 10 and 11, launched 4 and 5 years earlier, are the first four spacecraft to escape the gravity of our Solar System on their journeys into the Milky Way. Due to better launch dates and a speed advantage, the Voyagers are now outdistancing the Pioneers and achieving certain milestones first. Voyager 1 crossed Pluto's orbit in 1988 before Pioneer 10 at about 29 astronomical units (AU), when Pluto's orbit was inside Neptune's. Although Pioneer 11 crossed Uranus' orbit just before Voyager 2's 1986 encounter, Voyager 2 encountered Neptune in 1989 before Pioneer 11 crossed Neptune's orbit.

The Voyagers, depicted in Fig. 3-2, each carry the following instruments:[3]

- Plasma spectrometer (PLS) measures velocity, density, and pressure of plasma ions

---

[2] The NASA Strategic Plan is available at http://science.nasa.gov/about-us/science-strategy/. The 2000 Strategic Plan cited in this chapter is no longer accessible to the public. A link to the 2011 Plan is http://www.nasa.gov/pdf/516579main_NASA2011StrategicPlan.pdf

[3] Figure 3-2 shows these instrument locations on the spacecraft. For more information on the instruments and experiments, see http://nssdc.gsfc.nasa.gov/nmc/spacecraftDisplay.do?id=1977-084A in the National Space Science Data Center [4].

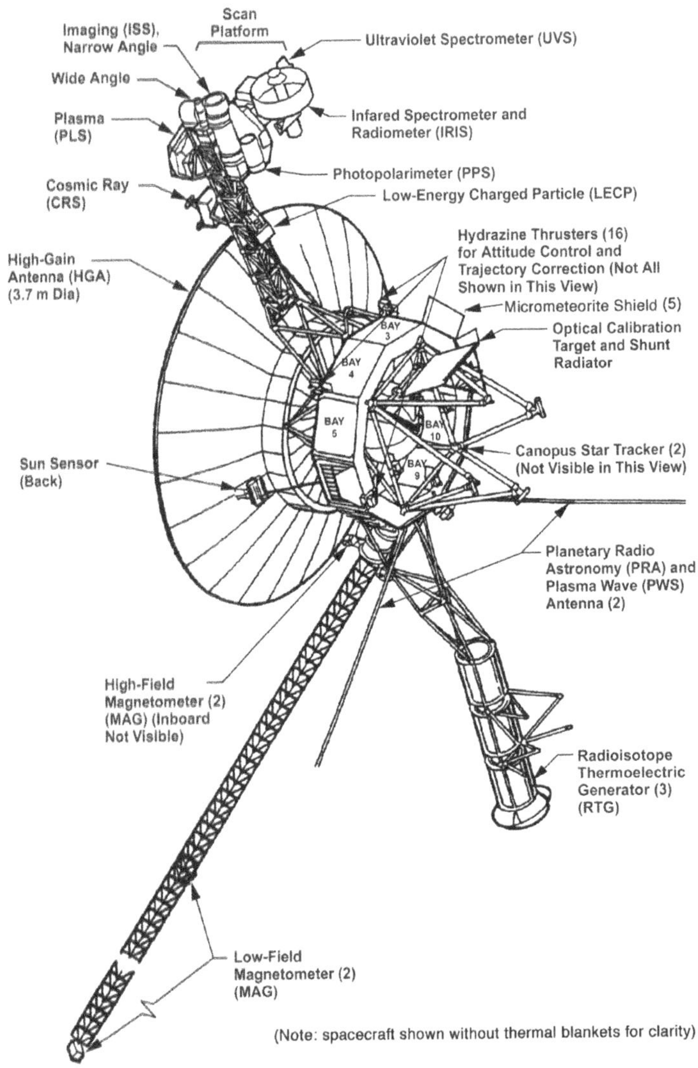

Fig. 3-2. Voyager spacecraft and science instruments.

- Low-energy charged particles (LECP) experiment measures electrons, protons, and heavier ions in the tens of kilo-electron volts (keV) to mega-electron volts (MeV) range
- Cosmic ray system (CRS) measures cosmic ray electron and nuclei energies in the 3 to 30 MeV range
- Triaxial fluxgate magnetometer (MAG) measures the strengths of planetary and interplanetary magnetic fields
- Plasma wave system (PWS) observes low-radio-frequency electron-density profiles and plasma wave-particle interactions

- Planetary radio astronomy (PRA) experiment studied radio-emission signals from Jupiter and Saturn
- Ultraviolet spectrometer (UVS) measures atmospheric properties in the ultraviolet spectrum
- Imaging science system (ISS) includes one narrow-angle, long-focal-length camera and one wide-angle, short-focal-length camera
- Photopolarimeter system (PPS), to collect emission intensity data, includes a polarizer and a filter for one of eight bands in the 220- to 730-nanometers (nm) spectral region
- Infrared interferometer spectrometer (IRIS) and radiometer measures local and global energy balance and vertical temperature profiles of the planets, satellites, and rings.

The spacecraft and instruments are generally in good health. With two exceptions, the instruments work well and all have the sensitivity to continue observations in the environments expected beyond the termination shock and heliopause. The PLS on Voyager 1 no longer returns useful data. The Voyager 2 MAG experiment has had a continuing problem with noise generated by the spacecraft and other instruments making reliable analysis very difficult, but the increase in magnetic field strength as solar maximum approached in 2001 and again in 2013 has made that problem more tractable.

The VIM consists of three distinct phases: termination shock, heliosheath exploration, and interstellar exploration. The two Voyager spacecraft began the VIM operating in an environment controlled by the Sun's magnetic field with the plasma particles being dominated by those contained in the expanding supersonic solar wind. This is the characteristic environment of the termination shock phase. At some distance from the Sun, the supersonic solar wind is held back from further expansion by the interstellar wind. The first feature encountered by a spacecraft as a result of this interstellar wind/solar wind interaction is the termination shock where the solar wind slows from supersonic to subsonic speed and large changes in plasma flow direction and magnetic field orientation occur.

Passage through the termination shock ended the termination shock phase and began the heliosheath exploration phase. Voyager 1 crossed the termination shock at 94 AU in December 2004, and Voyager 2 crossed at 84 AU in August 2007. After passage through the termination shock, each spacecraft was operating in the heliosheath environment, which is still dominated by the Sun's magnetic field and particles contained in the solar wind. The thickness of the heliosheath had been uncertain, estimated to be tens of astronomical units thick, taking several years to traverse.

The heliosheath exploration phase ends with passage through the heliopause which is the outer extent of the Sun's magnetic field and solar wind. Voyager 1 has completed its passage through the heliopause [5], thus starting the interstellar exploration phase with the spacecraft operating in an interstellar wind dominated environment. This interstellar exploration is the ultimate goal of the Voyager Interstellar Mission.

Voyager 1 has been escaping the Solar System at a speed of about 3.6 AU per year, 35 degrees (deg) out of the ecliptic plane to the north, in the general direction of the Solar Apex (the direction of the Sun's motion relative to nearby stars). Voyager 2 is also escaping the Solar System at a speed of about 3.3 AU per year, 48 deg out of the ecliptic plane to the south.

Both Voyagers were expected to cross the heliopause 10 to 20 years after reaching the termination shock. The crossing has been determined to be 2012 for Voyager 1 and should be within the span of ~2017–2027 for Voyager 2. In late-2013, Voyager 1 was announced as the first human-made object to venture into interstellar space [6]. "We believe this is mankind's historic leap into interstellar space," said Ed Stone, Voyager project scientist based at the California Institute of Technology, Pasadena. "The Voyager team needed time to analyze those observations and make sense of them. But we can now answer the question we've all been asking – 'Are we there yet?' Yes, we are."

Voyager 1 is just outside the solar bubble, where some effects from our sun are still evident. Figure 3-3 is an artist's concept of the outer environments or regions that were being explored by the VIM at about the time of the Voyager 1 entry into interstellar space.[4]

---

[4] A summary of the "solar bubble" (heliosphere) boundaries and regions defined in Figure 3-3 is in http://www.jpl.nasa.gov/news/news.php?feature=3838.

In 2004, Voyager 1 passed through the termination shock into the slow-down region, as it first detected the increased pressure of interstellar space on the heliosphere.

In 2010, it then passed into the "stagnation region" where the outward velocity of the solar wind slowed to zero and sporadically reversed direction. In Fig. 3-3, in the slow-down and stagnation regions, the prevalence of low-energy charged particles from the heliosphere jumped dramatically and is indicated by the dots in the two regions.

On August 25, 2012, Voyager 1 entered the depletion region, where the magnetic field allows energetic ions from inside the heliosphere to escape out, and cosmic rays from interstellar space zoom in. This outer region is also called the magnetic highway for the magnetic field and its effect on the ions and cosmic rays. Additional details are in a section of [1] that also has references to mid-2013 scientific papers.

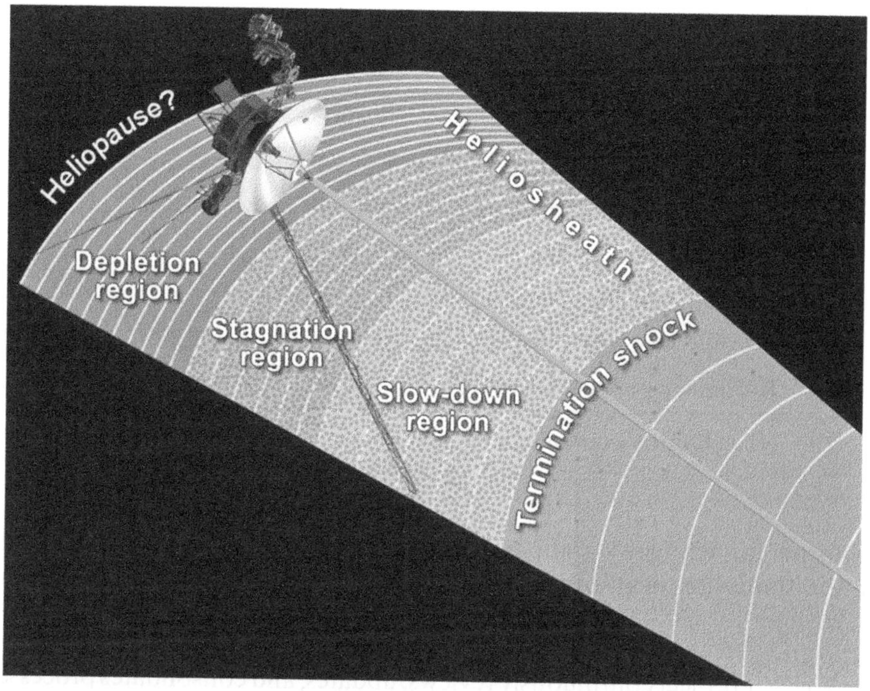

Fig. 3-3. Voyagers 1 and 2 exploration of the outer regions of the solar bubble as of 2013.

The Voyagers have enough electrical power and thruster fuel to operate at least until 2025. By that time, Voyager 1 will be 15.5 billion miles (24.9 billion kilometers [km]) from the Sun and Voyager 2 will be 13.0 billion miles (20.9 billion km) away. Eventually, the Voyagers will pass other stars. In about 40,000 years, Voyager 1 will drift within 1.6 light years (9.3 trillion miles) of AC+79 3888, a star in the constellation of Camelopardalis. In some 296,000 years, Voyager 2 will pass 4.3 light years (25 trillion miles) from Sirius, the brightest star in the sky. The Voyagers are destined—perhaps eternally—to wander the Milky Way.

The two Voyagers were the first operational spacecraft to reach the heliopause and to return the science observations from that region. The duration of the VIM will be limited primarily by the decreasing spacecraft electrical power from the two radioisotope thermoelectric generators (RTGs) and telemetry link capability. Table 3-1 provides life estimates for electrical power, telecommunications, and hydrazine (for attitude control).

Voyager 1 can be tracked by stations at all three sites. With Voyager 2 far south of the ecliptic, it is not visible from the northern hemisphere stations so the telecommunications link is only through Canberra. The table shows telemetry data rate limits for two Deep Space Station sizes at Goldstone,

Table 3-1. Spacecraft lifetime estimates in calendar years.

	Voyager 1	Voyager 2
Electrical power	2023	2023
Telemetry link capability		
7200 bps, 70-/34-m HEF[a] array	1994	1998
1400 bps, 70-m antenna	2007	2011
160 bps, 34-m HEF antenna	2024	2029
40 bps, 34-m HEF antenna	2050	2057
Hydrazine for attitude control	2040	2048

[a] High Efficiency (antenna)

California for Voyager 1 and near Canberra, Australia for Voyager 2. Limits for the third site, near Madrid, Spain, are similar to those at Goldstone for Voyager 1.

The Voyager project continuously reviews, updates, and consolidates processes in order to increase efficiency and improve its return on public investment. During VIM, Voyager has reduced its flight team staffing by 97 percent, from approximately 300 in 1989 to 10 in 2002. Reduced staffing since then has constrained VIM in the areas of non-routine activity planning, execution and analysis, and anomaly response.

The allocations of VIM telemetry rate to types of data are as follows. (At 160 bits per second (bps) or 600 bps, the different data types are interleaved.)

- Playbacks of data recorded at 7200-bps or 1400-bps on the tape recorder
- 160-bps real-time fields, particles, and waves; UVS subset; engineering
- 40-bps real-time engineering data.

3.2 Overview of Telecom Functional Capabilities

This section describes telecom system capabilities that existed at launch. Figure 3-4 shows the functions of the spacecraft and the DSN telecom system. Some functions, such as S-band downlink and the spacecraft low-gain antenna (LGA), are no longer used. Section **3.7**, Operational Scenarios of the Voyager Interstellar Mission, describes the combinations of capabilities being used in the Voyager Interstellar Mission (VIM).

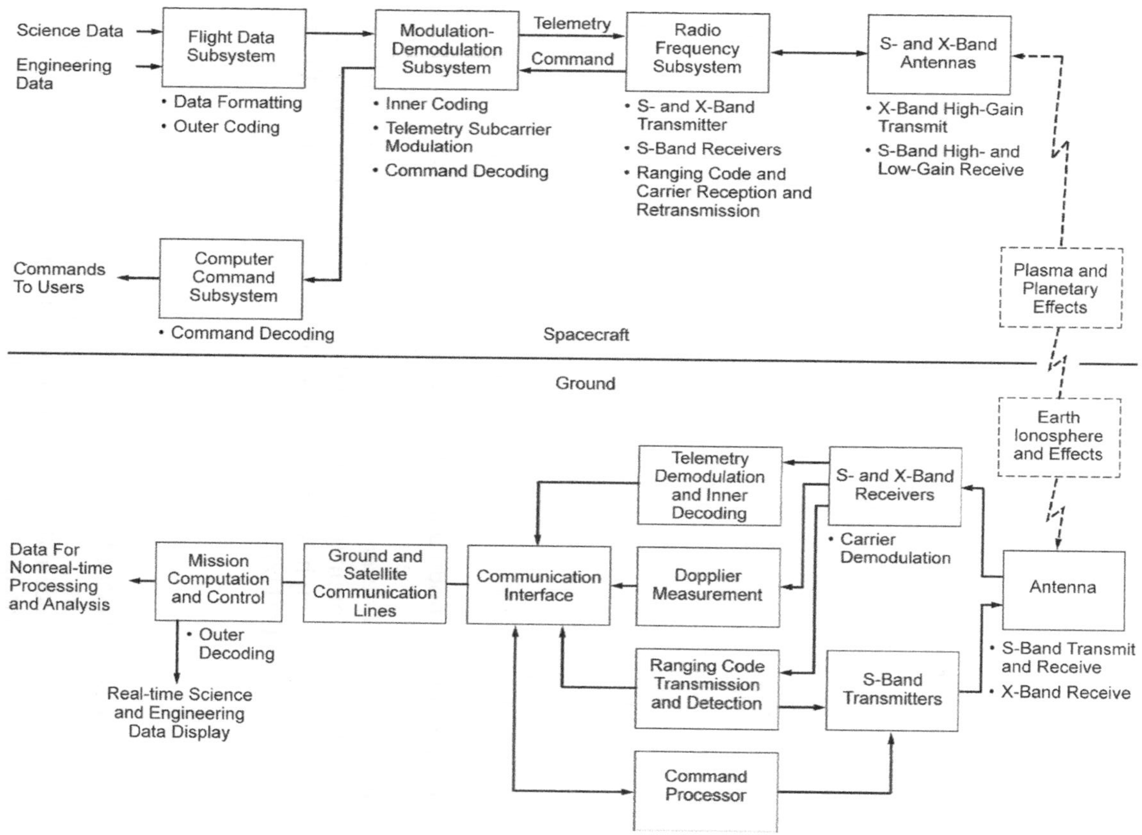

Fig. 3-4. Overview of spacecraft and ground telecommunications functions for Voyager.

3.2.1 Uplink

3.2.1.1 Uplink Carrier. Each Deep Space Station (DSS) transmits an uplink carrier frequency[5] of 2114.676697 megahertz (MHz) to Voyager 1 and 2113.312500 MHz to Voyager 2. The carrier may be unmodulated or modulated with command (CMD) or ranging (RNG) data or both. Phase lock to the uplink carrier is provided. When the transponder[6] receiver (RCVR) is phase locked, its voltage-controlled oscillator (VCO) provides a frequency reference to the exciter to generate a downlink carrier that is two-way coherent with the uplink.

3.2.1.2 Ranging Modulation. Voyager uses standard DSN turnaround sequential ranging modulation. "Turnaround" means the ranging modulation on the uplink carrier is demodulated by the spacecraft receiver and remodulated on the downlink carrier. "Sequential" means that a series of ranging codes are transmitted one after the other, allowing for both sufficient resolution in range and elimination of ambiguity in range. (The DSN ranging modulation is described in Module 203, Sequential Ranging, of the *DSN Telecommunications Link Design Handbook* [7].)

The spacecraft transponder has the capability to demodulate the uplink ranging data from the uplink carrier and modulate it on the S-band[7] downlink carrier, the X-band downlink carrier, or both downlink carriers simultaneously. For the ranging acquisitions to be valid, the transponder must be configured (set) for two-way coherent operation.

3.2.1.3 Command Demodulation. Voyager receives and demodulates the command signal from the uplink carrier. (DSN command modulation is described in the *DSN Telecommunications Link Design Handbook* [7], module 205, 34-m and 70-m Command.) The signal consists of 16-bps, Manchester-encoded commands, biphase modulated onto a squarewave subcarrier frequency of 512 hertz (Hz).

---

[5] These frequencies are DSN Channel 18 and Channel 14, respectively. The specific values are the defined channel center frequencies. The DSN channels are defined in Module 201, Frequency and Channel Assignments, in the *DSN Telecommunications Link Design Handbook* [7].

[6] A transponder includes a receiver and an exciter. An exciter is the part of a radio transmitter that produces the downlink carrier frequency.

[7] For spacecraft in the deep space frequency bands, S-band refers to an uplink frequency of about 2115 MHz and a downlink frequency of about 2295 MHz. X-band refers to a downlink frequency of about 8415 MHz.

Table 3-2. Voyager 1 and Voyager 2 downlink frequencies and channels.

Spacecraft	Coherent Downlink Frequency (MHz)	Channel	Non-Coherent Downlink Frequency (MHz)	Channel
Voyager 1	2296.481481	18	2295.000000	14
Voyager 2	2295.000000	14	2296.481481	18
Voyager 1	8420.432097	18	8415.000000	14
Voyager 2	8415.000000	14	8420.432097	18

3.2.2 Downlink

3.2.2.1 Downlink Carriers. When the transponder is set to the two-way coherent tracking mode and is locked to an uplink carrier, the received carrier frequency is used to generate phase and frequency coherent downlink carriers. The ratio between downlink frequency and uplink frequency is 240/221 for the S-band downlink and 880/221 for the X-band downlink.

The transponder may also be set to a mode in which the receiver may be locked to an uplink, but the downlink carrier is not coherent with that uplink carrier.[8] In this mode, or when the receiver is not locked to an uplink carrier, an onboard frequency source generates the downlink carrier frequencies.

3.2.2.2 Transmit Frequencies. Table 3-2 contains the downlink carrier frequencies and associated DSN channel numbers that Voyager 1 and Voyager 2 produce in the coherent and non-coherent modes.

3.2.2.3 Downlink Polarizations. Table 3-3 defines the downlink polarization produced at S-band (from either power amplifier) and X-band (from the selected traveling wave tube amplifier (TWTA).

[8] The described mode is "two-way non-coherent on," or "TWNC on." Voyager is one of many JPL Deep Space missions that have two transponder modes called "TWNC on" and "TWNC off." (Beginning in the late 1990s, the term "coherency disabled" replaced "TWNC on" and "coherency enabled" replaced "TWNC off".) TWNC is pronounced "twink." The TWNC on mode means the downlink frequency cannot be coherent with an uplink frequency. The TWNC off mode means the downlink will be coherent with the uplink when the transponder's receiver is in lock.

Table 3-3. S-band and X-band downlink polarizations.

Link	Polarization
S-band	Right circular
X-band TWTA-1	Left circular
X-band TWTA-2	Right circular

3.2.2.4 Telemetry Modulation. The telemetry comes to the telemetry modulation unit (TMU) separately as a "low-rate" channel and a "high-rate" channel. Low rate is 40 bps only and its routing through the TMU is such that it can only be downlinked as uncoded bits. High rate, one of a set of rates between 10 bps and 115.2 kilobits per second (kbps), is downlinked as coded symbols. The TMU encodes the high-rate data stream with a convolutional code having constraint length of 7 and a symbol rate equal to twice the bit rate ($k = 7$, $r = 1/2$).

Either modulator can biphase-modulate the telemetry symbols onto either a 22.5-kilohertz (kHz) or a 360-kHz subcarrier. A subcarrier is a symmetrical square wave signal derived from a TMU crystal oscillator that has a nominal frequency of 2.88 MHz. The 360-kHz subcarrier is required for bit rates greater than 7.2 kbps. The TMU has one modulator for the S-band downlink and another for X-band downlink. The modulated subcarrier goes to the S- or X-band exciter.[9]

3.3 Spacecraft Telecom System Design

3.3.1 Spacecraft Telecom System Overview

The telecom system consists of three subsystems, as detailed in Table 3-4. The table shows the subsystem acronyms for reference.

Note: In Table 3-4, the pairs of components, such as S-RCVRs, are identical to one another, providing redundancy for the function. In each case, one (and only one) of the components is operative (powered on) at a time.

[9] DSN telemetry data bit, symbol, and subcarrier waveform requirements are defined in the *DSN Telecommunications Link Design Handbook* [7], Module 207, 34-m and 70-m Telemetry Reception.

Table 3-4. Voyager spacecraft telecom subsystems and their components.

Subsystem/Component Quantity	Quantity
Radio frequency subsystem (RFS)	
S-band receiver (S-RCVR)	2
S-band exciter (S-EXC)	2
X-band exciter (X-EXC)	2
S-band traveling wave tube amplifier (S-TWTA)	1
S-band solid-state amplifier (SSA)	1
X-band traveling wave tube amplifier (X-TWTA)	2
Ultrastable oscillator (USO)	1
Modulation demodulation subsystem (MDS)	
Command detector unit (CDU)	2
Telemetry modulation unit (TMU)	
S/X-band antenna subsystem (SXA)	
High-gain antenna (HGA)	1
Low-gain antenna (LGA)	1

Figure 3-5 is a functional block diagram of the Voyager telecom system.[10] The telecom system is housed in equipment bays 1, 9, and 10 of the spacecraft bus. The bus is a decagonal structure, with each of the ten sides making up the external surface of one equipment bay, as shown in Figure 3-2.

The radio frequency subsystem is designed to perform as a command receiver, a phase-coherent ranging transponder, and a telemetry transmitter. Final X-band amplification is provided by redundant traveling-wave tube amplifiers; final S-band amplification by a traveling-wave tube amplifier or solid state amplifier, working as a redundant pair. Microwave components provide radio frequency (RF) filtering and switching for connecting the transmitters and receivers to the high-gain or low-gain antenna.

The modulation demodulation subsystem has redundant command detector units and telemetry modulation units.

[10] The numbered triangular markers in the block diagram indicate capabilities that are no longer available in VIM, including those lost due to hardware failures or other circumstances. The Voyager 2 receiver problems (flags 3 and 4) still require special procedures as described in Section **3.7**, Operational Scenarios.

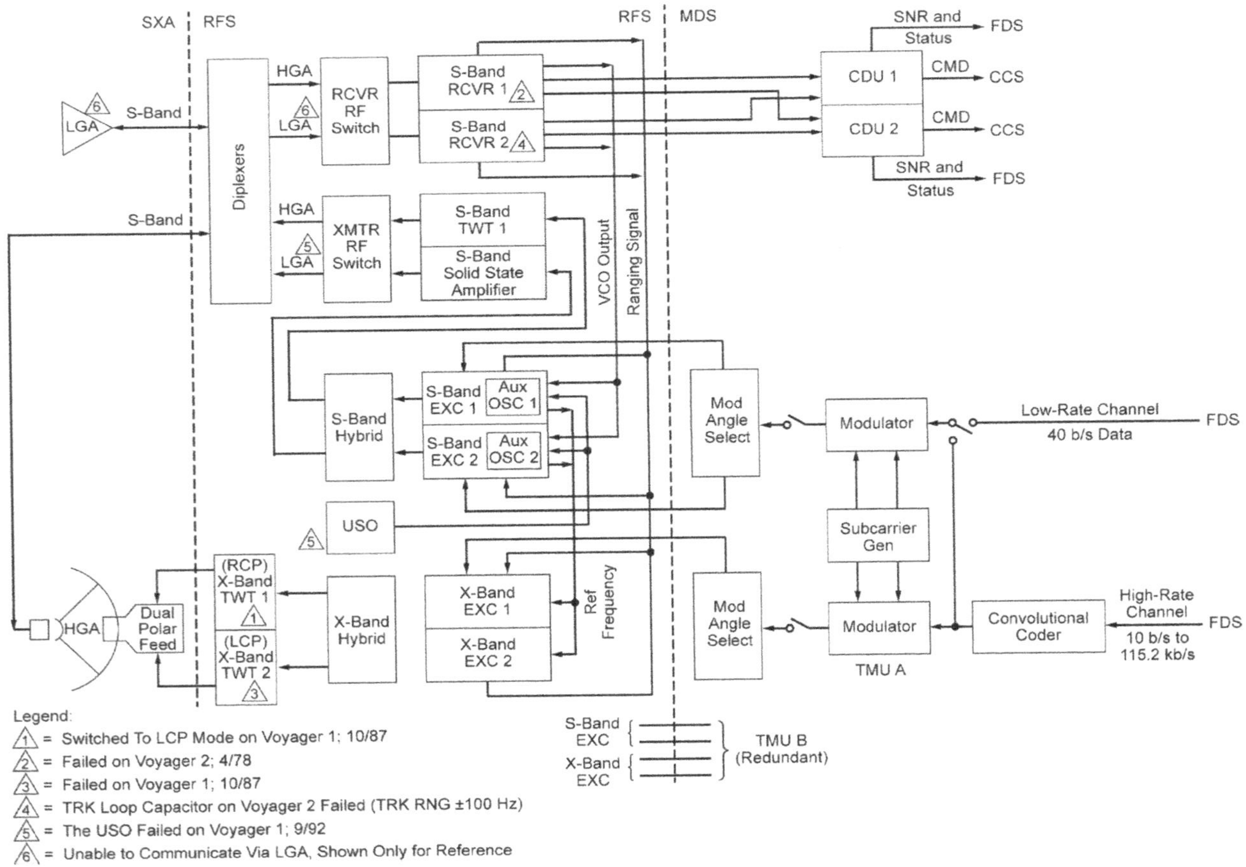

Fig. 3-5. Voyager telecommunications functional block diagram.

The S-/X-band antenna subsystem consists of an LGA and an HGA. The subsystem receives S-band signals and transmits S-band and X-band signals to and from the Deep Space Network.

The telecom system receives control instructions from the computer command subsystem (CCS) and the flight data subsystem (FDS) to select its operating modes. The primary modes are:

- S-band TWTA/SSA high power
- S-band TWTA/SSA low power
- S-band ranging on
- S-band ranging off
- X-band TWTA high power
- X-band TWTA low power
- X-band ranging on
- X-band ranging off
- HGA select for transmitting and receiving
- LGA select for transmitting and receiving
- TWNC on
- TWNC off
- USO on
- USO off

In Table 3-5, S-band bit rates are convolutionally coded, except for 40 bps which is uncoded. X-band planetary cruise bit rates 10–2560 bps are convolutionally coded, while all VIM rates and planetary playback rates (7.2–115.2 kbps) are coded with a concatenation of convolutional and Golay or convolutional and Reed-Solomon coding.

3.3.2 Modulation Demodulation Subsystem

3.3.2.1 Command Detector Units. The CDU demodulates the command subcarrier from the radio frequency subsystem (RFS) receiver, synchronizes its internal clock to the received command bit rate, and detects the command bits.

The CDU outputs to the CCS the detected command bits and a clock signal derived from the command bit rate. The CDU outputs status signals to the FDS.

3.3.2.2 Telemetry Modulation Units. The TMU can receive both high-rate and low-rate, non-return-to-zero (NRZ), serial digital data from the FDS. By using control input from the CCS, shown as switches in the TMU-A section

Table 3-5. Displays typical configurations used by Voyager for each mission phase.

Mission Phase	Transmitter Power		Antenna	Ranging		Subcarrier Freq (kHz)		Link Data Rates and Coding (bps)		RFS Tracking Configuration
	S	X		S	X	S	X	S	X	
Launch	Lo	Off	LGA	Off	Off	22.5	22.5	1200[a]	Off	1-way
1st 80 days	Hi	Off	LGA	Off	Off	22.5	Off	10–2560[a]	Off	2-way coherent
Planetary cruise	Off	Low	HGA	Off	On	Off	22.5	Off	10–2560[a]	2-way coherent
Planetary playback	Lo	Hi	HGA	On	On	22.5	360	40[c]	7.2 k–115.2 k[b]	2-way coherent
VIM cruise	Off	Lo	HGA	Off	Off	Off	22.5	Off	160[b]	1-way
VIM playback	Off	Hi	HGA	Off	Off	Off	22.5	Off	1.4 k–7.2 k[b]	1-way

[a] Convolutionally coded

[b] Convolutionally coded with Golay or Reed-Solomon

[c] Uncoded

of Figure 3-5, the TMU selects for a set of modes to process telemetry data through the TMU. CCS control inputs determine low-rate/high-rate data routing for S-band, subcarrier frequency selection, modulation index value, and the input of the modulated subcarrier to the S-band and X-band exciter. The low-rate data is not coded; the high-rate data is convolutionally coded. High-rate data is always available for the X-band downlink.

3.3.3 Radio Frequency Subsystem

3.3.3.1 Receivers. The receiver is a narrow-band, double-conversion, superheterodyne, automatic-phase-control design. The receiver has a coherent amplitude detector that detects and measures received-signal strength and provides the receiver with an automatic gain control (AGC) function. Receiver AGC is telemetered as a primary uplink performance parameter.

When phase locked to an uplink signal, the receiver's phase detector will

1) control the phase and frequency of the transmitted downlink carriers if in two-way coherent mode,

2) demodulate the composite command signal, if present, and

3) demodulate the ranging signal if present. The ranging signal level is controlled by the ranging AGC.

The received carrier frequency controls the generation of the coherent downlink at both S-and X-band, at transmit/receive frequency ratios of 240/221 and 880/221, respectively (two-way tracking). The ranging signal is provided by independently switched paths to the S- and X-band exciters for modulation of the downlinks. The S- and X-band ranging channels are controlled by discrete commands from the spacecraft CCS, regardless of which receiver is powered.

3.3.3.2 S-Band Exciters. The S-EXC provides RF drive to the S-band power amplifier and a frequency reference to the X-EXC. One or the other S-EXC must be powered on at all times. The X-band downlink requires an S-EXC, an X-EXC, and an X-TWTA.

The S-EXC also phase modulated the S-band downlink carrier present in earlier mission phases with the composite telemetry signal (modulated subcarrier) from the TMU and with the ranging signal detected by the receiver when the S-band ranging channel was controlled on.

Each S-EXC has a crystal-controlled auxiliary oscillator (aux osc) that can generate the downlink (D/L) in the TWNC-on mode or when there is no uplink (U/L). Although the USO (Section 3.3.3.6) is preferred as the D/L source in these cases, the aux osc can take over if the USO fails[11] or is switched off.

3.3.3.3 S-Band Power Amplifiers. Only one S-band power amplifier, either the S-TWTA or the SSA, may be powered at a time, or both may be off when no S-band downlink is required. Both power amplifiers have two RF output power levels available.[12] A CCS control instruction (high power/low power) establishes the power level in both units, with a separate control input to turn the selected power amplifier on or off.

3.3.3.4 X-Band Exciters. The X-band exciter converts the frequency at the output of the S-band exciter to X-band to drive the X-band TWTA. Comparable to the S-band exciter function, the X-band exciter phase modulates the RF

---

[11] The Voyager 1 USO failed in September 1992, as flagged by legend item 5 in Figure 3-5. The exciter aux osc has generated the 1-way downlink since then. Use of the less stable aux osc restricts Voyager 1 to transmitting downlink in the residual carrier mode only.

[12] The low-power and high-power RF levels to the HGA for the S-band TWTA are 6.5 W and 19 W. For the S-SSA, they are 6 W and 15 W [8].

signal with the composite telemetry signal from the TMU and, if the X-band ranging channel is on, the ranging signal detected by the receiver.

3.3.3.5 X-Band Power Amplifiers. Only one X-TWTA can be powered at a time. Further, a control input from the CCS ensures that the X-TWTA is powered off when the X-exciter is off. As is the case for S-band, whether powered on or not, the X-TWTA power level is selected to either of two levels by CCS control input. (The low-power and high-power RF levels to the HGA for the X-TWTA are 12 W and 18 W.)

In October 1987, the Voyager 1 X-TWTA-2 failed, as annotated in Figure 3-5, legend item 3. The primary downlink was switched to X-TWTA-1. In November 1998, Voyager 2 switched from X-TWTA-2 to X-TWTA-1.[13] For both spacecraft, legend item 1 in the figure flags the changes from X-TWTA-2's right-hand circular polarized (RHCP) downlink to X-TWTA-1's left-hand circular polarized (LHCP) downlink. (The relationship between X-TWTA selection and the resultant polarization of the X-band downlink is described in the SXA Section **3.3.4** that follows.)

3.3.3.6 Ultrastable Oscillator. The RFS has one USO. The USO provides the most stable frequency reference available for the downlink in the non-coherent mode of operation. The USO has short-term stability of 12×10^{-12} MHz and lifetime stability of $\pm 2 \times 10^{-6}$ MHz. The USO on/off state is via control input from the CCS. When the USO is turned off, the aux osc in the powered S-band exciter is consequently turned on, and vice versa.

3.3.4 S/X-Band Antenna Subsystem

The S/X-band antenna (SXA) consists of an S-band low-gain antenna (LGA) and an S-band and X-band high-gain antenna (HGA). For the HGA, the SXA has a 3.66-m diameter main reflector, a frequency selective surface (FSS) subreflector, and S-band and X-band feeds. Axially in front of the main reflector are, in order, the X-band feed, the FSS, the S-band feed, and the LGA (farthest from the spacecraft).

The main reflector, X-band feed, and the FSS (reflecting at X-band) form a Cassegrain radiator. The main reflector, FSS (transparent at S-band), and S-band feed form a prime focus radiator. The LGA is mounted on the back of the S-band feed structure. The SXA also includes an X-band waveguide, an S-band coaxial cable, and RF power probes for each frequency.

[13] The switch to the backup X-TWTA is in the status report section of the Voyager mission status web page, http://voyager.jpl.nasa.gov/news/voyager_status.html [1].

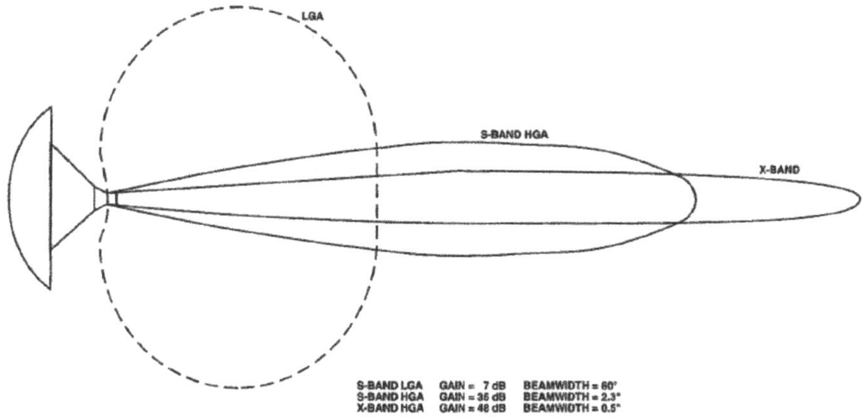

Fig. 3-6. Voyager SXA patterns and beamwidths.

Figure 3-6 sketches the relative patterns of the HGA and LGA, with the angles not to scale. The figure also lists the gain and beamwidth values. As the figure indicates, the LGA and HGA boresights are aligned with each other. The LGA has a broad S-band pattern about its boresight; the HGA has narrower S-band and X-band patterns as determined by the main reflector's diameter.

3.3.4.1 High-Gain Antenna. In the VIM, communication to and from the spacecraft is through the HGA. The HGA consists of S- and X-band feeds backed by a circular parabolic reflector. S-band gain is approximately 36 dBi; X-band gain is approximately 48 dBi.

The HGA is right-hand circularly polarized at S-band. At X-band it uses a dual-polarized Cassegrain feed that produces a right-hand or left-hand circularly polarized wave, depending on which of the two X-TWTAs is driving the feed. A left-hand circularly polarized downlink comes from X-TWTA-1; a right-hand circularly polarized downlink comes from X-TWTA-2.

3.3.4.2 Low-Gain Antenna. The spacecraft also carries an S-band-only LGA. The LGA consists of a right-hand circularly polarized radiator. The radiation pattern is approximately a cardioid of revolution. The gain of the LGA is approximately 7 dBi.

The LGA, used immediately following launch, was also available for emergency communications until margin for LGA links was exhausted in the 1980s.

3.3.5 Telecom System Input Power and Mass

Table 3-6 summarizes the steady-state spacecraft input power to the major telecom system units for both high-power and low-power modes, as applicable. The table also summarizes the masses of components of the system.

Table 3-6. Voyager spacecraft input power and mass summary.

	No. of Units	Input Power (W)[a]	Mass (kg)[b]
RFS			**44.0***
Transponder	2		4.7
Receiver		4.3	
S-Band Exciter		2.4	
ACIS[c]		0.9	2.5
S-TWTA	1	33.0/86.4[a]	5.1
S-band SSA	1	35.7/91.2[a]	5.0
X-TWTA	2	48.3/71.9[a]	5.8
USO	1	2.7	2.0
Diplexer	2		1.4
Receiver RF switch			1.2
Transmitter RF switch			0.9
Other microwave			3.5
Cabling			2.3
MDS			**8.4***
TMU	2	5.7	2.2
CDU	2	5.4	2.0
SXA			**53.0***
SXA, SXA coax, SXA waveguide			2.1
SXA structure, including main reflectors			50.9
Mass Total			**105.4**

[a] Low power/high power values do not include turn-on or turn-off transients.

[b] The stated mass is for one unit; for example, each X-TWTA weighs 5.8 kilograms (kg).

[c] Antenna control and interface system.

* The bold numbers 44.0, 8.4 and 53 are subtotals.

3.4 Telecom Ground System Description

The DSN is an international network of ground stations (antennas, transmitters, receivers, and associated systems) that operated intensively only at S-band and X-band during the first decades of the Voyager mission, with a Ka-band capability being developed in the 1990s.[14] The DSN supports interplanetary spacecraft missions and radio and radar astronomy observations for the exploration of the Solar System and beyond. The DSN consists of three deep-space communications complexes located approximately 120 deg from each other at Goldstone, in California's Mojave Desert; near Madrid, Spain; and near Canberra, Australia. Each complex has one 70-m antenna, two or more 34-m antennas, and one 26-m antenna (not used for Voyager).

Specific DSN numerical parameters for Voyager are defined in DSN Operations Plan for the Voyager Interstellar Mission [9]. The *Deep Space Mission Systems Telecom Link Design Handbook* [7] includes functional capability descriptions of each antenna type for the purpose of modeling link capability between a spacecraft and station.[15]

3.4.1 Uplink and Downlink Carrier Operation

Voyager uses an S-band uplink, X-band primary downlink, and S-band secondary downlink.[16] Command uplinks and maneuver and tape-recorder-playback downlinks require the 70-m antennas. The 34-m antennas are limited to reception of the relatively low-rate 160 bps cruise data.

Figure 3-7 shows the antenna and microwave sections of a 70-m station. The following paragraphs describe Voyager-related functions of that type of station. Refer to the *DSN Telecommunications Link Design Handbook* [7] for corresponding figures and descriptions of the other types of DSN stations.

3.4.1.1 Uplink. The uplink signal produced by the 20-kilowatt (kW) S-band transmitter goes through an S-band diplexer, orthomode junction, and polarizer to the S-band feed. The signal then passes through an S-/X-band dichroic reflector, subreflector, and main 70-m reflector before radiation to the spacecraft. Voyager has no plan to use the 400-kW S-band transmitter.

---

[14] A link to the home page of the DSN is http://deepspace.jpl.nasa.gov/dsndocs/810-005/ which has an "About the DSN" section with brief descriptions of the functions and history of the Network, as well as information about the tracking station complexes in Australia, Spain, and California.

[15] 810-005 (Rev. E) [7], was initially released January 2001 as a paper document. Modules in Rev E are updated as needed and are now maintained online at the link http://deepspace.jpl.nasa.gov/dsndocs/810-005/. Though the Voyager spacecraft was originally designed to work with ground systems defined in previous versions of the Handbook, the Rev E systems continue to support the Voyagers. Also, see http://www.jpl.nasa.gov/basics/bsf18-3.html for a general description of uplink and downlink data flow at a Deep Space Communications Complex.

[16] For Neptune encounter and later, the X-band downlink was designated "primary" to return science data because it has greater telecom performance than the S-band link via the HGA. The S-band link was designated "secondary" because it would be selected by system fault protection if the X-band link were to fail.

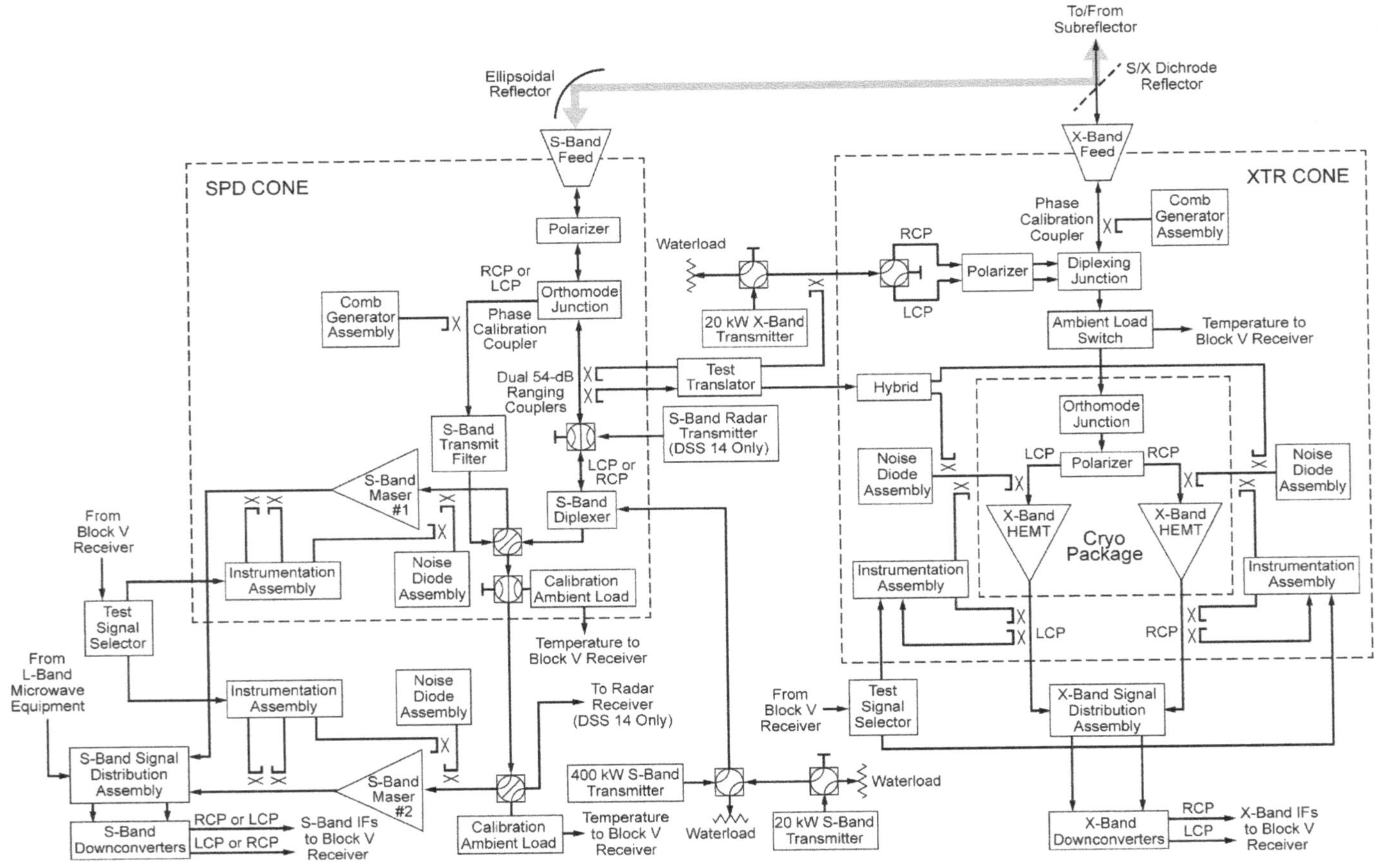

Fig. 3-7. DSS-14 and DSS-43 microwave and transmitter block diagram.

3.4.1.2 Downlink. The X- and S-band downlinks from the main (70-m) reflector and the subreflector are both focused at the S/X dichroic reflector. A dichroic is reflective at one frequency band and transparent at another, thus allowing S-band frequencies to be separated from X-band frequencies. This dichroic reflector reflects the S-band (on the path shown by the thick line in Figure 3-7) to the S-band feed and passes the X-band through to the X-band feed with very low loss.

From the diplexing junction, the X-band signal goes to a polarizer that selects the right circular polarization output for both spacecraft. The output from the polarizer is amplified by the X-band high-electron-mobility field-effect transistor (HEMT) preamplifier and frequency-downconverted for input to the block V receiver (BVR).

The S-band downlink signal shares a common path with the uplink signal between the feed and the S-band diplexer. The diplexer routes the downlink to the S-band low-noise preamplifier (S-band maser). After further routing and downconversion similar to the X-band downlink, the S-band downlink is input to the BVR.

3.4.2 Command Processing

Voyager command files are transferred to the station minutes in advance of transmission in a store-and-forward system. At the station, standards and limits tables the command processor assembly (CPA) and the command modulator assembly (CMA) clock out the command bit stream, modulate the command subcarrier, and provide the modulated subcarrier to the station's exciter for modulation of the RF uplink carrier. The command bit rates, the command subcarrier frequency, and the command modulation index (suppression of the uplink carrier) are controlled through standards and limits tables.

The JPL Voyager Spacecraft Mission Controller, referred to as the ACE, operates the multimission command system from a workstation in the mission support area. Just prior to a command session, the ACE directs the station to turn command modulation on and selects the 16-bps command rate and a calibrated "buffer" in the station's CMA. The CMA produces the command subcarrier, which produces a 512-Hz square wave to match the subcarrier-tracking-loop best-lock frequency in the Voyager CDU. As the ACE sends the spacecraft commands, the CMA modulates the command-bit waveform onto the subcarrier. When finished, the ACE directs the station to turn command modulation off.

3.4.3 Telemetry Processing

Two BVRs are assigned to a project's tracking pass. Each BVR has phase-locked loops for acquiring and tracking the carrier, telemetry subcarrier, and telemetry symbol stream. Voyager generates a 22.5-kHz subcarrier for use with bit rates less than or equal to 7.2 kbps and a 360-kHz subcarrier for use with bit rates greater than 7.2 kbps. In the residual carrier mode, the X-band carrier modulation index settings vary from 51 deg for the lowest data rate (10 bps) to 80 deg for the highest (115.2 kbps).[17]

The BVR delivers telemetry symbols to the maximum likelihood convolutional decoder (MCD). Voyager can use either the Block 2 or Block 3 MCD (MCD2 or MCD3)[18] to process the (7,1/2) convolutional code. The MCD outputs decoded telemetry bits to the frame synchronizer subsystem (FSS).

An MCD/FSS pair makes up a telemetry channel assembly (TCA). The telemetry group controller governs the operation of TCA1 (with MCD3) and TCA2 (with MCD2). After the MCD achieves lock, the FSS requires recognition of a minimum of two successive frame-sync words to output telemetry to the project. Validation requires recognition of a third sync word. The number of allowable miscompares (between received and expected bit values) in each frame-sync word recognized by the synchronizer can be set in the software.

3.5 Sample Telecom System Performance

The Voyager spacecraft receives an S-band uplink from the Earth and transmits S-and X-band downlinks to the Earth, compatible with DSN station configurations and performance defined in the *DSN Network Operations Plan for VIM* [9] and the *DSN Telecommunications Link Design Handbook* [7].

The telecommunications system is capable of simultaneous commanding, telemetry processing, and radiometric tracking using any combination of the available uplink and downlink frequency bands. See the *Voyager Telecommunications Design Control Document* [10] for planned telecom configurations and predicted uplink and downlink performance during Voyager's prime mission. (This section is limited to a summary of the telemetry performance during VIM.)

The Voyager communication link margins are computed using the link budget techniques and statistical criteria defined in *Deep Space Telecommunications Systems Engineering* [11].

The four VGR telecom functions are carrier tracking (Doppler), command, telemetry, and ranging. The performance of each function is expressed as a signal-to-noise ratio (SNR), as shown in Table 3-7.

---

[17] A modulation index of 90 deg puts all of the power in the sidebands and therefore produces a suppressed carrier mode. Suppressed carrier mode is used during VIM to extend Voyager 2 playback data rate capability. See Section **3.6**, New Telecom Technology.

[18] See Module 208, Telemetry Data Decoding, in the *DSN Telecommunications Link Design Handbook* [7] for a description of the Block 2 and Block 3 MCDs. Block 3 refers to a later DSN equipment implementation than Block 2, and has been available for operational use since 1997.

Table 3-7. VGR telecom link functions and signal-to-noise ratios.

Function	SNR Definition
Carrier	P_c/N_0
Command	E_b/N_0
Telemetry	E_s/N_0
Ranging	P_r/N_0

Each SNR is expressed in terms of N_0, which is noise spectral density. The "signal" part of the SNR is P_c (carrier power), E_b (energy per command bit), E_s (energy per telemetry symbol), or P_r (downlink ranging power). Each function has a minimum SNR, the threshold, at which the quality of the link meets the criteria defined by the project.

3.5.1 Design Control Tables

Link performance is book-kept using a design control table (DCT), sometimes called a link budget. When used for planning future capability, Voyager link predictions are based on a criterion of positive margin under the two conditions of Command at mean minus 3-sigma and Telemetry at mean minus 2-sigma. Sigma refers to the standard deviation of the command E_b/N_0 and telemetry E_s/N_0.

A DCT includes numerous link parameters and their tolerances, but it applies to only one point in time. For planning and analyzing performance during flight, the project may prefer tabulations or plots of key quantities versus time.

The original JPL Deep Space Communications and Navigation Systems (DESCANSO) article from which this chapter is based [3] contains DCTs for the S-band uplink (carrier and command channel) and the X-band downlink (carrier and telemetry channel).

3.5.2 Long-Term Planning Predicts

Often, plots are more compact and useful for displaying link performance than either a series of DCTs or a set of tabulations. Figure 3-8 summarizes predictions of downlink P_t/N_0[19] at one station (DSS-43) and one day of the year (January 30) for Voyager 2 from 1995 to 2020, one year per curve. The

[19] The quantity downlink P_t/N_0 is a convenient link parameter to plot for telemetry links. There is a fixed value of P_t/N_0 that represents threshold for each data rate. Performance of bit rates that extend over orders of magnitude (600 bps to 7.2 kbps in Fig. 3-8 and 160 bps to 1.4 kbps in Fig. 3-9) can be shown on a scale of 10 dB per decade. The difference between the thresholds for the 1.4-kbps data rate for Voyager 2 (Fig. 3-8) and Voyager 1 (Fig. 3-9) represents the additional 0.8 dB gain achieved by using suppressed carrier for that rate on Voyager 2 only. See Section 3.4.3 Telemetry Processing and Section 3.6.3 Ground System Performance Improvements for additional information regarding the suppressed carrier downlink.

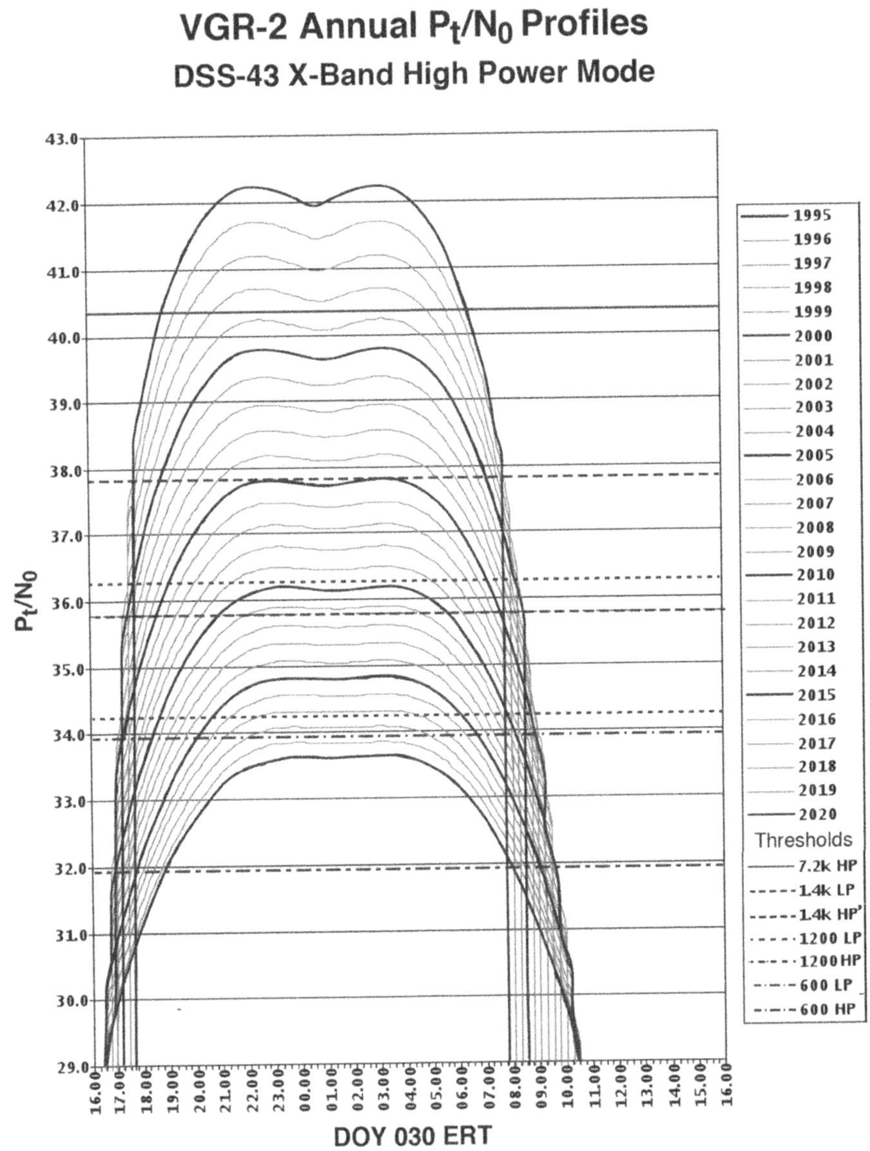

Fig. 3-8. 25 Years of Voyager 2 telecom performance predictions for DSS-43. Please visit www.wiley.com/go/deepspacecommunications to access the color version of this figure.

horizontal axis is a 24-hour period. The main shape of any of the curves is due to the increasing DSS-43 elevation angle to Voyager 2 (from 17:00 to 21:00) and the decreasing elevation angle (from 05:00 to 09:00). All of the Fig. 3-8 curves have similar shapes because the Earth's orbit around the Sun repeats from year to year. Though they have the same general shape, the other curves are offset from each other vertically because the communications distance between Voyager 2 and Earth increases steadily from year to year. Telemetry thresholds are displayed as horizontal lines per the legend on the right. The plot is used for long-term mission planning purposes.

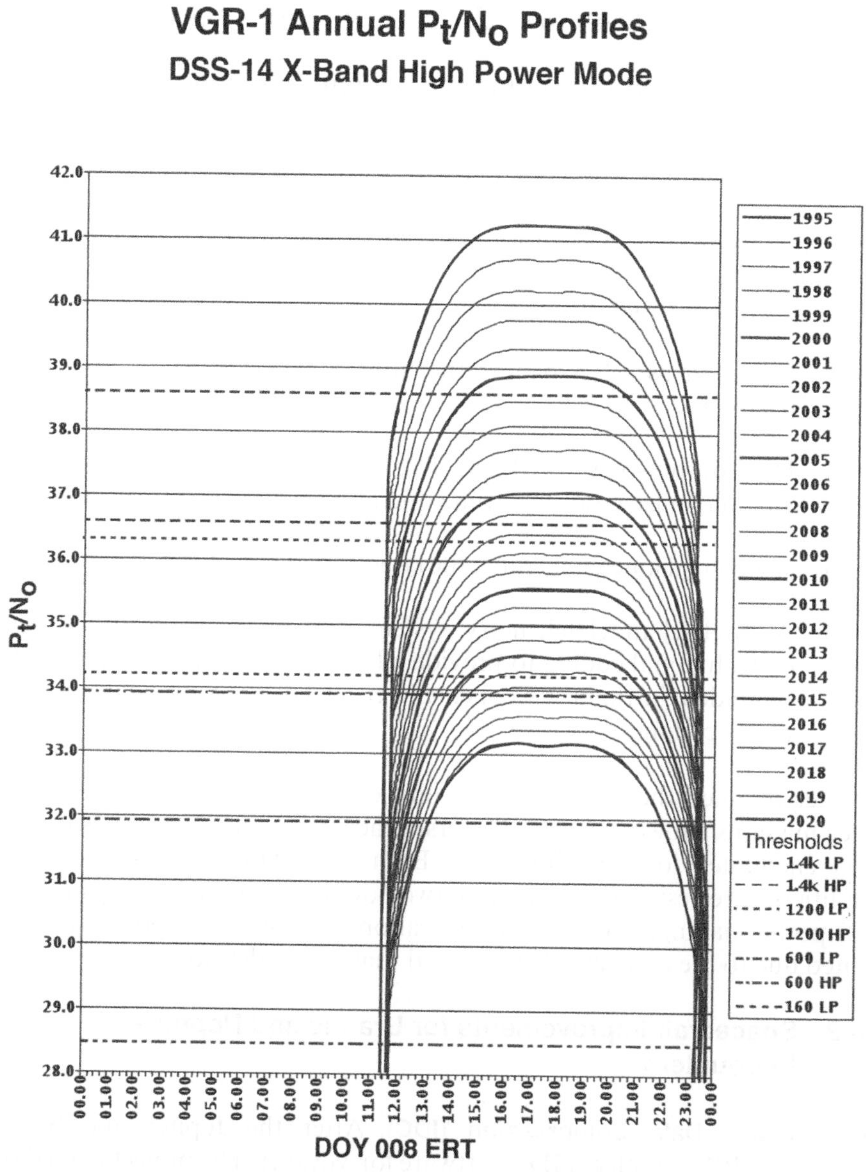

Fig. 3-9. 25 years of Voyager 1 telecom performance predictions for DSS-14. Please visit www.wiley.com/go/deepspacecommunications to access the color version of this figure.

Telemetry link margin exhaustion dates can be estimated within about half a year for each spacecraft by comparing the annual performance predictions with the bit-rate thresholds on each chart. Threshold lines for 160 bps and 40 bps are not included because they fall below the vertical scale, except for Voyager 1 at 160 bps in low-power mode. There is plenty of link margin at those rates to operate well beyond 2020. The project will continue to plan data rate usage and scheduled station support. In the same manner, Fig. 3-9 displays 25 years of Voyager 1 day-of-year (DOY) 008 performance predictions and bit-rate thresholds at DSS-14.

3.6 New Spacecraft and Ground Telecom Technology

3.6.1 Spacecraft and Telecom Link Design Compared with Previous Missions

The Voyager telecom system design was heavily influenced by the telecom system designs for Mariner-Venus-Mercury (1973 launch) and Viking Orbiter (1975 launch). Both of these prior missions flew primary S-band uplink and downlink systems and performed X-band experiments. Key Voyager design improvements consisted of:

1) First-ever use of X-band rather than S-band for primary downlink telemetry
2) Dual-output-power X-band TWTAs, designed to minimize mass and maximize efficiency while operating more than 50,000 hours
3) A 3.66-m diameter antenna, the largest solid reflector flown as of 1977
4) A single-channel telemetry system with concatenated Golay and convolutional coding to provide efficient transmission of data, later upgraded in-flight to concatenated Reed-Solomon and convolutional coding.

Voyager retained the S-band uplink and downlink similar to earlier deep space missions. However, it was the first spacecraft to use X-band as the primary encounter downlink frequency. Both the S-band and X-band power amplifiers were designed to operate at two power levels for flexibility in spacecraft power loading. Simultaneous operation of both at high power was prohibited due to the excessive thermal load that would develop.

3.6.2 Spacecraft Improvements for Uranus and Neptune Encounters

3.6.2.1 Image Data Compression (IDC). After the Jupiter and Saturn encounters, JPL completed IDC software for Voyager. The project loaded the software into the backup flight data subsystem (FDS) computer that was reconfigured to handle just that task [12,13].[20] Uncompressed Voyager images

[20] The *Voyager Neptune Travel Guide* [12] describes the specific IDC algorithm implemented on Voyager. In *Channel Coding and Data Compression System Considerations* [13], Rice discusses Voyager image data compression in context with other aspects of error-correcting coding for the deep space channel. These include convolutional (Viterbi) and Reed-Solomon codes, interleaving, and frame synchronization.

contain 800 lines, 800 dots (pixels) per line, and 8 bits per pixel (to express one of 256 gray levels). However, much of the data content in a typical planetary or satellite image is dark space or low-contrast cloud features. By counting only the differences between adjacent pixel gray levels, rather than the full 8-bit values, image data compression reduced the number of bits for the typical image by 60 percent without unduly compromising the information. This reduced the time needed to transmit each complete image from Uranus and Neptune to Earth by the same 60%.

3.6.2.2 Error-Correcting Coding. Like other deep space links, the Voyager telemetry link is subject to noise in the communications channel changing the values of bits transmitted over the channel—in other words, causing bit errors. Error-correcting coding reduces the rate of errors in the received information that is output. Such coding increases the redundancy of the signal by increasing the number of bits transmitted relative to the information bit rate.[21] The Golay encoding algorithm used at Jupiter and Saturn required the transmission of one overhead bit for every information bit transmitted (100 percent overhead). Voyager carried an experimental Reed-Solomon data encoder, expressly for the greater communication range of the Uranus and Neptune phase of the mission. The new Reed-Solomon encoding scheme reduced the overhead to about one bit in five (20-percent overhead) and reduced the bit-error rate in the output information from 5×10^{-3} to 10^{-6}.

3.6.3 Ground System Performance Improvements

The capability of the DSN 64-m stations, as it existed during the Voyager Jupiter and Saturn encounters, allowed for maximum downlink rates of 115,200 bps at Jupiter (in 1979) and 44,800 bps at Saturn (in 1981 for Voyager 2). Prior to Voyager 2's Uranus and Neptune encounters (1986 and 1989), several major enhancements described in this section were made to the ground receiving system used for Voyager.

The comparison in Table 3-8 provides an overview of the effectiveness of the upgrade in capabilities. The comparison is between what the maximum downlink rates actually were at each of the four planetary encounters and what they would have been at Uranus and Neptune without the upgrades. The

[21] The total channel data rate can be considered apportioned between the original information rate and the redundancy bits as an "overhead". The bits of coded data transmitted over the channel are often referred to as symbols. Because of overhead, the symbol rate is higher than the information bit rate. The power of an error-correcting code is that the reduced effect of noise on the signal allows a higher information rate, a lower bit error rate, or a lower transmitter power, or a desirable combination of these.

Table 3-8. Voyager 2 ground system performance improvements (DSN with 1979–1981 capability).

Encounter	Inverse Square	Expected Rate (bps) from Inverse Square	Achieved Maximum Rate (bps)	Factor of Improvement
Jupiter	1/1	115,200 (baseline for inverse square bit rates)	115,200	—
Saturn	1/4	~29,000	44,800	×1.5
Uranus	1/13	~9,000	29,900	×3.3
Neptune	1/36	~3,200	21,600	×6.8

comparison is approximate because of the finite set of Voyager downlink rates available and differences in mission priorities and margin criteria at each encounter.

With other factors constant, communications capability is inversely proportional to the square of the distance from the spacecraft to the Earth. At the encounters, Jupiter-Earth distance averaged 5.2 AU, Saturn-Earth averaged 10 AU, Uranus-Earth was 19 AU, and Neptune-Earth was 30 AU. With no ground upgrade the communications capability at Saturn, Uranus, and Neptune would have been 1/4, 1/13, and 1/36 that at Jupiter, respectively.

3.6.3.1 DSN 64-m to 70-m Upgrade. The most significant DSN upgrade benefiting Voyager was the upgrade of the 64-m antennas to 70 m. The 70-m upgrade was accomplished by removing the old metallic surface plates and structural outrigger beams, then installing a totally new outer support structure along with precision surface plates that could be adjusted to sub-millimeter accuracy. Holographic alignment techniques were introduced that permitted sharp focusing of the X-band radio signals [14]. Together, the larger surface area and alignment and calibration techniques yielded an improvement in signal strength averaging 1.4 dB for each 70-m antenna.

3.6.3.2 Arraying with DSN Antennas. The second-most significant DSN upgrade benefiting Voyager was the installation of baseband combiner technology for arraying multiple antennas. Baseband combining added another 0.8 dB to the 70-m performance by arraying the 70-m antenna with a 34-m high-efficiency (HEF) antenna, and it added 1.2 dB by arraying the 70-m antenna with two 34-m antennas.

3.6.3.3 Arraying with Non-DSN Antennas for Neptune Encounter. The Voyager Project called upon ground resources beyond the NASA/JPL-operated DSN for data acquisition at the Neptune encounter. As had been done for the

Uranus encounter, the DSN again teamed with the Australian government's Parkes 64-m radio astronomy antenna operated by the Commonwealth Scientific and Industrial Research Organization (CSIRO). The 70-m antenna and a 34-m antenna of the DSN facility in Canberra were arrayed with the Parkes antenna, connected by a 320-km (200 mi) microwave link.[22]

By simultaneously tracking Voyager from these three antennas during the Neptune encounter period, the DSN and Parkes radio observatory achieved an increase in the combined signal strength roughly proportional to the combined surface areas of the arrayed antennas. Other factors being the same, the DSN–Parkes array provided double the bit-rate capability of a single 70-m antenna.

By far the greatest signal strength improvement for Neptune resulted from arraying the twenty-seven 25-m dishes of the National Radio Astronomy Observatory's (NRAO) Very Large Array (VLA) near Socorro, New Mexico with the 70-m DSN antenna at Goldstone, California. The received signal power (or data rate capability) with the VLA arrayed with the 70-m DSN antenna was nearly triple that of the 70-m antenna by itself. An array of a 70-m antenna, two 34-m antennas, and the VLA increased the downlink capability by 5.6 dB relative to the 70-m antenna alone, almost a factor of four in bit rate.

Last, a cooperative venture with the Japanese space agency permitted use of its 64-m Usuda antenna on encounter day for non-real-time combining of radio science data.

3.6.3.4 Block V Receiver. During the early 1990s the DSN developed a software receiver, the BVR. Among other benefits, the BVR offered Voyager the capability to operate in the suppressed-carrier mode. By changing the spacecraft exciter's phase modulation index to 90 deg, there is no separate carrier, and all of the power goes into the modulated telemetry subcarrier. With the BVR in suppressed-carrier mode, the 7200-bps tape recorder playback capability was extended for Voyager 2 by approximately two years

---

[22] Voyager 2's closest approach with Neptune was on August 25, 1989. Arrays with the Parkes antenna were used as early as March. Parkes was used more days than not during June, July, and August. Closer in time to encounter, Voyager 2 received nearly continuous downlink using arrays at all three sites on most days. At Canberra, the array for a particular pass included as many as three stations scheduled from among: DSS-43, DSS-45, DSS-42, and DSS-49 (the numerical designator for Parkes). At Madrid, the array consisted of DSS-63 and DSS-65. At Goldstone, it usually consisted of DSS-14 and DSS-15, with occasional inclusion of DSS-19 (the numerical designator for the VLA).

beyond the capability using traditional residual carrier mode. (Voyager 1 exhausted its 7200 bps capability before the BVR became available.)

3.6.3.5 Improvements in System Noise Temperature. During 2000 to 2001 the DSN replaced the high-maintenance maser preamplifiers with HEMT technology and decreased the preamplifier system noise temperature at the 70-m stations.[23] The two upgrades produced approximately 0.5-dB performance increase for downlink telemetry. From Fig. 3-8 or 3-9, the spacecraft recedes from the Earth by the equivalent of 0.5 dB per year in the late 1990s, falling to 0.2 dB per year nearer 2020. Thus, the two upgrades have the effect of prolonging the bit-rate capability (as compared to that with no upgrade) at any time by another year or two.

3.6.3.6 Additional 34-m Stations and Full-Spectrum Combining for Array. The DSN has upgraded the array capability at all three communication complexes from baseband to full-spectrum combining[24] and has installed more 34-m stations. Full spectrum combining allows Voyager to extend the use of science playbacks beyond termination shock, heliosheath, and heliopause into the interstellar space beyond. The availability of more 34-m stations reduces competition among any given number of projects for scarce DSN resources. In 2013, Voyager routinely arrays a pair of 34-m antennas for daily 160 bps cruise telemetry.

3.6.4 Ground Display and Operability Improvements

Prior to VIM, the flight team viewed real-time spacecraft telemetry produced by the Voyager implementation of the Test and Telemetry System (TTS) [15]. The TTS was a 1960s-era system of Univac 1530, and Univac 1219, and Modcomp

---

[23] The overall efficiency of a receiving system is sometimes expressed as *G*/*T*, where *G* is antenna gain and *T* is the system noise temperature. The 70-m upgrades included the X-band transmit receive (XTR) cone shown in Fig. 3-7. Besides providing X-band transmit capability, use of the XTR cone results in an X-band system noise temperature that is lower by the equivalent to 0.5 dB at higher elevation angles for Voyager. See *DSN Telecommunications Link Design Handbook*, Module 101, 70-m Subnet Telecommunications Interfaces [7]. The XTR cone's feed design includes a diplexing junction to inject the transmitted signal directly into the feed. This eliminates the need for a waveguide diplexer and a common path for the received and transmitted signals. As a result, much of the received path can be cryogenically cooled with a significant reduction in operating system temperature.

[24] See "New Spanish Dish Will Aid Interplanetary Communications" at http://www.jpl.nasa.gov/news/news.php?feature=553 for discussion of the 2002–2003 construction of an additional 34-m antenna near Madrid. See http://www.techbriefs.com/component/content/article/1264-ntb/tech-briefs/electronics-and-computers/7394 "Full-Spectrum Arraying of Receiving Radio Antennas" describing the DSN's 34-m array project with full spectrum combining.

II computers that ran a Viking operating system and Voyager-specific applications. Flight team inputs to operate the TTS were submitted on punch cards. Fixed format output was viewed either on small black and white monitors (called DTV, digital television) or "green-bar" fan-fold printer paper.

At the start of VIM in 1990, Voyager was the second project (after Magellan) to adopt the JPL Advanced Multimission Operations System (AMMOS) [16]. Use of AMMOS leapfrogged Voyager to networked Unix workstations, including such improvements over TTS as color graphical user interfaces, real-time "on-the-fly" charting, laser printers, and much greater file storage capacity. AMMOS technology contributed significantly to flight team efficiency gains.

3.7 Operational Scenarios of the Voyager Interstellar Mission

When planning VIM in the late 1980s, it was recognized from the outset that planning thirty-year missions (1990 to 2020) for spacecraft that had already flown for 12 years (1977 to 1989) might appear unreasonably optimistic. The fundamental goal for this long-term mission is to return significant science data from environments not yet well understood. The approach was to establish a basic operational framework with the inherent flexibility to respond to new situations or reformulated goals, as necessary. Conservative spacecraft operation practices were generally planned to maximize lifetime for the various subsystems [17].

3.7.1 Tracking Coverage

The planned VIM tracking requirement was set at 16 hours per day for each Voyager spacecraft. Due to competing needs from other spacecraft, the actual coverage in 2000 and 2001 was roughly 12 hours per day. By late 2012, the coverage had been reduced to 2 to 6 hours per day for each Voyager due to contentions with another project in the same part of the sky in that year as well as simultaneous downtimes of several months at two stations. The coverage for each Voyager returned to the norm by the end of the downtimes.

3.7.1.1 Termination Shock, Heliosheath, and Heliopause. The requirements on spacing and duration of DSN tracking passes depend somewhat on the abruptness of the termination shock and heliopause. The 2013 plan was for one or more passes per day, with as long as 10 hours total per day.[25]

---

[25] The information from Voyager 1 as it left the heliosphere was unique. To ensure scientists would catch the change – its timing and duration unknown in advance – when it happened, the project requested as much tracking coverage as possible [18].

3.7.1.2 Uplink. Because of the large spacecraft–Earth distance, 70-m stations are used to transmit uplinks to each Voyager to meet the following periodic requirements:

- Weekly: transmission of a Command Loss timer reset command (see Section **3.7.3.2** for an explanation of the Command Loss timer)
- Every 4 months: loading a command sequence
- Annually: transmission of a computer command subsystem (CCS) clock calibration, timing test, and memory refresh.

3.7.1.3 Downlink. The Voyager primary cruise data rate is 160 bps. Either a 34-m beam waveguide (BWG) or 34-m HEF station provides sufficient capability for cruise data.

Two 6.6-hour digital tape recorder (DTR) playbacks are received per year per spacecraft, plus occasional special playbacks requested by scientists. These playbacks require a 70-m downlink station. Near the start of VIM, playbacks were downlinked at 7200 bps. As the 7200-bps link margin approached exhaustion at the 70-m stations, 70-m/34-m HEF arrays were employed to gain a 0.8-dB performance improvement for approximately two more years of capability. When array link margin was exhausted, Voyager reduced the playback data rate to 1400 bps and returned to the use of 70-m stations standing alone. The mission has needed arrays again to capture 1400-bps playbacks downlinked from Voyager 1 beginning in 2007 and from Voyager 2 in 2011.

Four 7-hour and two 0.5-hour attitude control calibration maneuvers are performed per spacecraft every year, each requiring 70-m station downlink coverage to ensure uninterrupted downlink telemetry.

Once per year, a 70-m downlink is required to capture a 1200-bps CCS Timing Test from each spacecraft.

A 70-m station is required to capture 600-bps ultraviolet science downlink telemetry, when requested by scientists.

3.7.2 RFS Strategies

3.7.2.1 X-Band TWTA High-/Low-Power-Level Drivers. Selection of a power level is a function of such considerations as amount of ultraviolet and cruise science data coverage, periodic general science and engineering (GS&E) telemetry and DTR playbacks, and TWTA lifetime relative to heliopause attainment. Both spacecraft are operated in X-band low power, except when high power is needed to receive the selected data rate at the required level of confidence and at the required bit error rate with the available tracking support. Power-level choices derive from 1) the fact that the high-power-based science data are generally of significant value, 2) the risk that the spacecraft may fail first from other causes, and 3) the knowledge that low-power versus high-power operation can affect lifetime by at most 25 percent.

3.7.2.2 X-Band TWTA Power-Level Switching Cycles Minimized. The X-band TWTA power level is switched from high to low power whenever high power is not needed for more than two weeks. Fewer TWTA power switches will maximize TWTA lifetime.

3.7.2.3 X-Band TWTA On/Off Switching Not Planned. There are no TWTA on/off switches planned unless there is a spacecraft anomaly. On/off switching is thought to degrade the TWTAs faster than low-/high-power mode switching.

3.7.2.4 S-Band Downlink Not Required. For both spacecraft, the S-band system will not be turned on again unless there is a spacecraft anomaly because VIM has no requirement for S-band, and the electrical power load is considerably higher than for X-band. If the last X-band TWTA failed or a spacecraft attitude anomaly degraded the pointing accuracy required for X-band, a decision could be made to turn on the S-band system.

3.7.2.5 Two-Way Coherent Tracking Not Required. Voyager has no plan to deviate from trajectories established prior to the start of VIM. Planetary encounter navigational exactitude is not necessary for VIM. Requirements for two-way coherent Doppler and ranging were eliminated from VIM to reduce cost.

3.7.2.6 Voyager 2 Procedures to Compensate for Voyager 2 Receiver Problem. On April 6, 1978, a fault-protection algorithm onboard Voyager 2 automatically switched from the prime to backup receiver. However, the backup receiver's tracking-loop capacitor[26] was found to have failed sometime previously. Soon after returning to the prime receiver by ground command, that receiver failed, leaving the spacecraft uncommandable. Seven days later, the algorithm switched back to the degraded backup receiver, forever thereafter requiring special detailed uplink procedures to command Voyager 2.

For a command pass, the DSN offsets the Voyager 2 uplink frequency to compensate for the predicted Doppler. The failed tracking loop necessitates

---

[26] The tracking-loop capacitor is in the receiver phase-lock loop circuitry to facilitate uplink acquisitions and track Doppler-induced frequency shifts resulting from changes in relative velocity between the spacecraft and the DSN antenna. For Voyager, these velocity changes are due primarily to the Earth's rotation, so they occur on every pass.

that the uplink signal be received within 100 Hz of the best-lock frequency (BLF)[27] to maintain lock. The BLF changes by approximately 100 Hz with each 0.25-deg receiver temperature change. Major temperature changes are caused by spacecraft configuration changes. However, even with a constant configuration, seemingly random temperature variations affect the BLF significantly. Two special procedures (BLF test and command moratorium), described in the next two paragraphs, are required to reliably command Voyager 2 in the presence of the spacecraft thermal conditions.

Standardized BLF tests, performed about twice per week, tune the uplink signal through the last-known range of the BLF. By observing the resulting spacecraft-receiver signal-level peak time in downlink telemetry and subtracting the round-trip light time, the flight-operations staff determines the DSN uplink frequency that produced the peak. The frequency that is determined becomes the latest BLF and is used for commanding and centering the next BLF test.

The flight team plans a Voyager 2 "command moratorium" (a period with no command uplinking permitted) for as long as 3 days following configuration changes that affect the spacecraft thermal profile. The command moratorium is to provide a period of stabilization of the thermal profile before the next BLF test in preparation for commanding.

3.7.3 Spacecraft Fault Protection

The CCS has five fault-protection algorithms (FPAs) stored in memory, as summarized in Table 3-9. The two algorithms most directly related to the telecommunications system are named RF Loss and Command Loss [19].

3.7.3.1 RF Loss. RF Loss provides a means for the spacecraft to automatically recover from an S- or X-band exciter or power amplifier degradation or failure affecting the unit's RF output. The CCS monitors the output RF power at four points in the RFS: the S-band exciter and S-band power amplifier and the X-band exciter and X-TWTA. If the output RF power from one or more powered-on units drops below a threshold level, the algorithm will attempt to correct the problem by switching to the redundant unit.

---

[27] The term "best-lock frequency" in a phase-locked loop refers to the natural oscillation frequency of the loop with no input. When the receiver loop is receiving an uplink carrier exactly at BLF (the center of its bandwidth), the loop indicates a zero static phase error telemetry measurement. The bandwidth of a healthy Voyager receiver is about 100 kHz, as compared with 100 Hz for the receiver with the failed tracking loop.

Table 3-9. Voyager fault-protection algorithms.

FPA Name	Description
RF Loss	Monitors S- and X-band exciter and transmitter hardware and switches to redundant unit if a failure is detected
Command Loss	Switches to redundant command reception hardware units in an effort to re-establish command reception capability if a command has not been received within a specified interval
AACS[a] Power Code Processing	Monitors AACS status information and issues preprogrammed recovery responses in the event of AACS anomalies
CCS Error	Responds to critical anomalous CCS hardware and software conditions. The response typically stops any on-going sequence activities, places the CCS in a known quiescent state, and waits for ground action
Power Check	Responds to CCS tolerance detector trip or spacecraft under-voltage power condition by switching to redundant hardware in an attempt to isolate an electrical fault and then eliminating power loads in a predetermined manner if required

[a] Attitude and articulation control subsystem.

3.7.3.2 Command Loss. Command Loss provides a means for the spacecraft to automatically respond to an onboard failure resulting in the inability to receive or recognize ground commands. If a period of time set in the flight software goes by without the spacecraft recognizing a valid uplinked command, the Command Loss timer expires. The algorithm responds to the presumed spacecraft failure[28] and attempts to correct that failure by systematically switching to redundant hardware elements until a valid command is received. Command Loss will be executed four consecutive times if command reception is not successful. After four unsuccessful executions, the CCS will disable Command Loss and activate a set of sequences of commands named the backup mission load (BML) and described below.

3.7.3.3 Backup Mission Load. In the event of permanent loss of command reception capability, a BML command sequence stored onboard each spacecraft is programmed to continue controlling the spacecraft and achieving fundamental VIM objectives. The BML will begin execution two weeks after the first execution of Command Loss and continue until the spacecraft stops

[28] A ground system procedural error or station problem that results in failure to transmit the Command Loss timer reset command can also result in the algorithm tripping. A Command Loss timer reset command is usually transmitted to each spacecraft weekly. If a period of time set in the flight software goes by without the spacecraft recognizing a valid uplinked command, the Command Loss timer expires.

operating. It will transmit cruise science and engineering telemetry, store science observations on the tape recorder, and downlink playbacks regularly.

The BML will configure the downlink to X-band high-power, with S-band remaining off. The basic cruise data rate is 160 bps, with playbacks at 1400 bps. The BML was designed for 34-m station supports, except during playbacks when 70-m or 70-m/34-m HEF array support is assumed.

At the beginning of VIM, HGA pointing information was uplinked to an on-board table that will provide accurate attitude control through the planned end of mission in 2025.

References

Much of the telecom design information in this chapter comes from original Voyager prime mission design documentation. These include the design control document for the telecommunications links [11], the functional description of the telecommunications system [8], and the hardware design requirement for the modulation demodulation subsystem (MDS) [20]. Much of the mission and operational information was obtained from the Voyager Operational Handbook [21], the Voyager Neptune Travel Guide [12], and the Voyager project public website [1].

[1] *Voyager The Interstellar Mission*, project public website, Jet Propulsion Laboratory, California Institute of Technology, Pasadena, California. http://voyager.jpl.nasa.gov/ (accessed October 30, 2014)

[2] E. C. Stone, J. D. Richardson, and E. B. Massey, *The Voyager Interstellar Mission Proposal to Senior Review 2010 of the Mission Operations and Data Analysis Program for the Heliophysics Operating Missions*, Jet Propulsion Laboratory, California Institute of Technology, Pasadena, California, March 2010, http://voyager.jpl.nasa.gov/mission/interstellar.html (accessed November 18, 2013)

[3] R. Ludwig and J. Taylor, *Voyager Telecommunications, DESCANSO Design and Performance Summary Series*, Article 4, Jet Propulsion Laboratory, California Institute of Technology, Pasadena, California, March 2002. http://descanso.jpl.nasa.gov/DPSummary/summary.html (accessed October 30, 2014)

[4] [Voyager experiments web page], NASA Data Center website, National Aeronautics and Space Administration. http://nssdc.gsfc.nasa.gov/nmc/spacecraftDisplay.do?id=1977-076A in the National Space Science Data Center (accessed December 30, 2013)

[5] D. A. Gurnett, W. S. Kurth, L. F. Burlaga, and N. F. Ness, "In Situ Observations of Interstellar Plasma with Voyager 1," *Science*, vol. 341, no. 6153, pp. 1489–1492, September 2013. http://www.sciencemag.org/content/341/6153/1489.full (accessed October 30, 2014)

[6] *NASA Spacecraft Embarks on Historic Journey into Interstellar Space*, September 12, 2013, NASA public website, http://www.nasa.gov/mission_pages/voyager/voyager20130912.html#.UozsERAeIsK (accessed November 20, 2013)

[7] *DSN Telecommunications Link Design Handbook*, 810-005, Rev. E, Jet Propulsion Laboratory, Pasadena, California. http://deepspace.jpl.nasa.gov/dsndocs/810-005/ (accessed October 30, 2014)

[8] C. R. Paul, *Voyager Telecommunications System Functional Description,* 618-822 (internal document, Jet Propulsion Laboratory, California Institute of Technology, Pasadena, California, June 1980.

[9] E. Batka, *Deep Space Network Operations Plan, Voyager Interstellar Mission (*VIM*) Project* 618-700 (internal document), Jet Propulsion Laboratory, California Institute of Technology, Pasadena, California, April 1995.

[10] *Voyager Telecommunications Design Control Document*, 618-257 (internal document), Jet Propulsion Laboratory, Pasadena, California, January 15, 1988.

[11] J. H Yuen, editor, *Deep Space Telecommunications Systems Engineering*, Plenum Press, New York, 1983.

[12] C. Kohlhase, *The Voyager Neptune Travel Guide*, 89-24 (internal document), Jet Propulsion Laboratory, California Institute of Technology, Pasadena, California, June 1, 1989.

[13] R. F. Rice, *Channel Coding and Data Compression System Considerations for Efficient Communication of Planetary Imaging Data*, JPL Technical Memorandum 33-695 (Rev. 1) (also NASA-CR-140181), Jet Propulsion Laboratory, California Institute of Technology, Pasadena, California, September 1, 1974.

[14] D. J. Rochblatt, Chapter 8, "Microwave Antenna Holography," *Low-Noise Systems in the Deep Space Network*, M. S. Reid, editor, John Wiley & Sons, Inc., Hoboken, New Jersey, 2008. Also http://descanso.jpl.nasa.gov/monograph/mono.html (accessed October 30, 2014)

[15] *Voyager Telemetry System User's Guide*, Volume I, *Introduction and Overview*, 618-686 (internal document), Jet Propulsion Laboratory, California Institute of Technology, Pasadena, California, March 1977.

[16] *AMMOS User Guides*, Vol. I, MGDS User's Overview, document D-6057, (internal document) Jet Propulsion Laboratory, California Institute of Technology, Pasadena, California, April 1994.

[17] *Voyager Mission Design Guidelines and Constraints, Interstellar Mission*, 618-123, Vol. V (internal document), Jet Propulsion Laboratory, California Institute of Technology, Pasadena, California, March 29, 1991.

[18] A. Witze, "Voyager: Outward bound," *Nature*, May 23, 2013, http://www.nature.com/news/voyager-outward-bound-1.13040 (accessed November 25, 2013)

[19] C. E. Presley, *Voyager Computer Command Subsystem Flight Software Design Description, Assembly Language Listings*, 618-235, Vol. 2, Rev. K, (internal document, Jet Propulsion Laboratory, Pasadena, California, September 15, 2005.

[20] *Design Requirement – Mariner Jupiter/ Saturn 1977 Flight Equipment Modulation Demodulation Subsystem*, Design Requirement MJS77-2003-1 (internal document), Jet Propulsion Laboratory, California Institute of Technology, Pasadena, California, April 9, 1974.

[21] *Voyager Operational Handbook*, 618-804, document (internal document), Jet Propulsion Laboratory, California Institute of Technology, Pasadena, California, August 1989.

Additional Resources

[1] T. Ferris, "Voyagers' Never-Ending Journey," *Smithsonian magazine*, May 2012. http://www.smithsonianmag.com/science-nature/Timothy-Ferris-on-Voyagers-Never-Ending-Journey.html (accessed November 25, 2013)

[2] M. Wall, "NASA's Voyager 1 Probe Enters New Realm Near Interstellar Space," *Space.com*, June 27, 2013. http://www.space.com/21751-voyager-spacecraft-nears-interstellar-space.html (accessed November 25, 2013)

[3] "Interplanetary Monitoring Platform (IMP 8) Completes 28-year Observing Marathon," Spaceref.com, Oct. 30, 2001. http://science1.nasa.gov/missions/imp-8/ and "Interplanetary Monitoring Platform (IMP 8) Completes 28-year Observing Marathon," Goddard Space Flight Center, Greenbelt, Maryland, October 30, 2011, http://www.spaceref.com/news/viewpr.html?pid=6420 (both accessed October 30, 2014)

[4] WIND spacecraft, NASA websites (both accessed October 30, 2014): www.nasa.gov/mission_pages/sunearth/news/wind-slamswaves.html

[5] *Advanced Composition Explorer (ACE)*, website, California Institute of Technology, http://www.srl.caltech.edu/ACE/ (accessed November 18, 2013)

[6] "SAMPEX (Solar Anomalous and Magnetospheric Particle Explorer) spacecraft, web page, University of Colorado, Boulder. http://lasp.colorado.edu/home/sampex/ (accessed November 18, 2013)

[7] *Unisys History Newsletter*, Unisys website. https://wiki.cc.gatech.edu/folklore/index.php/Main_Page (accessed November 18, 2013)

[8] *Modcomp Systems & Solutions*, website, Modcomp, Inc., Deerfield, Florida. http://www.ddp116.org/modcomp/modcomp.pdf (accessed November 18, 2013)

[9] "Voyager Experiment and Instruction Descriptions," *National Space Science Data Center*, website, National Aeronautics and Space Administration, http://nssdc.gsfc.nasa.gov/nmc/spacecraftDisplay.do?id=1977-076A (accessed November 18, 2013)

[10] *NASA Science Missions*, website, National Aeronautics and Space Administration. http://spacescience.nasa.gov/missions/ (accessed November 18, 2013)

[11] *NASA history*, website, National Aeronautics and Space Administration. http://history.nasa.gov/ (accessed November 18, 2013)

[12] *DSN Telecommunications Link Design Handbook*, 810-5, Handbook Glossary, Module 901, March 12, 2012, http://deepspace.jpl.nasa.gov/dsndocs/810-005/901/901F.pdf (accessed November 18, 2013)

[13] *Basics of Space Flight*, website, Jet Propulsion Laboratory, California Institute of Technology, Pasadena, California, http://www.jpl.nasa.gov/basics/ (accessed November 18, 2013)

[14] T. T. Pham, A. P. Jongeling, and D. H. Rogstad, "Enhancing Telemetry and Navigation Performance with Full Spectrum Arraying," presented at the *2000 IEEE Aerospace Conference*, Big Sky, Montana, March 2000. http://ieeexplore.ieee.org/stamp/stamp.jsp?tp=&arnumber=879875 (accessed October 30, 2014)

[15] "Science Strategy," *NASA Science*, NASA website. http://science1.nasa.gov/about-us/science-strategy/ (accessed November 14, 2013)

Chapter 4
Galileo Telecommunications

Jim Taylor, Kar-Ming Cheung, and Dongae Seo

4.1 Mission and Spacecraft Description

This chapter describes how the Galileo orbiter received and transmitted data with the Deep Space Network (DSN). The relay communications subsystems and the link between the Galileo probe and the orbiter are also described briefly. The chapter is at a functional level, intended to illuminate the unique mission requirements and constraints that led to both design of the communications system and how the mission had to be modified and operated in flight.

Augmenting the spacecraft downlink design and the supporting ground system for science return with only the low-gain antenna (LGA) was a particular challenge for the Galileo planetary mission.

The Galileo orbiter was designed and built at the Jet Propulsion Laboratory (JPL) in Pasadena, California, and the Galileo probe was designed and built at the NASA Ames Research Center (ARC) in Sunnyvale, California. The orbiter flight team was located at JPL, as was the probe flight team during that portion of the mission.

4.1.1 The Mission

The Galileo spacecraft was launched in 1989 aboard the Space Shuttle Atlantis (STS [Space Transportation System]-34). Its primary objective was to study the Jovian System. The Galileo launch delay after the Challenger Space

Some of the grey scale figures in this book have color versions in the companion website www.wiley.com/go/deepspacecommunications. Please refer to this website where the text refers to colors.

Deep Space Communications, First Edition. Edited by Jim Taylor.

Shuttle accident in 1986 necessitated a change in the strategy to get Galileo to Jupiter.[1] The original strategy was a relatively direct flight to Jupiter with a single gravity assist at Mars. The new mission plan had to work with less propulsion, so it made use of a longer, much less direct flight, with gravitational assists from Venus once and Earth twice, to give the spacecraft enough energy to get to Jupiter. During the cruise phase of the mission,[2] the Galileo spacecraft took the first close-up images of an asteroid (Gaspra) in October 1991, and discovered the first known moon (Dactyl) of an asteroid (Ida) in August 1993 [2]. During the latter part of the cruise, Galileo was used to observe the collisions of fragments of Comet Shoemaker-Levy 9 with Jupiter in July 1994.

The Galileo primary mission (1995–1997) involved

- Penetration of Jupiter's atmosphere by the probe that returned a Jovian "weather report" on temperature, pressure, composition, winds, clouds, and lightning
- Initial orbiter flyby of the Jovian satellite Io and passage through the Io torus
- Jupiter orbit insertion (JOI)
- A two-year "tour" of the major satellites by the orbiter that returned images, radio science, and data on fields and particles.

The probe descended through an unusually dry spot in Jupiter's top cloudy layer, and probably melted in the hot atmosphere somewhere below the clouds.

The orbiter had six scientific instruments on one section that spun (at 3 revolutions per minute, rpm) for pointing stability and for collecting three-dimensional fields and particles data near the spacecraft. The "de-spun" section used gyros to point the four remote-sensing instruments at a target to obtain images, composition, surface structure, and temperature data.[3] The orbiter's umbrella-like high-gain antenna (HGA) did not deploy fully, so Galileo's computer was reprogrammed to compress and record the data taken during Jovian satellite flybys to the on-board tape recorder. The data was returned to Earth during the remainder of each orbit using the low-gain antenna (LGA) and modifications to the ground receiving systems of the Deep Space Network (DSN).

---

[1] The last planetary launch before Galileo in 1989 was Pioneer Venus in 1978. Galileo remained in "new mission" status for these years while the launch vehicle was changed four times. Each change, none of them due to the Galileo spacecraft itself, necessitated a complete redesign of the mission with corresponding changes to the requirements for tracking and data acquisition support by the DSN [1].

[2] Refer to http://solarsystem.nasa.gov/missions/galileo/indepth [3] for more on the Galileo interplanetary mission design. (accessed January 10, 2013)

[3] The last remote sensing data from the orbiter was received in March 2002.

The orbiter was powered by two radioisotope-thermoelectric-generators (RTGs). It used its 400 newton (N) main engine to go into Jupiter orbit, but maintained pointing and fine-tuning of each new orbit with clusters of 10-N thrusters.

The prime-mission tour consisted of 11 different elliptical orbits around Jupiter, with each orbit (except one) involving a close flyby and gravity assist at Jupiter's moons Ganymede, Callisto, or Europa. The major scientific returns from the primary mission included data on

- Jupiter's storms and rings
- Hot, active volcanoes on Io
- Strong evidence for a possible ocean on Europa
- Ganymede's own magnetic field
- Evidence suggesting the possibility of liquid saltwater oceans beneath the surfaces of Ganymede and Callisto surface.

After completing its primary mission, Galileo began a two-year extended mission called the Galileo Europa Mission (GEM) on December 8, 1997. GEM was a 14-orbit, low-cost extension of Galileo's exploration of the Jovian system. This mission was divided into three main phases:

- The Europa Campaign (December 1997–May 1999), which searched for further signs of a past or present ocean beneath Europa's icy surface;
- The Jupiter Water/Io Torus Study (May 1999–October 1999), which focused on detailed storm and wind patterns in Jupiter's atmosphere; and
- The Io Campaign (October 1999–December 1999), which obtained, from two flybys, high-resolution images and a compositional map of Io with a sample of a volcanic plume.

At the end of the GEM, December 31, 1999, the orbiter started another mission called the Galileo Millennium Mission (GMM). This mission originally was planned for completion within approximately 14 months but was extended to 2003. The GMM mission plan originally consisted of two phases, the first named Io and the second Cassini.[4] The 2003 mission extension included plans for the final disposition of the orbiter.[5] During the GMM, the orbiter made additional close flybys of all four large moons, including four encounters of Io from 2000 through 2002. The spacecraft studied Io's extensive volcanic

[4] For more on the Galileo/Cassini 2001–2002 cooperative mission refer to http://www.jpl.nasa.gov/jupiterflyby/ and to http://www.jpl.nasa.gov/news/fact_sheets/galileo0309.pdf (both accessed January 10, 2013)

[5] Refer to http://www.jpl.nasa.gov/news/fact_sheets/galileo0309.pdf [5] for more information on the GMM and a table showing the dates and flyby altitudes of all of Galileo's satellite encounters. (accessed January 10, 2013)

activity and the magnetic environment at high resolution. It also observed Europa's ionosphere, generated by ultraviolet radiation from the Sun and interaction of the charged particles from the Jovian magnetosphere. In the Cassini phase, the spacecraft performed cooperative measurements with the Cassini spacecraft as Cassini received its own gravity assist from Jupiter in December 2000. Galileo was also relatively near Jupiter at that time. Galileo collected data from Jupiter's inner magnetosphere, the dusk side of the magnetosphere, and the solar wind.

In November 2002, Galileo's orbit took it closer to Jupiter than ever before, flying less than 1000 kilometers (km) over the moon Amalthea,[6] which is less than one-tenth the size of Io and less than half as far from Jupiter. Measurements of changes in Galileo's radio signal frequencies during the flyby were used refine the mass and density of Amalthea. This passage produced information on dust particles as Galileo flew through Jupiter's gossamer rings as well as new information on magnetic forces and energetic charged particles close to the planet. Galileo's final orbit took an elongated loop away from Jupiter. Then on September 21, 2003, came an intentional mission-ending plunge into Jupiter's atmosphere to ensure against the possibility of impact and Earthly contamination of any Jovian satellites. Eight years after probe entry, the orbiter also made a direct impact with Jupiter, vaporizing as it plowed into the dense atmosphere.

Figure 4-1 is a graphical representation of the sizes and orientations of the orbits around Jupiter for the prime mission and the GEM [4,32]. For clarity, only some of the orbits are labeled with the alphabetic character indicating the targeted Jovian satellite (Callisto, Ganymede, Europa, or Io) for that orbit, and the two-digit number representing the orbit number.

4.1.2 The Spacecraft

The Galileo spacecraft (Fig. 4-2) had two main components at launch, the 6.2-meter (m) tall orbiter, and the 0.9-m long probe.[7] The orbiter's launch mass was 2,223 kilograms (kg), including a 118-kg science payload and

[6] Amalthea averages 189 km in diameter (270 × 166 × 150). Amalthea was the nymph who nursed the infant Zeus with goat's milk in Greek mythology.
http://en.wikipedia.org/wiki/Amalthea_%28moon%29 (accessed January 10, 2013)

[7] Figure 4-1 and Figure 4-2 come from predecessors to https://solarsystem.nasa.gov/galileo/mission/index.cfm
This link to the Galileo legacy website [3] includes brief descriptions of the Galileo mission, spacecraft (orbiter and probe), and mission operations. (accessed May 19, 2016)

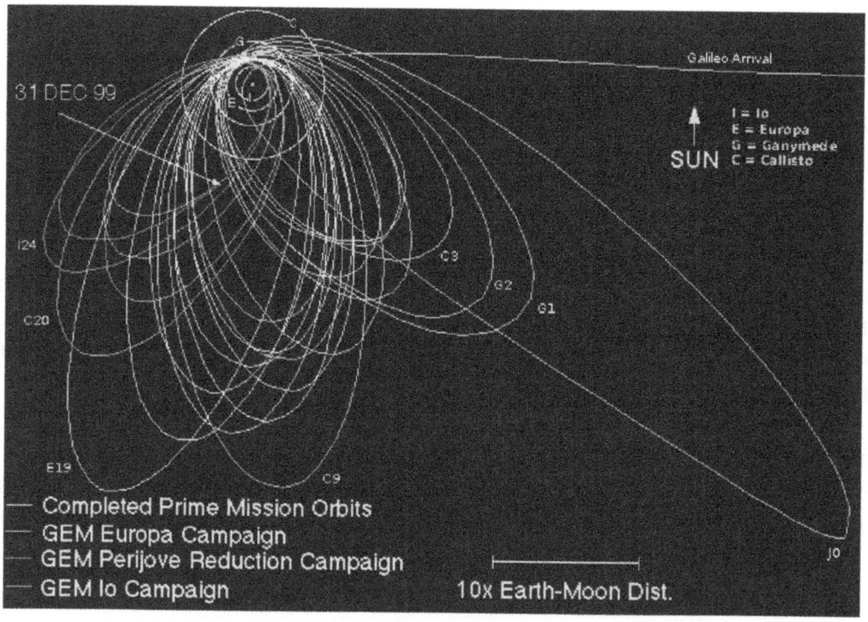

Fig. 4-1. Galileo Europa mission and prime mission tours. Please visit www.wiley.com/go/deepspacecommunications to access the color version of this figure.

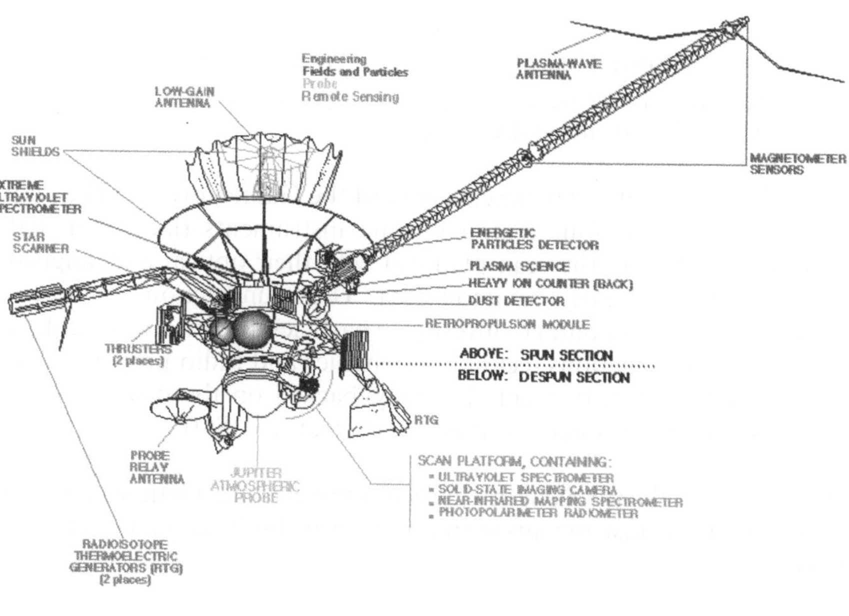

Fig. 4-2. Galileo spacecraft. Please visit www.wiley.com/go/deepspacecommunications to access the color version of this figure.

925 kg of usable propellant.[8] The probe's total mass was 339 kg: the probe descent module was 121 kg, including a 30-kg science payload.

4.1.2.1 Galileo Orbiter. The Galileo orbiter combined features of spinner spacecraft (the Pioneers and Ulysses) and three-axis-stabilized spacecraft (the Voyagers). The orbiter incorporated an innovative "dual-spin" design. Part of the orbiter (including the telecom electronics and antennas and some instrument booms) rotated while another part (containing an instrument platform) remained fixed in inertial space. The orbiter was a good platform for fields and particles experiments that perform best when rapidly gathering data from different directions. The orbiter was also a good platform for remote sensing experiments that require accurate and steady pointing.

The orbiter used two RTGs to supply electrical power to run the spacecraft's devices. The radioactive decay of plutonium produced heat that was converted to electricity. The RTGs produced about 570 watts (W) at launch. The power output decreased at the rate of 0.6 W per month and was 493 W when Galileo arrived at Jupiter.

The attitude and articulation control subsystem (AACS) was responsible for determining the orientation of the spacecraft in inertial space, keeping track of the spacecraft orientation between attitude determinations, and changing the orientation, instrument pointing, spin rate, or wobble of the spacecraft. Software in the AACS computer performed the calculations necessary to do these functions. As part of the S-band (2 to 4 gigahertz; GHz) mission (described in Section 4.4 of this chapter), the AACS software was updated to include the ability to compress imaging and plasma wave data down to as little as 1/80th of their original volume.

There were 12 scientific experiments aboard the Galileo orbiter. The despun section was home to four remote-sensing instruments (labeled in red in Fig. 4-2), mounted on the scan platform with their optical axes aligned so that they viewed a nearly common area. The spun section contained six instruments (labeled in blue) to investigate particles and magnetic fields. Two radio-science investigations (celestial mechanics and radio propagation) did not have individual instruments but piggybacked on the orbiter's telecom system, including the system's ultrastable oscillator (USO).

Figure 4-3 shows the wavelength ranges of the electromagnetic spectrum that the remote-sensing instruments monitored during both encounters and cruise periods.

---

[8] Propellant made up 41 percent of the orbiter's launch mass. Most of the propellant was consumed at JOI.

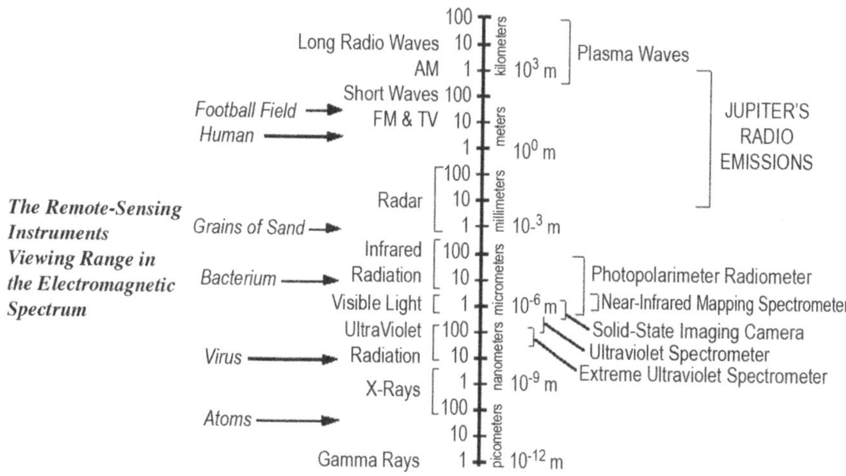

Fig. 4-3. Galileo orbiter's remote sensing instrument wavelength ranges in the electromagnetic spectrum.

4.1.2.2 Galileo Probe. The probe consisted of two main parts, the deceleration module and the descent module.[9] The deceleration module was required for the transition from the vacuum and cold of interplanetary space to the intense heat and structural loads incurred during a hypersonic entry into a planetary atmosphere—and from a speed of tens of kilometers per second to a relatively placid descent by parachute. The descent module carried the scientific instruments and supporting engineering subsystems that collected and transmitted scientific data to the orbiter, which was flying overhead.

The probe did not have an engine or thrusters so it could not change the path set for it by the orbiter at separation. The probe was spin-stabilized, achieved by spinning the orbiter up to 10.5 rpm before release. There was no communication between orbiter and probe during the coast to Jupiter because the probe had no capability to receive radio signals. During atmospheric entry, the probe stored no data, collecting and transmitting it in real time.

The probe's entry into the Jovian atmosphere generated temperatures of 14,000 K. The materials used for the probe's descent module heat shields—carbon phenolic for the forebody shield and phenolic nylon for the afterbody shield—have also been used for Earth re-entry vehicles.

---

[9] The probe description comes from http://solarsystem.nasa.gov/galileo/mission/journey-probe.cfm [3]. (accessed January 10, 2013)

See Section 4.6 of this chapter for more detail on the probe-to-orbiter relay link.

Parachutes were used for two key functions, separating the deceleration and descent modules and providing an appropriate rate of descent through the atmosphere. Before deployment of the main chute, a smaller, pilot parachute was fired at 30 meters per second (m/s) by a mortar to start the deployment process. The deployment occurred in less than 2 s, pulling away the aft cover and unfurling the main chute. The main parachute's diameter was 2.5 m. The canopy and lines were made of Dacron and Kevlar, respectively. Once the main chute was fully deployed, the forebody shield (aeroshell) was jettisoned.

To save weight, the Galileo descent module, carrying six scientific instruments, was not sealed against the influx of the Jovian atmosphere. However, the two relay radio systems were hermetically sealed within housings designed to withstand pressures up to 20 bars and tested to 16 bars (2 and 1.6 megapascals, MPa).

4.2 Galileo Spacecraft Telecommunications System

The Galileo telecommunications system (Fig. 4-4) was on the spun section of the dual-spin orbiter. The system consists of four hardware subsystems:

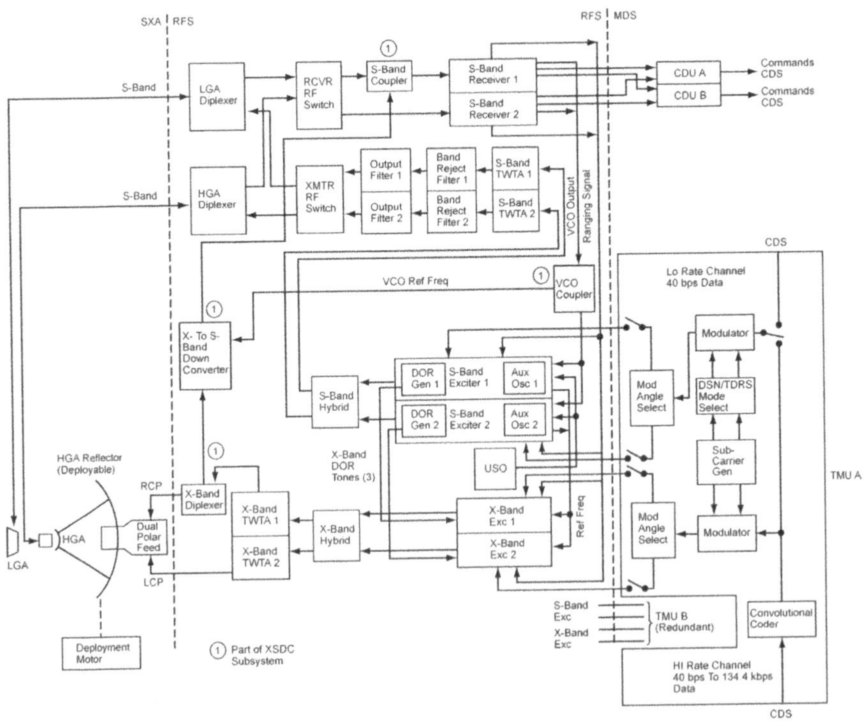

Fig. 4-4. Galileo orbiter telecom system.

1) Radio frequency subsystem (RFS)
2) Modulation demodulation subsystem (MDS)
3) S-/X-band[10] antenna (SXA) subsystem
4) X- to S-band downconverter (XSDC).

4.2.1 Galileo Telecommunications Functions and Modes

The Galileo telecommunications system[11] enabled the orbiter to provide: (a) uplink carrier tracking and downlink carrier generation, (b) command detection, (c) telemetry encoding and modulation, and (d) radiometric communications with the Deep Space Network (DSN).[12] For interplanetary cruise, Galileo originally planned to use a ground station operated by the German Space Operations Center (GSOC)[13] as well as those of the DSN. During the prime mission, antenna arrays included the Parkes antenna operated by the Australian Commonwealth Scientific and Industrial Research Organization (CSIRO).[14]

4.2.1.1 Uplink. Depending on the required uplink mode, the carrier could be unmodulated, modulated with a command subcarrier or ranging modulation, or both.

4.2.1.1.1 Uplink Carrier. The spacecraft receiver could acquire an uplink carrier arriving close enough in frequency, and then maintain phase-lock on

[10] For Galileo, S-band refers to carrier frequencies of about 2.1 GHz (uplink) and 2.3 GHz (downlink). X-band refers to carrier frequencies of about 7.2 GHz (uplink) and 8.4 GHz (downlink).

[11] Sections 4.2 and 4.3 describe the Galileo orbiter telecom system and the ground system as they were originally intended for use. The orbiter no longer exists, and the ground system has evolved away from S-band. References to uplink or downlink at X-band assume the availability of a fully deployed high-gain antenna (HGA). As described in Section 4.4, the HGA did not deploy. The X-band parts of the RFS were verified operational in short tests in 1991 and 1993. The S-band parts of the telecom system that used the LGA functioned as designed.

[12] The terms "radiometric communications" or radiometric data in this chapter refer collectively to one-way or two-way Doppler, turnaround (sequential) ranging, and differential one-way ranging (DOR).

[13] Current information about GSOC is available at http://www.dlr.de/iss/en/desktopdefault.aspx/tabid-1412/2072_read-3536/

[14] Current information about the Parkes antenna is available at http://www.atnf.csiro.au/ (both links accessed January 10, 2013)

that carrier as long as it was present. The telecom system was able to operate in the following uplink modes:

- With an uplink or with no uplink
- With the uplink at S-band or at X-band
- With the uplink modulated or unmodulated
- With the uplink transmitted from the either the DSN or GSOC.

4.2.1.1.2 Command Detection. The RFS S-band receiver (S-RCVR) and the command detector unit (CDU) received and demodulated the command waveform from either an S-band or an X-band uplink carrier, and send it to the hardware command decoder in the command data subsystem (CDS). The command waveform could be present alone or simultaneously with ranging modulation.

4.2.1.2 Downlink. The downlink carrier could be unmodulated, modulated with a telemetry subcarrier or ranging modulation, or both.

4.2.1.2.1 Downlink Carrier. The RFS exciters (EXC) and power amplifiers made up the transmitters that gave the orbiter the capability to generate, modulate, and transmit downlink carriers. With or without an uplink carrier present, the RFS was able to generate and transmit an S band downlink carrier alone, an X-band downlink carrier alone, or both simultaneously. With either an S-band or X-band uplink carrier present, the RFS had the capability to use the uplink carrier to generate downlink S-band or X-band carrier frequencies or both. The S-band and X band downlink carriers were always coherent with each other. Depending on RFS mode, the downlink carriers both were coherent with the uplink carrier or both were noncoherent.[15]

4.2.1.2.2 Telemetry Encoding and Modulation. The MDS's telemetry modulation unit (TMU) and the RFS's S-band exciter (S-EXC) and X-band exciter (X-EXC) processed the telemetry "low rate" and "high-rate" data-bit streams[16] from the CDS

[15] Galileo was one of many JPL deep-space missions having two downlink modes called "TWNC on" and "TWNC off." TWNC (two-way non-coherent) is pronounced "twink." The TWNC-on mode means the downlink frequency cannot be coherent with an uplink frequency. The TWNC-off mode means the downlink will be coherent with a received uplink when the transponder's receiver is in lock to the uplink carrier.

[16] As more fully described in Section 4.2.3, the CDS continuously output to the TMU both a 40 bits-per-second (bps) low-rate data stream and a high-rate data stream. The bit rate of the high-rate stream was set between 10 bps and 134.4 kbps, so in most cases its bit rate was higher than the fixed 40 bps of the low-rate channel.

into modulated telemetry subcarriers that phase modulated the downlink carriers. The TMU provided two telemetry modes: Tracking and Data Relay Satellite System (TDRSS)[17] and DSN.

4.2.1.3 Radiometric Data. Radiometric communications are those that are required of the telecom system to meet project navigation and radio science data quantity and accuracy requirements. Radiometric data used with Galileo included two-way or one-way Doppler, turnaround ranging, and differential one-way ranging (DOR).

4.2.1.4 Probe Relay. The L-band[18] relay link from the probe to the orbiter, active for about one hour on December 7, 1995, used equipment entirely separate from the orbiter's S-band and X-band uplink and downlink. Section 4.6 describes the relay link and its telecom-related results in more detail.

The orbiter-mounted relay receiving hardware (RRH) received the L-band signal from the probe. Though mounted on and in the orbiter, the RRH antenna and receivers were designated part of the probe system. To eliminate single-point, catastrophic mission failures, the relay link system configuration included two nearly identical RF links with dual electrical and electronic probe transmitting and orbiter receiving systems. Two parallel and simultaneous data streams went from the probe's scientific instruments to the orbiter. One of the data streams used a USO for transmission to the orbiter. The probe Doppler wind experiment used variations in the frequencies of the carrier signals received at the orbiter to deduce the wind speeds in the atmosphere [6].

4.2.2 Radio Frequency Subsystem

The RFS had the major components listed in Table 4-1. The table includes the subsystem acronyms for reference.

Where there were duplicated units in a pair (such as S-RCVRs), generally each could provide full functionality. The units were cross-strapped but with

---

[17] In the TDRSS mode, the TMU convolutionally coded a 1200-bps data stream received from the CDS. The symbol stream phase-modulated the RFS S-band RF carrier at 90±3 deg, without use of a telemetry subcarrier. This mode was used only for the immediate post-launch phase while the spacecraft was still attached to the Inertial Upper Stage (IUS). See 4.5 Telecom Operational Scenarios in this chapter.

[18] L-band refers to frequencies between 390 megahertz (MHz) and 1550 MHz. The probe-to-orbiter relay link carriers were 1387.0 MHz and 1387.1 MHz, chosen to provide the best link performance through Jupiter's atmosphere.

Table 4-1. RFS components.

Element	Number of Units
S-band receiver (S-RCVR)	2
S-band exciter (S-EXC)	2
X-band exciter (X-EXC)	2
Ultrastable oscillator (USO)	1
S-band traveling-wave-tube amplifier (S-TWTA)	2
X-band traveling-wave-tube amplifier (X-TWTA)	2
Differential one-way ranging (DOR) generator	1
S-band antenna switches (LGA-1/LGA-2 switches, HGA/LGA switch)	2,1
Microwave routing and interface elements	N/A

only one unit powered at a time.[19] For example, either S-RCVR could drive either S-EXC, with the powered receiver driving the powered exciter. Similarly, either S-EXC could drive either S-TWTA. Additional functional redundancy was built into the RFS in the sense that the 1-way downlink frequency source could be either the exciter's auxiliary oscillator or the USO, and (when using the HGA), the downlink could be at either S-band or X-band.

4.2.3 Modulation Demodulation Subsystem

The MDS consisted of two TMUs and two CDUs, with one CDU and one TMU powered at a time. The CDU was responsible for the detection (demodulation) of uplink command data for decoding by the CDS, and the TMU was responsible for the modulation of telemetry data for downlink transmission.

Because of the critical functions performed by the CDU and TMU, each had a large amount of hardware redundancy and cross-strapping with the

[19] The term "cross-strapped" refers to the interconnections at the unit input or output. Because RCVR-1 and RCVR-2 were cross-strapped with S-EXC-1 and S-EXC-2, the transponder could operate with RCVR-1 driving either S-EXC-1 or S-EXC-2, or RCVR-2 driving either S-EXC. "Generally" means there were exceptions, required either in hardware design or flight rule, or a factor in selecting configuration [5]. RFS exceptions to "generally": Only RCVR-1 was connected to the XSDC; RCVR-2 operated on the same channel as the USO, so potential frequency interference was one factor in launching with RCVR-1 selected; and the XSDC received too much noise when X-TWTA-2 was on, so the two could not operate together.

interfacing RFS elements. The two CDUs were identical to each other, and the two TMUs were nearly identical to each other.[20] The TMU and the RFS exciter were fully cross-strapped. Likewise, the CDU pair was cross-strapped with the RFS receiver pair and with the hardware command decoder of the CDS.

The TMU received two serial data streams from the telemetry formatter of the CDS. The use of the uncoded 40-bps low-rate data was reserved for when the spacecraft entered safing. The 10-bps to 134.4-kbps high-rate stream was convolutionally encoded[21] by the TMU. The TMU could modulate either the low-rate bit stream or the high-rate symbol stream on either a 22.5-kHz subcarrier or a 360-kHz subcarrier from an internal TMU oscillator. TMU-B could provide the symbol stream directly to the exciter (TDRSS mode).

The CDU received a modulated 16-kHz command subcarrier from the RFS receiver. Depending on the ground station command mode, the subcarrier could be unmodulated, modulated with a bit-synchronization (bit-sync) waveform equivalent to an all-zeroes command data stream, or with both bit-sync and command bits. The CDU demodulated the command subcarrier. It provided three separate outputs to the CDS command decoder:

1) A CDU in-lock or out-of-lock indicator

2) The 32-bps command-bit timing ("clock")

3) The command bits.

---

[20] The TMUs were almost identical. They differed as follows: TMU-A had an experimental "coder-2" that could produce a (15,1/4) convolutional code for 115.2 kbps and 134.4 kbps and modulate the coder-2 symbols on a 720 kbps subcarrier. TMU-B had a "TDRSS mode" wherein symbols from the (7,1/2) coder directly modulated the S-band carrier (no subcarrier). Adding the experimental coder less than 2 years before launch was a result of the delay in launch date from 1982 to 1989 and consequently the prime mission period to 1995–1997. The decrease in output from the already-fueled RTG power supply during the delay meant the TWTA would likely operate only in the low-power mode. Part of the communications shortfall was to be made up by using the more efficient (15,1/4) code; the remainder by planning an array of the DSN's 70-m antenna with the Very Large Array (VLA) radio science antenna system in New Mexico for the critical encounter data [1]. See Ref. [7] for a description of the VLA and [8] for the use of the VLA as an arrayed antenna resource during the Voyager mission.

[21] See Section 4.3, Galileo S-band Mission, for a description of the "concatenated coding" used from 1996 until the end of the mission.

4.2.4 S-/X-Band Antenna Subsystem

The SXA consisted of an HGA[22] and two LGAs (LGA-1 and LGA-2). The two LGAs only worked at S band. The HGA was designed to work at S-band and X-band. Because the LGAs had no X-band capability, uplink or downlink at X-band required the HGA. The spacecraft could be configured (via real time or sequenced commands) to receive and transmit S band on the HGA, on LGA-1, or on LGA-2. The same antenna had to be used for both reception and transmission of S-band at a given time. Galileo S-band antennas were right circularly polarized (RCP), simplifying the task of configuring the DSN. The X-band downlink polarization was RCP or left circularly polarized (LCP) depending which of the X-TWTAs was powered on.

The S-band antennas operate at a nominal uplink frequency of 2115 megahertz (MHz) and a nominal downlink frequency of 2295 MHz. The actual frequencies were DSN channel 18 for RFS receiver 1 (and for a two-way coherent downlink with that receiver) and channel 14 for RFS receiver 2 (and its coherent downlink) or a USO-generated downlink [5].

Because the most prominent part of the HGA was a main reflector 4.8 m in diameter, it looked like a single antenna. However, the HGA had two separate feed systems, one for S-band and the other for X-band. In its functions, the HGA could in many ways be considered as two distinct antennas (S-HGA and X-HGA). The X-band and S-band boresights (direction of maximum gain) were co-aligned in the direction of the LGA-1 boresight, which was the –z axis.

Though this was not the original plan, LGA-1 was selected for most of the mission. LGA-2, with its boresight aligned in the opposite direction from the LGA-1 boresight, was only used at specific times when the trajectory geometry required: Venus flyby and Earth-1 flyby (see Section 4.5, Operational Scenarios).

In addition to LGA-1, LGA-2, and HGA, the orbiter also had two other antennas that were not considered parts of the orbiter telecom system. These were the relay receiving antenna (RRA) for the Galileo probe-to-orbiter relay link and the plasma wave spectrometer (PWS) antenna, part of a science instrument.

---

[22] The Galileo HGA did not deploy fully and therefore was never functional for use in the mission. The antenna description in Section 4.2.4 is of the system as built and *intended* for use. See Section 4.3, Galileo S-Band Mission, for the workarounds developed during flight to enable a successful mission.

4.2.5 X- to S-Band Downconverter

The Galileo project always considered the orbiter's single XSDC as an experimental subsystem, meaning that use of an X-band uplink was not essential for receiving commands or other critical mission functions. The XSDC parts were

- Downconverter
- S-band coupler
- X-band diplexer
- Low pass filter
- Voltage-controlled oscillator (VCO) coupler.

The X-band diplexer allowed simultaneous X-band reception (via the XSDC) and transmission (via the X-band TWTA) through the HGA. The S-band coupler connected the down-converted X-band modulated carrier to RFS S-RCVR-1. Working with the RFS receiver and transmitter, and depending on the controlled configuration, the XSDC provided the telecom system with non-coherent or two-way coherent carrier operating modes with an X-band uplink.

4.2.6 Telecom Hardware Performance during Flight

The orbiter was launched with the following elements active: S-RCVR-1, S-EXC-1, S-TWTA-1, CDU-A, and TMU-A. The USO was turned on a few weeks after launch. As of the end of the Galileo primary and extended missions, these originally selected telecom units were all still selected and generally operating without a problem.

The telecom hardware problems that occurred during flight were

- HGA failure to deploy (discussed separately in Sections 4.4 and 4.6)
- RFS receiver "wandering VCO anomaly"
- Unexpected CDU lock-count changes
- USO frequency drift rate changed by radiation.

4.2.6.1 "Wandering VCO" RFS Receiver Incident. Several days after Ganymede-2, the second of the 10 Jovian satellite encounters of the prime mission, the orbiter receiver failed to acquire a routine uplink from the Madrid tracking station on September 11, 1996.[23] The spacecraft turned, 90 degrees

[23] A Galileo uplink acquisition required the station to turn on its 100-kW S-band transmitter and perform a frequency sweep. A standard sweep varied the frequency at a specified rate over a range of ± 12 kHz about a center frequency, returning to the center frequency for the rest of the pass. Even with Doppler over a pass, this center frequency reached the spacecraft near enough to the "best lock frequency," the frequency at which the receiver VCO oscillated without an uplink.

(deg) off Earth and back, several hours before the incident. However this change in geometry could not be confirmed as a contributing factor in the subsequently modeled hardware problem. Examination of the VCO volts telemetry by telecom showed the measurement had deviated by as much as −8 kHz and +24 kHz from best lock prior to the acquisition attempt [9]. The station acquired the uplink using a resweep frequency range 2.5 times as wide as standard. Over the next several days, tracking stations had to use sweep frequency ranges as much as 7.5 times the standard.

The receiver returned to normal operation after a radio frequency subsystem tracking-loop capacitor (RFSTLC) test. RFSTLC tests, conducted periodically through the prime mission, required a station to sweep its transmitter in frequency to pull the VCO to either +65 kHz or −65 kHz from the best-lock frequency (BLF), then to turn the transmitter off. The VCO frequency then relaxed back to BLF; the time it took to do so provided a measure of the time constant of the resistor-capacitor network in the tracking loop. In the first RFSTLC test after the incident, the frequency-change-vs.-time signature and the loop time constant were not normal. However, the VCO wandering stopped, and subsequent RFSTLC tests were normal in all respects. This receiver operated normally through the remainder of the prime and extended missions.

Ground testing and analysis focused on the receiver tracking-loop integrator, an LM108 operational amplifier. A model [10] that involves ionic contamination by sodium ions (Na+) fits the inflight data well, including the evident self-healing ("annealing"). No further on-board receiver problems occurred.

4.2.6.2 Unexpected CDU Lock-Count Changes. As part of its normal operation, the CDU incremented a software counter each time it changed to or from out-of-lock to subcarrier-lock or subcarrier-lock to bit-sync-lock. The spacecraft telemetered the count as engineering data periodically. The CDU passed command data to the command decoder in the CDS only when it was in bit-sync lock. The number of lock counts for each session of planned commanding was known. If the count exceeded the predicted number, this was defined as an "unexpected CDU/CDS lock-count change." These unexpected lock counts have also occurred on Voyager and other projects, and they have never caused any problem with commanding. The several unexpected lock-change events per year on Galileo placed no restrictions on commanding activities. With greatly reduced engineering telemetry sample rates and telecom staffing in GEM and GMM, unexpected lock-count events were no longer analyzed, though they presumably still occurred. There was no evidence of any change in receiver or CDU command performance after launch.

On Galileo, prime-mission unexpected lock counts occurred from a variety of station configuration and operational problems [10]. Also, one repeatable spacecraft cause is known, having been verified in ground tests of a Galileo receiver and CDU in the early 1990s. During the test, one pair of lock changes occurred as a result of the combination of (a) the uplink tuning rate at the initial uplink acquisition by the unmodulated carrier at the start of a pass, (b) the uplink signal level of the unmodulated carrier, and (c) the relative frequency rate between the sweep and the receiver VCO. The effective frequency rate in the test included the combined effect of Doppler frequency at the receiver and the "random walk" frequency of the VCO.

The "U/L ACQ (uplink acquisition) sweep" mechanism resulted in a waveform momentarily at the RFS receiver output to the CDU, which the CDU in turn interpreted (in error) as bit sync command modulation. This waveform occurred when the frequency difference was about 512 Hz, the same as the Galileo command subcarrier frequency. The CDU sent an "in-lock" signal to the CDS, which recorded it as a lock-count change. After a moment, the RFS receiver output waveform was different, the CDU no longer interpreted it as a command signal, and it sent an "out-of-lock" to the CDS. The CDS made another lock count, for a total of two.

4.2.6.3 USO Radiation-Induced Frequency Offset and Rate Change. The USO was of Voyager project inheritance. Though each S-EXC had an internal auxiliary oscillator (aux osc), the USO had been the predominant non-coherent downlink carrier frequency source since it was turned on December 5, 1989. The USO was turned off for a few tens of days in late 1991 and once again in early 1992 in support of the anomaly investigation of the HGA failure to deploy.

The frequency of the crystal oscillator in a USO changes with time (ages). The multi-mission navigation team accounted for the relative velocity between station and spacecraft in their orbit determination and predictions of one-way Doppler frequency. Frequency shifts not accounted for in the navigation orbit-determination process were used to ascertain other effects, such as the crystal aging or the effects of radiation.

It was also known, from Voyager's one-way Doppler profiles before and after the Voyager spacecraft flybys of Jupiter in 1979, that the Galileo USO frequency rate would be affected by the radiation dose at each planetary encounter. For the prime mission, the effect was qualitatively expected to be greatest at JOI because the largest radiation dose occurred there. However, the Voyager experience could not be confidently carried over to make a quantitative prediction of the radiation-induced USO frequency change for

Galileo. Based on the Voyager experience, the project and the DSN coordinated, as part of the overall JOI telecom strategy, to search for and quickly find the one-way downlink.

The radiation-induced USO frequency changes continued through the Jupiter encounters of the prime, GEM, and GMM missions. The offset changes were usually fairly small (less than 5 Hz at S-band over a couple of days) at each encounter.[24] The pre-encounter drift (aging) rate was observed to resume as the orbiter returned to greater distances from Jupiter. To ensure rapid lockup by the ground receiver, the DSN sent out periodic USO frequency update messages (known as TFREQ updates) for use in tracking operations whenever the USO frequency (referenced to S-band) changed by more than 0.5 Hz.[25]

4.2.7 Orbiter Input Power and Mass Summary

When operating, each telecom system element had a single power input mode except for the TWTAs, which had both high-power and low-power modes. Table 4-2 summarizes the steady-state spacecraft input power and the RF output power for both high-power and low-power modes, as applicable. The table also summarizes the masses of components of the system.

---

[24] On November 5, 2002, the Galileo orbiter flew past the satellite Amalthea at a distance of less than 1 R_J above Jupiter's cloud tops. The radiation level experienced during this flyby was significantly greater than that during JOI. The total dose was estimated to be 2.5 times that incurred during a typical Io flyby. The project configured the orbiter's flight software (including fault protection algorithm updates) to accept the probable effects of the radiation environment. These fault-protection response changes balanced swapping redundant-element swapping or mode changes that would significantly reduce the continuity or quantity of critical science data against mode changes essential to continuing the orbital mission.

[25] On December 11, 2001 and again on January 27, 2002, the received downlink frequency in the one-way mode exhibited rapid and unexpected variations of several tenths of a hertz. These fluctuations, thus far unexplained, occurred over more than one 70-m station. They each resulted in the loss of several frames of telemetry data when the station receiver carrier loop was unable to follow the rapid frequency changes. After a period of several hours, the downlink frequency became stable again, though at an offset of several tenths of a hertz from before each episode. Because the Sun–Earth–spacecraft angle was greater than 90 deg and the Sun was not unusually active, solar effects on the S-band downlink are ruled out. The episodes did not occur when the orbiter was in a high-radiation region. By elimination of other possible causes, the circumstances point to the USO or its control circuitry as a source of the frequency fluctuations, but not to a specific cause.

Table 4-2. Galileo orbiter input power and mass summary.

	Number of Units	Input Power (W)[a, b]	Output Power (W)[c]	Mass (kg)[a, d]
RFS				53.5
Transponder	2			
Receiver		4.5		
S-band exciter		2.6		
X-band exciter		3.5		
Antenna control and interface subsystem (ACIS)[e]	1	0.7		
USO	1	2.7/4.5		
X-band TWTA	2	46.9/72.4	11.6/20.0	
S-band TWTA	2	34.9/87.1	4.9/14.8	
DOR		1	0.5	
XSDC	1	3.1		2.5
MDS				9.8
CDU	2	4.4		
TMU	2	5.5/5.8		
SXA				
Deployment motor		2	12.0	
Antenna[f]				8.1

[a] Mass is from Ref. [5], module GLL-3-230; input power is from Ref. [5], module GLL-3-250.

[b] For TWTAs, the smaller power value is for low-power mode, the larger for high-power. For USO and TMU, the lower value is near-Earth, and the larger is at Jupiter.

[c] RF power defined as design value at RFS/SXA interface (LGA-1 for S-TWTA and HGA for X-TWTA).

[d] The stated mass is the total for the subsystem (for example, 9.8 kg for the MDS includes two TMUs and two CDUs).

[e] Antenna control and interface system.

[f] Mass does not include antenna structural elements. The entire orbiter structure is 255.5 kg.

4.3 Galileo S-Band Mission

4.3.1 Overview

The Galileo project had intended to use the Deep Space Network as a means to transmit conventional uplink signals to the spacecraft via S-band and X-band carriers, and to receive conventional downlink signals from the spacecraft. In brief, these signals were

- Uplink carrier
- Command subcarrier and data stream
- Downlink carrier
- Telemetry subcarrier and data stream
- Two-way Doppler data from downlink carrier (with coherent uplink carrier in lock)
- Turnaround ranging modulation on the uplink carrier and downlink carrier(s)
- Delta-DOR modulation on the downlink carrier(s)

The conventional capabilities of the DSN, planned for Galileo use [11], as they existed in the 1990s and early 2000s, are described in Chapter 4. Because of a failure of the spacecraft high-gain antenna (HGA) to deploy fully and consequent loss of any meaningful X-band capability, the support provided to the Galileo "S-band mission" by the DSN was anything but conventional.

This section describes the extraordinary collaboration effort between a reconstituted Galileo project software development team, the Galileo flight team, and the DSN technology development team that saved the Galileo mission from the HGA failure and eventually led to the overall success of the Galileo mission. This effort was made the subject of a DSN Advanced Systems Program[26] case study [12].

During its early cruise phase, the Galileo orbiter communicated with Earth using the S band signals from the LGA. As designed for thermal control, the HGA "umbrella" antenna with X band capability was to remain furled until the Sun–spacecraft range became and remained greater than 0.9 astronomical units (AU) before the second flyby of Earth. On April 10, 1991, about 1-1/2 years after launch and with the thermal constraint lifted, the orbiter was commanded to turn the HGA deploy motors on to unfurl its HGA. The

[26] The DSN Advanced Systems Program sponsored a number of improvements in capability during the Galileo era (1980s and 1990s). For the Galileo LGA mission, station arraying improvements and new coding/image data compression techniques were the most significant.

antenna failed to fully deploy. Analysis of telemetry data and pre-launch design and test data pointed to a scenario that 3 of the 18 ribs of the umbrella antenna remained stuck to the antenna's central tower. Several unsuccessful attempts were made to free the stuck ribs. Because the reflector had not achieved a parabolic shape, the antenna was not functional. The only way to continue communicating was through the use of the Earth-facing LGA-1. And if the then-current configuration (ground and spacecraft) remained unchanged, the telemetry data rate would decrease to 10 bps by JOI. The originally planned X-band HGA downlink data rate was 134.4 kbps.

For over a year, much thought was expended in ground testing and analysis, leading to multiple efforts to free the stuck ribs. Most attempts involved turning the spacecraft toward and away from the Sun, in the hope that warming and cooling of the antenna assembly would free the ribs through thermal expansion and contraction. These attempts were unsuccessful. Other analysis suggested that turning the antenna deployment motors on and off repeatedly ("hammering") might deliver enough of a jolt to free the sticking and open the antenna. Unfortunately this effort failed also. Other approaches were tried, but none of them worked. For example, the X-band downlink and uplink were operated through the partly deployed HGA to compare end-to-end capability with the S-band LGA-1 capability.[27]

In parallel with the efforts to unstick the HGA, the JPL Flight Projects Office (Galileo Project), the JPL Telecommunication Division, and the JPL Tracking and Data Acquisition Office supported a study from December 1991 through March 1992 to evaluate various options for improving S-band telemetry capability through LGA-1.[28] The study assumed that image and instrument data, as well as spacecraft calibration and monitoring data, would have to be heavily edited and compressed using the Galileo's onboard processors, which had severe limited computation and memory resources. The study

[27] Use of the HGA was found not viable. The test showed the X-band downlink, near the HGA boresight, had about 2-dB improvement relative to LGA-1 S-band downlink. The pattern had numerous deep nulls, suggesting that keeping the antenna sufficiently pointed would be a major operational challenge. Further, it wasn't known if the nulls might change position with time due to temperature changes or mechanical movement.

[28] As early as October 1991, the TDA Office chartered a 1-month study to identify a set of options to improve the telemetry performance of the Galileo mission at Jupiter, using only the LGA. At the end of the study, the four options recommended for further evaluation (arraying of ground antennas, data compression, advanced coding, and suppressed carrier downlink) eventually were all applied in the S-band mission [1]. Note: at the time of the study, the DSN was part of the Telecommunications and Data Acquisition (TDA) organization at JPL.

also presumed significant science and mission replanning and major ground system improvements would also be necessary.

The Galileo S-band mission was formally approved and funded in January 1993. The concept involved substantial changes to both the spacecraft and the DSN. Some key communications technologies used were

1) Intra-site and inter-continental antenna arraying, to increase the effective aperture by combining signals from up to six antennas
2) S-band "ultracone" feed and low-noise amplifier at the Canberra 70-m station, to provide a receive-only very low system temperature
3) Suppressed carrier tracking with the BVR, to improve modulation efficiency
4) Advanced channel coding, to reduce the operating signal-to-noise ratio (SNR) threshold
5) Low-complexity lossless and lossy data compression and image editing schemes, to reduce the onboard data volume without compromising the science objectives.

The first four items together, it was estimated, would increase the supportable downlink data rate by one order of magnitude, from 10 to about 100 bps on average.[29] Including the fifth, data compression, would provide another order-of-magnitude increase in performance.[30]

These expectations were achieved. With the improved S-band downlink, the orbiter was able to complete 70 percent of the objectives of the original primary mission. With continued use of the improved S-band downlink (except

---

[29] Arraying would improve the downlink by as much as 4 decibels (dB). Depending on which antennas were used, the ultracone would improve the downlink by another 1.7 dB, suppressed carrier modulation by 3.3 dB, and advanced coding by 1.7 dB. Together with the corresponding spacecraft modifications including data compression, the ground enhancements met the S-band mission Project objective to return one full tape recorder load of data after each satellite encounter, as well as satisfying the Project requirement to receive continuous engineering data and low-rate science [1].

[30] Data compression reduces the transmission and storage bandwidth required by removing intrinsic redundancy in the source data, but leaving the transmitted data more vulnerable to communication channel errors. Error correction coding introduces structured redundancy to the data to reduce the effects of channel errors, incidentally increasing channel bandwidth. Data compression and coding, used together as in the Galileo S-band mission, can produce a large improvement in the end-to-end system efficiency.

for arraying) in GEM and GMM, the orbiter returned significant amounts of science data.

4.3.2 Ground System Improvements for Galileo S-Band Mission

The DSN changes involved various enhancements to the three DSCCs that could provide a factor of 10 increase in data return from the Galileo spacecraft as compared with the data return that would result from use of the existing DSN configuration receiving S-band via the spacecraft LGA-1 only. The Galileo conceptual design is shown in Fig. 4-5. It included the addition of the Deep Space Communications Complex (DSCC) Galileo Telemetry (system) (DGT), a new telemetry subsystem to serve as a signal processor, specifically designed to handle the Galileo low-signal conditions.[31] The S-band mission packet-telemetry, suppressed-carrier DGT mode (known in Galileo spacecraft flight software as "Phase-2") began in May 1996.[32]

4.3.2.1 DSCC Galileo Telemetry. The DGT was installed in parallel with the existing Block V Receiver (BVR) and telemetry channel assembly (TCA), which formed a part of the DSN telemetry subsystem. The BVR and TCA continued to provide for Doppler extraction and spacecraft

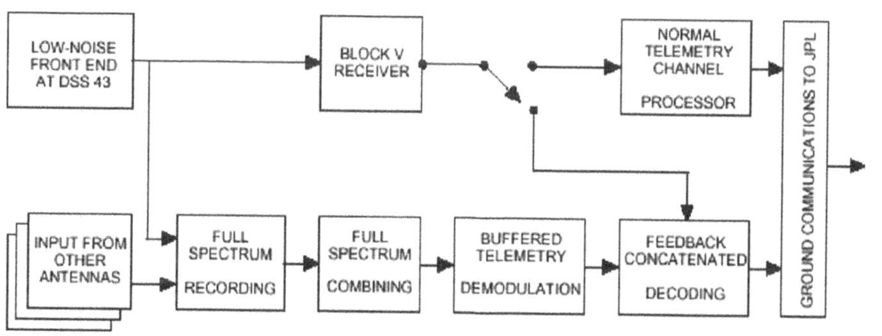

Fig. 4-5. Conceptual form of the DGT for Galileo.

---

[31] Material in this paragraph comes largely from Ref. [9]. The DGT included a full-spectrum recorder (FSR), a full-spectrum combiner (FSC), the buffered telemetry demodulator (BTD, a receiver with phase-locked loops for carrier, subcarrier, and symbols), and a feedback concatenated decoder (FCD). The DGT implementation was specifically for Galileo and was decommissioned in 2000; therefore, it is no longer a DSN capability.

[32] To reduce the risk to the Galileo one-chance-only events from schedule slips in the new ground system development, Galileo planned the critical December 1995 Probe data return and JOI activities to operate using the existing spacecraft Phase-1 software and the existing telemetry system only. Section 4.5.6 describes Phase-1 and Phase-2.

emergency support.[33] In 1995 the BVR was a new digital receiver for multi-mission support that was used for Galileo at Jupiter encounter. The BVR was capable of acquiring and tracking the spacecraft carrier in a residual or suppressed-carrier mode and of demodulating carrier, subcarrier, and symbols. For the Galileo S-band mission (Phase-2), the BVR delivered symbols to either the DGT's feedback concatenated decoder (FCD) in the packet mode or to the TCA's maximum likelihood convolutional decoder (MCD) in the time-division multiplexing (TDM) mode. Figure 4-5 shows the BVR interface to the FCD as developed for non-arrayed operation.

4.3.2.2 Ultracone at Canberra 70-m Station. In addition to the DGT, an ultra-low-noise receive-only feed system was added to the Canberra 70-m antenna (DSS-43) to reduce the S-band system noise temperature to 12.5 kelvin (K), excluding atmospheric effects. Prior to installation of this so-called "ultracone," Galileo operations had been conducted with an S-band polarization diversity (SPD) feed cone having system noise temperatures at zenith of 19.9 K in the diplexed transmit/receive mode and 15.6 K in the receive-only mode, both excluding the effects of the atmosphere. The ultracone met its system noise temperature (SNT) design objective. It continued to be used in the GMM, with a total system noise temperature (including atmospheric effects) of about 15 K at high elevation angles in good weather.

4.3.2.3 Arraying Ground Antennas. Further enhancement of the Galileo downlink signal was obtained through the following antenna-arraying techniques at the Canberra CDSCC:

- Intercontinental arraying of the 70-m antenna at the Goldstone, California (DSS-14), with the 70 m antenna near Canberra, Australia (DSS-43), during mutual view periods
- Addition of two of the three 34-m antennas at Canberra (DSS-34 and DSS-45) into the array with the 70-m antennas at Canberra and Goldstone
- Addition of the Australian 64-m radio telescope at Parkes into the array. Parkes, called Deep Space Station 49 (DSS 49), for DSN identification, supported the Galileo mission as an additional element of the Canberra array.

The timeline for arraying was generally: (a) begin the array, as Canberra "rises", by adding the Canberra 70-m and two 34-m antennas to the

[33] Through GEM and GMM, the S-band mission safemode continued to produce a residual carrier downlink, modulated by 40 bps "high-rate" TDM data with (7,1/2) convolutional encoding.

Goldstone 70-m antenna already tracking, (b) add Parkes about 2 hours later,[34] (c) then delete the Goldstone 70-m antenna as it sets, and (d) finally, delete Parkes as it sets about 2 hours before the Canberra array sets.

NASA provided several enhancements to the Parkes radio telescope to increase its contributions to the array.

The overall network configuration used to support this phase of the Galileo mission is shown in Fig. 4-6. At each antenna, the S-band signal from the spacecraft was converted to a 300-MHz intermediate frequency (IF) by an open-loop downconverter. The IF outputs went simultaneously to the BVR channel and the DGT's FSR channels.

4.3.3 Data Compression

The objective of data editing and compression for both imaging and non-imaging data was the same: to reduce the number of information bits to be stored on the Galileo tape recorder and transmitted to the DSN.[35] One

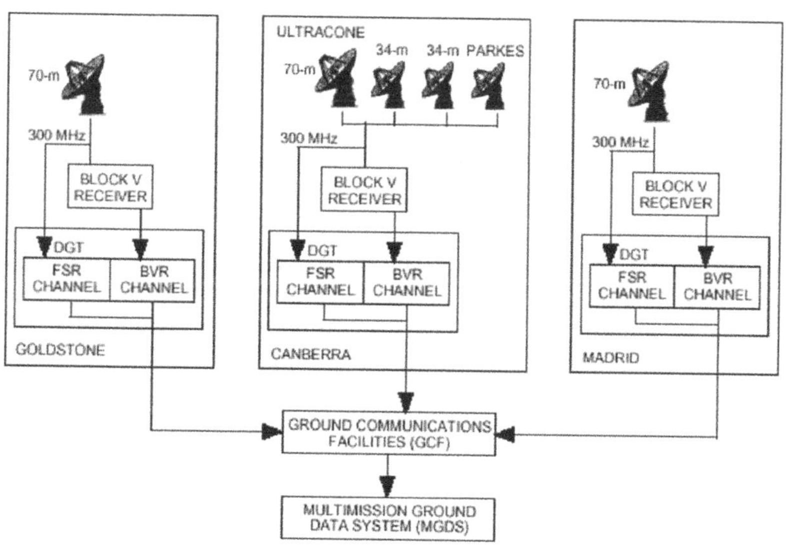

Fig. 4-6. DSN configuration for Jupiter orbital operations.

[34] The Parkes antenna, with a minimum operating elevation angle of 30 deg, has a later rise time and an earlier set time than the Canberra antennas, with their minimum elevation angles of 6–8 deg.

[35] The material in this subsection came largely from Ref. 13.

uncompressed Galileo image consisted of 800 lines of 800 picture elements (pixels), with each 8-bit pixel defining one of 64 grey-scale levels.

The development of data compression for the S-band mission included several steps:

- Selection of a compression scheme
- Evaluation of acceptability of scheme by the scientists
- Development of compression ratio estimates for mission planning
- Post-processing techniques to remove artifacts without compromising accuracy.

Because the bulk of Galileo's data volume was imaging data, the following description of data compression uses imaging data as an example.

4.3.3.1 Compression Scheme. The candidate scheme chosen for detailed evaluation and eventual implementation was the integer cosine transform (ICT) scheme for lossy image compression. ICT can be viewed as an integer approximation of the discrete cosine transform (DCT) scheme, regarded as one of the best transform techniques in image coding. Its independence from the source data and the availability of fast transform algorithms make the DCT an attractive candidate for this and other practical image-processing applications.

Data compression was to be accomplished in Galileo's onboard processors prior to the compressed data being recorded on the tape recorder. The processors were severely limited in computation and memory resources. The specific Galileo scheme used an 8 × 8 ICT. The integer property reduced the computational complexity by eliminating real multiplication and real addition. The relationship between the ICT and DCT allowed the use of efficient (fast) techniques that had been previously developed for DCT. Simulation of the Galileo ICT produced similar rate distortion results as a standard DCT scheme.[36]

4.3.3.2 Scientist Evaluation. Because the prime mission images and other Galileo data were expected to be of much higher resolution than data from the Voyager flybys of Jupiter, it was essential for the lossy data compression to preserve the scientific accuracy (validity) of the data. Two methods were

[36] Rate distortion theory is used to compute the minimum bit rate required to transmit a given image, for a specified amount of distortion. The results can be obtained without consideration of a specific coding scheme. A summary of rate distortion theory is available in http://www.stanford.edu/class/ee368b/Handouts/04-RateDistortionTheory.pdf (accessed 01/10/2013)

used to achieve and maintain the required accuracy. First, the Galileo principal investigators (PIs) and other planetary scientists evaluated the effects of compression on the best previously available images. Second, small portions of images (named "truth windows") were to be stored and transmitted without lossy compression. The scientist-evaluation process, named "PI-in-the-loop visual evaluation," was done in collaboration with the Remote Payload Systems Research group and the Vision group at the NASA Ames Research Center. The experiment, using sets of monochromatic astronomical images, converged rapidly on an acceptable set of customized quantization tables and verified the existence of compression/distortion tradeoffs acceptable for scientific evaluation [13].

4.3.3.3 Truth Windows. To ensure adequate accuracy, the concept of an addressable truth window (TW) was built into the image data compression. The TW was a fixed 96 × 96 pixel region that could be located anywhere in the 800 × 800 pixel image. To conserve onboard memory, the TW was losslessly compressed using the Huffman encoding module of the ICT compression algorithm, thus not requiring any additional onboard software. The PI could use the TW both to preserve important details and as a statistical reference to the rest of the image following application of image restoration techniques.

4.3.3.4 Compression Ratio Prediction Techniques. These techniques facilitated science and mission planning. For the Galileo fixed-to-variable compression scheme, an algorithm was given to the scientists. The algorithm predicted the compression ratio from a lookup table, based on the known statistics of the camera, the type of image expected, and its estimated entropy. The entropy, in terms of adjacent pixel differences, was modeled with a generalized Gaussian function with parameters based on previously available planetary images.

4.3.3.5 Post-Processing. Image restoration techniques had previously been used in other applications to remove the undesirable blockiness and checkerboard effects inherent in the output decompressed images produced by block-based transform compression schemes. However, the Galileo scientists' concern was that, while these techniques might make the image "look better," this was at the expense of introducing distortions that reduced detail and thus compromised scientific accuracy. With this in mind, the Galileo post-decompression restoration techniques worked first in the frequency domain, then in the spatial domain. Frequency coefficients were adjusted within the range of possible original values. Linear filtering was then performed with the constraint that frequency coefficients stay within their range of possible original values, creating a restored image that could be acceptably close to the original image.

4.3.4 Galileo Encoding and Feedback Concatenated Decoding

4.3.4.1 Overview. The Galileo S-band mission was supported by a coding system that used an inner convolutional code concatenated with outer Reed-Solomon (RS) codes having four different redundancies.[37] To reduce the effects of error bursts, the interleaving depth was 8. Contrast this signal design with the original Galileo signal design for the HGA mission as defined in [5]. In that original design, the solid state imaging (SSI) imaging data was coded by a (255,241) RS code, with an interleaving depth of 2, and the output of that code was convolutionally coded by the TMU.[38]

For the S-band mission, the staggered RS redundancy profile was designed to facilitate the novel feedback concatenated decoding strategy. Figure 4-7 is a block diagram of the Galileo FCD. The S-band mission decoding process

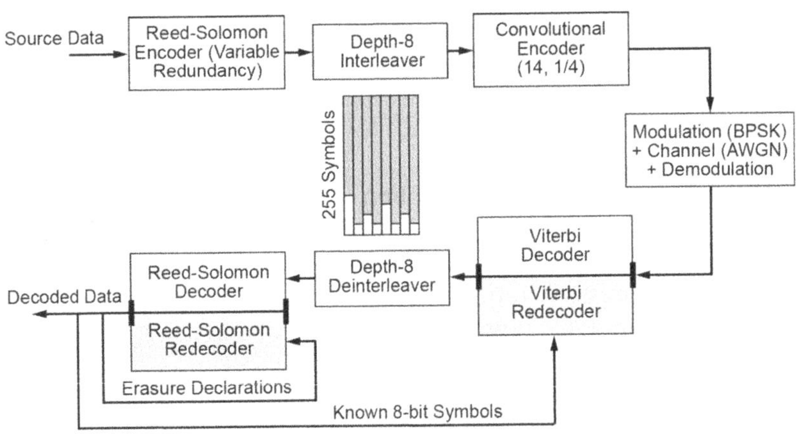

Fig. 4-7. Galileo encoding and feedback concatenated decoder (FCD).

---

[37] The material in this section came largely from Refs. 13 and 14. The Galileo S-band mission error-correction coding scheme used a (14,1/4) convolutional code as the inner code and a (255,*k*) variable redundancy RS code as the outer code. The RS codewords were interleaved to depth 8 in a frame. The redundancy profile of the RS codes was (94, 10, 30, 10, 60, 10, 30, 10). The generator polynomial, in octal, of the (14,1/4) code is (26042, 36575, 25715, 16723).

[38] From the *Galileo Orbiter Functional Requirements* [5], module 3-300, *Telecommunications*, the orbiter was launched with two kinds of convolutional encoders. Besides the standard (7,1/2) encoder in each TMU, TMU-A also had an experimental (15,1/4) encoder. This coder could not be used for the LGA S-band mission because it was designed to operate only at 115.2 kbps or 134.4 kbps.

proceeded in four distinct stages of Viterbi decoding, each followed by Reed-Solomon decoding. The RS decoders used a time-domain Euclid algorithm to correct errors and declare erasures.[39] In each successive stage, the Reed-Solomon decoder tried to decode the highest redundancy codewords not yet decoded in previous stages, and the Viterbi decoder redecoded its data utilizing the known symbols from all previously decoded Reed-Solomon codewords.

The (14,1/4) convolutional code used for the Galileo mission is the concatenation of a software (11,1/2) code and the existing (7,1/2) code in the TMU hardware. The choice of this convolutional code was constrained to use the existing (7,1/2) code and by the processing speed of the ground decoder.

The Viterbi decoder portion of the FCD was implemented in software in a multiprocessor workstation with shared memory architecture. The use of a software decoder was possible in the 1990s due to the low downlink rate from the Galileo orbiter. The advantages of a software-based decoder for Galileo were that its development cost was relatively low, and that it provided the flexibility necessary for feedback concatenated decoding. To exploit parallel processing in multiple processors, the Viterbi algorithm used "round-robin" frame decoding. In effect, this consisted of running several complete, independent decoders for several frames in parallel. Compared with other approaches considered, the round-robin required minimum synchronization and communication because each processor was an entity independent of the others.

4.3.4.2 Orbiter Coding and Modulation. An RS-encoded data block was interleaved to depth 8 and then encoded by the (14,1/4) convolutional encoder. The RS codewords could have four different levels of redundancies, as depicted by the lightly shaded areas at the bottom of the code block in Fig. 4-7. In the spacecraft, the encoded symbols were modulated on a subcarrier that modulated the downlink carrier. The deep space communications channel was characterized as additive white Gaussian noise (AWGN).

4.3.4.3 Ground Decoding and Redecoding. At the station (Fig. 4-6), the downlink carrier and subcarrier were demodulated in two parallel paths, either in the standard BVR or in the FSR DGT's own receiver, the buffered telemetry demodulator (BTD). We had parallel BVR and FSR paths, based on the premise that this was a one-time chance to obtain precious data. The

---

[39] The definition of an RS(n,k) code is one that accepts as input k data bytes and produces as a code word n bytes, where $n > k$. An RS(n,k) code can correct t errors and s erasures if $2t + s \leq n-k$. The Galileo codes are referred to as RS(255,161), RS(255,195), RS(255,225), RS(255,245).

FSR/DGT path included the ability to recover data through reprocessing of data recoded by the FSR. BTD reprocessing included forward and backward (in time) tracking and tinkering with loop parameters. The FCD processed the data once each time the BTD produced the output from reprocessing. Reprocessing increased the Galileo science data return by about 1–3 percent.[40]

As shown in Fig. 4-7, the channel symbols to the FCD first went to a Viterbi decoder. After deinterleaving, the codeword or set of code words with the highest redundancy was decoded by the RS decoder. If decoding of the first codeword was successful, the results (the "known 8-bit symbols" in Fig. 4-7) were fed back for Viterbi redecoding, as described subsequently in this section.

Redecoding facilitates Viterbi decoding. A correctly decoded RS bit forced the add-compare-select operation at each state to select the path that corresponded to the correct bit. The Viterbi decoder was thus constrained (when decoding again, or "redecoding") to follow only paths consistent with known symbols from previously decodable RS code words. The Viterbi decoder was much less likely to choose a long erroneous path because any path under consideration was pinned to coincide with the correct path at the locations of the known symbols.

The RS-Viterbi decoding-redecoding process repeated for as many as four times if necessary. In the first pass, only the first (strongest) code word RS(255,161) was decoded.[41] The symbols in the code words decoded by the RS decoder were fed back to assist the Viterbi decoder to redecode the symbols in weaker code words. At this and each successive stage, the output of the Viterbi redecoder was deinterleaved. In the second pass, the fifth codeword RS(255,195), which has the second highest redundancy, was decoded. The newly decoded symbols were fed back to further assist the Viterbi redecoder. The process was repeated twice more. In the third pass, the third and seventh code words RS(255,225) were decoded, and finally in the fourth pass, the second, fourth, sixth, and eighth (weakest) code words RS(255,245) were decoded.

Figure 4-7 also shows a shorter feedback loop entirely within the RS decoder using erasure declarations.[42] If an RS byte error was detected but the byte could not be decoded, it could still be declared an erasure for future RS

---

[40] Personal communication, Timothy Pham (JPL), July 7, 2014

[41] RS code words are made up groups of eight bits, each called a "byte" or an RS symbol. RS symbols are not the same as the soft quantized communication channel symbols that are input to the FCD from the BTD or the BVR.

[42] This loop was implemented in the FCD but was not used operationally for Galileo.

redecoding attempts. RS redecoding using erasure declarations based on error forecasting was worth about 0.19 dB when used in conjunction with one-stage decoding of the Galileo LGA convolutional code, shrinking to 0.02 dB with two-stage Viterbi decoding and almost nil with four-stage decoding [14].

Occasionally decoding remained unsuccessful even after four stages with two parallel FCDs, and the affected telemetry frame was declared lost.[43]

4.3.4.4 Control of Interaction between Data Compression and Decoding Performance. By definition, data compression reduces the inherent redundancy in the source data. Loss of any packets of the compressed data from failure to decode causes a phenomenon called error propagation. How the error propagates depends on the compression schemes being used. The compressed Galileo data had to be safeguarded against catastrophic error propagation.

The ICT scheme for Galileo imaging data included a simple but effective error containment strategy. The basic idea was to insert sync markers and counters at regular intervals in the onboard data to delimit uncompressed data into independent blocks.[44] In case of ground packet loss or other anomalies, the decompressor could search for the sync marker and continue to decompress the rest of the data. For an 800-line × 800-pixel image before compression, the interval was eight lines. This error-containment strategy guaranteed that error propagation would not go beyond 1 percent of the lines in an image.

---

[43] The open-loop downlink data (prior to BTD demodulation) was recorded to tape by the FSR. If high-value telemetry frames could not be decoded in real time, the FSR tapes were returned to JPL for labor-intensive non-real-time processing. Sometimes these frames could be successfully decoded after repeated attempts with different BTD or FCD parameter settings.

[44] The Galileo image error containment scheme worked as follow. Every eight-line block of camera output was compressed into a variable-length compressed data block. The DC (steady-state bias) value was reset to zero at the start of every eight lines, thus making every eight lines independent. A 25-bit sync marker and a seven-bit modulo counter were inserted at the beginning of every eight lines. The chosen sync marker minimized the probability of false acquisition to 10^{-8} in a bursty channel environment. In the ground decompressor, the error detection/sync software checked the prefix condition of the Huffman codes to detect any anomaly. When an anomaly was detected, decompression resumed from the next sync marker, and the reconstructed blocks were realigned using the modulo counter. The undecodable portion of the data was flagged and reported.

4.3.4.5 Concatenated Coding/Decoding Performance. Verification of the actual performance of the concatenated codes and interleaving that had been chosen by analysis required building the DGT. Because the orbiter packet-mode flight software was still in development, the DGT was tested with ground-generated signals during the year before deployment.[45] That testing verified an expected bit-energy-to-noise spectral density ratio (E_b/N_0) threshold of +0.6 dB (Viterbi decoder output), equivalent to a symbol-energy-to-noise spectral density ratio (E_s/N_0) threshold of –5.4 dB (BTD symbol output).

Downlink performance analysis in the GMM continued to show that the DGT decodes successfully at these levels. Empirically, the station monitor data shows that telemetry frames were lost (not decoded successfully in four passes through the decoder) rarely when the E_s/N_0 averaged –6 dB or greater. After this verification, data rate planning for the S-band mission was based on making data rate changes when the equivalent of the mean value of E_s/N_0 was at a level of –5.4 dB.

4.4 Telecom Link Performance

This section summarizes the uplink and downlink predicted performance for the orbiter from December 1997 through August 2002.[46] Communication link margins are computed using the link budget techniques and statistical criteria defined in *Deep Space Telecommunications Systems Engineering* [15]. Link performance was book-kept using a design control table (DCT), an orderly listing of parameters from transmitter to decoder. The Galileo DCT included favorable and adverse tolerances for each parameter that are used to determine a mean value and statistical variance for that parameter. As required by JPL link design policy[47] [16], overall performance was established in terms of the mean and the standard deviation (sigma).

---

[45] Data available at the time of the S-band mission studies in 1991–1992 included Ref. 18 published in 1988. That paper referenced the effects of interleaving depth on concatenated system performance, including some test data for the (7,1/2) code. There was no in-depth analysis from which to extrapolate to the case of the (15,1/4) code. Simulation of concatenated system performance with the (15,1/4) code had not been feasible because of the amount of data needed to verify bit-error rates (BERs) even in the 10^{-5} to 10^{-6} range. One (15,1/4) simulation would have taken 30 hours of Sun-3/260 CPU time per 100,000 decoded bits.

[46] See the Galileo article in the Design and Performance Summary series in http://descanso.jpl.nasa.gov/DPSummary/summary.html for the complete design control tables and plots of predicted uplink and downlink performance discussed in this section. The Galileo DCTs are similar in content to the Deep Space 1 DCTs in Chapter 5. (accessed January 10, 2013)

[47] The link policy itself is posted at http://descanso.jpl.nasa.gov/telecompolicy/policy.html (accessed October 30, 2014).

Table 4-3. Galileo orbiter telecom link functions and SNR criteria.

Function	SNR Definition	Galileo Criterion (designated multiple of sigma)
Carrier	P_c/N_0	Mean minus 3-sigma (UL), minus 2-sigma (DL)
Command	E_b/N_0	Mean minus 3-sigma
Telemetry	E_s/N_0	Mean minus 2-sigma
Ranging	P_r/N_0	Mean minus 2-sigma
DOR	P_{tone}/N_0	Mean minus 2-sigma

DL = downlink UL = uplink

Five link functions were used during the mission: carrier tracking (Doppler), command, telemetry, turnaround ranging, and DOR. The functions that continued to be used in the GMM were Doppler, command, and telemetry.

The performance of each function was expressed as an SNR as shown in Table 4-3. The "noise" part of the SNR was expressed in terms of N_0, which is noise spectral density. The "signal" part of the SNR was P_c (carrier power), E_b (energy per command bit), E_s (energy per telemetry symbol), P_r (downlink ranging power), or P_{tone} (power in DOR tone). Each function had a minimum SNR, the threshold, at which the quality of the link meets the bit-error rate (BER) or other criteria defined by the project. The predicted SNR at all times was required to exceed the threshold SNR by a designated multiple of the standard deviation (sigma).

4.4.1 Design Control Tables

Predicted telecom performance at a particular time during the mission is defined in a link budget, also known as a design control table or DCT. A design control table is an orderly listing of geometric quantities, carrier channel performance, and the performance of each data channel.

Geometric (usually input as a trajectory file and a pointing file):

- station-to-spacecraft separation in kilometers (km) or astronomical units (AU)
- angle between the station antenna boresight and the spacecraft
- angle between the spacecraft antenna boresight and the spacecraft
- elevation angle of the station antenna
- weather model

Carrier channel:

- frequency band
- spacecraft antenna used

- station antenna size (or station identification, ID)
- transmitter power
- receiver system noise temperature
- RF circuit losses in the transmitter and receiver
- modulation index of each data channel
- bandwidth of the receiver's carrier lock loop
- receiver threshold

Data channel

- type of data (command, telemetry, ranging, delta-DOR)
- type of coding (for command or telemetry)
- bit rate or integration time
- modulation index of each data channel
- threshold criterion (such as "mean minus 2-sigma")
- threshold (as a signal to noise ratio or a bit error rate)

The Galileo DCTs displayed in the original Deep Space Communications and Navigation Systems Center of Excellence (DESCANSO) article were produced by the telecom forecaster predictor (TFP) [19,20]. TFP is a multimission tool for link performance prediction built upon Matlab. The Galileo TFP adaptation used standard "common models" for station parameters, and Galileo spacecraft models.

4.4.2 Long-Term Planning Predicts

For planning spacecraft data-rate sequencing, TFP can produce tabulations or plots. While a DCT is a snapshot of many link parameters at one point in time, the tabulation (when read into a spreadsheet) can represent a whole series of snapshots. The rows represent successive points in time, and the columns represent values of individual parameters. Parameters can also be displayed as plots over a period of time.

For detailed data-rate planning, tabulations or plots can cover one station pass (8 to 12 hours) with points every 10 to 20 minutes. At another extreme, reasonably sized tabulations or plots can reach over spans of years with data spacing every 10 to 20 days [17].

4.4.2.1 Uplink Quantities During a Single Pass. The uplink carrier power P_c and the command channel signal-to-noise ratio E_b/N_0 each vary much less over the same range of elevation angle than the downlink P_c/N_0 and E_s/N_0 even though the S-band uplink and downlink are not that far apart in frequency.

As elevation angle changes, two uplink and three downlink values change. Variation of atmospheric attenuation and station antenna gain (affected by structural deformation) are similar on uplink and downlink. Station system noise temperature, the largest downlink contributor, is not a factor for uplinks.

4.4.2.2 Downlink Quantities During a Single Pass. The dominant quantity causing the variation is the station SNT which varies considerably with elevation angle. The Galileo S-band mission used a telemetry modulation index of 90 deg, producing a suppressed carrier downlink. The station's BVR is configured with a Costas loop for receiving the suppressed carrier downlink.

4.4.2.3 Range and One-Way Light Time During GEM and GMM. For long-term predictions during GEM and GMM, TFP was set to predict for a constant elevation angle of 25 deg, with a data point plotted once every 20 days.

Galileo was in orbit around Jupiter. With a negligibly small error in the resulting performance in decibels, the spacecraft-Earth range could be assumed the same as the Jupiter–Earth range. The range varied from just greater than 4 AU to just greater than 6 AU with a periodicity of about 13 months as the planets moved in their orbits about the Sun. The difference in performance was proportional to 20 × log ($range_{max}/range_{min}$), or about 3.6 dB.

The one-way light time (OWLT), Fig. 4-8, is proportional to the station-spacecraft range.

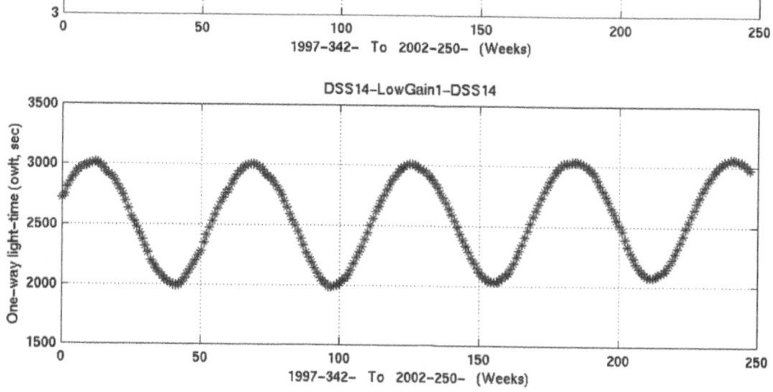

Fig. 4-8. Earth-to-Jupiter range in AU (top) and one-way light-time (bottom).

4.4.2.4 LGA-1 Boresight Angle and Sun–Earth–Craft Angle During GEM and GMM. The angle from the LGA-1 boresight to the station depends on the size and timing of spacecraft turns that keep the antenna oriented. The project goal was to keep the offpoint angle smaller than 4 deg to minimize downlink performance losses. However, this required attitude reference stars available to the AACS at the desired inertial attitudes.

The Sun–Earth–craft (SEC) angle[48] was the driving factor in solar conjunction planning, as described in Section 4.5.7. Conjunctions occurred about every 13 months, when the angle was small. Modulation index and bandwidth reconfiguration were made for SEC angles smaller than 22 deg, and commanding was prohibited for angles smaller than 7 deg.

4.4.2.5 Downlink and Uplink Carrier Performance During GEM and GMM. The station receiver's Costas loop remained above threshold except when the range was near maximum and the offpoint angle was larger than usual. Additional variation would be apparent if the full range of station elevation angles were considered in the predictions rather than only 25 deg. Near the time of solar conjunction, additional degradation occurred.

The uplink received power had somewhat less variation because the LGA 1 pattern was somewhat broader at the uplink frequency. Also the carrier SNR is nonlinear with total received downlink power, indicating greater carrier tracking difficulty at low levels.

4.4.2.6 Telemetry and Command Performance During GEM and GMM. The telemetry symbol signal-to-noise ratio E_s/N_0 mimics the downlink carrier performance, and the command E_b/N_0 (bottom plot) mimics that of the uplink carrier power.

For (14,1/4) coding, the telemetry threshold of +0.6 dB E_b/N_0 was –5.4 dB referenced to E_s/N_0. Predictions showed that the 60-bps rate could be supported at 25-deg elevation at the smaller ranges. Using a non-diplexed mode, or the DSS-43 ultracone, resulted in the availability of 80 and 120 bps at smaller ranges. At larger ranges, rates down to 8 bps were required, especially inside the 22-deg SEC angle conjunction limit.

The command E_b/N_0 showed that 32-bps command capability was never a problem for a 100 kW transmitter. Because solar noise is not Gaussian,

---

48 In most DESCANSO articles, this angle is called the Sun–Earth–probe (SEP) angle, its traditional name at JPL. Because the Galileo mission included a probe spacecraft, the term SEC angle is used in this chapter. The SEC angle most commonly is used in planning solar conjunction communications.

commanding was not scheduled inside 7 deg SEC angle. When commanding inside the 7-deg limit was occasionally required because of a ground or spacecraft problem, use of the full 400-kW power capability and multiple transmissions were successful (to about 4-deg SEC angle, the smallest angle attempted).

4.5 Telecom Operational Scenarios

4.5.1 Planned and Actual DSN Coverage

The pre-launch plan was for Galileo to use primarily a set of three 34-m stations, one each at Goldstone, Canberra and Madrid. Communications assumed the use of the orbiter's HGA (for S band uplink and X-band downlink) when the HGA could be Earth-pointed and one of the LGAs (for S-band uplink and downlink) during maneuvers or at other times the HGA could not be Earth-pointed. The 34-m tracking station at Weilheim, Germany, operated by GSOC was an integral part of the planned cruise mission.

Upon failure of the HGA to deploy in 1991, the project converted the station coverage plan for most of the remaining 4 years of the interplanetary cruise to the 70-m net only, with S-band uplink and S-band downlink via the LGA.

In addition, with the S-band mission in place for a substantial portion of the prime orbital mission, the DSN was operated in an arrayed configuration for Galileo passes. Both intra-site and inter-site (intercontinental) arraying were used to increase the effective receiving antenna aperture and therefore the supportable downlink rate. This arraying involved real-time combining of the spacecraft signals from the DSN 70-m and 34-m antennas at Canberra with those from the 70-m and 34-m antenna at Goldstone. The combined signals were enhanced further by the addition of the signal from the Australian 64-m radio astronomy antenna at Parkes. The array gain, relative to a single 70-m station, varied from about 1 dB (one 70-m station with one 34-m station) to almost 4 dB (full array, including Parkes).

4.5.2 Launch Phase

The Space Shuttle Atlantis, on Mission STS-34, launched the Galileo spacecraft (the orbiter and the attached probe) on October 18, 1989[49] [21].

---

[49] Originally the Galileo science mission was planned and the orbiter and probe spacecraft designed for a 1982 launch. Changes in launch vehicle and the Space Shuttle Challenger accident delayed Galileo's launch from 1982 to 1986 to 1989. The redesign for a 1986 launch put the spacecraft in the shuttle bay with a Centaur booster capable of a direct launch to Jupiter. The post–Challenger redesign for the eventual 1989 launch required replacing the powerful, Lox/ LH_2-burning Centaur with the weaker, but safer solid rocket motor (SRM) IUS and longer, complex, gravity-assisted trajectory.

Prior to the separation of the spacecraft and the attached Inertial Upper Stage (IUS) from the shuttle, communications for most activities on STS-34 were to be conducted through the TDRSS, a constellation of three communications satellites in geosynchronous orbit 35,900 km above the Earth. A minimum amount of IUS/spacecraft telemetry data was sent via the shuttle downlink. During the first minutes of flight, three NASA Spaceflight Tracking and Data Network (STDN) ground stations received the downlink from the shuttle. Afterwards, the TDRS-East and TDRS-West satellites provided communications with the shuttle during 85 percent or better of each orbit.

The spacecraft/IUS were deployed from the shuttle about 6 hours into flight. The Galileo orbiter's receiver and command detector had remained on through launch. Command "discretes" from the IUS activated the orbiter command and data subsystem (CDS) commands to select the LGA (LGA-1), turn the S-band exciter on, and then turn the S-band TWTA on in the low-power mode. After a 5-minute TWTA warm-up, the orbiter telecom system was ready to support the first independent downlink from the orbiter: 1200 bps engineering data in the "TDRSS mode" (no subcarrier, direct carrier modulation at 90-deg modulation index).

Prior to the first DSN acquisition, the CDS commanded the TMU to the "DSN mode," which was used for the rest of the mission. The 1200-bps rate continued, but from that time on the 22.5-kHz subcarrier and with a different modulation index.[50] The CDS also commanded the RFS to the TWNC off mode, so the downlink carrier could provide two-way Doppler data for initial trajectory determination.

Approximately 9 hours after the orbiter separation from the IUS, the CDS stored sequence switched the data rate from 1200 bps to 28.8 kbps, on the 360-kHz subcarrier and with a modulation index change. The launch-phase spacecraft data from the orbiter tape recorder was played back at this rate. Following playback, routine real-time engineering telemetry resumed at the 1200-bps rate.

4.5.3 Cruise Phase

The "VEEGA" (Venus-Earth-Earth gravity assist) cruise to Jupiter included a flyby of Venus about 90 days after launch, followed by two flybys of

[50] The orbiter, like other JPL missions tracked by the DSN, had a specific subcarrier frequency and modulation index setting for each data rate. During the first DSN pass, rates of 1200 bps and 28.8 kbps were planned. These rates are associated with 22.5-kHz and 360-kHz subcarrier frequency, and with 68-deg and 80-deg modulation index, respectively.

Earth.[51] Following the second flyby of Earth, the spacecraft passed the orbit of Mars and went through the asteroid belt, the orbiter finally reaching Jupiter December 7, 1995. Release of the probe from the orbiter was in July 1995, with the probe entering Jupiter's atmosphere on the same day as the orbiter went into orbit around Jupiter [3].

The Venus flyby occurred on February 10, 1990 at an altitude of 16,000 km, with data playback scheduled the following October when the spacecraft would be closer to the Earth. The orbiter was originally designed thermally for operation only between Earth and Jupiter, where sunlight is 25 times weaker than at Earth and temperatures are much lower. The VEEGA mission exposed the spacecraft to a hotter environment in the region between Earth and Venus. Engineers devised sunshades to protect the craft. For the shades to work, the –Z axis had to be aimed precisely at the Sun, with the HGA remaining furled for protection from the Sun's rays until after the first Earth flyby. The original plan was to deploy and begin using the HGA within 2 months of launch. The VEEGA mission necessitated a wait until the spacecraft was close to Earth to receive a high volume of recorded Venus data at rates up to 134.4 kbps, transmitted through the LGA.

The first Earth flyby (Earth-1) occurred on December 8, 1990 at an altitude of 960 km and the second on December 8, 1992 at an altitude of 305 km. Between the two Earth flybys was a flyby of the asteroid Gaspra on October 29, 1991. On the final leg to Jupiter was a flyby of the asteroid Ida in August 28, 1993. The Gaspra flyby altitude was 1,600 km at a flyby velocity of about 30,000 km/hr. The Ida flyby altitude was about 2,400 km at a velocity of nearly 45,000 km/hr relative to Ida. The second Earth flyby included an optical communications experiment: the Galileo SSI detected laser pulses transmitted via a telescope at Table Mountain, California [1]. The experiment yielded good data in support of theoretical studies and encouraged the further development of the technology for optical communications.

Within a few days of launch, the S-band TWTA was switched to its high power mode, where it generally remained from that time until end of mission. During cruise, the orbiter communicated via either the primary LGA-1 or the aft-facing LGA-2. LGA-1 was boresighted in the same direction as the

---

[51] The change to a 1989 shuttle launch also required redesign of the interplanetary cruise trajectory to include a flyby of Venus (and the two of Earth) for enough energy to reach Jupiter. Flying inward toward the Sun resulted in the need for redesign of the spacecraft's thermal control and the addition of LGA-2 to maintain communications with Earth on the Venus leg. See http://solarsystem.nasa.gov/galileo/mission/journey-cruise.cfm for more information on the cruise phase. (accessed January 10, 2013)

HGA. LGA-2 was added to the spacecraft when the mission was redesigned to include a Venus flyby. Because of the flyby geometry relative to the tracking stations, LGA-2 was also required for about two days at Earth-1 and could have been used for a similar period at Earth-2. The project's antenna selection tradeoff during planning for Earth-2 was reduction in risk (two fewer antenna switches) at a small cost in decreased communications capability.

Cruise telemetry data rates were either 1200 bps or 40 bps, using the (7,1/2) convolutional code. The lower rate was always required for trajectory correction maneuvers at large LGA offpoint angles from Earth and at the larger Earth-spacecraft ranges. The single Galileo command rate was 32 bps, uncoded. When more command link performance was required, this was achieved by use of the high power (100 kW) transmitters at the 70-m stations. During the early portion of cruise, turnaround ranging was possible via the LGA. Around the time of Earth-1, the delta-DOR tones were also transmitted on the S-band downlink carrier [22] and used to verify the navigation solution for the Earth-1 flyby.

The Galileo probe was turned on and tested, using the S-band orbiter-DSN links, during cruise.

4.5.4 HGA Deployment Attempts

The orbiter HGA was a very close derivative of the unfurlable TDRS antennas and was built for Galileo by the same manufacturer. The 4.8-m parabolic reflector was gold-plated molybdenum wire mesh attached to 18 graphite epoxy ribs. Each rib rotated about a pivot at the base. A ballscrew on the centerline, driven by redundant motors, raised a carrier ring attached to the ballnut. A pushrod connected each rib to the carrier. As the carrier rose, the ribs were intended nominally to rotate symmetrically into position [23].

The HGA deployment phase began when the temperature control constraints permitted Earth-pointing of the HGA. On April 11, 1991, the orbiter began to deploy the HGA under computer-sequence control [24]. The antenna had been furled and protected behind a small sunshade for almost 18 months since launch, in which the spacecraft spent a time closer to the Sun than to the Earth. Communications, including Venus and Earth-moon science data return, had been using the LGAs.

Within minutes, Galileo's flight team, watching spacecraft telemetry from 37 million miles (60 million kilometers) away, could see that something was wrong: The deployment motors had stalled, something had stuck, and the antenna had opened only part way.

Within weeks, a "tiger team" had thoroughly analyzed the telemetry, begun ground testing and analysis, and presented its first report. They attributed the problem to the sticking of a few antenna ribs due to friction between their standoff pins and their sockets. In addition to the planned 18 months from launch to scheduled deployment, the launch itself had been delayed from 1986 to 1989 with the antenna in its furled launch configuration. The first remedial action was taken—turning the spacecraft to warm and expand the central tower, in hopes of freeing the stuck pins.

Beyond thermal cycling, the tiger team developed other ideas to loosen the stuck ribs. These ideas, generally seconded by the review board and workshop experts, included producing a small vibration and shock by retracting the second low-gain antenna (on a pivoting boom), pulsing the antenna motors, and increasing the spacecraft spin rate to a maximum of 10 rpm (normally about 3 rpm). The deploy motor pulsing was called "hammering." On December 28, 1992, a warming turn produced maximum tower extension from thermal expansion, but no rib released. The next day over 2,000 pulses were applied. The ballscrew rotated about 1.5 turns (about the amount predicted after ground tests of the spare HGA at JPL) before stalling again after a few hundred pulses. Eventually, more than 13,000 hammer pulses were applied through January 19, 1993 [25].

A special HGA Deploy Anomaly Review Board, mostly made up of experts from outside JPL, met with the project and its tiger team monthly. In June 1992, a comprehensive two-day workshop was held at JPL, attended by nearly fifty technical specialists from outside JPL, reviewing the work to date and seeking new ideas [26]. A follow-on review, ground testing of mechanisms, and investigation [27] by the cognizant engineers concluded that cold welding and galling were credible failure causes. Thus, vacuum, temperature or rubbing of parts–or some combination of these mechanisms was suspected, but it could not be proven. A significant contributing factor considered was wear on the deploy touch points that occurred during the multiple trips that the HGA assembly in its shipping container endured to and from KSC.[52] No definitive cause for the in-flight failure of the HGA to deploy was documented.

Based on the investigative work, the appearance of the partly-deployed HGA has been depicted as shown in Ref. 28. To see if the partly deployed antenna was of any use for communications, the flight team operated the X-band TWTA downlink and the XSDC X-band uplink through the HGA to assess link capability in the stuck position. Although the orbiter received the X-band uplink and the DSN received the X-band downlink, the downlink capability was only slightly greater than available at S-band through the LGA. The project considered the sequencing complexity to maintain HGA pointing to Earth as too risky in trade for the small improvement.

[52] Personal communication, Mark Gatti, JPL, July 3, 2014.

A two-year campaign to try to free the stuck ribs, including seven heating or cooling cycles, failed to release any more ribs. The project concluded there was no longer any significant prospect of deploying the HGA. One last attempt was made in March of 1996 [24]. When that also was unsuccessful, the project continued to devote its resources to completing the implementation of the S-band mission, using only LGA-1.

4.5.5 Probe Separation, Jupiter Cruise, and Jupiter Orbit Insertion

This Jupiter cruise phase began four months before and ended two months before Jupiter encounter. In addition to the actual separation of the probe from the orbiter on July 13, 1995, this phase included probe turn-on and final checkout as well as the preliminary positioning of the orbiter-mounted relay radio hardware (RRH) antenna. This articulated antenna was repositioned several times during probe descent.

In addition, the orbiter's tape recorder malfunctioned October 11, 1995, less than two months before JOI. The tape recorder failed to stop rewinding as expected after recording some imaging data. Commands were sent to halt the tape recorder immediately upon discovery of the problem, but by that time it had been trying to rewind with the tape stuck in one position for 15 hours. The flight team investigated the problem using an identical recorder on the ground. They also began redesigning the encounter sequence in case the recorder could not be used again. Within a week, the project had a plan to return all of the planned probe relay data as well as 50 percent of orbiter science data planned for the S-band mission, even without the tape recorder. On October 20, the recorder was tested and proved to be still operational. Though the recorder was considered to be unreliable under some operating conditions, the ground tests showed the problem to be manageable. Periodic "tape conditioning" sequences to avoid further tape sticks were instituted, and the recorder continued to work through the prime and extended missions.[53]

---

[53] The last time the flight team restored the recorder's capability was in mid-2002. During a standard tape conditioning activity on April 12, 2002, fault protection in the flight software's tape manager tripped, locking out subsequent tape commands. This type of fault trip is caused by a failure of the tape rate to properly synchronize with an internal timing reference. On May 7, a test confirmed that the recorder's motor was operating as expected and that the motor current was consistent with the tape being stuck to one or more heads. On June 8, the tape recorder was successfully unstuck during a high rate slew. The tape pulled free shortly after the slew command was issued and behaved normally during a subsequent short playback slew. Over the next several weeks a series of tape motions to condition the tape and reduce the possibility of future hard sticks was begun. Ground tests combined with a revised empirical model were then used to define future tape operating strategies. There were no more recorder incidents during the remainder of the mission.

The JOI and probe relay phase was the most complex and scrutinized phase of the mission.[54] This mission phase began two months before JOI and ended a month after JOI. Activities included two approach trim maneuvers, a close flyby of the Jovian satellite Io, probe entry and data relay, JOI, and a post-JOI orbit trim maneuver. The orbiter passed through the most intense radiation environment of the prime mission during the Io flyby at a distance of 4Rj (Jupiter radii). (There were six subsequent Io flybys in GEM and GMM, and an even more intense radiation environment occurred around the Amalthea flyby in November 2002.) About two hours after the first signal was received from the probe, the orbiter's 400 newton (N) main engine fired for 49 minutes to achieve JOI. For telecom, the Doppler variation through the closest-approach station pass was several times the amount observed in interplanetary cruise or orbital cruise passes.

Continuous DSN coverage was required throughout this phase for navigation and telemetry. Unique coordination with the DSN was required to ensure the proper sequence of bandwidth settings in the station's Block V Receiver (BVR). Also, unique uplink acquisition and tuning profiles were coordinated to minimize the Doppler variation through the close encounter pass. Additional telecom factors included planned loss of downlink at the end of JOI (due to Doppler) and a solar conjunction with loss of data expected about one week after JOI. Within these constraints, the orbiter and ground telecom systems were configured for the maximum supportable downlink data rate via the LGA, with probe data being the highest priority.

Section 4.6 describes the probe-orbiter relay system and links in some detail. In summary, the probe returned data through the RRH for about one hour. The radio signal from the probe ended 61.4 minutes into the entry when the high atmospheric temperatures caused the probe's radio transmitter to fail [29]. For a "first look" of probe data, the orbiter transmitted from CDS memory the highest-priority 40 minutes of probe data by December 13. The orbit plan also included multiple playbacks of all probe data from the tape recorder. The playback campaign began on January 3, 1996 (after solar conjunction) and ended April 15 after three full or partial playbacks.

4.5.6 Orbital Operational Phase

The orbiter's prime mission included 11 orbits of Jupiter, with flybys of one or more Jovian satellites on 10 of these encounters.[55] The prime mission was defined to end December 7, 1997 (two years after JOI), at which time the

[54] Section 4.6 describes the requirements, implementation, and performance of the probe-to-orbiter relay link.

[55] The fifth encounter, in January 1997, occurred during a solar conjunction. No satellite close-approach was planned, and this phasing orbit for subsequent encounters was sometimes referred to as J5, for Jupiter 5.

GEM began. The GEM ended December 7, 1999, at which time the GMM began. That mission continued into 2003.

As defined by the flight software, two major downlink spacecraft telecom configurations existed during the prime mission, "Phase-1" and "Phase-2" [9]. Phase-1 downlinked the same TDM data that had been used through cruise. The TDM telemetry data was (7,1/2) convolutionally coded, modulating the 22.5-kHz subcarrier, and with the subcarrier modulating the carrier in a residual carrier mode (modulation index 72 deg maximum). The S-band TWTA operated in its high-power mode. LGA-1 was kept pointed as close to Earth as possible. On the ground, the link was supported with the BVR.

Phase-2, which became operational in June 1996, involved significant reprogramming of the CDS to produce a packet-formatted telemetry stream, to partially code the stream for input to the TMU, and to set a data rate at one of a small set of rates between 8 bps and 160 bps (the TMU and RFS were not reprogrammable in flight). The Phase-2 downlink used a concatenation of block-length 255 Reed-Solomon coding, interleaved to a depth of 8, and (14,1/4) convolutional encoding. The packet-mode symbol stream modulated the 22.5 kHz subcarrier as in Phase-1, but the subcarrier modulated the carrier at a 90-deg modulation index, producing a suppressed carrier.

In support of Phase-2, the ground system implemented the full spectrum recorder/full spectrum combiner (FSR/FSC), the buffered telemetry demodulator (BTD) and the feedback concatenated decoder (FCD). The FSR/FSC enabled the use of efficient local and intercontinental station-arraying with signal-combining at IF (intermediate frequency). For redundancy, the station operated with two FCDs, the second one receiving demodulated channel symbols from a BVR. The BVR also produced two-way Doppler for navigation.

To maximize the downlink "bits-to-ground" data volume return per pass, orbiter sequencing system software was upgraded to incorporate the data-rate capability file (DRCF) prediction into an automated telemetry-rate generator (TLMGEN) to create the series of commands to change the downlink rate in coordination with the downlink configuration and allocated station passes. Routinely, the rate would be set so that that residual (defined as the actual achieved symbol SNR minus the predicted symbol SNR) remained in the 0.5-dB to 1-dB range. The DRCF/TLMGEN rate accounted for diplexed versus low-noise configuration of the station, the changing elevation angle, and the particular stations that were assigned to the array at a given time. It also sequenced "fill data" (defined as that which could be lost without penalty) at times the downlink was likely to be out of lock due to a one-way to two-way transition or other defined spacecraft conditions such as turns for trajectory correction maneuvers.

The telecom analyst maintained plots of residual (observed minus predicted) values of symbol SNR and SNT for representative data rates during each 70-m station pass. Individual data plots could be displayed by such criteria as

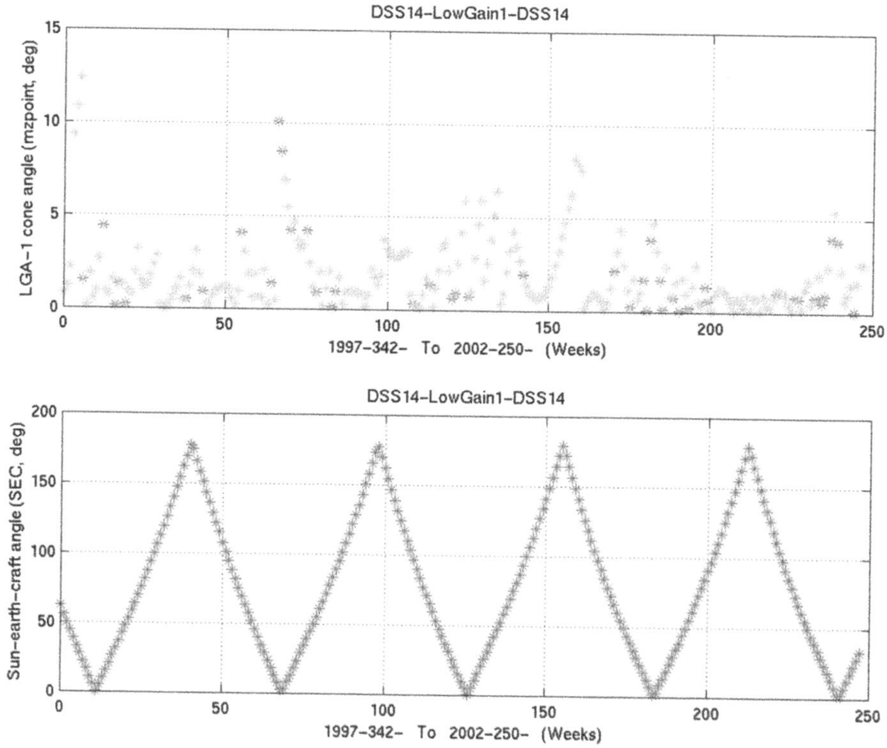

Fig. 4-9. LGA-1 off-Earth angle (top) and Sun–Earth–Spacecraft angle (bottom).

station ID or diplexer mode. This allowed the project to determine if one station or one operating mode became degraded relative to others. By superimposing a plot of LGA off-Earth angle, for example one like the top portion of Fig. 4-9 on the residuals, it was possible to assess the quality of the pre-launch antenna pattern modeled in the prediction software. During the prime mission, the pattern was updated for the DRCF/TLMGEN software.

4.5.7 Solar Conjunction

Solar conjunction was difficult for communications because the radio signals traveling between the Earth and Jupiter passed through regions of high and variable charged-particle density close to the Sun.[56]

[56] A superior solar conjunction (like Galileo's) occurs when the Sun is between the spacecraft and the Earth. Planning for superior conjunction effects on deep-space links at JPL currently takes into account only the carrier-frequency band and the SEC angle. Solar activity varies in cycles, and the 11-year solar cycle near a maximum in 2000–2001. The effects on a link, caused by charged particles from the Sun producing amplitude and phase scintillation, may also be highly variable over periods of a few minutes to a few hours. Coronal-mass ejections (CMEs) of charged particles that crossed the ray path between Earth and the spacecraft sometimes degraded Galileo S-band links even when the SEC angle was greater than 90 deg.

Based on the observed Phase-2 performance in the latter half of 1996 and the experience gained in receiving the Phase-1 downlink in December 1995–January 1996, the telecom analysts devised a strategy to configure the links for the January–February 1997 conjunction. The strategy involved a succession of steps going in to the smallest SEC angle, then reversing these coming back out: (a) changing from suppressed carrier to residual carrier downlink mode, (b) reducing the telemetry modulation index, (c) reducing the data rate, and (d) increasing the carrier-loop bandwidth to larger than normal values.

The purpose for each of these changes was as follows:

- A suppressed carrier waveform requires a Costas loop for carrier tracking; the Costas loop is significantly more susceptible than a normal phase-locked loop to half-cycle slips resulting from the solar disturbances to signal amplitude and phase.
- Reducing the telemetry modulation index puts more of the power into the carrier, increasing the ability of the carrier-tracking loop to hold lock.
- Reducing the data rate makes up for the reduced amount of power available in the data sidebands relative to the carrier.
- Increasing the carrier-loop bandwidth (CLBW) reduces the loop SNRs, but it permits the loop to remain in lock through a wider spectrum of (non-Gaussian) solar fluctuations.

The same strategy was successfully used for subsequent conjunctions, which occurred every 13 months. The configuration changes were based solely on the SEC angle. Independent of the solar cycle or short-term solar fluctuations, the size of the SEC angle proved to be the single best predictor of solar effects on Galileo S-band communications. The following specific strategy was first used in 1997 and worked well for each conjunction subsequently:

- At 22 deg SEC angle inbound (decreasing SEC angles), transitioned from the standard loop bandwidth mode (0.3 Hz CLBW) to 0.4 Hz CLBW. Conversely, outbound from conjunction at approximately 22 deg SEC, returned to the standard mode.
- At 18 deg SEC angle, transitioned from suppressed carrier mode (90 deg modulation index) to a residual carrier mode (60 deg modulation index) and a still wider CLBW of 0.6 Hz.
- At 12 deg SEC angle, raised the data rate thresholds used for DRCF/TLMGEN downlink sequencing.
- At 9 deg SEC angle, transitioned to a lower modulation index residual carrier mode and a wider bandwidth (51 deg mod index, 0.8 Hz CLBW).

- Within 7 deg SEC angle, imposed a "moratorium" on commanding. Because of this constraint, all planned commanding, including resetting of the command loss timer, occurred outside 7 deg SEC.
- Within 6 deg SEC angle, expect significant loss of telemetry data. Because of this expectation, the project elected to place only lower-value "fill data" on the downlink during the time the SEC angle was within 6 deg.

4.5.8 Galileo Europa Mission and Galileo Millennium Mission

These extended mission segments spanned the periods of December 1997–December 1999 and December 1999–September 2003, respectively. GEM encounters began with Europa-12 and ended with Io-25. The GMM encounters commenced with Europa-26 and continued through Amalthea-34, followed by a final plunge into Jupiter's atmosphere [3]. The Io-24 and Io-25 flybys again subjected the spacecraft to the more intense regions of the Jovian radiation belts. Patches to the flight software had been made to minimize the effects of radiation-induced power-on reset flags that had halted some of the previous encounter sequences.

The same Phase-2 telecom mode that began in 1996 during the prime mission was continued through GEM and GMM except that station arraying was no longer scheduled. All uplink and downlink was scheduled through 70-m standalone passes. With reduced staffing in the telecom analysis area, routine generation of residual data had ceased by this time. However, predictions for every station pass continued to be made. These were used to generate as-needed performance analysis for passes in which telemetry data is lost for "reasons unknown" or "low link performance."

4.6 Probe-to-Orbiter Relay-Link Design

4.6.1 Overview

During interplanetary cruise, four in-flight probe checkout tests took place as the orbiter and probe traveled together.[57] Separation of the probe from the orbiter was completed on July 12, 1995. On December 7, 1995, both spacecraft arrived at Jupiter. As the probe entered Jupiter's atmosphere, the orbiter flew past Jupiter's satellite Io, received the relay data from the probe, and fired its main engine for orbit insertion around Jupiter.

The strategy for returning the probe data to Earth took into account several factors. The loss of the use of the orbiter high-gain antenna (HGA) prevented

---

[57] This probe operations overview and relay link performance summary sections come from Ref. 29.

the downlink of real-time probe data during the encounter, leaving the orbiter's tape recorder to collect the data. To restore the redundancy implied by both real-time transmission and later playback of the probe data, a method was devised to store a reduced set of probe symbols in the spacecraft memory. Also, the frequency data (to detect Doppler shifts resulting from wind) from the probe receiver onboard the orbiter was stored in orbiter memory. An anomaly with the orbiter tape recorder on October 11, 1995 resulted in more tweaks to the strategy to minimize the risk of loss of probe data from the orbiter before it could be returned to Earth. Finally, because solar conjunction would cut communications from the orbiter to the Earth about a week after arrival, the strategy included playing back only the highest priority symbol set before conjunction, leaving the remaining playback until after conjunction, in January 1996.

4.6.2 Link Requirements and Design

Figure 4-10 is a block diagram of the probe-to-orbiter relay link [30], with the bottom showing elements housed in the probe and the top showing those in the orbiter.[58] The probe instruments and flight software created two data streams called "A-string" and "B-string."

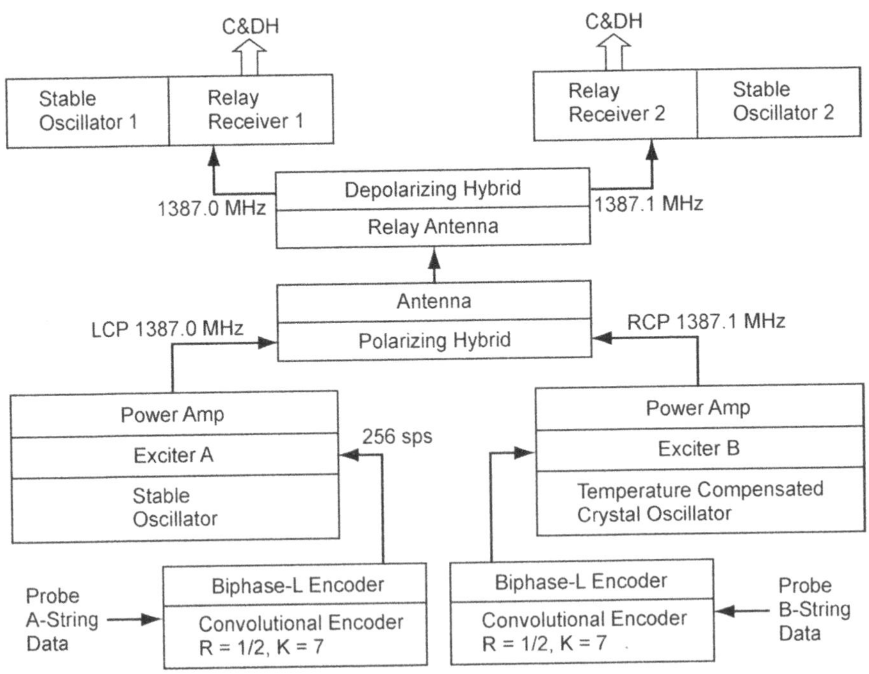

Fig. 4-10. Probe/RRH communications block diagram.

[58] The probe relay link requirements come from Refs. 30 and 31.

These represent two separate RF channels that were differentiated only by frequency and circular polarization sense. Each channel carried identical symbols, had a data rate of 128 bps, and was coded with a (7,1/2) convolutional code. The A-string symbol stream went to exciter A to biphase modulate the 1387.0 MHz L-band carrier, and similarly the B-string stream went to exciter B to biphase modulate the 1837.1 MHz carrier.

A stable oscillator provided a 23-MHz frequency reference for the 1387.0-MHz carrier, and a (less stable) temperature-compensated crystal oscillator for the 1387.1-MHz carrier. The stable oscillator used a quartz crystal frequency source and was housed within a double-proportioned control oven. The 23-MHz oscillator outputs were frequency multiplied to the final carrier frequencies. The 1387.0-MHz carrier had the stability required for radio science, the stability being in the range of 10^{-10} (due to pressure variations) to 10^{-9} (due to motion).

Each power amplifier in the probe output an RF level of 23 W to the antenna, one carrying A-string data and the other B-string data. The carriers first passed through a polarizing hybrid to make the 1387.0 MHz downlink LCP and the 1387.1 MHz downlink RCP. Each active unit (encoder, exciter, power amplifier) was enclosed in a sealed pressurized container, designed to survive to a pressure of 20 bars (2 megapascals, MPa).

The probe antenna was a crossed-dipole cup. For both frequencies, the antenna gain was 10 dBi, with a beamwidth of 56 deg between the half-power points. The antenna was fixed to the aft end of the probe, with its boresight intended to remain generally aligned to the local vertical throughout descent.

After leaving the probe antenna, the RF signals traversed a portion of the Jovian atmosphere, suffering absorption by ammonia and by clouds that were anticipated to exist in the region between 2.5 to 6.3 bars (0.25 to 0.63 MPa) pressure. The signals also suffered fading due to scintillation in Jupiter's ionosphere. Predicted link performance was based on a relay communications range between 214,000 km at entry to 229,000 km an hour later.

At the orbiter, the RRA (a 1.1-m parabolic dish) received the carriers. The RRA gain was 21.0 dB peak, with a 25-deg beamwidth between the half-power points. The RRA also received background noise from Jupiter's disk and from synchrotron activity in the Jovian magnetosphere. The two carriers were separated by a depolarizing hybrid that output them to each of the two RRH receivers. Each receiver had a USO[59] of essentially the same design as the one in the probe.

---

[59] These USOs are distinct from the USO associated with the orbiter's S-band downlink in the one-way mode.

Each receiver acquired, tracked, and demodulated one of the channels. When in the phase-locked mode, the receiver provided estimates of the (signal + noise) amplitude and the noise amplitude for downlink in the probe data from the orbiter. The receiver also provided the numerically controlled oscillator (NCO) control word for use in estimating the signal frequency and changes in signal frequency. More receiver detail, including a description of software algorithms, is in the *Galileo Probe Operations Manual* [30].

The detected symbols were output from a three-bit soft-quantizer, each symbol thus providing a sign and a relative level. The detected symbols were not further decoded, but rather stored onboard the orbiter for later transmission to the Earth on the S-band downlink.

The orbiter targeting, articulation of the relay radio antenna (RRA), and near-JOI sequence of events were required to allow acquisition of at least 60 minutes of data from the probe, with up to 75 minutes if other constraints allowed. The orbiter was to provide the RRA with a minimum unobstructed field of view of 12 deg half cone angle, from the edge of the antenna, for all required pointing directions.

The orbiter's sequence of activities was to include returning at least the first 39 minutes of the relay data in real time. This is based on an assumed knowledge accuracy of 75 s, at 99 percent confidence, for probe entry time. The Jupiter arrival date and geometry were chosen to avoid solar conjunction interference with the return of the initial probe data and to avoid having the relay signal pass through Jupiter's rings.

Relay link performance at 10 bars (1 MPa) atmospheric pressure was based on achieving a BER of less than 1/1000, at 128 bps. The link was required to have positive margin relative to 99 percent adverse environmental tolerances plus the root sum square of the 99-percent system performance tolerances. Probe link lockup was to occur about 70 s after parachute (chute) deployment at about 0.8 bars (0.08 MPa) atmospheric pressure.

4.6.3 Summary of Achieved Relay-Link Performance

The initial downlink from the orbiter memory readouts (MRO) showed both A-string and B-string downlinks had locked up. The quality bits attached to the probe symbols were all "high," indicating the communications link was solid. Orbiter telemetry verified that the RRA successfully went through its four commanded repositionings to maintain the communications link. The first (pre-conjunction) MRO provided an overall look at the probe mission. The B-string data lasted to entry + 51.2 minutes (approximately 13 bars (1.3 MPa) pressure in the worst-case model, and the A-string to entry + 61.4

minutes.[60] A "coast timer" had begun counting down at separation of the probe from the orbiter. This timed out accurately, resulting in successful acquisition of pre-entry data. The MRO also showed all probe science instrument were working and returning data.

The signal level at the RRH was more than sufficient to maintain lock for the entire mission. The reported signal level was an average of 1.5 to 2 dB lower than predicted. Possible causes for the discrepancy were analyzed, including RRA mispointing, changes in hardware performance since launch, and calibration errors (less likely). The carrier to noise ratio (P_c/N_0) was in the range of +35 to +40 dBm for both A-string and B-string from entry to entry + 50 minutes. A-string fell to 28 dBm before recovering to 31 dBm just before it went out of lock. During the major part of descent, the Probe link P_c/N_0 was well above the threshold of ~26 dB, and no bit errors occurred.

The MRO data from both RRH strings showed that loss-of-lock was preceded by a sudden drop in the transmitted power. The temperature of the probe communications equipment was about 115°C, higher than expected and well above the 60°C qualification temperature.

4.7 Lessons Learned

Do not fly a complex system that has single points of failure if simpler systems can provide sufficient performance.

As expressed in JPL's principles for flight systems [32]:

> "Designs shall employ a 'keep-it-simple' philosophy (straightforward designs) to reduce risk/cost, to enable easy implementation, design verification and flight operational usage."

Use of "complex" design implementations shall be avoided. Added complexity shall be justified to be essential to meet mission requirements or constraints.

The deployable Galileo HGA was 4.8 m in diameter, as compared with the 3.7-m solid Voyager HGA. Disregarding all factors other than planned communication capability, this is a difference of about 2 dB. The GLL HGA was based on, but not identical to, the TDRS deployable antenna.

---

[60] See Ref. 29, from which these times came. Times of 48.3 minutes (B string) and 58.5 minutes (A string) are also given, relative to a reference of major/minor frame zero (MF 0). Entry was 166 s prior to MF 0.

Back up critical spacecraft functions.

The cost of including backup (redundant) hardware is spacecraft mass and perhaps complexity. The risk of not including it is the loss of the mission. There have been two situations involving telecom functions that made us glad to have a backup.

Receiver-A developed the "wandering VCO volts" anomaly (VCO = voltage controlled oscillator), which eventually cleared. A working receiver is essential to continuing a mission, and Galileo continued with Receiver-B in reserve.

The ultrastable oscillator (USO) frequency, because it is so stable, was observed to be affected by the radiation environment near Jupiter. In 2001–2002 there may have also been a transient condition in which abrupt frequency shifts occurred, severe enough to cause loss of downlink lock. An onboard RF frequency source, while possibly not essential to mission continuation, is certainly reassuring. An aux osc in each S-EXC, provides a separate means to generate a one-way downlink in the absence of the single USO.

Model and handle telecom link margin wisely.

The S-band mission required Galileo to make use of link margin aggressively though not recklessly. Repeated inflight measurements of SNR and SNT resulted in changes to the modeled LGA-1 antenna pattern and updates to the DSN interface document[61] [33]. This work also established that spacecraft and station performance was stable and accurately modeled in the prediction tool. The project established a margin policy for data rate sequencing. The policy was that the transition point for switching up or down in data rate is at a time when the predicted mean P_t/N_0 is 0.5 dB higher than the threshold of the higher data rate. This 0.5-dB margin is much lower than on the typical deep space mission. It was established as an optimum level that resulted in the loss of only occasional telemetry frames but that prudently maximized the sequenced bits-to-ground data volume for every pass.

---

[61] In the Galileo era, Revision D of 810-5 ("Deep Space Network/Flight Project Interface Design Handbook") was in effect. Revision D was a paper document that contained modules very similar in form to Revision E [33], which superseded it in 2001 and is accessible at http://deepspace.jpl.nasa.gov/dsndocs/810-005/ (accessed October 30, 2014)

Individual modules in 810-5 at this link are updated for significant configuration or performance capabilities.

The S-band mission performance improvement techniques are reusable.

Development of the onboard data compression and advanced error-correcting coding while Galileo was in flight, and the concurrent development of intercontinental arraying of ground stations and the feedback concatenated decoder were necessary to save the mission. The Galileo mission was very costly in terms of DSN tracking time required.

A lesson learned from the success of this development has been applied to reduce tracking time of deep-space missions after Galileo. As expressed in [34]:

> "To [accommodate] limited DSN tracking pass capability, the information system design shall consider significant use of data editing, data compression, and improved data encoding techniques to meet downlink telemetry data requirement."

References

[1] D. J. Mudgway, *Uplink-Downlink: a History of the Deep Space Network 1957–1997*, National Aeronautics and Space Administration, Washington, D.C., 2001.

[2] P. E. Beyer, R. C. O'Connor, and D. J. Mudgway, "Galileo Early Cruise, Including Venus, First Earth, and Gaspra Encounters," *The Telecommunications and Data Acquisition Progress Report 42-109*, Jet Propulsion Laboratory, Pasadena, California, pp. 265–281, May 15, 1992, http://ipnpr.jpl.nasa.gov/progress_report/42-109/109T.PDF (accessed January 10, 2013)

[3] "Galileo Legacy Site," *Solar System Exploration*, website, National Aeronautics and Space Administration (formerly Galileo Project Website), Jet Propulsion Laboratory, California Institute of Technology, Pasadena, California. http://www.jpl.nasa.gov/news/fact_sheets/galileo0309.pdf (accessed October 30, 2014)

[4] "Galileo Mission to Jupiter," *NASA Facts*, website, JPL, National Aeronautics and Space Administration, February 2003. http://www.jpl.nasa.gov/news/fact_sheets/galileo0309.pdf (accessed January 10, 2013)

[5] *Galileo Orbiter Functional Requirements*, Galileo Project Document 625-205 (internal document), Jet Propulsion Laboratory, Pasadena, California, June 1989. Its Level 3 and Level 4 modules below were updated until the 1989 launch, as required.

- Equipment List and Mass Allocations, module GLL-3-230
- Power Profiles and Allocations, module GLL-3-250
- Galileo Telecom Requirements, module GLL-3-300

- Modulation Demodulation Subsystem Requirements, module
- GLL-4-2003
- Radio Frequency Subsystem Requirements, module GLL-4-2004
- S/X Band Antenna Subsystem Requirements, module GLL-4-2017

[6] *Probe Mission*, Galileo Legacy Site, National Aeronautics and Space Administration. http://solarsystem.nasa.gov/galileo/mission/journey-probe.cfm (accessed January 10, 2013)

[7] *Welcome to the Very Large Array!* (website), National Radio Astronomy Observatory, National Science Foundation, http://www.vla.nrao.edu/ (accessed January 10, 2013)

[8] R. Ludwig and F. H. Taylor, *Voyager Telecommunications, DESCANSO Design and Performance Summary Series*, Article 4, Jet Propulsion Laboratory, California Institute of Technology, Pasadena, California, March 2002. http://descanso.jpl.nasa.gov/DPSummary/summary.html (accessed October 30, 2014)

[9] P. E. Beyer, B. G. Yetter, R. G. Torres, and D. J. Mudgway, "Deep Space Network Support for the Galileo Mission to Jupiter: Jupiter Orbital Operations from Post-Jupiter Orbit Insertion Through the End of the Prime Mission," *The Telecommunications and Mission Operations Progress Report 42-133, January–March 1998*, Jet Propulsion Laboratory, Pasadena, California, 23 pp., May 15, 1998. http://ipnpr.jpl.nasa.gov/progress_report/42-133/133A.pdf (accessed January 10, 2013)

[10] W. Mayo and F. H. Taylor, *Galileo Final Report Telecom Sections*, JPL D-68095 (internal document), Jet Propulsion Laboratory, Pasadena, California, December 17, 1997.

[11] *Network Operations Plan for Galileo Project*, Vol. II, Rev. D, *Operations Manual*, DSN 870-7, JPL D-1997 (internal document), Jet Propulsion Laboratory, California Institute of Technology, Pasadena, California, June 26, 1996.

[12] J. W. Layland and L. L. Rauch, "The Evolution of Technology in the Deep Space Network: A History of the Advanced Systems Program," *The Telecommunications and Data Acquisition Progress Report 42-130, April–June 1997*, Jet Propulsion Laboratory, California Institute of Technology, Pasadena, California, August 15, 1997. http://ipnpr.jpl.nasa.gov/progress_report/42-130/130H.pdf (accessed October 30, 2014)

[13] K.-M. Cheung, K. Tong, and T. Chauvin, "Enhancing the Galileo Data Return Using Advanced Source and Channel Coding," *Proceedings of the NASA Technology 2004 Conference*, Washington D.C., September 1994.

[14] S. Dolinar and M. Belongie, "Enhanced Decoding for the Galileo Low-Gain Antenna Mission: Viterbi Redecoding with Four Decoding

Stages," *The Telecommunications and Data Acquisition Progress Report 42-121, January–March, 1995*, Jet Propulsion Laboratory, California Institute of Technology, Pasadena, California, pp. 96–109, May 15, 1995, http://ipnpr.jpl.nasa.gov/progress_report/42-121/121Q.pdf (accessed January 10, 2013)

[15] J. H. Yuen, editor, *Deep Space Telecommunications Systems Engineering*, Plenum Press, New York, New York, 1983.

[16] J. H. Yuen, *A Practical Statistical Model for Telecommunications Performance Uncertainty*, JPL Technical Memorandum 33-732, Jet Propulsion Laboratory, Pasadena, California, June 15, 1975.

[17] C. T. Timpe, *Galileo Orbiter Deep Space Network Telecommunications Design Control Document*, Galileo Project Document 625-257, D-713 (internal document), Jet Propulsion Laboratory, Pasadena, California, March 1983.

[18] K.-M. Cheung and S. J. Dolinar, Jr., "Performance of Galileo's Concatenated Codes With Non-ideal Interleaving," *The Telecommunications and Data Acquisition Progress Report 42-95*, July-September 1988, Jet Propulsion Laboratory, Pasadena, California, pp. 148–152, November 15, 1988. http://ipnpr.jpl.nasa.gov/progress_report/42-95/95N.PDF (accessed January 10, 2013)

[19] R. H. Tung, *User's Guide, Telecom Forecaster Predictor/Unified Telecom Predictor (TFP/UTP)*, V2.1, DSMS No. 887-000036 (internal document), Jet Propulsion Laboratory, Pasadena, California, February 21, 2002.

[20] K. K. Tong and R. H. Tung, "A Multimission Deep-Space Telecommunications Analysis Tool: The Telecom Forecaster Predictor," *The Telecommunications and Mission Operations Progress Report 42-140, October–December 1999*, Jet Propulsion Laboratory, Pasadena, California, February 15, 2000. http://ipnpr.jpl.nasa.gov/progress_report/42-140/140C.pdf (accessed January 10, 2013)

[21] STS-34 [website], Kennedy Space Center, Cape Canaveral, Florida. http://solarsystem.nasa.gov/missions/dsn/indepth; http://solarsystem.nasa.gov/galileo/docs/STS-34_Press_Kit.pdf (both links accessed October 30, 2014)

[22] D. L. Gray, "ΔVLBI Data Performance in the Galileo Spacecraft Earth Flyby of December 1990," *The Telecommunications and Data Acquisition Progress Report 42-106, April-June 1991*, Jet Propulsion Laboratory, Pasadena, California, pp. 335–352, August 15, 1991. http://ipnpr.jpl.nasa.gov/progress_report/42-106/106X.PDF (accessed October 30, 2014)

[23] W. J. O'Neil, "Project Galileo Mission Status," IAF-91-468, *42nd Congress of the International Astronautical Federation*, Montreal, Canada, October 5–11, 1991.

[24] T. V. Johnson, "The Galileo Mission," Scientific American, vol. 273, no. 6, pp. 44–51, Dec. 1995.

[25] W. J. O'Neil, N. E. Ausman Jr., T. V. Johnson, M. R. Landano, and J. C. Marr, "Performing the Galileo Jupiter Mission with the Low-Gain Antenna (LGA) and an Enroute Progress Report," IAF-93.Q.5.411, *44th Congress of the International Astronautical Federation*, Graz, Austria, October 16–22, 1993.

[26] W. J. O'Neil, N. E. Ausman Jr., T. V. Johnson, and M. R. Landano, "Galileo Completing VEEGA—a Mid-Term Report," IAF-92-0560, *43rd Congress of the International Astronautical Federation*, Washington, D.C., August 28–September 5, 1992.

[27] G. Levanas, M. Johnson, *High Gain Antenna Deploy Failure Extended Investigation Final Report for the Galileo Mission*, JPL D-15345 (internal document), Jet Propulsion Laboratory, California Institute of Technology, Pasadena, California, December 31, 1997.

[28] F. H. Taylor, K. Cheung, and D. Seo, *Galileo Telecommunications, DESCANSO Design and Performance Analysis Series, Article 5*, Jet Propulsion Laboratory, California Institute of Technology, Pasadena, California, July 2002. http://descanso.jpl.nasa.gov/DPSummary/summary.html (accessed October 30, 2014)

[29] *Galileo Probe Mission Operations Final Report*, Hughes Space and Communications Company, Contract NAS2-10000, September 1996.

[30] *Galileo Probe Operations Manual, Vol. 1, Probe/RRH System Description*, Hughes Aircraft Co., E1812, SCG 850509R, undated but released prior to launch roughly 1995.

[31] L. E. Bright, *Galileo Probe-Orbiter Relay Link Integration Report*, 1625-145, Rev. A (internal document), Jet Propulsion Laboratory, California Institute of Technology, Pasadena, California, January 16, 1984.

[32] M. R. Landano, "Design, Verification/Validation and Operations Principles for Flight Systems," JPL D-17868, Rev. A, (JPL internal document), Jet Propulsion Laboratory, California Institute of Technology, Pasadena, California, November 15, 2000.

[33] *Deep Space Network Telecommunications Link Design Handbook*, JPL 810-005, Rev. E, Jet Propulsion Laboratory, California Institute of Technology, overall release dated April 1, 2010, subject to individual module updates at this link: http://deepspace.jpl.nasa.gov/dsndocs/810-005/ (accessed October 30, 2014) Note: the document now has the organizational name Deep Space Network (DSN). In the early 2000s it was Deep Space Mission Systems (DSMS) to reflect the organization's name then.

[34] "The Journey to Jupiter: Galileo Orbital Tour and Extended Mission Tours," *Solar System Exploration*, website, National Aeronautics and Space Administration. Selected from pages within http://solarsystem.nasa.gov/galileo/mission/index.cfm (accessed October 30, 2014)

Chapter 5
Deep Space 1

Jim Taylor, Michela Muñoz Fernández, Ana I. Bolea-Alamañac, and Kar-Ming Cheung

This chapter describes how the Deep Space 1 (DS1) spacecraft and the Deep Space Network (DSN) ground systems received and transmitted data[1]. The signal to the spacecraft was at X-band, and the signal to the ground was at X-band or Ka-band or both together. The description is at a functional level, intended to illuminate the unique DS1 mission requirements and constraints that led to the communications-system design, and how the spacecraft was operated in flight.

DS1 was the first project at JPL that used the Small Deep Space Transponder (SDST), the spacecraft radio that has become standard hardware for many deep space missions using X-band. In the DS1 era, the DSN acquired many of its current characteristics. This chapter includes descriptions of the DSN systems used for carrier tracking, radiometric data, command transmission, and telemetry reception as DS1 used them through 2001.

[1] This chapter describes the DS1 spacecraft as it operated from 1998–2001. The mission ended December 18, 2001. Though the functions remain the same, some details of the 2001-era DSN and project ground systems that supported DS1 differ from the current versions. Chapter 1 has a current description of DSN operations.

Some of the grey scale figures in this book have color versions in the companion website www.wiley.com/go/deepspacecommunications. Please refer to this website where the text refers to colors.

Deep Space Communications, First Edition. Edited by Jim Taylor.

5.1 Mission and Spacecraft Description

5.1.1 Technology Validation

DS1 was the first of the New Millennium Program (NMP) deep-space technology-validation missions. The development of DS1 was led by JPL, with Spectrum Astro, Inc. as the industry partner for spacecraft development.

DS1's payload consisted of 12 advanced technologies for deep space that flew for the first time. With the three involving telecom listed first, the technologies demonstrated by DS1 are:

1) Small Deep Space Transponder (SDST) for X-band uplink and X- and Ka-band downlink

2) Ka-band solid-state Power Amplifier (KaPA) and associated experiments in Ka-band carrier tracking, telemetry demodulation, and turnaround ranging

3) Beacon Monitor Operations Experiment (BMOX) for autonomous onboard health and status summarization and request for ground assistance

4) Miniature Integrated CAmera Spectrometer (MICAS), a panchromatic visible imager and infrared and ultraviolet imaging spectrometers

5) Solar-electric propulsion (SEP) technology, implemented as the Ion-Propulsion System (IPS)

6) Autonomous onboard navigation (AutoNav)

7) Solar-Concentrator Arrays, using Refractive Linear Element-Technology (SCARLET)

8) Integrated ion-and-electron spectrometer, known as the Plasma Experiment for Planetary Exploration (PEPE)

9) Remote Agent eXperiment (RAX) architecture for autonomous-onboard planning and execution

10) Set of Low-Power Electronics (LPE)

11) High-packaging-density smart power switch, known as a Power-Actuation and Switching Module (PASM)

12) Multi-Functional Structure (MFS) experiment combining electronics and thermal control in a structural element.

Although there were 12 advanced technologies on DS1, the rest of the spacecraft payload was composed of components that were already current, low-cost, and tested on other missions when DS1 was designed. For example, the high-gain antenna (HGA) was a flight spare from the Mars Pathfinder program, and the flight computer was based on that used by Mars Pathfinder [2].

This approach—combining new technologies with tried-and-true components—was used because the New Millennium Program (1995–2008) focus was to prove that certain advanced technologies work in space, not to build a spacecraft out of advanced but unproven components.

5.1.2 Mission Overview

DS1 was launched October 24, 1998 [3] and completed its extended mission on December 18, 2001 [4]. The DS1 primary-mission design and execution focused exclusively on the validation of the 12 new technologies [5]. Technology testing was completed two weeks before the encounter with Asteroid 9969 1992KD (renamed Braille shortly before the encounter) on July 29, 1999 [3]. As a bonus to its technology-validation mission, DS1 collected a wealth of science data. The MICAS instrument recorded pictures and spectra of Mars, Jupiter, and selected stars. PEPE recorded extensive solar-wind data, some in collaboration with the Cassini spacecraft.

The primary mission concluded, having met or exceeded all of the mission success criteria, on September 18, 1999.

The extended mission's goal, in contrast to technology validation, was to return science data from the encounter with comet Borrelly in September 22, 2001.[2] The primary challenge in the extended mission was working around the failure in November 1999 of the star tracker, or stellar-reference unit (SRU). By June 2000, the flight team had devised a major revision of the flight software to use the science camera (MICAS) as a substitute for feeding star data to the attitude-control system. Since then, project-mission planning also accommodated a solar conjunction (spacecraft on the opposite side of the Sun from Earth) in November 2000 and another flight-software update in March 2001 to improve the probability of acquiring remote-sensing data during the Borrelly encounter. The risky encounter with comet Borrelly went well on September 22, 2001 [6], and the spacecraft used all four of its instruments.

---

[2] The original plan for the DS1 extended mission was for a flyby of the comet Borelly [8]. During technology validation, as we learned how, and how well, the spacecraft worked, we added a flyby of comet Wilson-Harrington for early 2001. The stellar-reference unit (SRU) failure and the recovery from that failure resulted in an extended period without IPS thrusting and a consequent replanning of the mission for a Borelly flyby only.

For telecom operations, the DS1 flight team initially responded to the onboard SRU failure by substituting from the ground in near-real time, downlink carrier-signal observation, telecom analysis, and uplink control. The replaced functionality achieves pointing the body-mounted HGA to within an acceptable angle of Earth. The "HGA activity" [7] described more fully later, was labor-intensive and exacting in timing requirements. The RTLT (round-trip light time) delay in the HGA activity's downlink signal monitor and corrective-command transmission process was manageable. The RTLT was about 30 min in early 2000, 40 min at solar conjunction, 34 min during the March 2001 software update, and about 25 min during the Borrelly encounter.

Following the successful flyby of the comet Borrelly, DS1 began what was colloquially named the hyperextended mission. This final mission phase, which included some additional technology validation of the IPS and the KaPA, ended with the spacecraft's downlink being shut off on December 18, 2001 [4, 8].

5.1.3 Telecom Subsystem Overview

By project policy, and like other parts of the spacecraft, the DS1 telecom subsystem was "single string" (without block redundancy). The subsystem elements include a transponder (receiver-transmitter in which the downlink can be phase-coherent with the uplink), power amplifiers for X- and Ka-band, and selectable directive and wide-beamwidth antennas. See Figs. 5-1 and 5-2 in Section 5.3.

The primary communication link was on Channel 19 at X-band (7.168-gigahertz [GHz] uplink and 8.422-GHz downlink). The SDST included the X-band receiver, command-detection and telemetry-modulation functions, and X- and Ka-band exciters.[3] The X- and Ka-band solid-state power amplifiers (XPA and KaPA) provided 12 W of RF power at X-band, and 2.2 W at Ka-band.

The Ka-band downlink carrier, phase coherent with the X-band downlink carrier, was also on Channel 19 (32,156 megahertz [MHz]). The Ka-band carrier can be unmodulated, or modulated with telemetry or ranging data like the X-band carrier.

[3] "Exciter" is a generic term for the portion of a radio transmitter that produces the carrier frequency. The SDST had two exciter functions, one for X-band and the other for Ka-band. Besides generating the output carrier frequency, each exciter also had a phase modulator and the modulation index control for each kind of downlink modulation.

DS1 had four X-band antennas. The high-gain antenna had a half-power beamwidth[4] of about ±4.0 degree (deg), and ±4.5 deg on the downlink and uplink, respectively. The three low-gain antennas were pointed along different spacecraft axes and had beamwidths of about ±35 deg for both downlink and uplink. As controlled through waveguide-transfer switches, the X-band uplink and downlink were always on the same antenna.

The Ka-band downlink was transmitted from the KHA (Ka-band horn antenna), which had a half-power beamwidth of about ±3.5 deg.

5.2 Telecom Subsystem Requirements

The DS1 development-phase project policies and top-level requirements led to a number of high-level directives regarding subsystem implementation. DS1 was intended as a capability-driven—as opposed to science-driven—mission.

"Science-driven" means the requirements that define a scientific mission govern the design of the spacecraft, its mission design, and its ground system. "Capability-driven" means that the requirements placed on the spacecraft, etc., follow from (rather than determine) the definition of hardware and software systems that are available.

Deep Space 1 spacecraft- and ground-system designs were driven strongly by existing hardware, software, and system capabilities in order to meet cost, schedule, and risk constraints:

- Capability-Driven Design: High-level requirements could be renegotiated (requirements reduced) if they conflicted with understood capabilities of existing hardware
- Single-String Implementation: The project policy identified that a single-string design was to be employed unless an existing design already incorporated redundancy.

For telecom, these constraints resulted in flying a single unit of each of two advanced technology subsystems: the SDST and the KaPA. The SDST was a flight-engineering model (FEM), as project development and test schedules precluded a full flight model.

---

[4] The direction of maximum gain of an antenna is called the boresight. The half-power beamwidth is defined in terms of the angle from boresight at which the antenna would have the capability to transmit (or receive) half as much power as at the boresight. In this chapter, to avoid ambiguity, the half-power beamwidth is expressed in terms of ± deg from boresight. A half-power beamwidth of ±4 deg would be a total beamwidth of 8 deg.

Except where functional redundancies already existed (for example, telemetry available on either X- or Ka-band downlink, and X-band downlink available via either high- or low-gain antenna), project policy precluded "conventional" backups for these functions. Furthermore, it was project policy to employ single-string design, and avoid cross-strapped redundancy unless existing designs (off-the-shelf or advanced technologies) already had it, and it was cost-effective to retain it.

Unlike a traditional science-driven mission, DS1 imposed fewer absolute link-performance requirements (such as minimum downlink rate vs. time) that the telecom subsystem had to meet. Nevertheless, a number of issues imposed requirements on the telecom subsystem. Sources of these system requirements were:

- Project policies
- Mission-coverage needs
- Technology-validation goals
- End-to-end information-flow considerations
- Interoperability with the DSN
- Spacecraft-architecture constraints
- Radiometric-tracking accuracy.

The above considerations led to the definition of the flight-system (spacecraft) telecom requirements. Top-level telecom-subsystem capabilities and link design to meet the requirements were defined in the DS1 Project Requirements/TMOD Support Agreement (PR/TSA) [9]. SDST parameter values measured during prelaunch testing were in the "Telecommunication FEM SDST/DSN compatibility and performance Motorola test report" [10].

5.3 Telecom System Description

The DS1 telecom subsystem provided X-band uplink and X/Ka-band downlink capabilities to handle all RF communications between the DS1 spacecraft and the DS1 mission operations team via DSN. The telecom subsystem received and demodulated uplink commands, transmits science- and engineering-telemetry data on either an X-band or a Ka-band downlink or both, and provided coherent two-way Doppler and range-measurement capabilities using the X-band uplink, and the X- or Ka-band downlink. Figure 5-1 is a block diagram of the telecom-system functional elements.

The DS1 spacecraft had four antennas for X-band (one HGA and three LGAs) and one Ka-band antenna (the KHA). Figure 5-2 shows the antenna locations on the spacecraft. Each DS1 antennas had a direction of maximum gain, often called the boresight. The boresights of the HGA, LGAX, and KHA were parallel to the +x-axis. The boresights of LGAZ+ and

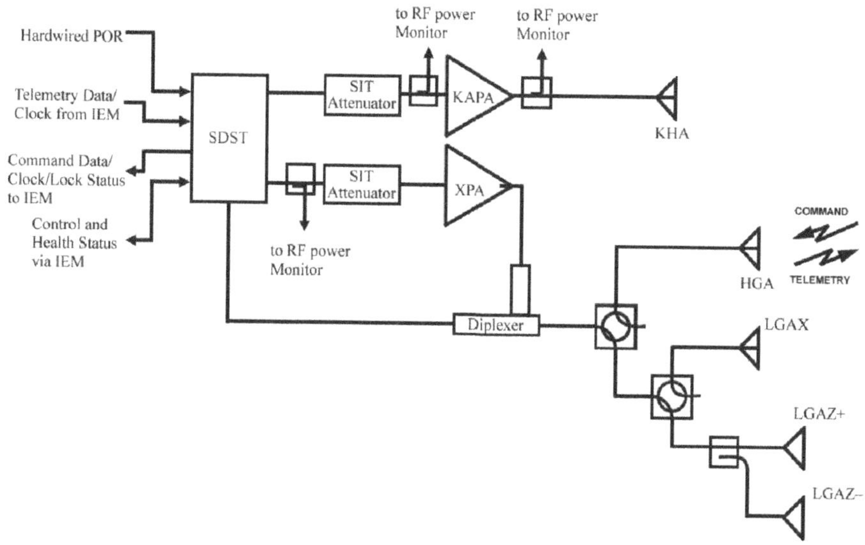

Fig. 5-1. DS1 spacecraft telecom system functional block diagram.

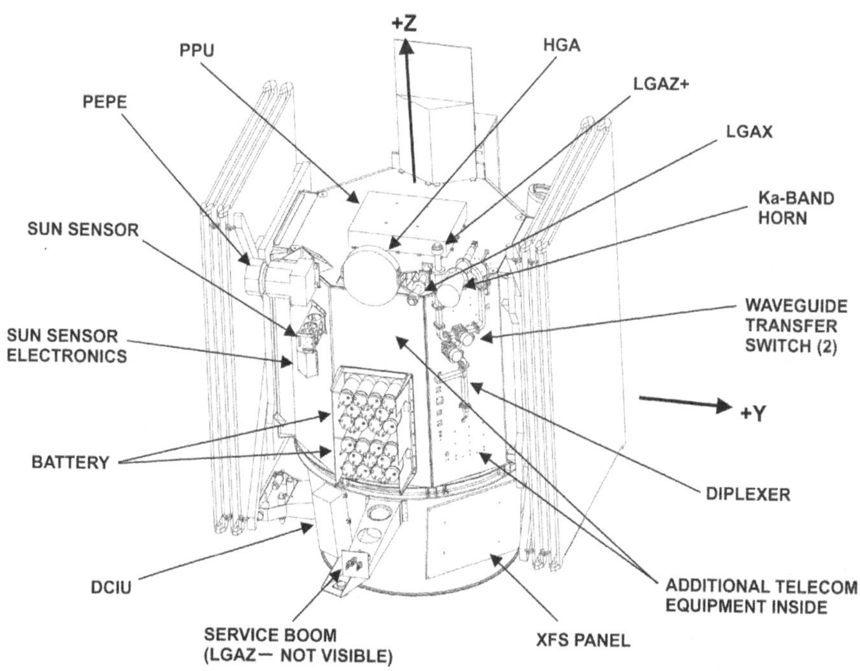

Fig. 5-2. Launch mode configuration with telecom subsystem components.

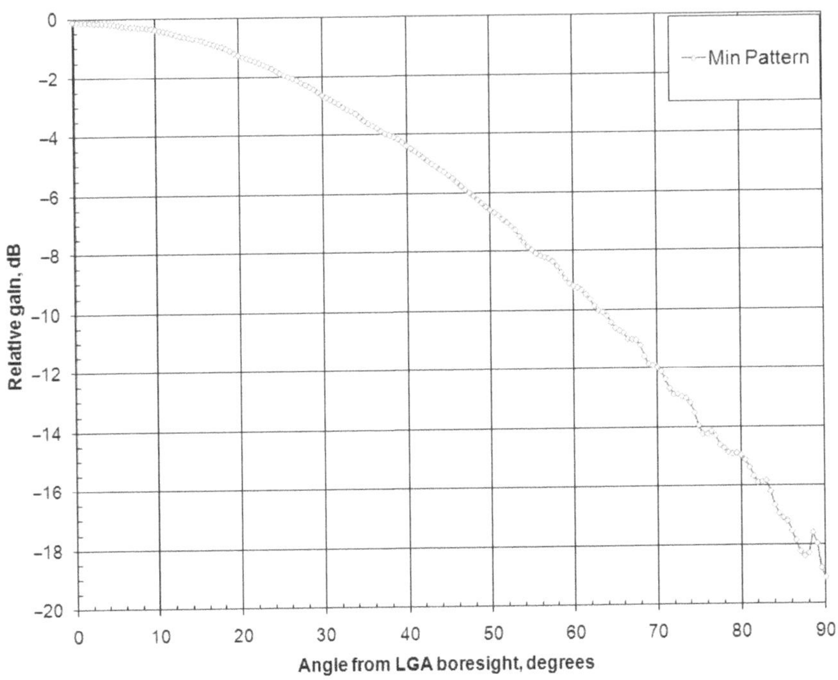

Fig. 5-3. LGA downlink pattern (relative gain as a function of angle from boresight).

LGAZ– were parallel to the +z-axis and the –z-axis, respectively. Orienting the DS1 spacecraft so that the +x-axis pointed at Earth maximized the performance of links using the HGA, LGAX, or KHA. All antennas were right-circularly polarized (RCP) except LGAZ–, which was left-circularly polarized (LCP).[5]

Figure 5-3 shows the downlink pattern of each LGA; Fig. 5-4 shows the X-band downlink pattern of the HGA. The X-band uplink patterns were similar, but slightly broader because of the lower-uplink frequency. Figure 5-5 shows the KHA pattern.

[5] The LGAZ– antenna element was a duplicate of LGAZ+, mounted midway out on the service boom, with its boresight oriented along the –z-axis. LGAZ– was added to the spacecraft late in the development phase, less than one year before the planned launch date. The need for the antenna was in the first weeks after launch, when the range was small (strong signals) but Earth-spacecraft geometry would result in blockage of signal paths to LGAX or LGAZ+. Antenna-system design needed to preserve the capability of LGAZ+ as much as possible, and at the same time to disturb the existing configuration and spacecraft system interfaces as little as possible. These needs led to the choice of a passive vs. an active-coupling system, and to a 25 percent/75 percent power split between the LGAZs.

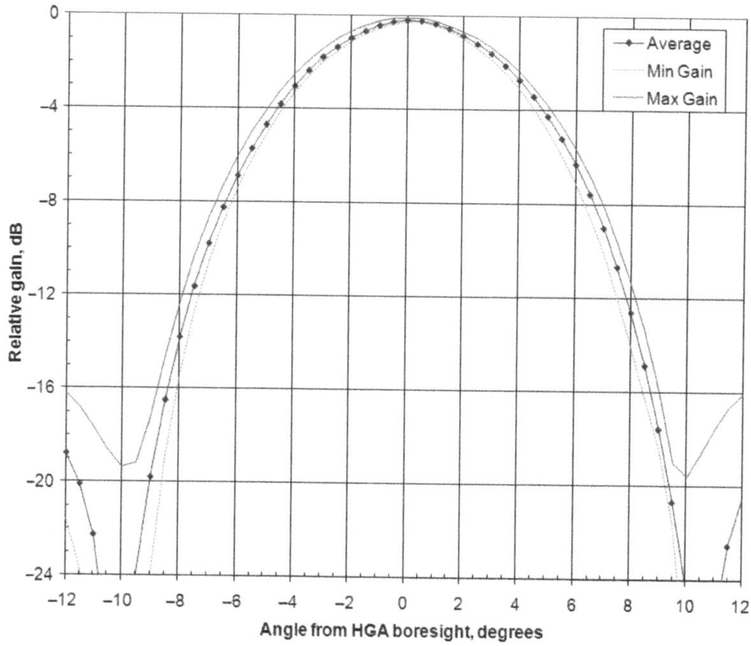

Fig. 5-4. HGA downlink pattern (relative gain as a function of angle from boresight).

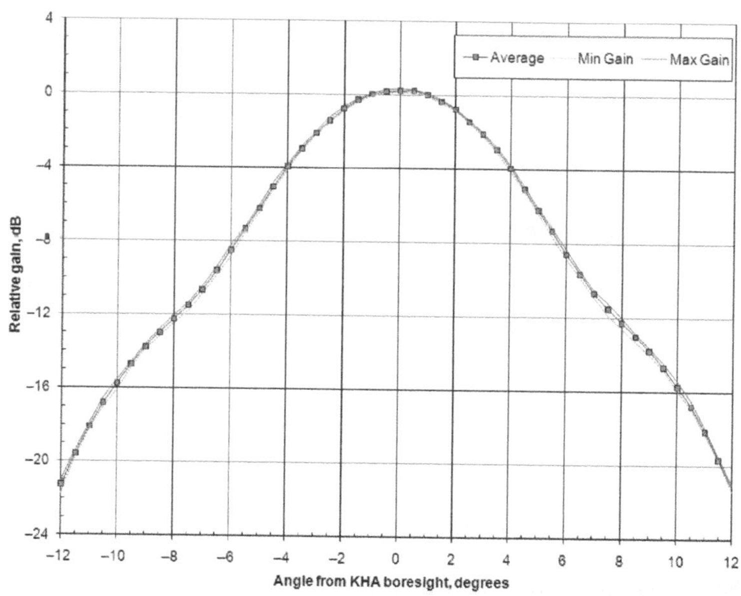

Fig. 5-5. KHA pattern (relative gain as a function of angle from boresight).

The SDST provided the detected-command bits for decoding and an in-lock/out-of-lock indicator to the Integrated Electronics Module (IEM) of the avionics subsystem. The IEM could send a power-on-reset (POR) signal to the SDST to activate a relay to remove spacecraft power from the SDST for 3 s, and then restore power. The SDST received a serial stream of telemetry-data bits and a clock signal from the IEM.

The amount of RF power input to the XPA from the SDST X-band exciter was established by a "select in test" (SIT) attenuator. Similarly, a SIT attenuator established the KaPA's input RF-power level. A 6-dB passive coupler connected the two z-axis LGAs, making both LGAZ+ and LGAZ– active when "the LGAZs" were selected for X-band. This means that (on the downlink) RF energy radiated out of both antennas when the LGAZs were selected, with the 6-dB coupler sending 25 percent of the energy to LGAZ–.

The HGA had a larger on-boresight gain, but also a narrower pattern. When the spacecraft x-axis could be kept within 6 deg of Earthline, the HGA was selected (it had 15 dB more gain than LGAX). Otherwise, the spacecraft was commanded or sequenced to operate on either LGAX (aligned with the +x-axis) or on the system of LGAZ+ and LGAZ– (aligned with the +z- and –z-axis, respectively).

The three LGAs all had the same patterns of gain as a function of angle from boresight. Because of different circuit losses between the SDST and each antenna, LGAZ+ had an effective gain about 1.5 dB lower than LGAX, and LGAZ– about 7 dB lower than LGAX. Much of the in-flight telecom analysis involved what uplink- or downlink-data rates would be available for different conditions of spacecraft pointing and antenna selection.

5.4 DS1 Telecom Technology

The three telecom-related technologies [5] demonstrated during the DS1 prime mission were:

1) Small Deep Space Transponder

2) Ka-band

3) Beacon Monitor Operations Experiment (BMOX).

5.4.1 Small Deep Space Transponder (SDST)

The design of the SDST (Fig. 5-6) facilitated command, telemetry, and radiometric communication between mission control and the spacecraft. The SDST combined the spacecraft receiver, command detector, telemetry modulator, turnaround-ranging channels, exciters, and control functions

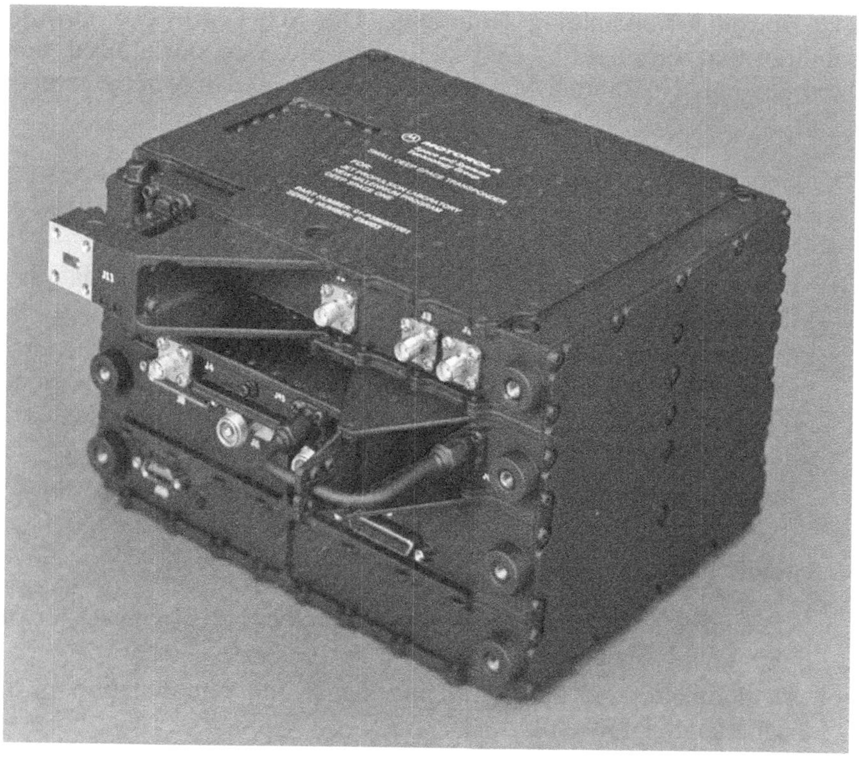

Fig. 5-6. The small deep-space transponder (SDST). Please visit www.wiley.com/go/deepspacecommunications to access the color version of this figure.

into one 3-kg package, about 18 × 11 × 16 cm in size. Developed by Motorola, Inc. (now General Dynamics, Inc.), Scottsdale, Arizona, under funding from NASA's Jet Propulsion Laboratory, the DS1 SDST provided a spacecraft terminal for X- and Ka-band telecoms with the NASA DSN, allowing X-band uplink, and X- and Ka-band downlink. It also provided coherent and noncoherent operation for radio-navigation purposes. This compact, low-mass transponder design was enabled by the use of advanced GaAs (gallium arsenide) monolithic microwave-integrated circuits.

As the heart of the telecom subsystem, SDST performed the following key functions.

5.4.1.1 Uplink-Receiving Functions

- Reception and demodulation of the X-band-uplink carrier
- Provision of an uplink AGC (automatic gain control) function for receive-power control and measurement
- Reception and demodulation the command subcarrier and data stream.

5.4.1.2 Downlink-Transmitting Functions. The SDST provided downlink capabilities that were used by DS1 and others that were not utilized. Some capabilities, noted in italics below, were not used or used only in the extended mission.

- Generation of a noncoherent downlink with either the SDST auxiliary oscillator *or an external ultrastable oscillator (USO)*. DS1 used the SDST aux osc only.
- Performing convolutional encoding[6] and subcarrier modulation of the downlink telemetry.
- Modulation of X- and Ka-band carriers with telemetry subcarriers *or with telemetry symbols directly onto the carrier*. The DS1 telemetry downlink symbol rate was low enough to allow for use of subcarrier modulation only.
- Independent control of X- and Ka-band modulation-index values.

5.4.1.3 Radiometric Functions

- Generation of two-way coherent downlink carriers by phase locking with uplink signal.[7]
- Demodulation of uplink-ranging signal and remodulation of the signal on the downlink.
- *Generation of differential one-way ranging (DOR) tones for downlink.* The SDST DOR tone capability was checked out but not used for navigation during the prime mission technical validation. Late in the extended mission, to improve the navigational knowledge of the flyby past Borrelly, the project scheduled the DSN operational delta-DOR equipment and transmitted DOR tones from the SDST twice in the week prior to the Borrelly encounter.

5.4.1.4 SDST Performance Monitoring and Spacecraft Data Interfaces

- Acceptance of control signals from the Integrated Electronics Module (IEM)
- Generation of analog-engineering status within the subsystem
- Providing status and performance parameters to the IEM

---

[6] See Section 5.6 for a description of the telemetry-transfer frame, which was convolutionally encoded by the SDST.

[7] DS1 operated on DSN channel 19, with frequencies as defined in the PR/TSA (Project Requirements/TMOD Support Agreement) [9] and in JPL document 810-005 [11]. The defined X-band-downlink frequency (8.422 GHz) is 880/729 times the defined X-band-uplink frequency (7.168 GHz). The defined Ka-band-downlink frequency (32.156 GHz) is 3360/749 times the X-band-uplink frequency.

- The SDST design accommodated interfaces with spacecraft avionics via a MIL-STD-1553 [12], a MIL-STD-1773 [13], or an RS422 [14] serial bus using the 1553 protocol. The DS1 SDST Command and Data Handling (C&DH) communication was via the 1553, and the data interface uses the RS422 [5].

Technology-validation, link-performance tests for SDST (and the KaPA, below) included transmitting each of the 19 DS1 telemetry rates simultaneously over X- and Ka-band to verify that the station could lock up to and decode data at each rate. The ranging channel was operated at low- and high-modulation index values, and the received-range delay compared between the two bands. Frequency-stability and carrier-noise levels (both affecting Doppler data quality) were compared between the bands. The SDST DOR modulation was turned on briefly to verify its operability.

As a result of DS1's success in proving the SDST design in flight, numerous other missions have since used the SDST, including the Mars orbiter and Mars rover missions described in this book. Typically, several missions pool their resources with "group buy" SDSTs. For instance, the Mars Science Laboratory (Chapter 8) and Juno both launched in 2011 with Group Buy III SDSTs.

5.4.2 Ka-Band Solid-State Power Amplifier (KaPA)

5.4.2.1 KaPA and Ka-Band Overview. At DS1 launch, the KaPA (Fig. 5-7) was the highest-power deep-space solid-state Ka-band amplifier yet flown. The KaPA, developed by Lockheed Martin Communication and Power Center, operated at 32 GHz and weighed 0.7 kg. As established during in-flight-technology validation, the KaPA amplified the RF output from the SDST Ka-band exciter to 2.2 W with an overall efficiency of 12 percent [5].

The DS1 SDST was the first to include an internal Ka-band exciter. A later mission, MRO (Chapter 6), carried SDSTs to generate and modulate carriers that could be used to generate Ka-band downlinks external to the SDST.

Ka-band offers a potential link-performance advantage for deep-space communications. With future improvement of ground facilities and spacecraft hardware, assuming similar power efficiencies and spacecraft antenna sizes, Ka-band holds a potential four-fold increase in data rate compared to X-band. This fact alone is obviously important as in the end it means reduced project cost. Ka-band offers greater available bandwidth as NASA and other agencies move away from lower frequencies shared with personal communications systems and other emerging information-technology ventures.

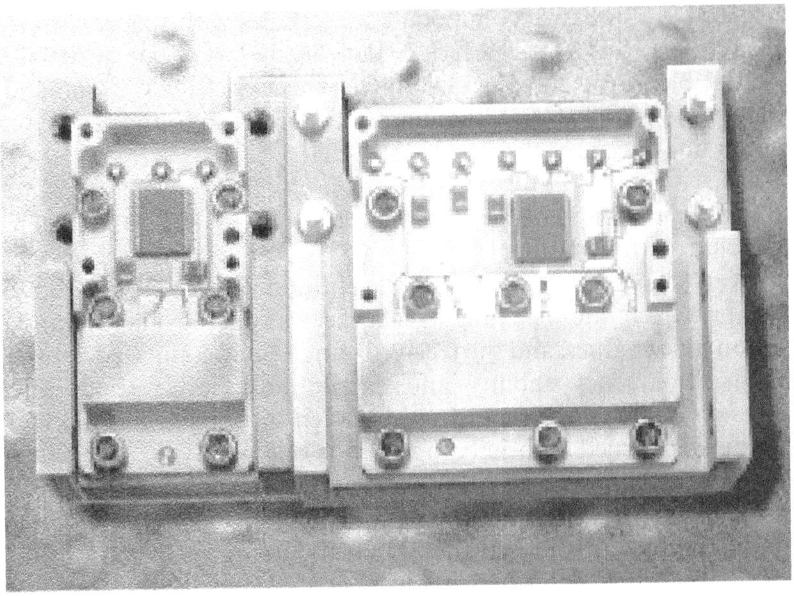

Fig. 5-7. Ka-band power amplifier (KaPA) RF module.

On the debit side, the need for a KaPA/Ka-band technology demonstration on DS1 spoke to the relative immaturity of flight systems at this frequency, in contrast to X-band. Ka-band link performance is also more sensitive than X-band to clouds and rain, which continues to be a challenge to designing reliable deep-space Ka-band links. Arrays that take into consideration different seasonal weather patterns at each DSN longitude (for example, California and Arizona) can increase link reliability. Once necessary Ka-band ground systems are in place, a higher data rate requires fewer ground resources and less mission-operation support per spacecraft-operation week.

5.4.2.2 KaPA and Ka-Band In-Flight Technology Validation. As part of the technology validation, DS1 first successfully demonstrated the KaPA in flight less than two months after launch. On December 9 and 10, 1998, the SDST Ka-band exciter and the KaPA were first powered on in flight. During two passes, the Ka-band link functions were methodically verified. These functions (also tested with the X-band downlink) included coherent and noncoherent downlink carrier tracking, turnaround ranging, and telemetry decoding at all DS1 downlink rates.

KaPA engineering-telemetry measurements were confirmed as nominal during these tests. Internal to the KaPA were temperature sensor, gate current, and gate-voltage telemetry measurements. External to the KaPA were other temperature sensors, as well as RF power detectors to monitor both input and output RF power. From these, RF gain could be deduced. At the same time, the SDST collected internal and external diagnostic-telemetry signals that could isolate (to the SDST RF output, the intervening telecom-system components, or the KaPA) the location of any potential degradation of performance. This ability to isolate problems was part of the DS1 technology-validation plan, as SDST and KaPA came from different industrial partners.

Besides characterizing the KaPA operation and link during the primary mission, DS1 subsequently provided Ka-band modulated and unmodulated signals for DSN performance-verification, and improved ground-system design and network-component upgrades to operational use of Ka-band. The lifetime of the KaPA was proven through hundreds of hours of reliable operation through the end of the mission.

5.4.3 Beacon Monitor Operations Experiment (BMOX)

5.4.3.1 Beacon System Concept Description. Beacon-monitor technology allows a spacecraft to report its status without transmitting telemetry on the downlink. The status provides information the ground system requires to intervene by scheduling a telemetry-downlink or command-uplink session.

The Mars landing missions (such as the Mars Exploration Rover in Chapter 7 and the Mars Science Laboratory in Chapter 8) employ a form of beacon operations during their critical Entry-Descent-Landing phase. These beacons, called "semaphores" or (multiple frequency shift keying) "MFSK tones" are based on the same principles as the DS1 BMOX, though their intent is to signal successful completion of a series of activities rather than to signal a problem.

The main appeal of a beacon system is that when DSN resources are scarce and spread out among many missions, it is cheaper to build and deploy small stations at different locations with tone-detection capability only. The beacon monitor concept behind the DS1 experiment envisioned dedicated antennas 3 to 10 m in diameter [2]. Noncoherent tone does not require phase-locked receivers, and detection is possible at a lower total-received power than for telemetry at even low-bit rates. Figure 5-8 shows beacon monitoring-system elements as they were envisioned at the time of the DS1 BMOX activities.

The onboard monitoring subsystem for a typical, beacon-equipped spacecraft [15, 16] would consist of flight software and part of the telecom subsystem, and be responsible for:

- Analyzing the engineering data to determine spacecraft health
- Reducing health status to one of a few (perhaps the four implemented in DS1) monitoring states, also known as beacon states or tone states
- Mapping current monitoring state into an appropriate monitoring signal
- Transmitting the monitoring signal to the ground.

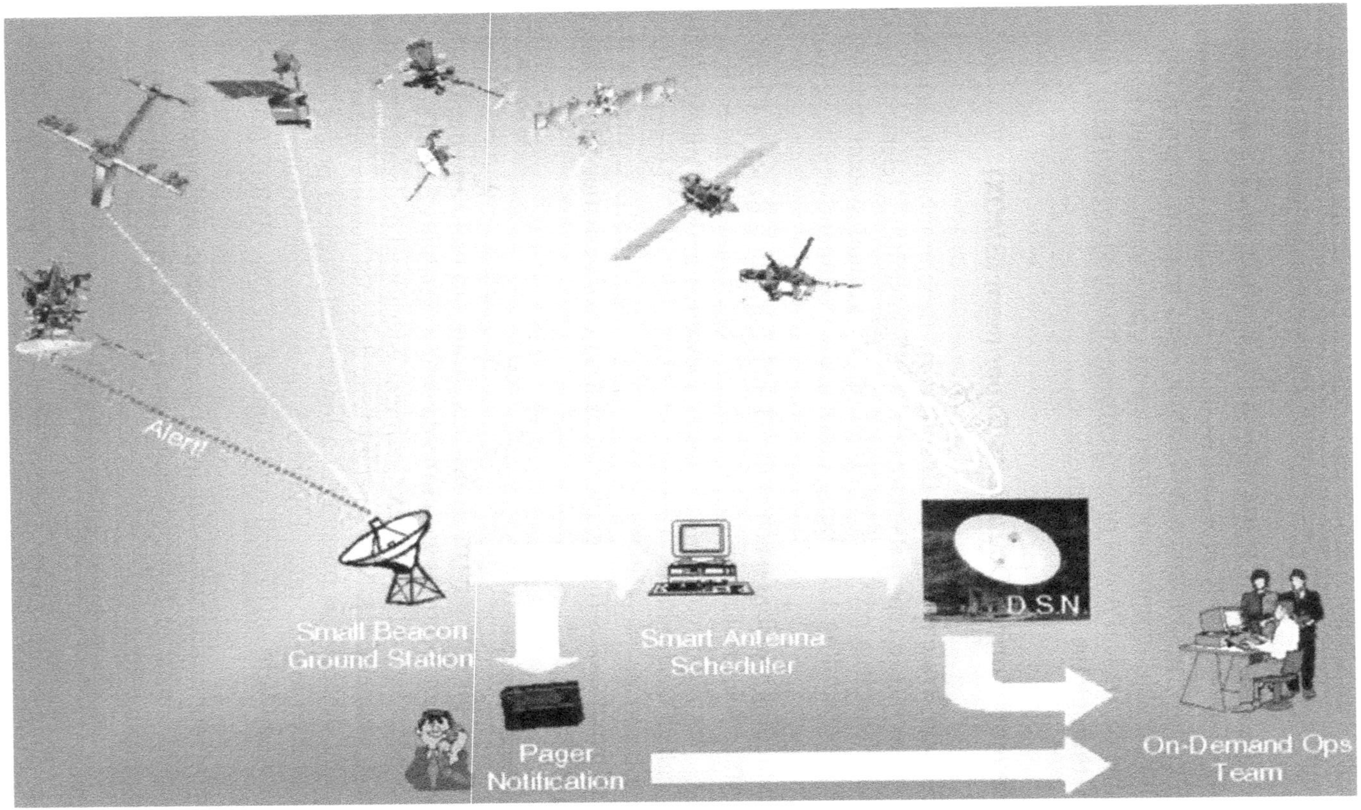

Fig. 5-8. Beacon monitoring system elements building on DS1 BMOX demonstration concepts.

During DS1, system-ground components were envisioned as a set of separate ground stations (not currently implemented) and a coordination computer. A beacon system would also include support by project-operations teams and DSN-station scheduling, prediction, and operation systems.

On board, the concept envisioned translating the overall spacecraft health and status into one of four general states. Using the spacecraft's radio, the software would direct the radio to create one of four subcarrier frequencies (without any telemetry modulation on the subcarrier) that would then be modulated onto the downlink carrier. The tone frequency indicated the spacecraft state. A so-called "green" tone indicated that the spacecraft was operating within acceptable conditions. An "orange" tone indicated that an anomaly was resolved by the spacecraft but conditions were acceptable. A "yellow" tone indicated a desire to send data to the ground or to request help with a problem that might escalate to jeopardize the mission. Finally, a "red" tone indicated that the spacecraft had a critical anomaly it could not resolve and required urgent assistance from the ground.

A beacon-monitoring station would detect the monitoring signal using the schedule and predictions from the coordination computer, and then send the result back to the computer. The computer would interpret the beacon message based on rules established by the project. It would maintain a monitoring schedule for all spacecraft, and it would make pass requests for a 34-m or 70-m antenna and notify the project when needed. It would also initiate urgent responses when triggered by an urgent message. The DSN prediction systems would provide carrier-frequency and antenna-pointing predictions to the computer, which would send these to the monitor station. The DSN would be responsible for scheduling 34-m or 70-m antenna passes in response to the computer requests, as triggered by the detected messages. This future beacon-monitoring system is complemented with the DSN's larger antennas to track spacecraft and send telemetry data to the projects in accordance with the DSN schedule.

When operating in monitoring mode, each spacecraft would maintain a continuous ability to receive commands from the ground. It would transmit its monitoring signal continuously or on a scheduled basis if constrained by spacecraft power or other factors. In the scheduled case, a pre-agreed communication window could be established for monitoring purposes.

During a spacecraft emergency, the DSN would work directly with the project-operations teams as usual, bypassing the coordination computer. When intensive interaction is needed between the spacecraft and the ground, the monitoring mode could be terminated by a ground command, or by the onboard computer. If onboard fault-protection software detected a condition requiring rapid ground intervention, the spacecraft would revert to safe mode and transmit low-rate telemetry to the ground.

5.4.3.2 The DS1 Beacon Monitor Operations Experiment (BMOX). The DS1 BMOX new technology consisted of flight software to control existing SDST subcarrier-frequency modes to provide two functions:

1) Problem- or condition-detection and tone transmission—instead of routinely sending spacecraft-health data, the spacecraft evaluated its own state and transmitted one of four beacon tones to reveal how urgent it would be to send high-rate health data

2) Data summarization—when telemetry tracking was required, the data summarization function created and transmitted "intelligent" summaries of onboard conditions to the ground instead of bulk-telemetry data.

The tone-generation function was validated first. Stored-command sequences controlled the SDST directly, producing, over a period of several hours, an unmodulated carrier, and a suppressed carrier that was successively modulated by subcarrier frequencies of 20, 25, 30, and 35 kHz. The subcarrier frequencies served as the tones. In a fully functional beacon system, the particular tone would indicate a "nominal," "interesting," "important," or "urgent" condition.

The spacecraft-technology validation's tone-transmission also checked the station-predict function, the BMOX station-control software, and the station's ability to detect weak X- and Ka-band tone-modulated carriers. Subsequently, the BMOX flight software did not depend on just a stored sequence; it controlled which SDST subcarrier would be produced. The experiment also depended on use of the existing DSN 34-m stations that have full tracking, command, and telemetry capabilities.

In early 2000, weekly tone-transmission tests over scheduled 34-m stations were sequenced to complete the station BMOX-operations automation. The end-to end-process completed with e-mailed reports indicating the time and frequency of the tone received.

Tone-transmission capability was first used operationally in mid-2000, not for BMOX but as part of the overall spacecraft-pointing algorithm[8] and IPS-thrusting operation. At this time, the tones conveyed one piece of information:

[8] The attitude control system (ACS) pointing algorithm developed after the SRU failure depended on the software maintaining lock to a reference star. A "tone detection" sequence that was activated during selected tracking passes would cause a 35-kHz frequency to modulate the downlink carrier if star-lock status had remained normal since the last check. It would modulate the downlink with a 20-kHz frequency if star-lock had been lost for more than a preset time—1.5 hours.

the pointing algorithm's star-lock history. Transmission of star-lock time information served a real operational purpose and did not involve the BMOX software for detection. The BMOX data-summarization function matured later in both the prime and the extended mission.

5.4.4 Telecom System Mass and Input Power

For comparison with similar functions in other spacecraft, Table 5-1 shows values of mass and spacecraft power for major elements of the DS1 telecom hardware discussed in Sections 5.3 and 5.4. The mass values and some power values come from the technology validation reports [5] and pre-launch project reports (JPL internal documents). Where available, the power values were taken from in-flight engineering telemetry. The telemetry confirmed there was negligible drift in power usage by the receiver, exciters, or power amplifiers from launch through the end of mission.

5.5 Telecom Ground System Description

While not duplicating the current general information in Chapter 2, this section includes brief descriptions of the DSN systems as they existed in 1998–2001 to provide carrier tracking, radiometric data (Doppler and ranging) collection, command uplinking, and telemetry reception and decoding for DS1.

Table 5-1. DS1 telecom system mass and power summary.

System Unit	Input Power (W[a])	Mass (kg)	Dimensions (cm)
Receiver	11.8		
X-band exciter, 2-way	1.8		
X-band exciter, 1-way	2.3		
Ka-band exciter	3.9		
SDST		3.1	17.9 × 11.2 × 16.4
XPA	52.5	1.6	
KAPA	16.9	0.7	14.2 × 15.2
HGA		1.2	
KHA		0.8	
LGAX		0.4	
LGAZ+		0.4	
LGAZ–	0.4		

[a] Based on in-flight telemetry data

Specific DSN numerical parameters for DS1 were defined in the PR/TSA [10] and the DS1 Project Network Operations Plan [16].

5.5.1 Uplink and Downlink Carrier Operation

During the DS1 era (1998 through 2001) the 34-m stations had X-band uplink capability. In 2001, the transmitter power was 4 kW. The 70-m stations were X-band downlink only at DS1 launch in October 1998; by the end of mission in December 2001, all three had X-band uplink capability also.

For DS1, the station transmitter was set to RCP except during passes that the spacecraft LGAZ– (which was LCP) was scheduled to be in view of Earth.

Of the operational 70-m and 34-m-BWG stations, only DSS-25 had Ka-band-downlink capability throughout the DS1 mission.

5.5.2 Radiometric Data (Doppler and Ranging)

As with other deep space missions, the DS1 uplink-and-downlink carriers provided a means of measuring the station-to-spacecraft velocity as a Doppler shift. In addition, ranging modulation applied to the uplink was turned around by the SDST to modulate the downlink to provide a means of measuring the station-to-spacecraft distance. Together, Doppler-and-ranging data provided radio navigation inputs to the project. Both radio navigation and optical navigation were mainstays for DS1 orbit determination. Radio navigation also played a part in the technology validation of "auto-nav." In the final 30 days of the spacecraft's approach to the asteroid Braille, auto-nav collected optical navigation images and conducted trajectory correction maneuvers at increasing frequencies to control the targeting of the final encounter.

The metric data assembly (MDA) at the tracking station processed DS1 Doppler data. The sequential ranging assembly (SRA) and the MDA together processed ranging data. The Fig. 5-9 block diagram shows the MDA in context with other major station elements and the spacecraft. The Fig. 5-10 block diagram adds the SRA and also shows the routing of radiometric data at JPL to the DS1 project navigation. Both diagrams are from the Network Operations Plan [17].

5.5.2.1 Doppler Data. The Doppler-sample rate for DS1 was normally 10 samples/s. Doppler-integration times were sometimes made longer to counter weak downlink levels.

5.5.2.2 Ranging Data. Figure 5-11 shows the end-to-end turnaround ranging data flow in the DS1 era. The block diagram is from Ref. [17]. The ranging uplink was transmitted from the station antenna to the spacecraft, and the

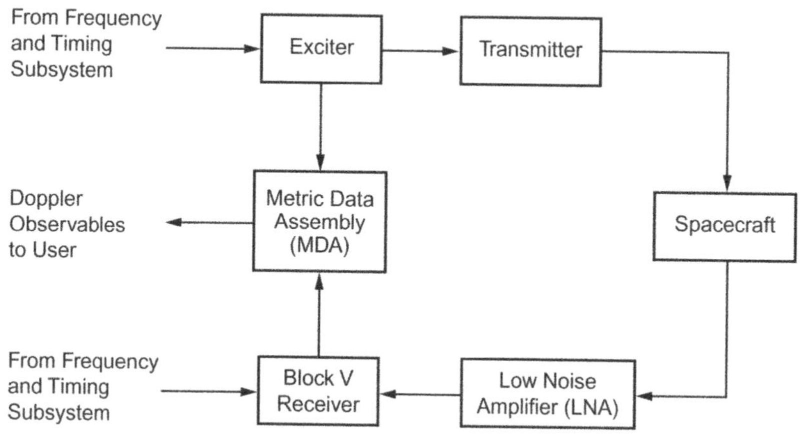

Fig. 5-9. DSN Doppler system.

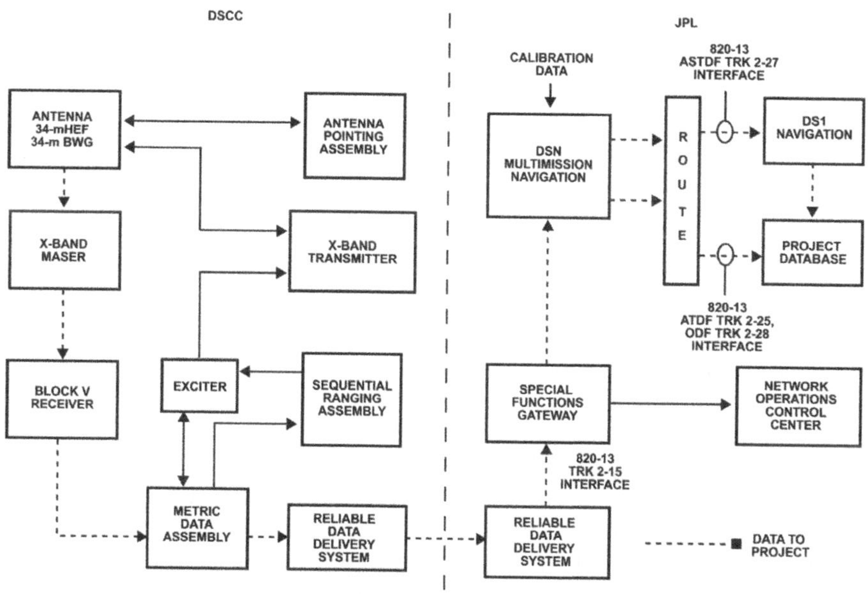

Fig. 5-10. DSN end-to-end radiometric data-flow diagram.

ranging downlink from the spacecraft was received via the same-station antenna for two-way ranging or at another station for three-way ranging.

For the DS1 mission, the highest frequency ("clock") provided fine resolution in range. The remaining components are known as ambiguity-resolving components. The number of ambiguity-resolving components that should be used in the ranging sequence for any particular range measurement was determined by the required ambiguity-resolving capability for that measurement. To accommodate the lower-link margins in the extended

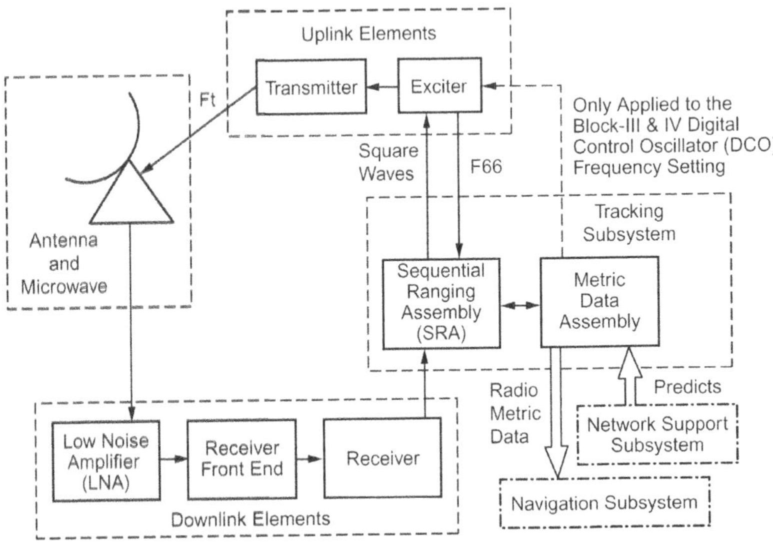

Fig. 5-11. DSN ranging system.

mission, DS1 used a 300-s integration time for the clock component and 20-s integration time for the lower-frequency components. Standard DS1 ranging used components 4 through 20 [17] for ambiguity resolution.

5.5.2.3 Ground Processing of Doppler and Ranging Data. At JPL, the Radiometric Data Conditioning Group, part of the Multimission Navigation function, processes and delivers the Doppler and ranging data to project navigation. DS1 navigation sometimes did further processing of the delivered Doppler and ranging files in the trajectory-determination process by, for example, weighting[9] the values of data from specific passes relative to other passes. The DS1 project relied on radio navigation data to plan the interplanetary trajectory and the use of the ion thruster system. The radio-navigation data sets were also used to generate P-files for delivery back to the DSN, for use in creating the frequency and pointing predicts for subsequent tracking passes. Frequency predicts were input to the BVR to assist in locking the receiver to expected periods of one-way, two-way, or three-way data.

[9] Weighting is an art in navigation-orbit determination, in which the available datasets (or even individual-ranging points) are assigned relative value (importance) relative to other datasets. Weighting may involve such factors as the amount of scatter between successive points, the agreement between the range and Doppler points within a pass, and how well the points from one pass "fit" into the solution model, as determined from previous passes. Orbit determination for DS1 was a challenging process because of the extensive periods of low-level thrusting. The effects of thrusting have to be separated from other small forces, such as solar pressure.

Pointing predicts were used to drive the station antenna in elevation and azimuth angle during the pass. Pointing predicts were supplemented by several tables that were specific to the station type, the antenna coordinates on the Earth, and the general declination of the spacecraft. These supplementary tables include corrections for atmospheric refraction as a function of elevation angle and azimuth as well as for deformation of the antenna structures (and thus, changes in the beam direction) as a function of elevation angle.

5.5.3 Command Processing and Radiation

The following description is of the systems used to command DS1. Figure 5-12 shows the systems at JPL and the station that were involved in the commanding process [17] for DS1.

At the station, the command-processor assembly (CPA) and the command-modulator assembly (CMA) clocked out the command bit stream, modulated the command subcarrier, and provided the subcarrier to the exciter for RF-uplink carrier modulation. Bit rates, the command subcarrier frequency, and the command-modulation index (suppression of the uplink carrier) were controlled through standards and limits (S&L) tables.

At JPL, the DS1 ACE (call sign for project real-time mission controller) operated the multimission command system from a workstation in the DS1 mission-support area (MSA). Experience with DS1 critical-command

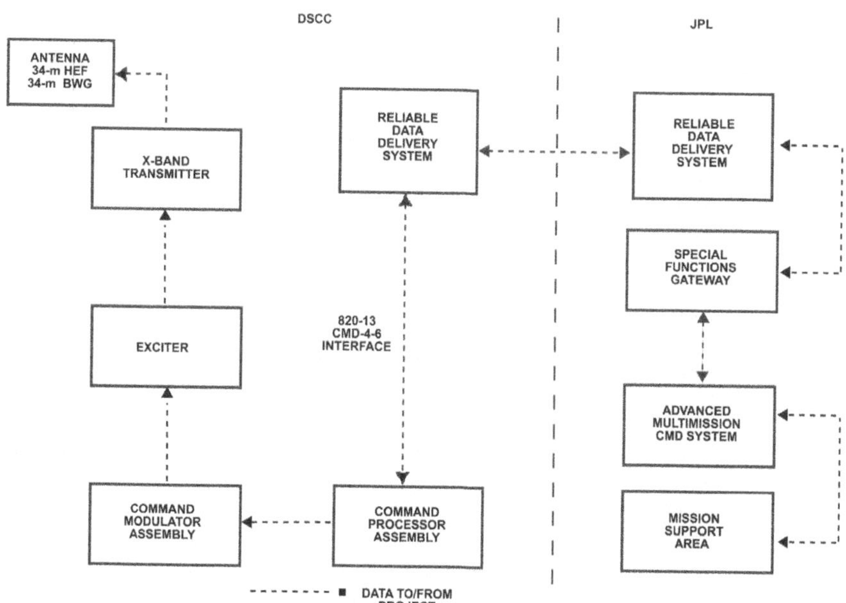

Fig. 5-12. DSN end-to-end command data-flow diagram.

timing, such as in the "HGA Activities" described in Section 5.7, showed that an ACE was able to activate command transmission within 2 s of the nominal time.

To begin or end a command session, the ACE requested the station to turn the command modulation on or off, respectively. The ACE selected a command rate the uplink would support, for example 125 bps. The selected rate was associated with one of four values of uplink-carrier suppression by command modulation (or modulation index). The carrier suppression was established by use of one of four calibrated "buffers" in the station's CMA. The CMA produced the command subcarrier at a nominal frequency of 16000.2 Hz to match the subcarrier tracking loop best-lock frequency in the DS1 SDST. The CMA also modulated the command-bit waveform onto the subcarrier.

The Reliable Data Delivery System transferred the command files to the station in the staging process, as well as the ACE directives for radiation of the staged commands. At the station, the command processor assembly performed the digital processing to create the command-bit stream from the command files as well as the activation signal.

5.5.4 Telemetry Demodulation, Decoding, Synchronization, and Display

Figure 5-13 is a block diagram of the station and JPL equipment that was involved in DS1 telemetry-demodulation and decoding [17].

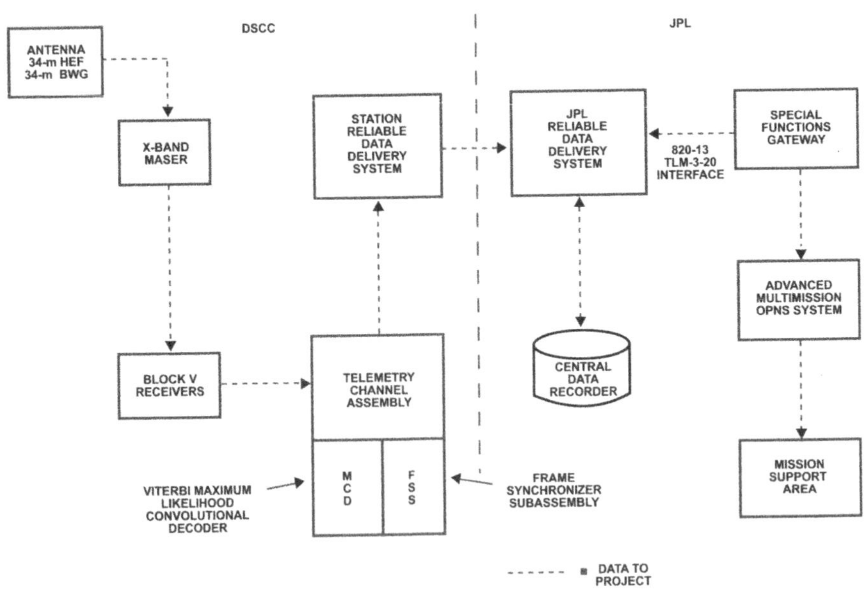

Fig. 5-13. DSN end-to-end telemetry-data-flow diagram (from 871-010-030 [11]).

The Advanced Multimission Operations System (AMMOS) processed telemetry in both near-real time (delays up to one minute) and in non-real time (as complete a record as possible, but with a delivery time guaranteed within 2 hours of the end of track. The non-real time version included retransmission of data lost between the station and JPL and replays from the central data recorder (CDR) as necessary.

Telemetry processing at JPL includes "channelizing" the data from the packets received, ordering the telemetry data that may have been transmitted in real time or from spacecraft storage, and time-tagging the data either by Earth-received time (ERT) or spacecraft-event time (SCET).

Station configuration and performance ("monitor") data were output by the Link Monitor and Control (LMC) at the station. Monitor data was channelized similarly to telemetry data and can be displayed or queried for telecom analysis.

Figure 5-14 shows in more detail the station equipment for DS1 telemetry demodulation and decoding. This equipment is still in use for some missions, though turbo codes are now standard for the newer missions.

Each redundant BVR has phase-locked loops for receiving (locking to) the carrier, the telemetry subcarrier, and the telemetry-symbol stream. DS1 generated a 375-kHz subcarrier for telemetry bit rates of 2100 bps or greater, and a 25-kHz subcarrier for bit rates lower than 2100 bps. DS1 X-band carrier-modulation index values range from 40 deg for the lowest data rate (10 bps) to 72 deg for the highest (19.908 kbps).

For DS1 the BVR delivered telemetry symbols to the maximum-likelihood convolutional decoder (MCD). The (15,1/6) convolutional code normally used by DS1 required the use of the MCD3. An MCD/FSS (Frame Synchronizer System) pair made up a telemetry-channel assembly (TCA). The telemetry-group controller (TGC) controlled the operation of TCA1 (containing the MCD3) and TCA2 (containing an MCD2)[10].

For DS1 the MCD output decoded telemetry bits to the frame-synchronizer (FS) subsystem. After the MCD declared lock, the FSS required recognition of a minimum of two successive frame-sync words to output ("flow") telemetry to the project. Validation required recognition of a third-sync word. The

[10] The MCD2 and MCD3 are distinct types of maximum-likelihood convolutional decoders. The MCD3, developed later, can handle both the $r = 1/2$ code and the $r = 1/6$ code. Though still in use by some deep space missions, the $r = 1/6$ code has been replaced by the turbo code, and MCD3s are becoming scarce through attrition.

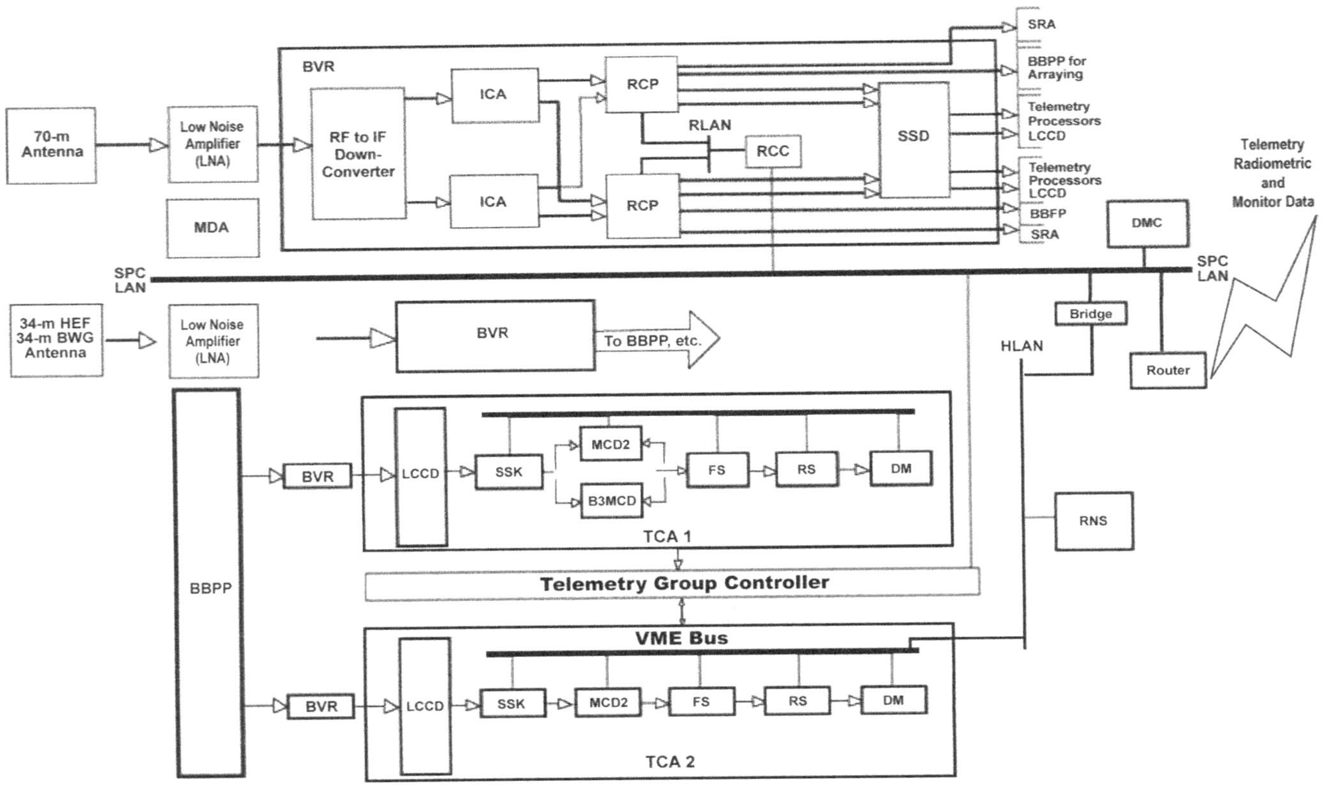

Fig. 5-14. DSN demodulation and production of telemetry data (from 871-010-030).

number of sync-word-allowable bit miscompares for recognition and validation could be set in the software.

In the DS1 MSA, the near-real-time data was "broadcast" to workstations, which could display them in the form of DMD (data monitor and display) pages. DS1 pages could be in list form, plots, or specially formatted "fixed" pages. Also, at their workstations, the DS1 analysts could query either the telemetry or the station monitor data. The query output was displayed as tabulations or plots on the screen, routed to a printer, or saved as a file for further processing.

5.6 Telecom Link Performance

As is the case for all modern deep space missions, the DS1 communication-link margins were calculated using statistical techniques to establish expected values from the mean and variance, and a further indication of variability from the tolerances and shape (uniform, Gaussian, etc.) [18]. When in-flight link performance was seen to differ significantly from the modeled predictions, link models (such as the LGAX antenna pattern, and the interaction of telemetry-and-ranging modulation in the SDST downlink) were updated from theoretical or pre-launch test data by additional or iterative assessment of available data.

The three DS1 link functions were command, telemetry, and ranging. Each had a minimum signal-to-noise ratio (called the threshold) at which the quality of the link meets a project-defined criterion.

Link performance is book-kept using a design-control table. In-flight DS1 operations were based on a criterion of positive-link margin under the following conditions: (a) command: mean minus three standard deviations (3σ), (b) telemetry: mean minus 2σ, and (c) ranging: mean minus 2σ. The command link did not have error-correcting coding, so data-stream bits were the same as channel symbols. The telemetry link had concatenated Reed-Solomon and convolutional coding. The parameter σ (spelled out as sigma) in the DCT refers to the standard deviation of the command E_b/N_0 (bit energy to noise-spectral-density ratio), the telemetry E_s/N_0 (symbol energy to noise-spectral-density ratio), or the downlink ranging P_r/N_0 (ranging power to noise-spectral-density ratio).[11] The quantity N_0 is the noise-spectral density; E_b is the energy per bit, E_s is the energy per symbol, and P_r is the ranging power.

---

[11] The ranging DCT defines mean and variance for P_r/N_0, as a bottom-line telecom-analysis quantity that can be compared against a like-named channel in the station-monitor data. Beyond this, navigation also defines a ranging "sigma" (computed as a function of P_r/N_0, but which is not included in DS1 DCTs) that is a prediction of the ranging-measurement scatter.

Tables 5-2, 5-3, and 5-4 are design-control tables (DCTs) containing predictions of DS1 telecom performance, generated by a software tool, the Telecom Forecaster Predictor (TFP). TFP, still in use for deep space telecom prediction, is a multimission tool for link-performance prediction built upon Matlab [19]. DS1 TFP used standard models for station parameters (the same for each project's TFP) adapted to include DS1 spacecraft models.

The three DCTs are all for a specific arbitrary instant in time, 2000-173/16:00 UTC (9 a.m. Pacific daylight time, June 21, 2000). At that instant in time, the DS1 spacecraft was scheduled to operate with the 70-m DSS-14 antenna at Goldstone. The spacecraft was configured for X-band uplink and downlink on the HGA. The command rate was 2000 bps, at an uplink-modulation index of 1.2 rad. The ranging modulation also suppressed the uplink, at a value of 3 dB. The downlink rate was 3150 bps, at a modulation index of 65.8 deg. The ranging also phase-modulated the downlink, at an index of 0.3 rad. The HGA was presumed to have its boresight misaligned from Earth by 2.5 deg.

TFP shows the time variation of link performance either as tabulations (columns of numbers to be read into a spreadsheet for formatting and printing) or as plot images for viewing or printing.

Performance of ranging and telemetry during the entire DSS-14 pass on June 21, 2000 is summarized in the two pairs of plots (Fig. 5-15 and Fig. 5-16) that come just before the three DCTs. The plots were created from the same computer run that produced the three DCTs. All plots contain values predicted once every 20 min, starting at the DCT time of 16:00 UTC and continuing to 04:00 UTC the next day. Quantities plotted for illustration are the mean values of the parameters.

The first plot-pair shows the downlink-ranging mean P_r/N_0 and its threshold of −10 dB at the top and the uplink command mean E_b/N_0 and its threshold of +9.6 dB at the bottom. The second plot-pair shows the station-elevation angle at the top and the downlink-telemetry mean E_s/N_0 with its threshold of −7.5 dB at the bottom. Figures 5-15 and 5-16 list, respectively, the downlink P_r/N_0 and uplink E_b/N_o; and station-elevation angle, and downlink-telemetry symbol SNR.

The top plot indicates how the ranging performance (predicted as downlink-ranging power to noise-spectral-density ratio) varies during a DSS-14 pass on June 21, 2000. Below a threshold of −10 dB, ranging quality would be unacceptable for navigation; below −5 dB, the quality would be marginal.

The bottom plot shows the 2000-bps command performance (predicted as uplink-command bit energy to noise-spectral-density ratio) for the same pass. Threshold is +9.6 dB for a bit-error rate of 10^{-5}.

Table 5-2. DS1 uplink (command and ranging) DCT.

Parameter	Value			
Basic Link Conditions				
Predict	2000-173T16:00:00 UTC			
Up- /Downlink	Two-way			
RF band	X:X			
Telecom link	DSS-14-HighGain. ConfigA-DSS-14			
Command Uplink Parameter Inputs				
Cmd data rate	2000 bps			
Cmd mod index	1.20 rad			
Cmd rngmod index	44.9 deg			
Operations mode	Nominal			
Mission phase	Launch phase			
DSN site	Gold-Gold			
DSN elevation	In view			
Percent probability of better weather	25			
Attitude pointing	Earth pointed			
External Data				
Range	(km)	3.0816e+08		
Range	(AU)	2.0599		
One-way light time (OWLT)	(hh:mm:ss)	00:17:07		
Station elevation(s)	(deg)	14.41		
DOFF: HGA, KHA	(deg)	2.50	2.50	
DOFF: LGA1, LGA2, LGA3	(deg)	2.50	92.50	87.50
Clk: HGA, KHA	(deg)	159.49	0.00	
Clk: LGA1, LGA2, LGA3	(deg)	159.49	0.00	0.00
Added s/c antenna pointing offset	(deg)	2.5		
DSN site considered	DSS-14/DSS-14			
At time	0.00 hours after the start time			

(continued)

Table 5-2. (Continued)

Link Parameter	Unit	Design Value	Fav Tol	Adv Tol	Mean Value	Var
Transmitter Parameters						
1. Total Xmitter power	dBm	73.01	0.00	–1.00	72.68	0.0556
2. Xmitter WG loss	dB	–0.41	0.05	–0.05	–0.41	0.0004
3. DSN antenna gain	dBi	72.45	0.20	–0.20	72.45	0.0133
4. Antenna pointing loss	dB	–0.10	0.10	–0.10	–0.10	0.0017
5. EIRP (1 + 2 + 3 + 4)	dBm	144.62	–0.80	–0.80	144.62	0.0710
Path Parameters						
6. Space loss	dB	–279.33	0.00	0.00	–279.33	0.0000
7. Atmospheric atten	dB	–0.14	0.00	0.00	–0.14	0.0000
Receiver Parameters						
8. Polarization loss	dB	–0.03	0.10	–0.10	–0.03	–0.0033
9. S/C ant pointing control loss	dB	–0.30	0.20	–0.20	–0.30	0.0133
10. Deg-off- boresight (DOFF) loss	dB	–0.44	0.43	–0.48	–0.47	0.0691
11. S/C antenna gain (at boresight)	dBi	20.10	0.50	–0.50	20.10	0.0417
12. Lumped circuit loss	dB	–1.79	0.30	–0.30	–1.79	0.0300
Total Power Summary						
13. Tot rcvd pwr (5 + 6 + 7 + 8+ 9 + 10 + 11 +12)	dBm	–117.34	–1.43	1.43	–117.34	0.2284
14. Noise spectral density	dBm/Hz	–172.22	–0.70	0.66	–172.23	0.0779
15. System noise temp.	K	434.75	–65.08	71.69	436.95	779.9427
16. Received P_t/N_0 (13–14)	dB-Hz	54.89	1.66	–1.66	54.89	0.3063
17. Required P_t/N_0	dB-Hz	50.60	0.00	0.00	50.60	0.0000
18. P_t/N_0 margin (16–17)	dB	4.29	1.66	–1.66	4.29	0.3063
19. P_t/N_0 margin sigma	dB	0.00	0.00	0.00	0.55	0.0000
20. P_t/N_0 margin-3 sigma (18–3*19)	dB	0.00	0.00	0.00	2.63	0.0000
Carrier Performance						
21. Recovered P_t/N_0 (16 + [AGC+BW])	dB-Hz	54.89	1.66	–1.66	54.89	0.3063
22. Command carrier suppression	dB	–3.46	0.20	–0.20	–3.46	0.0067

Table 5-2. (Continued)

Link Parameter	Unit	Design Value	Fav Tol	Adv Tol	Mean Value	Var
23. Ranging carrier suppression	dB	–3.00	0.10	–0.10	–3.00	0.0017
24. Carrier power (AGC)	dBm	–123.80	–1.46	1.46	–123.80	0.2367
25. Received P_c/N_0 (21 + 22 + 23)	dB-Hz	48.43	1.68	–1.68	48.43	0.3146
26. Carrier loop noise BW	dB-Hz	20.16	–0.20	0.15	20.13	0.0102
27. Carrier loop SNR (CNR) (25–26)	dB	28.30	1.71	–1.71	28.30	0.3248
28. Recommended CNR	dB	12.00	0.00	0.00	12.00	0.0000
29. Carrier loop SNR margin (27–28)	dB	16.30	1.71	–1.71	16.30	0.3248
Channel Performance						
30. Command data suppression	dB	–3.04	0.17	–0.18	–3.04	0.0051
31. Ranging data suppression	dB	–3.00	0.10	–0.10	–3.00	0.0017
32. Received P_d/N_0 (21 + 30 + 31)	dB-Hz	48.85	1.68	–1.68	48.85	0.3130
33. 3-sigma P_d/N_0 (32-3*sqrt [32var])	dB-Hz	47.17	0.00	0.00	47.17	0.0000
34. Data rate (dB-Hz)	dB-Hz	33.01	0.00	0.00	33.01	0.0000
35. Available E_b/N_0 (32–34)	dB	15.84	1.68	–1.68	15.84	0.3130
36. Implementation loss	dB	1.50	–0.50	0.50	1.50	0.0833
37. Radio loss	dB	0.00	–0.30	0.30	0.00	0.0300
38. Output E_b/N_0 (35–36–37)	dB	14.34	1.96	–1.96	14.34	0.4264
39. Required E_b/N_0	dB	9.60	0.00	0.00	9.60	0.0000
40. E_b/N_0 margin (38–39)	dB	4.74	1.96	–1.96	4.74	0.4264
41. E_b/N_0 margin sigma	dB	0.00	0.00	0.00	0.65	0.0000
42. E_b/N_0 margin-3 sigma (40–3*41)	dB	0.00	0.00	0.00	2.78	0.0000
43. BER (from 38) 8.5494e–14	None					

Table 5-3. DS1 downlink (telemetry and ranging) DCT.

Parameter	Value			
Basic Link Conditions				
Predict	2000-173T16:00:00 UTC			
Up-/Downlink	Two–way			
RF band	X:X			
Diplex mode	N/A			
LNA selection	LNA-1			
Telecom link	DSS-14-HighGain.ConfigA-DSS-14			
Telemetry Downlink Parameter Inputs				
Encoding	Reed-Solomon (255,223) concatenated with convolutional encoding (C.E.) (15,1/6)			
Carrier tracking	Residual			
Oscillator	Voltage-controlled oscillator (VCO)			
Subcarrier mode	Squarewave			
Phase-locked loop (PLL) bandwidth	1.00Hz			
Tlm usage	Engineering (ENG) - real time			
Tlm data rate/mod index	3150bps/ 65.80deg (38 DN)			
Operations mode	Nominal			
Mission phase	Launch phase			
DSN site	Gold-Gold			
DSN elevation	In View			
Percent probability of better weather	25			
Attitude pointing	Earth pointed			
External Data				
Range	(km)	3.0816e+08		
Range	(AU)	2.0599		
One-way light time (OWLT)	(hh:mm:ss)	00:17:07		
Station elevation(s)	(deg)	14.41		
DOFF: HGA, KHA	(deg)	2.50	2.50	
DOFF: LGA1, LGA2, LGA3	(deg)	2.50	92.50	87.50
Clk: HGA, KHA	(deg)	159.49	0.00	
Clk: LGA1, LGA2, LGA3	(deg)	159.49	0.00	0.00
Added s/c ant pnt offset	(deg)	2.5		
DSN site considered	DSS-14/DSS-14			
At time	0.00 hours after the start time			

Table 5-3. (Continued)

Link Parameter	Unit	Design Value	Fav Tol	Adv Tol	Mean Value	Var
Transmitter Parameters						
1. S/C transmitter power	dBm	40.97	0.50	–0.50	40.97	0.0417
2. S/C xmit circuit loss	dB	–1.91	0.30	–0.30	–1.91	0.0300
3. S/C antenna gain	dBi	24.60	0.60	–0.60	24.60	0.0600
4. Deg-off-boresight (DOFF) loss	dB	–0.98	0.21	–0.19	–0.97	0.0134
5. S/C pointing control loss	dB	–0.30	0.20	–0.20	–0.30	0.0133
6. EIRP (1 + 2 + 3 + 4 + 5)	dBm	62.39	1.19	–1.19	62.39	0.1584
Path Parameters						
7. Space loss	dB	–280.73	0.00	0.00	–280.73	0.0000
8. Atmospheric attenuation	dB	–0.14	0.00	0.00	–0.14	0.0000
Receiver Parameters						
9. DSN antenna gain	dBi	74.00	0.20	–0.20	74.00	0.0133
10. DSN antenna pointing loss	dB	–0.10	0.10	–0.10	–0.10	0.0033
11. Polarization loss	dB	–0.02	0.10	–0.10	–0.02	0.0033
Total Power Summary						
12. Tot rcvd pwr (6 + 7 + 8 + 9 + 10 + 11)	dBm	–144.61	–1.27	1.27	–144.61	0.1784
13. SNT (system-noise temperature) at zenith	K	18.39	–2.00	2.00	18.39	0.6667
14. SNT due to elevation	K	5.02	0.00	0.00	5.02	0.0000
15. SNT due to atmosphere	K	8.60	0.00	0.00	8.60	0.0000
16. SNT due to the Sun	K	0.00	0.00	0.00	0.00	0.0000
17. SNT due to other hot bodies	K	0.00	0.00	0.00	0.00	0.0000
18. System noise temperature (13 + 14 +15 + 16 +17)	K	32.01	–2.00	2.00	32.01	0.4444
19. Noise spectral density	dBm/Hz	–183.55	–0.28	0.26	–183.56	0.0082

(continued)

Table 5-3. (Continued)

Link Parameter	Unit	Design Value	Fav Tol	Adv Tol	Mean Value	Var
20. Received P_t/N_0 (12–19)	dB-Hz	38.95	1.30	–1.30	38.95	0.1866
21. Required P_t/N_0	dB-Hz	38.30	0.00	0.00	38.30	0.0000
22. P_t/N_0 margin (20–21)	dB	0.65	1.30	–1.30	0.65	0.1866
23. P_t/N_0 marg sigma	dB	0.00	0.00	0.00	0.43	0.0000
24. P_t/N_0 margin-2sigma (22–2*23)	dB	0.00	0.00	0.00	–0.22	0.0000
Carrier Performance						
25. Recovered P_t/N_0 (20 + [AGC+BPF])	dB-Hz	38.95	1.30	–1.30	38.95	0.1866
26. Theoretical tlm carrier sup	dB	–7.75	0.56	–0.61	–7.76	0.0570
27. Non-lin SDST tlm carr sup	dB	0.20	0.20	–0.20	0.20	0.0067
28. Total tlm carr sup (26 + 27)	dB	–7.56	–0.76	0.76	–7.56	0.0637
29. Theoretical rng carrier sup	dB	–0.26	0.04	–0.05	–0.26	0.0003
30. Non-lin SDST rng carr sup	dB	–0.54	0.20	–0.20	–0.54	0.0067
31. Total rng carr sup (29 + 30)	dB-Hz	–0.80	–0.25	0.25	–0.80	0.0070
32. DOR carrier suppression	dB	0.00	0.00	0.00	0.00	0.0000
33. Carrier power (AGC) (12 + 28 + 31 +32)	dBm	–152.97	–1.50	1.50	–152.97	0.2491
34. Received P_c/N_0 (25 +28 +31 +32)	dB-Hz	30.58	1.52	–1.52	30.58	0.2573
35. Carrier loop noise BW	dB-Hz	0.00	0.00	0.00	0.00	0.0000
36. Carrier loop SNR (CNR)	(34–35)	dB	30.58	1.52	–1.52	30.58
37. Recommended CNR	dB	10.00	0.00	0.00	10.00	0.0000
38. Carrier loop SNR margin (36–37)	dB	20.58	1.52	–1.52	20.58	0.2573
Telemetry Performance						
39. Theoretical tlm data sup	dB	–0.80	0.11	–0.12	–0.80	0.0023
40. Non-lin SDST tlm data sup	dB	0.00	0.20	–0.20	0.00	0.0067

Table 5-3. (Continued)

Link Parameter	Unit	Design Value	Fav Tol	Adv Tol	Mean Value	Var
41. Total tlm data sup (39 + 40)	dB	–0.80	–0.28	0.28	–0.80	0.0090
42. Theoretical rng data sup	dB	–0.26	0.04	–0.05	–0.26	0.0003
43. Non-lin SDST rng data sup	dB	–1.10	0.20	–0.20	–1.10	0.0067
44. Total rng data sup (42 + 43)	dB	–1.36	–0.25	0.25	–1.36	0.0070
45. DOR data suppression	dB	0.00	0.00	0.00	0.00	0.0000
46. Received P_d/N_0 (25 + 41 + 44 + 45)	dB-Hz	36.79	1.35	–1.35	36.79	0.2025
47. Two sigma P_d/N_0 (46–2*sqrt(46var))	dB-Hz	35.89	0.00	0.00	35.89	0.0000
48. Data rate	dB-Hz	34.98	0.00	0.00	34.98	0.0000
49. Available E_b/N_0 (46–48)	dB	1.80	1.35	–1.35	1.80	0.2025
50. Subcarrier demodulation loss	dB	0.01	0.00	0.00	0.01	0.0000
51. Symbol sync loss	dB	0.01	0.00	0.00	0.01	0.0000
52. Radio loss	dB	0.01	–0.00	0.00	0.01	0.0000
53. Output E_b/N_0 (49–50–51–52)	dB	1.78	1.35	–1.35	1.78	0.2025
54. Output SSNR (E_s/N_0)	dB	–6.00	–1.35	1.35	–6.00	0.2025
55. Required E_b/N_0	dB	0.30	0.00	0.00	0.30	0.0000
56. E_b/N_0 margin (53–55)	dB	1.48	1.35	–1.35	1.48	0.2025
57. E_b/N_0 margin sigma	dB	0.00	0.00	0.00	0.45	0.0000
58. E_b/N_0 margin–2sigma (56–2*57)	dB	0.00	0.00	0.00	0.58	0.0000
59. BER of conv decoder (from 53)	none	1.1063e-05				

The uplink-ranging carrier suppression was 3 dB, and the command-carrier suppression was –3.5 dB, both standard DS1 values in mid-2000 for 70-m station operation with the spacecraft HGA. The downlink ranging-modulation index was 17.5 deg (low), and the telemetry-modulation index was 65.8 deg. These are also standard DS1 values.

Table 5-4. DS1 ranging performance (uplink and downlink) DCT.

Parameter	Value			
Basic Link Conditions				
Predict	2000-173T16:00:00 UTC			
Up-/Downlink	Two-way			
RF band	X:X			
Diplex mode	N/A			
LNA selection	LNA-1			
Telecom link	DSS-14-HighGain.ConfigA-DSS-14			
Operations mode	Nominal			
Mission phase	Launch phase			
DSN site	Gold-Gold			
DSN elevation	In view			
Weather/CD	25			
Attitude pointing	EarthPointed			
Command Uplink Parameter Inputs				
Cmd data rate	2000 bps			
Cmd mod index	1.20 rad			
Cmd rngmod index	44.9 deg			
Telemetry Downlink Parameter Inputs				
Encoding	Reed Solomon (255,223) concatenated with C.E. (15,1/6)			
Carrier tracking	Residual			
Oscillator	VCO			
Subcarrier mode	Squarewave			
PLL bandwidth	1.00 Hz			
Tlm usage	Engineering (ENG) - real time			
Tlm data rate/mod index	3150 bps/ 65.80 deg (38 DN)			
Tlm rng/DOR mod index	Rng 0.3 rad / DOR 0.0 rad			
External Data				
Range	(km)	3.0816e+08		
Range	(AU)	2.0599		
One-way light time (OWLT)	(hh:mm:ss)	00:17:07		
Station elevation(s)	(deg)	14.41		
DOFF: HGA, KHA	(deg)	2.50	2.50	
DOFF: LGA1, LGA2, LGA3	(deg)	2.50	92.50	87.50
Clk: HGA, KHA	(deg)	159.49	0.00	
Clk: LGA1, LGA2, LGA3	(deg)	159.49	0.00	0.00
Added S/C ant point offset	(deg)	2.5		
DSN site considered:	DSS-14/DSS-14			
At time:	0.00 hours after the start time			

Table 5-4. (Continued)

Link Parameter	Unit	Design Value	Fav Tol	Adv Tol	Mean Value	Var
Uplink Turnaround Ranging Channel						
1. UL recovered P_t/N_0	dB-Hz	54.89	1.66	–1.66	54.89	0.3063
2. UL cmd ranging suppression	dB	–3.46	0.20	–0.20	–3.46	0.0067
3. UL ranging suppression	dB	–3.03	0.10	–0.10	–3.03	0.0033
4. UL Pr/Pt (2 + 3)	dB	–6.49	–0.30	0.30	–6.49	0.0100
5. UL filtering loss	dB	–0.91	0.20	–0.20	–0.91	0.0067
6. UL output P_r/N_0 (1 + 4 + 5)	dB-Hz	47.49	1.70	–1.70	47.49	0.3229
7. Ranging channel noise BW	dB-Hz	63.22	–0.43	0.20	63.14	0.0176
8. UL ranging SNR (6–7)	dB	–15.65	–1.75	1.75	–15.65	0.3406
Downlink Ranging Channel						
9. DL recovered P_t/N_0	dB-Hz	38.95	1.30	–1.30	38.95	0.1866
10. Theoretical telemetry suppression	dB	–7.75	0.56	–0.61	–7.76	0.0570
11. Non-linear SDST tlm suppression	dB	–0.57	0.20	–0.20	–0.57	0.0067
12. DL total tlm suppression	dB	–8.34	–0.76	0.76	–8.34	0.0637
13. Theoretical rng modulation loss	dB	–28.30	2.38	–2.46	–28.33	0.9756
14. Non-linear SDST rng mod loss	dB	0.00	0.20	–0.20	0.00	0.0067
15. DL total rng mod loss	dB	–28.33	–2.97	2.97	–28.33	0.9823
16. DL P_r/P_t (12 + 15)	dB	–36.66	–3.07	3.07	–36.66	1.0460
17. DL received P_r/N_0 (9 + 16)	dB-Hz	2.28	3.33	–3.33	2.28	1.2326
18. DL noisy ref loss	dB	0.00	0.00	0.00	0.00	0.0000
19. DL output P_r/N_0 (17 + 18)	dB-Hz	2.28	3.33	–3.33	2.28	1.2326
20. DL out P_r/N_0 sigma	dB-Hz	0.00	0.00	0.00	1.11	0.0000
21. DL out P_r/N_0 mean-2 sigma	dB-Hz	0.06	0.00	0.00	0.06	0.0000
22. DL required P_r/N_0	dB-Hz	–10.00	0.00	0.00	–10.00	0.0000
23. Ranging margin, mean (19–22)	dB	12.28	3.33	–3.33	12.28	1.2326
24. Ranging margin, mean-2 sigma (21–22)	dB	10.06	0.00	–0.00	10.06	0.0000

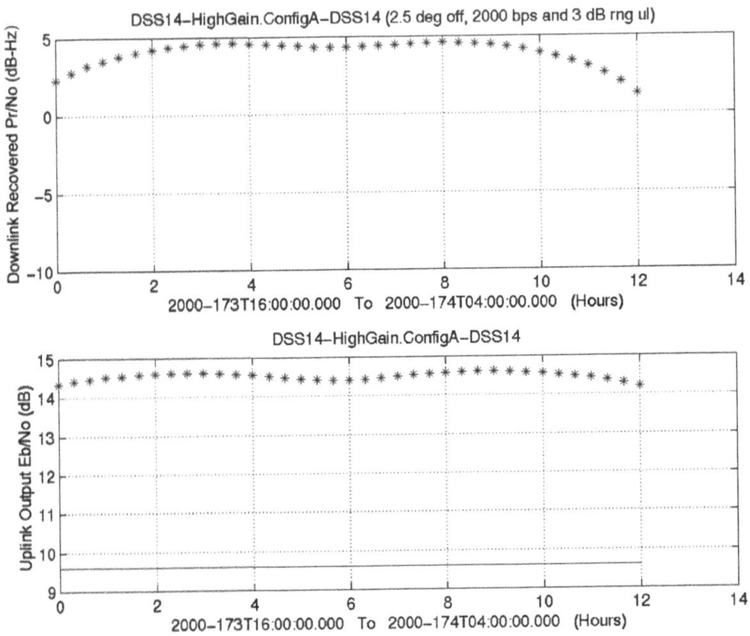

Fig. 5-15. Downlink P_r/N_0 and uplink E_b/N_0.

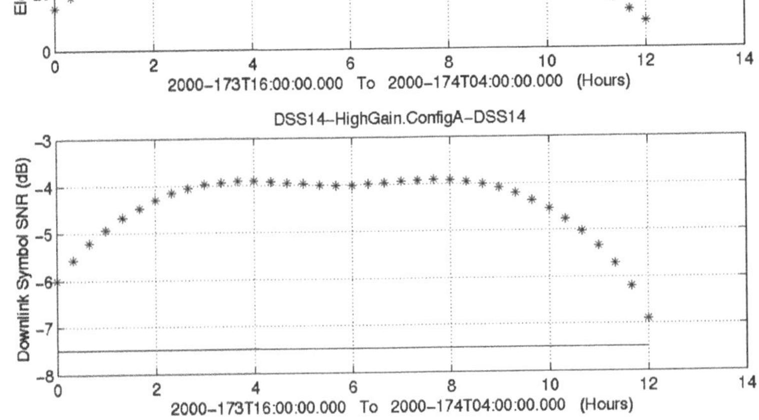

Fig. 5-16. Station elevation angle and downlink telemetry symbol SNR.

The top plot shows the variation of the DSS-14 elevation angle during the pass on June 21, 2000. Because the signal passed through more of Earth's atmosphere at lower elevation angles, the attenuation was larger and the system-noise temperature (SNT) was higher. Attenuation affected both uplink and downlink, while SNT affected downlink.

The bottom plot shows the predicted symbol signal-to-noise ratio (SSNR) of the 3150-bps telemetry. The downlink carrier also had ranging modulation at the "low" index. The telemetry-decoding threshold for the (15,1/6) concatenated code is –7.5 dB SSNR. DS1 experience showed that successful decoding by the MCD3 is improbable when the BVR produces an SSNR lower than –7.5 dB.

5.7 Operational Scenarios

The following scenarios describe the major telecom-subsystem operating modes in the context of supporting specific phases of the mission or major mission activities and modes.

5.7.1 Launch

The major prelaunch DS1 telecom decision was whether to launch with SDST in the coherent or the noncoherent mode. Noncoherent mode (coherency disabled) was chosen. The most important consideration leading to this choice was that noncoherent mode would provide an unambiguous downlink frequency for BVR acquisition regardless of whether an uplink was in lock. On the other hand, coherent mode with uplink in lock would have provided immediate two-way Doppler data to determine any corrections from errors in the launch trajectory.

The spacecraft launched with the LGAZ antennas selected, with LGAZ– at the smaller angle to Earth. One day after launch, LGAZ– was to be pointed within about 20 deg of Earth line. The uplink and downlink rates were to provide commandability and telemetry data via the selected LGA over a wide range of pointing errors. At launch, the command rate was 125 bps uncoded. The downlink rate was 2100 bps on a 25-kHz subcarrier, a 40-deg modulation index and (7,1/2) convolutional coding. During the initial acquisition pass, the SDST was commanded to go to the coherent mode ("TWNC [two-way non coherent] off" for the old timers), and to turn the X-band ranging channel on. Approximately one day after launch, the uplink rate was commanded to be 2000 bps, and a small sequence stored onboard before launch was activated to change the downlink rate to 19908 bps, the telemetry subcarrier frequency to 375 kHz, and the telemetry-modulation index to 65.8 deg. That configuration stayed the same for the first two weeks of the mission.

5.7.2 Safing

Safe mode normally occurs when the onboard fault-protection software detects a problem that requires unplanned ground intervention. (As on current spacecraft, a DS1 safe-mode configuration could also be commanded intentionally, such as when the flight software was "rebooted" after any software update.)

The original implementation of safe mode for DS1, depending on what fault occurs, would point the +x-axis either at the Sun or at Earth. After the late-1999 SRU failure, DS1 safe mode always pointed the +x-axis at the Sun and rotated the spacecraft about that axis at a rate of one revolution per hour. The system fault-protection software runs a "telecom script" (an unchanging series of commands, with defined intervals of time between the commands) to configure the SDST, XPA, and the antenna to provide the maximum degree of commandability and chance of a station receiving above-threshold telemetry. Until March 2001, much of the safe-mode telecom configuration (noncoherent mode, 7.8125-bps command rate, 40-bps telemetry rate, 25-kHz telemetry subcarrier frequency, (7,1/2) convolutional coding, ranging off, Ka-band downlink off) was similar to that of the launch mode.

Throughout the mission, the telecom script was updated as to which antenna it would select and what downlink-telemetry rate it would control. These updates matched downlink-performance changes caused by the changing Earth-DS1 distance and Sun-spacecraft-Earth angle. Telecom-script updates were through command-file uploads from the ground. As of March 2001 and through the end of the mission, the script selected LGAX, a telemetry rate of 79 bps, and the (15,1/6) coding.

5.7.3 Anchor Pass (at HGA Earth Point, High Rate)

The term "anchor" referred to the spacecraft stopping its mission activities to point the HGA at Earth and communicate. Anchor passes were scheduled approximately weekly to download telemetry data accumulated since the last anchor pass, to upload new command sequences, and to provide ranging and Doppler data. Prior to the start of the pass, the spacecraft may have been oriented toward a "thrust star" with the IPS thrusting. Since the +x-axis would be off-Earth, only minimal communication would be possible. After the SRU failure, the process of pointing to Earth for an anchor pass was based on the use of an onboard algorithm to control the spacecraft-pointing attitude without using the SRU. Before start of track, the spacecraft sequence stopped the thrusting, turned to an "Earth star" reference so the +x-axis was near Earth, selected the HGA, restarted the IPS thrust,[12] and the X-band

[12] Thrusting on Earth-point was not usually beneficial to the trajectory but conserved the very limited attitude-control propellant, hydrazine. The IPS was gimbaled in two axes so it could perform attitude control in the x- and y-axes. Thus, while Earth-point thrusting was at a lower level, hydrazine was only expended for attitude control about the third axis (z).

downlink (by turning the x-band exciter on). Depending on the amount of telemetry data to be downlinked, the ranging channel was sequenced on for either the entire pass (for less telemetry data) or part of the pass (for more data). One or two hours before the end of an anchor pass, these processes were reversed: the spacecraft returned to thrust attitude, the IPS was restarted, and (just before the scheduled end of track) the downlink was turned off.[13]

The flight team had developed several variations for starting an anchor pass, based on the degree of pointing certainty at the start of the pass. The most common variant was when the initial Earth-reference star was more than 5 deg from Earth, or there might have been a question about being locked to a star at all. If so, the flight team could elect to start the pass at a low telemetry rate (79 or 600 bps, depending on the uncertainty) with HGA selected but with a "lifeboat" sequence to reselect LGAX after three hours. The telecom analyst would compare telemetry E_s/N_0 with the value predicted for the expected off-Earth angle, then recommend a higher telemetry rate to be commanded in real time. If signal level was adequate, the flight team sent a command to "cancel" (deactivate) the lifeboat, and so remain on the HGA.

During sequence planning, the telecom analyst defined for each anchor pass the uplink and downlink rates, together with associated modulation-index values and subcarrier frequencies, and ranging-channel use. The analyst determined the supportable rates, using TFP. During the primary mission, these predicts were generated as a "data-rate capability file" intended to interface directly with sequencing software. Because confident long-range planning was less feasible when successively chosen Earth-stars were involved, the telecom analyst made predictions for each pass with the off-point angle input directly. In either case, data rates for each allocated 34- or 70-m pass entered the sequencing process via "service-package files."

5.7.4 Midweek Pass (at Thrust Attitude for IPS Operation)

Midweek passes alternated with anchor passes, which were usually scheduled for early in the week. During a midweek pass, the spacecraft was three-axis oriented for IPS thrusting through use of the Sun sensor, gyros,

[13] Until April 2001, IPS operation was nearly continuous and at a high-thrust level to reach Borrelly. This thrusting is called "deterministic," with the thrust level determined by the available spacecraft power and thrust direction as a function of time determined by reaching Borrelly at the planned time and miss distance. Subsequently, the mission moved into a period of lower-level "impulse thrusting" for continued hydrazine conservation, using successive orientations at intervals of one or two weeks toward a "north" and a "south" thrust-star, each near an ecliptic pole. Over a period of time, the impulses cancelled one another out. Like Earth-point thrusting, impulse thrusting was not at maximum thrust level and so allowed the X-band downlink to remain on between passes.

and science-camera data (assuming current operations without the failed SRU). During the prime mission, the LGAX or one of the LGAZ antennas dictated required-pointing direction, depending upon which antenna best supported communications on a particular day. For the rest of the extended mission, LGAX was best because thrust attitudes resulted in the +x-axis being 0 to 50 deg from Earth. If the angle was less than 7.5 deg, the HGA provided better performance than LGAX. In this case, no turn to an Earth-star was necessary, and HGA communications were possible for midweek as well as anchor passes. For larger angles, midweek passes via LGAX provided at least the beacon tone to indicate stability of pointing algorithm star-lock and two-way Doppler data to determine if thrusting was still continuing as planned.

As for anchor passes, the telecom analyst predicted midweek pass uplink and downlink performance as a function of time, and 34-m BWG, 34-m HEF, or 70-m station allocation. The spacecraft "backbone" sequence for the current period (about 2 to 3 weeks at a time) controlled the uplink-and-downlink rate to supportable values. The lowest command rate (7.8125 bps) was sequenced for 34-m BWG stations with their 4-kW, X-band transmitters. A rate of 125 bps was sequenced for either 34-m HEF stations or 70-m stations, both with 20-kW transmitters. For maximum IPS thrust capability, the X-band downlink was turned on shortly before the scheduled start of track and turned off a few minutes before the end of track. Observing the signal level change at turnoff validated that the sequence was operating. During midweek passes, the spacecraft was in the coherent mode, to provide two-way Doppler regarding the thrust.

Between tracks, the spacecraft was generally also left in a "distant-pass" configuration so that if an unscheduled pass should become necessary, DS1 was commandable. By the end of the mission, the distant-pass configuration was LGAX, 7.8125-bps command rate, and 40-bps telemetry rate.

5.7.5 High-Gain-Antenna Activity (January–June 2000, March 2001)

This section goes into some detail because it describes how the DS1 flight team creatively overcame the failure of a major onboard element of the attitude control system that was also single-string. After the failure, the team regained control well enough for short periods of HGA operation. This initial technique required exacting and labor-intensive real-time "ground in the loop" commanded pointing control. Over a little more than one year, flight software was updated to perform some of these functions on board using a science camera, then finally customized turn-size and rate commands were developed to provide pointing with sufficient accuracy for HGA communications, optical navigation, and comet science at Borrelly.

The DS1 spacecraft was launched with a Sun-sensor assembly (SSA), inertial measurement units (IMUs), and the previously mentioned SRU; together they provided three-axis control of spacecraft pointing.[14] During the prime mission and until November 11, 1999, when the HGA was required, it would be pointed at Earth within a normal dead-band tolerance of 1 deg.[15]. On that day, downlink performance was not consistent with the HGA at Earth-point. The spacecraft was found to be in safe mode (LGAX selected and x-axis pointed to the Sun) after a tracking station could not acquire the expected downlink. Subsequent telemetry analysis showed the SRU was inoperative. Afterwards, the spacecraft remained in safe mode for some months, and low-rate uplink and downlink communications were done via LGAX only.

To return valuable science data already stored onboard at the time of the failure, as well as moderately-voluminous engineering data concerning the failure itself, the project flight team invented a ground-in-the-loop method to point the spacecraft close enough to Earth to use the HGA. The name refers to the operation of feedback-control loops with delay, in which human analysts analyze spacecraft and station data in real time, then send corrective commands immediately in real time, all the while constrained by the tens of minutes of delay inherent in the light-time between Earth and the spacecraft. Fundamental to ground-in-the-loop control is the idea of moving the spacecraft from its +x-axis to the Sun attitude to a +x-axis near Earth. This was accomplished with a combination of inertial-control and attitude-control system inputs from the gyros and the SSA only. The first part of the process was to determine the position (stop the antenna); the second part was to maintain the position (keep the antenna pointed).

This special mode is described in some detail because it involved considerable use of the telecom-analyst's skills in monitoring and assessing the significance of variations in downlink-signal levels. The most basic measurement was of carrier performance as represented by the P_c/N_0 (carrier power-to-noise spectral-density ratio). The analyst assessed in real time what commands to send and when to send them. The commands were used to change basic spacecraft motion and pointing. The initial motion was called "coning," which was rotating the spacecraft around the line joining the Sun to the spacecraft, with the +x-axis at a fixed-offset angle from the Sun (equal to the Sun-spacecraft-Earth angle). The rotation (coning) rate was once

---

[14] The SSA was not used for three-axis control or knowledge in normal operation. Only the SRU and IMUs were used in normal operation.

[15] The term "dead band" comes from feedback-control theory. It refers here to an angle (1 deg) relative to the deviation of the actual pointing relative to each desired axis (x, y, z). When the difference between actual and planned pointing about an axis reaches the dead-band limit (say +1 deg), the control system fires a thruster in the negative direction. No corrective action occurs so long as the pointing error remains within the dead band.

per 45 min.[16] When coning was commanded to stop, the final motion was with the +x-axis pointed "near" Earth, under inertial control. Pointing control also commanded an indexing of the +x-axis by a selected number of degrees (from 2 to 8), to compensate for gyro drift during the hours of the pass.

5.7.5.1 Stopping the Antenna near Earth by Using the Planning Worksheet. Use of an Excel-planning spreadsheet requiring only simple and rapidly made inputs and providing simple and unambiguous-to-use outputs was essential to achieve the initial HGA pointing, starting from the +x-axis to the Sun condition. See Table 5-5 for a replica of the planning sheet developed by the DS1 project telecom analyst and used by the ACE to direct the times of specific station actions for the June 12, 2000 "stop coning" activity.

Table 5-5. Replica of "HGA Activity Planning Spreadsheet" for June 12, 2000 activity.

Action Plan (based on seeing first 2 peaks)

act_date 06/12/00

owlt_sec 1012
owlt= 0:16:52
rtlt= 0:33:44
time/rev= 0:45:00 (from 1st to 2nd observed peak)
typeover seeded items — Do **NOT** enable uplink station Conscan
type in observations — planned times/intervals

NO	coherency?
NO	ka-band dl?
NO	fsr recording?
NO	ul conscan?

station transmit	SCET	NOCC receive	QUERY receive	what	dB or time value	
		9:00:00	**0:00:00**	**observation: 1st HGA peak used in stopping HGA**		
		8:59:45		observation with delay removed		
		9:45:00		2nd HGA peak was expected	nominal = 0:45:00	
	observe-->	**9:45:00**	**9:45:00**	**observation: 2nd HGA peak used in stopping HGA**	**interval = 0:45:00**	(2nd - 1st peak)
		10:30:00	10:30:00	3rd HGA peak expected		
9:56:04		10:29:48	10:29:48	expected 3rd peak with delay removed		
9:48:00				Start Excel sheet update (2nd peak seen plus 1 min)	**worktime= 0:05:19**	
9:49:19 <--alert				Give ACE station drive_on time		
9:49:19				ACE verifies CMD buffer selected for 31.25 bps	nominal = 00:11:19	
9:54:19 <--action!			**9:54:19**	34m station's transmitter drive ON	**interval= 0:09:19**	(DrvOn - peak)
				start sweep (3 segments +/–10 kHz, 300 Hz/sec = 00:01:40 for sweep)		
9:55:59				Expected end of sweep (based on **ACQ** and nominal **ETX30XCN** duration of 00:01:40)		
9:56:04				Station turns command modulation ON at end of sweep		
9:56:11				ACE verifies command modulation ON	nominal = 0:13:24	
9:56:24 <--action!			**9:56:24**	"Stop coning" activate cmd bit1 (drive ON + 00:02:05)	**interval = 0:11:24**	(Bit1 - peak)
9:56:27				Actual radiation begin, including command system latency		
9:56:51	10:13:43			End radiation of activate command (for 31.25 bps only, excluding vc5 tail sequence)		
9:56:52	10:13:44			Sequence execution begins		
9:56:52	10:13:44			Sequence execution completes		
9:57:44	10:14:36	10:31:28		WAG: HGA stops.	0:01:40 after **real** 3rd peak	
		10:32:10	**10:32:10**	Stopped HGA expected	**00:02:10 nominal**	0:00:42 (stop - peak)
		10:37:30	**10:37:30**	HGA turn back expected	**00:05:20 nominal**	0:05:20 (back - stop)

[16] The coning rate was one rotation per 45 min in contrast to the safe-mode rotation rate of one revolution per 60 min. As part of the SRU recovery flight-operations redesign, the coning rate was made as rapid as possible while still providing sufficient time to get a "stop-coning" command to the spacecraft. The safe-mode rate was established before launch, and there was never any mission reason to change it.

Before the pass began, the telecom analyst customized the spreadsheet with the allocated start-and end-track times, the "start coning" and "final conditions" sequence start times, and the one-way light time. Table 5-5 shows one of three parts of the worksheet, with the times in the two "action" cells updating as soon as the analyst had filled in the "observe" time. The two action cells defined (a) the time the station would turn on its transmitter and begin an uplink-acquisition sweep, and (b) the time the DS1 controller (ACE) would radiate the "stop-coning" activate command.

The onboard "start-coning" sequence put the spacecraft into an attitude-control system mode that ensured the HGA would sweep its boresight past Earth periodically. The telecom system was switched from LGAX to HGA, telemetry modulation was removed from the downlink carrier, and the command rate was set to 31.25 bps. The +x-axis moved from the Sun, a distance equal to the Sun-spacecraft-Earth angle, and the spacecraft was sequenced to begin rotating about the Sun-spacecraft line at one rotation per 45 min.

Using the spreadsheet to determine the actual rotation rate, the telecom analyst observed and timed the occurrence of two sweeps of the HGA boresight past Earth, and then gave the ACE the "action" times. These two times determined when the uplink and then the command must be sent to reach the spacecraft, as the HGA was pointed near Earth the third time. Excluding station problems, 95 percent of the time this process stopped the antenna near Earth on the first attempt.

The March 2001 HGA activity was planned to stop the antenna, based on the analyst's seeing only one peak, on the assumption that the rotation was always close to once per 45 minutes and the OWLT shorter than about 20 minutes (giving 5 minutes of analysis, reaction time, and command transmission). This "single-peak" activity, which was checked against a similar one performed in June 2000, was also successful.

5.7.5.2 Keeping the Antenna Pointed Near Earth by Monitoring Signal Levels. The quantities monitored during this phase included the downlink P_c/N_0, telemetry E_s/N_0 (symbol energy to noise-spectral-density ratio), or uplink P_c (carrier power).

After the ACE had radiated the "stop-coning" command, the telecom analyst recorded the P_c/N_0 observed for each of the first two peaks from each of the two receivers. This single command activated a small stored sequence of commands that stopped the coning rotation, produced a turn-back in the opposite direction to return the HGA to near-perfect Earth-point, and reset the command and telemetry rates for further activity. The analyst compared the P_c/N_0 value against the mean value predicted by TFP that assumed the

HGA boresight was Earth-pointed with all telecom-link components operating at their expected values. The purpose of this assessment was to estimate the maximum downlink rate the link could support, taking into account link performance as well as the HGA-pointing control demonstrated during previous HGA activities.

Next, the station P_c/N_0 data showed the third peak, the subsequent halt in antenna motion, and the turn back to Earth. The telecom analyst directed the ACE to command the activation of a stored "telemetry-rate" sequence to establish the supportable rate, with its subcarrier frequency and modulation index. Thereafter, through the remaining hours of the pass, the analyst continued to monitor downlink P_c/N_0, E_s/N_0, and uplink P_c to determine if the antenna had drifted too far from Earth to support the rate. If so, the analyst would direct that a "corrective-turn" activate command be transmitted. The criterion for activating a "corrective-turn" sequence was if the E_s/N_0 first fell to –6 dB (with the threshold being –7.5 dB) fairly rapidly or sank to –6 dB twice but less rapidly.

Use of this labor-intensive and real-time process enabled the project to receive high-rate telemetry data from 14 passes. Through May 2000, we went from just sending a command to stop the rotation when we saw the signal from the HGA, evolving into a sequence that would stop the rotation and then turn back when OWLT became so large that the stop command would not arrive in time.[17] Finally, another 14 HGA activities in June 2000 enabled the project to reload several megabytes of flight software at a high rate. The new software brought to an end the routine use of the HGA activity described here because it used the science camera instead of the failed SRU to generate star data for onboard pointing control. The software update included changes to ensure near-Earth pointing for pointing relative to a thrust star for periods of ion engine thrusting. With full three-axis pointing capability restored for thrusting, DS1 resumed its science mission for a flyby of the comet Borrelly in September 2001 [7].

An activity in March 2001, using the new flight software for the Borrelly encounter [8], progressed in a manner similar to many HGA activities in 2000. For the first few hours after Earth-point was achieved, the HGA remained near Earth. However, the downlink performance became worse over a period of hours, with the likely cause being a slow drift in roll toward the HGA first null. To counter this, the DS1 attitude control analyst developed a new delta-turn "bump" sequence to mimic a fraction of a turn of the original-coning rotation. This bump restored HGA pointing, and was made standard for HGA activities required after any subsequent safe-mode events,

---

[17] Personal communications, Steve Collins and Tony Vanelli, JPL, July 8, 2014.

or flight-software updates. The flight team maintained a little "kit" of sunline turn commands to bump us forward or back by a few degrees. With this basic approach, the remaining uncertainty was in the gyro bias estimates. Incorrect estimates resulted in the spacecraft having a small remnant drift rate when the spacecraft attitude control system indicated that it was stopped.

Some art was required in choosing bump size and number of bumps without prior knowledge of which side of the HGA pattern the Earth had drifted to on a particular activity day. As we gained more experience with this technique, we got the idea of deliberately trying to "park" just to one side of the peak of the antenna pattern, with the idea that the gyro bias would cause us to drift along the antenna pattern where we would see the signal strength either increasing or decreasing. That change would tell us the sign of the required correction bump. Further, by observing a few revs at 1 rev/hour, by measuring the actual time between antenna signal peaks, we could also choose which side of the peak to "park" on.[18] The remaining mission was flown using this mode.

5.7.6 Solar Conjunction

Solar conjunction occurs when the spacecraft and the Sun are in the same angular region as viewed from the deep-space station [20]. The angular separation is the Sun-Earth-probe (SEP) angle. Effects on deep-space communications become more severe as the SEP angle becomes smaller.[19] For DS1, X-band up- and downlink, we considered an angle of 5 deg as the minimum at which to expect no degradation, and 3 deg as the minimum at which reliable communications could be planned. From October 20 to December 3, 2000, the angle was less than 5 deg, and from October 29 to November 25 it was less than 3 deg. The minimum angle was less than 0.5 deg during a scheduled pass on November 14, 2000. The 11-year solar cycle was near its maximum.

---

[18] Personal communications, Steve Collins and Tony Vanelli, JPL, July 8, 2014.

[19] A superior solar conjunction (like DS1's) occurs when the Sun is between the spacecraft and the Earth. Planning for superior conjunction effects on deep-space links at JPL currently takes into account only the carrier-frequency band and the Sun-Earth-spacecraft angle. Solar activity varies in cycles, with the 11-year solar cycle near a maximum in 2000–2001. The effects on a link, caused by charged particles from the Sun producing amplitude and phase scintillation, may also be highly variable over periods of a few minutes to a few hours. Coronal-mass ejections (CME) of charged particles that cross the ray path between Earth and the spacecraft have degraded Galileo low-margin S-band links even when the SEP angle is large (>90 deg). Apparent solar effects affected DS1 X-band up- and downlink signals during a pass on April 3, 2001 (at SEP ~31 deg), less than one day after a very large (class X20) solar flare occurred.

Both to minimize configuration changes and to use the HGA as much as possible, project navigation found a single-reference star with small x-axis-to-Earth angle throughout the conjunction period.[20] The HGA off-point from Earth varied from about 2.5 deg at period start to a minimum of 0.3 deg November 13, to about 2.8 deg at period end. The first scheduled post-conjunction pass that the project was able to receive telemetry was on November 20, with the Sun-Earth-spacecraft angle at 1.1 deg.

The DS1 project planned the solar conjunction as a single sequence with minimal configuration changes, to be loaded onboard for execution for the entire duration. The up- and downlink data rates were conservative, and the command-loss expiration was pushed to beyond the time the angle would again be greater than 3 deg. The command rate was made 7.8125 bps through the conjunction period.

Downlink strategy, as a function of Sun-Earth-spacecraft angle, was to:

- Sequence a downlink rate that was reduced by one or two data rates from normal between an angle of 3 and 5 deg. For example, instead of 4424 bps, sequence 3150 bps or 2100 bps. Instead of 790 bps sequence 600 bps or 420 bps. This increased the margin by about 1.2 to 3.1 dB
- Sequence 40 bps for passes with an angle less than 3 deg
- Modulate downlink carrier with only a subcarrier tone for passes with an angle less than 1 deg.

Link configurations were based on experience from previous solar conjunctions of Mars Global Surveyor and Cassini, as well as on the expert recommendations of the JPL Telecommunications Systems and Research Section. The sequence included operating both the X- and Ka-band downlinks. It also included periods with the downlink in the two-way coherent mode (SDST-coherency enabled and uplink in lock) and other periods with no planned uplink. The objective was to maximize the probability that at least one frequency band would be receivable during the scheduled weekly tracking passes. The strategy was successful both in monitoring spacecraft health and providing open- and closed-loop data for telecom analysis, and planning other project conjunctions [20].

---

[20] Part of the rationale in selecting the reference star used during conjunction was to yield good pointing relative to Earth for the November 20 pass. The star was also suitable for IPS thrusting which continued throughout conjunction. The 0.3-deg minimum Sun-Earth-spacecraft angle that occurred on November 13 compares with a solar radius of about 0.25 deg and meant the signal path from spacecraft to Earth was very nearly blocked by the Sun.

5.7.7 Ka-Band Downlink

Using the Ka-band downlink during the first months after launch was for technology validation of the KaPA. Ka-band was also used operationally during portions of the solar conjunction in November 2000. The Ka-band downlink was receivable via the KHA only when the +x-axis pointed at Earth and a station with Ka-band capability had been allocated. Sequencing of the Ka-band exciter of the SDST and the KaPA was through previously stored commands.

During the 1998–99 technology validation period, the Ka-band downlink carried telemetry and ranging data. Ranging channel on/off and modulation index (low/high) were individually controllable for X- and Ka-band, as was the telemetry subcarrier frequency and modulation index. However, only a single downlink rate was available at a time for use on both bands. Because the KHA and HGA gains were similar, but the KaPA had one-fifth of XPA's RF output, the Ka-band supportable telemetry rate was similarly reduced relative to X-band. Because DS1 was downlink-rate limited, the project generally chose the higher rate supportable on X-band. Subsequently through the end of the mission, except for special tests, the Ka-band downlink was left unmodulated.

5.8 Lessons Learned

5.8.1 Telecom-Related Lessons Learned

In December 1999, the telecommunications group at JPL presented lessons learned to DS1. This section is an update of that material [21], covering the development and testing of the SDST and KaPA, as well as the flight operations. Going on two decades since DS1 was active, many of recommendations have proved themselves in the testing and flight operations of other Deep Space missions. For future projects, the staffing-related experience has been incorporated into a task and level-of-effort model that is used during a project's mission operations design.

The section begins with some things we did mostly right and ends with other things that caused us difficulty.

5.8.1.1 Telecom Pre-Launch Testing. Telecom hardware testing by the manufacturer, telecom-development lab (TDL) subsystem testing, DSN-compatibility testing, and spacecraft-level prelaunch system testing was thorough enough on DS1 that the flight team found an untested SDST characteristic that was needed for in-flight planning or analysis. The DS1 telecom test plan should be used as a model for our in-development projects. Future test plans do not need to include repeats of development tests already done for unchanged components; however, the planning needs to be equally thorough in considering the telecom functions (command, telemetry, carrier tracking, including the presence of Doppler shifts, ranging) and mission plans.

5.8.1.2 Development-to-Operations "Handover." This was properly scheduled and well executed. The flight-team telecom analyst came onboard a year before the planned launch, and the development telecom-system people remained to lend a hand through the planned "40-day" technology-validation period that lasted six months. The intense period of telecom in-flight characterization (planning and execution) probably lasted half that long, with the formal SDST, KaPA, and BMOX technology-validation activities on the spacecraft occupying all or portions of about 15 passes. The telecom analyst was well seasoned by launch, including a sufficient familiarity with the DS1 spacecraft and specifically its subsystem. Even so, the telecom-development engineers played important roles in the formal-technology validation. Plans for our in-development projects need to include a development-to-operations handover of at least one year.

5.8.1.3 Flight Team Telecom-Support Level. The flight-team plan for telecom staffing was 1.0 full-time equivalent (FTE) in the prime mission, reducing to 0.5 during the extended mission. The reality was the need for 1.5 to 2.0 FTEs; through in-flight characterization, remaining at 1.0 through the SRU-anomaly resolution, and 0.75 by May 2001. Some causes: our software tools were not mature; there was more than expected hands-on flying of this dynamic spacecraft; and there was a continuing need to produce two forms of telecom-configuration definitions (a DSN keywords file and service packages). The DS1 experience taught us that we need a full-time, seasoned telecom analyst for each spacecraft, with moderately active telecom planning and execution, with tracking a half or more of the total time. This level can begin to be cut back once several projects have a common set of automation tools for telecom planning and analysis.

5.8.1.4 Effective Staffing Mix. On DS1, the senior-lead telecom analyst trained and mentored two junior analysts successively through the prime mission. We verified that routine or repetitive tasks such as performance-comparison runs do not require a senior analyst. A second, on-call senior analyst can step in for vacations, illness, etc. DS1 hoped to do the link-analysis task using a pool of qualified engineers. However, this "plug and play" approach did not work for DS1 because there were never enough members in the pool, and there were not enough projects subscribing to the concept. The implication here is that these problems have to be overcome to make a pure "service" approach work, and—even then—a senior analyst with continuity on each project is vital to mission success for a telecom-active mission. In the future, availability of more effective or integrated telecom software would allow for automation of the routine tasks, requiring only a review of the results by a less senior analyst.

5.8.1.5 Flight Team Co-location, near the MSA. Operating from a separate building by telephone and e-mail would not have worked during the highly interactive primary mission. Co-location reduced sequence-integration/review turnaround-time during iterations. The data was accessible only behind the firewall in the MSA. Interruptions were a resultant co-location cost in individual-analyst efficiency. These results underscore the lesson that the planning portions of the link-analysis task are project-dedicated flight-team functions, not a generic task.

5.8.1.6 Effective, Easy-to-Use Data Displays. DS1 made a formal agreement with the DSN to provide access in the MSA to operate (via graphical user interface) the workstations of the DSN's Network Operations Control-Center Real-Time (NOCC RT) System. This allowed a telecom analyst or the ACE to resolve configuration and bandwidth problems. DS1 demonstrated the need for rapid access to station configuration and performance data for telecom support of any deep space mission. To meet this need, the AMMOS system has subsequently developed DMD tabular and plot displays that access data from station monitor data (called "mon0158"). Telecom analysts on a variety of deep space missions use these displays when operating in the mission support area.

5.8.1.7 Querying Data. The AMMOS query system has limitations, particularly for even a few telemetry channels over long durations. Consequently, to complete the technology validation required 2-1/2 weeks of senior-analyst time for queries of KaPA performance and configuration back to the time of the in-flight characterization (December 1998). The set of tools, DMD/EZquery/Oplot is capable, but it works easily only for fairly recent data (of a few weeks' duration). (This problem has been eased in recent missions by use of alternative AMMOS query tools and improved yet again by use of the post-AMMOS data processors.)

5.8.1.8 Telecom-Sequencing Standardization. In DS1, configuration of the onboard telecom subsystem for the beginning of a station pass and after the end of a station pass was made standard by using sets of commands created by spacecraft blocks. Likewise, the configuration of the station to set up for and conduct a tracking pass was through the use of a DSN keywords file (DKF). Early in operations, DS1 analysts spent much time re-reviewing sequence products and even editing DKFs manually. Dual lessons here are to (a) publish sequence-generation guidelines early and stick to them, and (b) modularize for reuse the sequence elements at several levels higher than individual commands.

5.8.1.9 Need for an as-Flown Sequence. On DS1, we needed both a good one-page-per-day spaceflight operations schedule (SFOS) for planning and a good as-flown sequence of events (SOE) for problem analysis and technology validation. On subsequent projects, with standardized sequencing processes, there were fewer changes between plans and execution. We learned that the "as flown" products could be replaced by the excellent query and analysis tools.

5.8.1.10 Simultaneous Update. Selection of TFP (over the spreadsheet-based predictor that is still used for pre-launch telecom link design) was the correct choice for the solid tool needed for the operational environment, but it was new software. In the several months before and after launch, there were several deliveries of TFP as well as the need for individual "add path" telecom models.[21] As much as several hours of analyst time are needed to verify the correctness of each add-path update and several days to verify a formal TFP delivery before use in project-mission planning. The nonlinearity in the flight-engineering model SDST X-band modulator [21] required 1-1/2 months for users to determine how to model and implement in TFP. In a dozen projects since DS1, most of these TFP-development "growing pains" became less severe for in-development projects that adapted the same tool. The DS1 lesson is that the project-development plan needs to provide for a sufficient adaptation, debug, and shakedown effort for a new link-analysis tool ensemble.

5.8.1.11 Service Packages and DSN Keyword Files. As part of a ground-system "new technology," DS1 worked with the DSN to define the SP for both the telecom-configuration input to project-sequence planning and the project input to the new Network Planning and Preparation (NPP) system. An SP was a computer-generated listing of the services required of the DSN by a project. Services might include the station transmitting an uplink; commanding, transmitting, and receiving Doppler and ranging data; and receiving and decoding telemetry on the downlink. The Service concept was intended to replace an older station-configuration-management input called the DSN Keyword File (DKF).

A DKF is a time-ordered series of standardized information items or directive items. Each of these items has a defined stem (the "keyword") and may have parameters. A typical spacecraft information keyword is "S/C TLM

---

[21] See Ref. 19 for a description of TFP, including the "add path" capability. The name refers to maintaining the officially-delivered TFP spacecraft and station models in specific computer directories, then establishing updated models, as needed, in separate directories. TFP allows the user to specify that a TFP run use the updated model by adding its directory path (for short, an "add path" to the run instructions. For example, when many station performance parameters and models changed as the interface document 810-5 was updated to its current Rev. E in January 2001, the DS1 telecom analyst used an add path until an updated TFP could be delivered to the project.

For a list of 810-5 revisions from the first on-line version in 1996 to the present, see http://deepspace.jpl.nasa.gov/dsndocs/810-005/fm.pdf

Table 5-6. Example Dawn DKF keywords and definitions.

! 203 65 00299 209 163912 NMC D65	S/C TLM X-MOD, 40T6, 25.000, 53.0DG,
! 203 65 00300 209 163912 NMC D65	D/L ACQ, 2-WAY, X/CH. 29

From left to right, the fields in these typical keywords carry the values

Spacecraft ID = 203 (Dawn)

Tracking station ID = 65 (34m station at Madrid)

Row number in the DKF = 00299 (and 00300)

Day of year and UTC = 209 16:39:12

Tracking station ID = D65 (a redundancy)

Telemetry bit rate = 40 bps

Coding = Turbo 1/6 (T6)

Telemetry subcarrier frequency = 25 kHz

Telemetry modulation index = 53 deg.

Acquisition mode = 2-way

Acquisition band = X

DSN channel number = 29

X-MOD" with parameters that define the bit rate, subcarrier use and frequency, coding, and modulation index. A typical directive keyword is "D/L ACQ" with parameters specifying the frequency band, channel number, and tracking mode. These two examples refer to spacecraft status defining the X-band modulation conditions, and a directive to the station to acquire the downlink, respectively. Table 5-6 provides examples of these two keywords from a DKF for the Dawn spacecraft in 2014.

The NPP, intended as an automated means for station-configuration management, never became operational. As DS1 launch neared, significant concern about the NPP development resulted in a TMOD/project agreement for the project to use the older DKF station-configuration-management input. The project used SPs for telecom input to the sequencing process, as planned. Telecom time was lost before launch to develop and test the SP as an NPP input. With hindsight, a telecom interface for project sequence input only, could have been less complex than the SP. The late changeover to the DKF interface resulted in the need for several updates to the DKF-generation software through flight, and still a need for some hand-editing and a review of DKF outputs.

Numerous lessons that came from the DS1-SP experience were applied to later projects to make their data-rate-capability generation, configuration trade-off, and telecom-planning input processes and tools simple to use and easy to modify. On the DSN side, configuration codes (which existed for DS1) are an integral part of scheduling passes. The configuration codes define the services (such as transmitter, command, and telemetry) for each pass. At the other end of the sequence-generation process, DKF generation has become automated through project-sequencing tools and input to the DSN through a successor to the NPP.

5.8.2 Project-Level Lessons Learned

In December 2000, the former and current DS1 project managers and the spacecraft-development system manager presented the "lessons learned" to JPL [1]. The presentation included a mission summary, a discussion of the mission-success criteria, and the spacecraft-system-development schedule to place things into context.

Following one page titled, "Why did DS1 accomplish so much?" and another two titled, "What worked well?" there are 12 pages headed, "What didn't work well and why?"

Here, from that presentation, is a summary of the DS1 project-level lessons learned. Without the inclusion of the events that motivated each lesson, these can be taken more as a checklist for deep-space project operations in the future.

5.8.2.1 Project Management

- A project needs at least a year for Phase A/B, culminating in a review to ensure the mission concept is sound, the requirements are agreed to, and there are sufficient resources to do the job.
- During the early project phases, define phasing of funding, need dates for the launch vehicle, requirements and success criteria, etc., and do not proceed with further commitments until the entire project is better understood and agreement is in place with NASA Headquarters.

5.8.2.2 Organization-and-Team Dynamics

- The team is the most important factor in mission success.
- An unambiguous organization, adequate resources and the right environment are essential to allow the team to succeed. It is critical to have adequate resources to allow the team to do their job in a humane way.

5.8.2.3 Reviews

- Peer Reviews add the most value.
- Set up a peer-review plan early and get line-management support. Make sure the industry partner buys into the peer-review process.

5.8.2.4 Advanced Technologies

- Develop a technology plan during formulation that addresses risk-mitigation and technology readiness. Include meaningful technology-readiness gates to assess development progress, include clear-action plans if the gates are not met.
- Be cautious about having one technology rely on another for testing.
- If technologies are coupled, treat the independent technology as critical.

5.8.2.5 Communications and Data Transfer. Require data transfer to occur at the technical level, without intermediaries.

5.8.2.6 Assembly, Test, and Launch Operations. Include adequate margin in development schedules, particularly for technology development. Develop an Assembly, Test and Launch Operations plan that is resilient to late deliveries.

5.8.2.7 Operations

- Resources (personnel and schedule) need to be made available in order to allow spacecraft-and payload-team participation early in operations planning.
- Allocate time to allow development personnel to complete integration and test activities and to prepare for mission operations.

5.8.2.8 Contingency Procedures. Develop contingency procedures and update them during development and operations as new information makes them obsolete.

5.8.2.9 Operations-Test-Bed Environment

- If an activity is important and uncertain enough to test in a test bed, then require all subsystems with major involvement in it to review the test results.
- The test-bed configuration should be as flight-like as possible, and differences must be completely understood by the operations team.

5.8.2.10 Single-String Teams

- Build in human redundancy.
- Allocate funds for training and mentoring. Identify this as a major risk if the budget will not allow additional staffing.

5.8.2.11 External Communications. Define and maintain clear lines of communication to management and to the news media. Communicate the probable outcome of critical events and their impact clearly to JPL, Headquarters, and the media.

5.8.2.12 Science in a Technology-Validation Mission. Speak clearly as a project, with one voice, to ensure that external expectations match priorities. Project, JPL, and Headquarters must be in agreement on mission-success criteria.

References

[1] D. H. Lehman, L. Livesay, M. Rayman, and P. Varghese, *Experiences in Managing a "Faster, Better, Cheaper" Deep Space Project, PMI 2000 Symposium*, Houston, Texas, September 7, 2000, Project Management Institute, Inc., Newton Square, Pennsylvania. http://trs-new.jpl.nasa.gov/dspace/bitstream/2014/14339/1/00-0766.pdf (accessed February 21, 2013)

[2] Deep Space 1 project websites and web pages, Jet Propulsion Laboratory, California institute of Technology, Pasadena, California (all accessed October 30, 2014):The DS1 project web site, http://nmp.jpl.nasa.gov/ds1/

- Mission description, http://www.jpl.nasa.gov/missions/deep-space-1-ds1/
- Launch press kit, http://www.jpl.nasa.gov/news/press_kits/ds1launch.pdf
- Asteroid flyby press kit, http://www.jpl.nasa.gov/news/press_kits/ds1asteroid.pdf

[3] M. D. Rayman, P. Varghese, D. H. Lehman, and L. L. Livesay, "Results from the Deep Space 1 Technology Validation Mission," IAA-99-IAA.11.2.01, *50th International Astronautical Congress*, Amsterdam, The Netherlands, October 4–8, 1999. http://trs-new.jpl.nasa.gov/dspace/bitstream/2014/17743/1/99-1185.pdf (accessed October 30, 2014)

[4] "Deep Space 1 Says Farewell: News Release," *Near Earth Object* (website), Jet Propulsion Laboratory, California Institute of Technology, Pasadena, California. http://neo.jpl.nasa.gov/news/news124.html (accessed October 30, 2014)

[5] "New Millennium Program Technology Reports" (from DS1 and other spacecraft in the program), Jet Propulsion Laboratory, California Institute of Technology, Pasadena, California. http://ntrs.nasa.gov/search.jsp?R=20060032687 (accessed October 30, 2014)

[6] "Deep Space 1 Mission Status," *2001 New Releases* (website), Jet Propulsion Laboratory, California Institute of Technology, Pasadena, California, September 22, 2001. http://www.jpl.nasa.gov/releases/2001/release_2001_189.html (accessed February 21, 2013)

[7] J. Taylor and P. Ko, "Achieving and Maintaining Deep Space 1 Spacecraft High-Gain Antenna Pointing Control by Data Monitoring

and Immediate Corrective Commanding," *The Telecommunications and Mission Operations Progress Report 42–144, October-December 2000*, Jet Propulsion Laboratory, California Institute of Technology, Pasadena, California, February 15, 2001. http://ipnpr.jpl.nasa.gov/progress_report/42-144/144I.pdf (accessed February 21, 2013)

[8] M. D. Rayman, "The Deep Space 1 Extended Mission: Challenges in Preparing for an Encounter with Comet Borrelly," *Acta Astronautica*, vol. 51, no. 1-9, pp. 507–516, 2002. http://www.sciencedirect.com/science/article/pii/S009457650200070X (accessed February 21, 2013)

[9] *DS1 Project Requirements/TMOD Support Agreement (PR/TSA)*, Version 2, JPL D-13548 (internal document), Jet Propulsion Laboratory, California Institute of Technology, Pasadena, California, October 2, 1996.

[10] *Telecommunication FEM SDST/DSN Performance and Compatibility Motorola Test Report*, Revision A, Andre Makovsky, JPL D-19045 (internal document), Jet Propulsion Laboratory, California Institute of Technology, Pasadena, California, May 1998.

[11] *DSN Telecommunications Link Design Handbook*, 810-005, JPL D-19379, Rev. E, Jet Propulsion Laboratory, California Institute of Technology, Pasadena, California, continual updates. http://deepspace.jpl.nasa.gov/dsndocs/810-005/ (accessed October 30, 2014)

[12] *Digital Time Division Command/Response Multiplex Data Bus, MIL-STD-1553B*, United States Department of Defense, continual updates.

[13] *Fiber Optics Mechanization of an Aircraft Internal Time Division Command/Response Multiltiplex Data Bus*, MIL-STD-1773B, United States Department of Defense, continual updates.

[14] *RS422*, Wikipedia website. http://en.wikipedia.org/wiki/RS-422 (accessed February 21, 2013.)

[15] M. Muñoz Fernández, *Deep Space 1 Mission Support*, Master of Space Studies thesis, International Space University, Strasbourg, France, May 2000.

[16] A. I. Bolea Alamanac, *Deep Space 1 Operations: Lessons Learned*, Master of Space Studies thesis, International Space University, Strasbourg, France, May 2001.

[17] *Deep Space 1 (DS1) Project Network Operations Plan*, JPL D-14824, DSN 871-010-030 (internal document), Jet Propulsion Laboratory, California Institute of Technology, Pasadena, California, January 23, 2001.

[18] K.-M. Cheung, A. Makovsky, and J. Taylor, "Telecommunications Analysis for 'Faster, Better, Cheaper' Deep Space Flight Planning and Operations: The DS1 Experience," *Telecommunications and Mission Operations Progress Report 42-140*, Jet Propulsion Laboratory,

California Institute of Technology, Pasadena, California, February 15, 2000. http://ipnpr.jpl.nasa.gov/progress_report/42-140/140G.pdf (accessed February 21, 2013)

[19] K. K. Tong and R. H. Tung, "A Multimission Deep-Space Telecommunications Analysis Tool: The Telecom Forecaster Predictor," *The Telecommunications and Mission Operations Progress Report 42-140, October-December 1999*, Jet Propulsion Laboratory, Pasadena, California, February 15, 2000. http://ipnpr.jpl.nasa.gov/progress_report/42-140/140C.pdf (accessed February 21, 2013)

[20] D. Morabito, "Solar Corona-Induced Fluctuations on Spacecraft Signal Amplitude Observed during Solar Superior Conjunctions of the Cassini Spacecraft," *Radio Science*, vol. 42, RS3002, May 9, 2007. http://onlinelibrary.wiley.com/doi/10.1029/2005RS003425/pdf (accessed April 10, 2014)

[21] A. Makovsky, "New Telecom Loss Models for DS1," JPL interoffice memorandum 3312-98-09, D-93725 (internal document), Jet Propulsion Laboratory, Pasadena, California, June 18, 1998.

[22] D. H. Lehman, L. Livesay, M. Rayman, and P. Varghese, *Experiences in Managing a "Faster, Better, Cheaper" Deep Space Project, PMI 2000 Symposium*, Houston, Texas, September 7, 2000. http://trs-new.jpl.nasa.gov/dspace/bitstream/2014/14339/1/00-0766.pdf (accessed February 21, 2013)

Additional Resources

[1] M. I. Herman, S. Valas, W. Hatch, C.-C. Chen, S. H. Zingales, R. P. Scaramastra, L. R. Amaro, M. D. Rayman, "DS1 Telecommunication Development," SSC98-IV, *Utah Small Satellite Conference*, September 1, 1998.

[2] *Basics of Space Flight*, Jet Propulsion Laboratory, California Institute of Technology, Pasadena, California. This is a training module designed primarily to help JPL operations staff identify the range of concepts associated with deep-space missions and grasp the relationships these concepts exhibit for space flight. It also enjoys growing popularity among high school and college students, as well as faculty and those everywhere who are interested in interplanetary space flight. Its major sections are Environment, Flight Projects, and Operations: http://solarsystem.nasa.gov/basics/index.php (accessed February 21, 2013)

[3] Solar activity and space weather information:

- http://www.spaceweather.com/ (accessed February 21, 2013)
- http://www.nrl.navy.mil/accomplishments/solar-lunar-studies/lasco/ (accessed February 21, 2013)
- http://www.swpc.noaa.gov/communities/space-weather-enthusiasts (accessed February 21, 2013)

Chapter 6
Mars Reconnaissance Orbiter

Jim Taylor, Dennis K. Lee, and Shervin Shambayati

6.1 Mission Overview

The Mars Reconnaissance Orbiter (MRO) [1,2] has a suite of instruments making observations at Mars, and it provides data-relay services for Mars landers and rovers. MRO was launched on August 12, 2005. The orbiter successfully went into orbit around Mars on March 10, 2006 and began reducing its orbit altitude and circularizing the orbit in preparation for the science mission. The orbit changing was accomplished through a process called aerobraking, in preparation for the "science mission" starting in November 2006, followed by the "relay mission" starting in November 2008. MRO participated in the Mars Science Laboratory touchdown and surface mission that began in August 2012.

MRO communications has operated in three different frequency bands:

1) Most telecom in both directions has been with the Deep Space Network (DSN) at X-band (~8 GHz), and this band will continue to provide operational commanding, telemetry transmission, and radiometric tracking.

2) During cruise, the functional characteristics of a separate Ka-band (~32 GHz) downlink system were verified in preparation for an operational demonstration during orbit operations. After a Ka-band

Some of the grey scale figures in this book have color versions in the companion website www.wiley.com/go/deepspacecommunications. Please refer to this website where the text refers to colors.

Deep Space Communications, First Edition. Edited by Jim Taylor.

hardware anomaly in cruise, the project has not yet elected to initiate the originally planned operational demonstration (with yet-to-be-used redundant Ka-band hardware).

3) A new-generation ultra-high frequency (UHF) (~400 MHz) system was verified with the Mars Exploration Rovers in preparation for the successful relay communications with the Phoenix lander in 2008 and the later Mars Science Laboratory relay operations.

Lockheed Martin Space Systems, Denver, Colorado, is the prime contractor for MRO. They built the spacecraft and have provided flight operations support during the mission. The Jet Propulsion Laboratory (JPL), Pasadena, California, manages the project for the National Aeronautics and Space Administration (NASA), Washington, D.C. The Flight Team is located at both Lockheed and the Jet Propulsion Laboratory. Refer to the Mars Reconnaissance Orbiter home page [1,2] for current MRO information.

6.2 Mission Phases and Orbit Summary

6.2.1 Mission Objectives

The Mars Reconnaissance Orbiter (MRO) mission has the primary objective of placing a science orbiter into a low and near-circular Sun-synchronous Mars orbit to perform remote sensing investigations to characterize the surface, subsurface, and atmosphere of the planet and to identify potential landing sites for future missions. The MRO payload conducts observations in many parts of the electromagnetic spectrum, including ultraviolet and visible imaging, visible to near-infrared imaging spectrometry, thermal infrared atmospheric profiling, and radar subsurface sounding, at spatial resolutions substantially better than any preceding Mars orbiter.

The driving theme of the Mars Exploration Program (MEP) is to understand the role of water on Mars and its implications for possible past or current biological activity. The MRO is studying the history of water on Mars. Another Mars mission, the Mars Exploration Rover (MER), has shown that water flowed across the surface in Mars' history. The MRO is searching for evidence for when the water was on the surface and where it is now, and any indicators of whether water persisted on the surface of Mars long enough to provide a habitat for life.

In terms of telecommunications (telecom), the MRO mission

- Provides X-band (~8 GHz) uplink (command), downlink (telemetry), and navigation (two-way Doppler, turnaround ranging, and differential one-way ranging) with the Deep Space Network (DSN). The direct-from-Earth (DTE) uplink can also carry data intended for relay to a surface vehicle, and the DTE downlink can also carry data relayed to MRO from a surface vehicle.

- Provides ultra-high-frequency (UHF) data relay and navigation support services to landing MEP missions during their entry, descent, and landing (EDL) phase, and subsequently provide UHF two-way relay and navigation services to landed surface vehicles or to other orbiting spacecraft, for example, a sample-return canister waiting for pickup and return to Earth.
- Perform an operational demonstration of high-data-rate Ka-band (~32 GHz) downlink telecommunications and navigation services (using the X-band uplink) with the DSN.

6.2.2 The MRO Spacecraft

The MRO uses a new spacecraft bus design provided by Lockheed Martin Space Systems Company, Space Exploration Systems Division, in Denver, Colorado.

The X-band antennas for communication with the DSN are at the top in Fig. 6-1. Of the two low-gain antennas (LGAs) that are fixed-mounted to the high-gain antenna (HGA), LGA1 is called forward-facing because it is pointed in the same general direction as the gimbaled HGA. The other LGA, LGA2, points generally in the opposite direction. The UHF antenna that is used for communicating with surface vehicles is aligned with the +z axis,

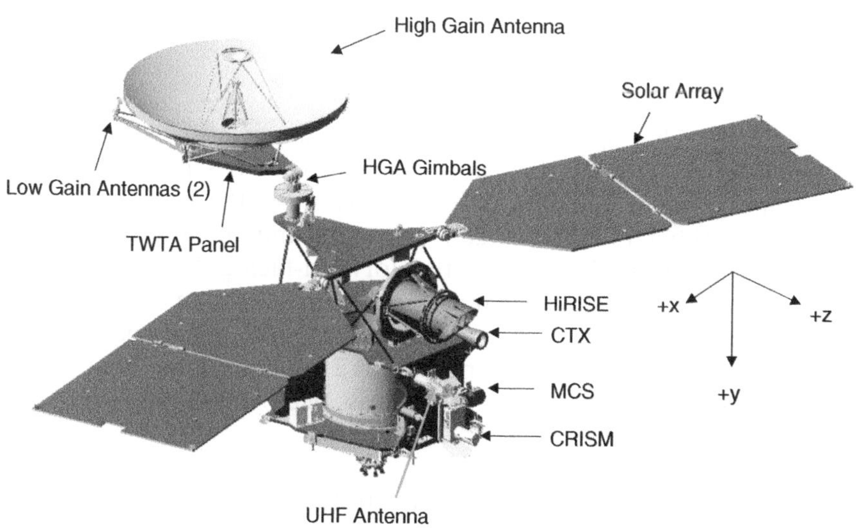

Fig. 6-1. Sketch of the MRO spacecraft with coordinate directions. Please visit www.wiley.com/go/deepspacecommunications to access the color version of this figure.

which is also the science instrument boresight. Throughout the science and relay operations phase, this axis is usually oriented vertical toward Mars.

The orbiter payload consists of six science instruments and three new engineering payload elements listed as follows:

- Science instruments
  - HiRISE, High Resolution Imaging Science Experiment
  - CRISM, Compact Reconnaissance Imaging Spectrometer for Mars
  - MCS, Mars Climate Sounder
  - MARCI, Mars Color Imager
  - CTX, Context Camera
  - SHARAD, Shallow (Subsurface) Radar
- New engineering payloads
  - Electra UHF communications and navigation package
  - Optical Navigation Camera Experiment (ONC)
  - Ka-band Telecommunications Experiment

Figure 6-1 is a sketch showing the major externally visible parts of the spacecraft.

6.2.3 Mission Phases

In order of occurrence, the six phases of the MRO primary mission include launch, cruise, approach and orbit insertion, aerobraking, primary science, and relay [3]. The following paragraphs provide overviews of the spacecraft activities in each phase. Specific telecom calibrations and activities are described in subsequent sections.

6.2.3.1 Launch. The spacecraft was launched on August 12, 2005. Approximately 58 minutes after launch, the spacecraft separated from the launch vehicle. About 4 minutes prior to separation, the X-band traveling-wave tube amplifier (TWTA) began warm-up, and the spacecraft began transmitting a downlink through the forward-facing low-gain antenna (LGA1) about 1 minute after separation. MRO remained in a single inertial attitude throughout the launch period. By 14 minutes after separation, the craft's solar panels finished unfolding. About 21 minutes after separation, in order to avoid interfering with solar array deployment, the HGA was deployed from the stow position.

The spacecraft established radio contact with Earth 61 minutes after launch and within 4 minutes of separation from the upper stage. Initial downlink-only contact came through an antenna at the Japan Aerospace Exploration Agency's Uchinoura Space Center in southern Japan.

When MRO came into view at the Goldstone, California, DSN site, a 34-m station established an X-band uplink with the MRO receiver. The uplink carrier provided a reference for two-way Doppler and turnaround ranging on the downlink, as well as establishing commandability.

The ultra-stable oscillator (USO) in the telecom subsystem was turned on within hours after launch so that the one-way downlink frequency would be stable prior to cruise phase activities. The USO has operated continuously except for a few hours during safe mode in January 2006.

6.2.3.2 Cruise. The cruise phase began about 3 days after launch and ended 60 days prior to Mars orbit insertion (MOI). The duration of the cruise phase was approximately 150 days.

Early cruise included an in-flight UHF antenna gain pattern measurement in conjunction with the 150-foot (46-m) diameter "Stanford Dish" radio telescope [23]. The receive and transmit gain measurements were made at the 400-MHz relay operations frequencies with the nadir deck (instrument and UHF boresight) pointed back toward the Earth.[1]

The Type-I interplanetary trajectories for MRO did not have transfer-angle constraints or other singularities, and so the trajectory correction maneuver (TCM) timing was based primarily on operational considerations and standard practice [22]. TCM-1 included firing the six main (170-newton, N) thrusters for 15 seconds on August 27, 2005. This engine burn followed a 30-second burn of six smaller (22 N) thrusters, which settled propellant in the craft's fuel tank for smoother flow. With communications on the LGA, MRO's orientation was adjusted prior to the burns to point the engines in the proper direction for the maneuver, and the spacecraft returned to cruise-phase attitude after the trajectory adjustment. Besides putting MRO on course for the Mars target point, TCM-1 checked out the engines that would be required for MOI.

Instrument payload calibrations began on August 30. The higher-resolution cameras were pointed at the Earth and the Moon as the spacecraft continued its flight to Mars.

---

[1] Pre-launch measurements could not be nearly so accurate because the spacecraft was not in its final flight configuration and because of antenna interactions outside the spacecraft. At 400 MHz, any electromagnetic conducting surfaces on the spacecraft parasitically couple with the UHF LGA forming a composite antenna pattern. This parasitic coupling would also occur on the ground with any metallic test equipment surrounding the spacecraft. The Stanford Dish test was a "first" in terms of an in-flight measurement of this type.

TCM-2, on November 18, 2005, used only the smaller TCM thrusters in a 20-second burn. Two other TCMs built into the mission plan [3] were not required.

6.2.3.3 Approach and Mars Orbit Insertion. Following the interplanetary cruise and Mars approach phases of the mission, the MRO achieved MOI on March 10, 2006. The MOI burn fired the craft's main thrusters for about 27 minutes to reduce velocity by about 20 percent as the spacecraft swung around Mars at about 5 kilometers per second (km/s), or about 11,000 miles per hour (mph).

The initial post-MOI orbit started from the MOI aim point of 360 km above the Martian surface, approaching from the south. As an example of how flight dynamics in the form of trajectory design works, Mars actually caught up to the slower moving MRO spacecraft at this point. After launch, MRO spiraled out from Earth to Mars, and Mars caught MRO from behind as the MRO radial motion around the Sun was slower than the motion of Mars at time of MOI.

After MOI and before the aerobraking phase began, the orbiter flew about 426 km (265 miles) above Mars' surface at the nearest point (periapsis) of each orbit, then swung out more than 43,000 km (27,000 miles) to the most distant point (apoapsis) before heading in again. The initial orbit period was about 35 hours.

After MOI, while preparing for aerobraking, the flight team tested several instruments, obtaining the orbiter's first Mars pictures and demonstrating the ability of its Mars Climate Sounder instrument to track the atmosphere's dust, water vapor, and temperatures.

6.2.3.4 Aerobraking. In aerobraking, the trajectory design deliberately causes the spacecraft to pass through the upper reaches of the Mars atmosphere on each periapsis pass. The atmospheric friction acts as a velocity brake, and each such pass lowers the apoapsis altitude at the other end of the orbit. After the correct apoapsis altitude is attained, the aerobraking phase is ended with a periapsis raise maneuver performed to bring periapsis out of the atmosphere. Aerobraking is made more complicated because the Martian atmospheric density as a function of altitude and latitude also varies with time. Infrared-sensing instruments and cameras on two earlier Mars orbiters (Mars Global Surveyor and Odyssey) were the main sources of information to the advisory team of atmospheric scientists, providing day-to-day data about variations in Mars' atmosphere as the aerobraking campaign continued. In addition, the Mars Climate Sounder instrument on MRO itself provided data to monitor changes in temperature that affected the atmosphere's thickness.

Aerobraking began on March 30, 2006 and ended August 30, 2006. An initial propulsive maneuver firing of the 22-newton (N) thrusters for 58 seconds at apoapsis put the MRO spacecraft into an active aerobraking orbit. That apoapsis maneuver lowered the periapsis altitude to 333 km (207 miles). The aerobraking phase required 445 orbits of carefully calculated dips into Mars' atmosphere. Aerobraking and a phasing maneuver on September 5 shrank its orbit from the post-MOI elongated ellipse to a more nearly circular orbit.

Aerobraking ended with MRO in a slightly elliptical low-altitude Sun-synchronous orbit, called the science orbit. After a successful final circularization, propulsive periapsis-raise, maneuver on September 11, 2006, the science orbit has a period of 1 hour and 52 minutes, with an apoapsis of 316 km over the North Pole and a periapsis of 250 km over the South Pole [1, 2].

Solar Conjunction: Between the end of aerobraking (with the primary science orbit established) and the start of the primary science mission phase was a solar conjunction. Defined as the time period when the Sun–Earth–Mars angle is 5 degrees (deg) or less, this first solar conjunction was from October 7 to November 8, 2006. The Ka-band communications demonstration was planned to conduct activities during conjunction to monitor and compare simultaneous X- and Ka-band telemetry downlinks. The DSN supported one 8-hr pass per day to a 34-m antenna during this period.

Solar conjunctions of Mars have a periodicity of about 26 months, and the Earth–Mars range is very nearly at maximum when the Sun–Earth–Mars angle is minimal at conjunction. The Sun–Earth–Mars geometry at conjunction causes several communications challenges. As the Sun–Earth–Mars angle decreases below about 5 deg, the communications signal passes through an increasing amount of solar plasma, which causes non-linear scintillation on the signal. In addition, the background noise from the Sun itself reduces the received signal-to-noise ratio (SNR) at the DSN. Finally, the Sun itself subtends 0.5 deg in the sky, as viewed from Earth, and can completely block the line-of-sight signal path to the MRO spacecraft if the Sun–Earth–Mars angle falls below 0.25 deg. Because conjunction is established by Mars–Earth–Sun geometry, all orbiters and landers at Mars have the same conjunctions. Table 6-1 gives the dates of Mars solar conjunction for 2004 through 2021. Proximity communications with surface vehicles would not be directly affected.

6.2.3.5 Primary Science Mission. During the science phase, MRO examined parts of the planet in detail and monitored the entire planet daily throughout a full cycle of Martian seasons, or 669 sols for one Mars year. The duration of this phase was about two Earth years (November 2006 to November 2008, starting with the end of aerobraking. The science experiments consisted of global mapping of Mars' surface, regional surveys for potential future Mars landing

Table 6-1. Mars solar conjunction dates and minimum SEP angles (2004–2021)*.

Date of Minimum SEP Angle	SEP Angle (deg)	Date of Minimum SEP Angle	SEP Angle (deg)	Date of Minimum SEP Angle	SEP Angle (deg)
09/15/2004	0.96	02/04/2011	1.08	07/27/2017	1.10
10/23/2006	0.39	04/18/2013	0.40	09/02/2019	1.08
12/05/2008	0.46	06/14/2015	0.62	10/08/2021	0.65

* Data received from Ref. 4 and personal communication from David Morabito, 12/08/2010.

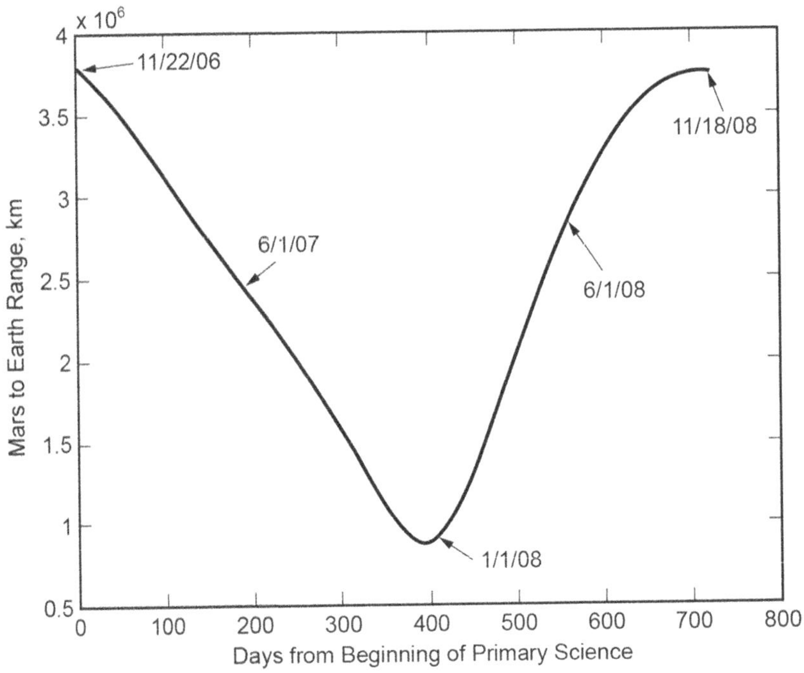

Fig. 6-2. Mars-to-Earth range during the primary science phase.

sites, targeted observations of areas of interest, and mapping of the Mars gravity field. The primary science mission ended with the onset of solar conjunction.[2] Figure 6-2 shows the Mars-to-Earth range in the primary science phase.

[2] While the primary science phase was planned to end in 2008 after one Martian year, NASA approved the continuation of science observations beyond the primary science phase until 2010, the end of the next major phase, the relay phase. NASA subsequently approved several additional mission extensions for science and relay operations. The latest MER extended mission continues into January 2016. http://mars.jpl.nasa.gov/mro/mission/timeline/

At the beginning of the science phase, Mars was about one-third of the way through a Northern Hemisphere summer. Throughout the phase, the orbiter generally kept its instruments pointed at Mars to collect data and its high-gain antenna pointed at Earth to send the data home. During this phase, conducting science observations was more complex than in previous Mars missions, because MRO had to coordinate three basic observation goals:

- Daily global mapping and profiling
- Regional surveys
- Globally distributed targeting of hundreds of specific sites.

Many targeted observations also involved nearly simultaneous, coordinated observations by more than one instrument.

During this phase, primary communications was through the HGA to a 34-m DSN station. Precise Doppler measurements were taken to aid the gravity science experiments. Several times a day, the orbiter pointed to an off-nadir target for high-resolution imaging for about 15 minutes. During these slews, the HGA pointing error increased, but communications with Earth were still possible.

During the primary science mission, the DSN allocation to MRO was two 34-m passes at X-band per day, plus three 70-m passes per week.[3]

6.2.3.6 Relay Mission. Beginning 6 months before the end of the primary science mission in December 2008 and continuing until the end of the MRO primary mission in December 2010, the Electra payload provided relay support to various Mars assets. During this relay phase, the Jet Propulsion Laboratory (JPL) Mission Management Office (MMO) was chartered to coordinate relay services between Martian surface assets and MRO. The coordination plan was based on a 4-week planning cycle for relay coordination, with weekly updates for ad hoc relay opportunity assignment. The relay mission continues the support of MER and also has supported two other spacecraft arriving at Mars and descending to the surface:

Phoenix: Launched in August 2007, the Phoenix Mars Mission was the first in NASA's Scout Program. Phoenix studied the history of water and habitability potential in the Martian Arctic's ice-rich soil. The solar-powered Phoenix lander operated for 2 months longer than its planned 3-month mission in the Martian arctic in 2008. During the surface mission, the MRO

---

[3] MRO has the highest rate Earth link telecom system for any planetary mission to date, going as high as 6 Mbps at closest distance Mars-Earth range. MRO has been collecting science data for over 9 years and so far has imaged about 2 percent of the planet surface with the high resolution camera.

mission plan stated that the Phoenix project would request two to three relay contacts daily with MRO's Electra at rates as great as 128 kilobits per second (kbps).

Mars Science Laboratory: The Mars Science Laboratory (MSL) is discussed in Chapter 8. MRO began has providing data for MSL site selection since 2009, and is scheduled to continue providing data through 2012. Since MSL landing in August 2012, MRO has received the bulk of MSL's science data via UHF relay and returned it to Earth by its X-band link.

At MSL arrival at Mars in 2012, MRO received one-way Doppler during entry, descent, and landing (EDL), and it received two-way Doppler for post-EDL reconstruction. After the MSL landing, MRO/Electra has been prime (with the Odyssey orbiter backup) for the surface-orbiter proximity communications relay, providing navigation and timing services, as well as forward- and return-link relay services.

For forward-link relay events, MRO has allocated space on the solid-state recorder (SSR) to store and forward up to 30 megabits per day (Mbits/day). For return-link events, the allocation is 5 gigabits per day (Gbits/day) for all landers. The MRO ground system has its own requirements for maximum data volume and data latency for data relayed from each lander during the primary science phase.

Figure 6-3 shows the activities planned to be performed during a typical relay session. Relay sessions between MRO and a surface asset are initiated by MRO. All information can be transferred via a reliable link—the Proximity-1 protocol (Prox-1). In outline, at the time of the overflight, MRO hails the surface asset. Once the surface asset has responded, the session begins. Once all the data has been transferred or the overflight is about to end, MRO terminates the link. If no scheduled termination time is forced, the link drops out due to geometric constraints, forcing a hard link termination.[4] The link session is later closed out by MRO Electra via the time out of a loss-of-lock event timer.

---

[4] In actual MRO/MSL operations, the geometric impact of lower signal levels or complete line-of-sight blockage is always the factor that ends data transfer. When data transfer halts, the MRO Electra continues to hail the MSL Electra to try to restore communications and pick up any last bits of information. Loss of carrier lock actually initiates this rehail procedure. This continues until the larger relay session itself times out. At this point the relay session is closed out, and the final block of relay data is passed from the Electra radio to the MRO SSR. There is a maximum rehail counter, but the threshold is set so high, 120 rehails at about 12 seconds per rehail cycle, that the effect is to have the Electra rehail until the programmed relay session ends, which is past the time that MRO sets below the horizon.

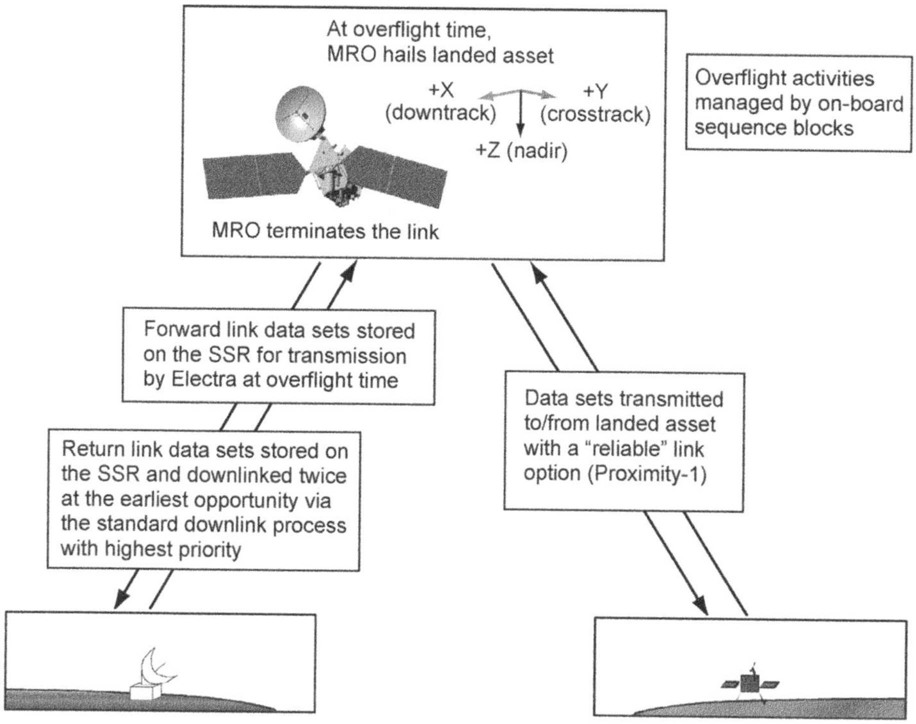

Fig. 6-3. A typical sequence of activities during a relay session.

Return-link data is downlinked at X-band from MRO to Earth at the earliest opportunity. The return-link data has the highest priority, and each frame will be sent twice.[5]

Transition from Primary Mission to Extended Mission: The nominal end of the MRO primary mission, which concluded with the extended science phase, was September, 2010. Through its mission extensions, MRO has continued both science and relay operations. The science goals include work on the nature and history of the Martian upper crust and on the polar caps and layered terrains and ice at all latitudes, as well as on atmospheric inter-annual variability. To support landing missions beyond MSL, in addition to providing UHF relay capability, MRO will provide data for landing site selection and for atmospheric characterization at the times of landing and surface operations [5].

6.2.3.7 Safe Mode. Safe mode provides a known, stable spacecraft configuration in case of a spacecraft anomaly. Safe mode may be entered via

[5] The MSL primary mission relay-data completeness requirements levied on MRO caused the project to respond by applying one retransmission of all relay data to Earth. The X-band downlink is not protected by a protocol like Proximity-1.

command (for example, for a flight software reboot) or from fault protection during any mission phase.

When MRO is configured in safe mode, LGA1 is boresighted at Earth, and the solar arrays are Sun-pointed. The spacecraft –y-axis tracks the Sun. Onboard Sun and Earth ephemerides that were loaded before launch are used to determine the Sun–probe–Earth (SPE) angle upon entry into safe mode and are used to point the HGA such that the forward-facing LGA1 boresight is pointed generally at Earth. The star trackers can be used to help with Sun acquisition if the spacecraft attitude knowledge is not good.

If the star trackers are not functioning and attitude knowledge is limited to that from Sun sensors, the spacecraft will rotate about its –y-axis (which in safe mode is pointed at the Sun) with a period of one hour for most mission phases. The rotation will cause the LGA1 boresight relative to the Earth to trace a cone of approximately half the SPE angle. As a result, the DSN station will observe a repeating power-level profile that depends on the SPE and LGA pattern.

In safe mode, the default USO is powered on. The X-band telecom transmit and receive paths are via LGA1. In safe mode, the command bit rate is set to 7.8125 bits per second (bps), and the X-band telemetry bit rate is set to 34.4 bps with (7,1/2) + Reed–Solomon (interleaving depth, $I = 1$) encoding. The short frame length reduces frame acquisition time at the station.

Further actions in safe mode ensure that the Ka-band TWTA is powered off and the small deep-space transponder (SDST) Ka-band exciter is turned off. The fault protection software also safes the Electra UHF transceiver (EUT).

6.2.4 The MRO Orbit and Its Relay Coverage for Surface Vehicles

MRO and Odyssey are the two NASA orbiters with Proximity-1 relay communications capability. Their orbits are Sun synchronous. Each time the orbiter crosses over the Martian Equator from south to north, the mean local solar time (LST) at the ground directly below is 3:00 p.m. (for MRO) or 5:00 a.m. (for Odyssey).

Table 6-2 [6] shows the orbit elements and related data for MRO and Odyssey. The MRO relay coverage defined in the three figures that follow is based on these values. Figures 6-4 through 6-6 define geometric coverage conditions between MRO and a surface vehicle as a function of the Martian latitude of the surface vehicle. The figures are based on composite statistics averaged over longitude and reflecting the maximum, average, or minimum over a 24-sol simulation using the Telecom Orbit Analysis and Simulation Tool (TOAST) [7].

Table 6-2. Orbit elements for MRO and Odyssey.

Orbit Element	MRO	Odyssey
Periapsis radius (km)	3624.4	3766.1
Apoapsis radius (km)	3691.1	3839.5
Semi-major axis (km)	3657.7	3802.8
Eccentricity	0.0091	0.0096
Inclination (deg)	92.6	93.1
Ascending node (deg)	–14.7	–159.8
Perigee argument (deg)	–78.8	–83.7
Time from perigee (s)	–1818.8	–1423.8
Epoch	2008-147T01:00:00	2008-147T01:00:00
Related data	**MRO**	**Odyssey**
Periapsis altitude/location	255 km/south pole	370 km/South Pole
Apoapsis altitude/location	320 km/north pole	444 km/North Pole
Mean LST, ascending node	3:00 p.m.	5:00 a.m.
Mean LST, descending node	3:00 a.m.	5:00 p.m.
Orbit period	1 hr 52 min	1 hr 58 min

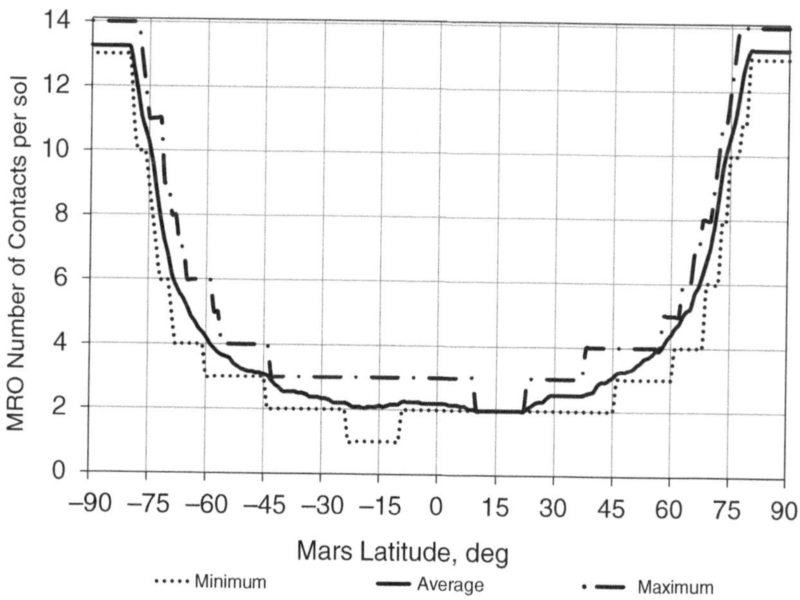

Fig. 6-4. Maximum, average, and minimum number of contacts per sol versus latitude of the lander for MRO orbit.

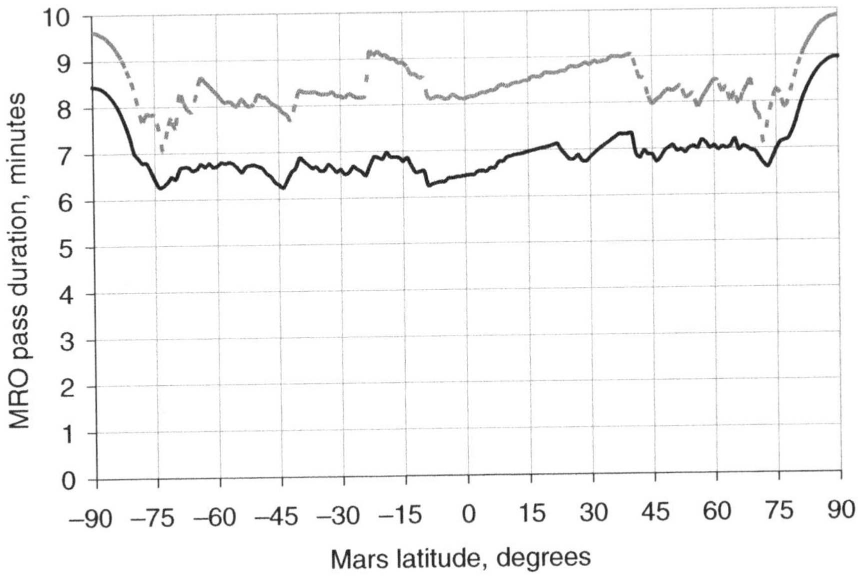

Fig. 6-5. Maximum (top) and average (bottom) pass duration versus Mars latitude for MRO orbit.

Figure 6-4 shows the number of contacts (lasting at least 1 minute above 10 deg). Figure 6-5 shows potential average and maximum MRO pass durations in minutes as a function of landed latitude, assuming a 10-deg minimum elevation angle from the surface. Pass duration is the time the orbiter appears above the minimum elevation angle.[6] Figure 6-6 shows the maximum gap times between potential contacts with MRO. A gap is the duration of time between geometric contact opportunities. In polar locations, for the 1-hour 52-min MRO orbit, the gaps would be about 1-3/4 hours. At some near-equatorial latitudes, there is one contact per sol, resulting in a gap longer than 24 hours.

6.2.5 MRO Orbit Phasing to Support Landing Vehicle EDL

To cover a critical event such as an arriving spacecraft's EDL, MRO can perform an orbit trim maneuver to adjust the orbit phasing (that is, adjust the true anomaly of the orbit). However, MRO does not have the propellant budget necessary to make an orbit plane change (that is, significantly shift the local solar time that MRO crosses the Equator). Orbit phasing moves the

[6] The minimum 10-deg elevation angle and assumed minimum 1-minute pass duration are for illustration. The figure omits minimum pass duration, which is generally not a useful statistic. For a near-circular Sun-synchronous orbit, there will always be a pass geometry that results in near-zero pass time except for surface locations near the poles.

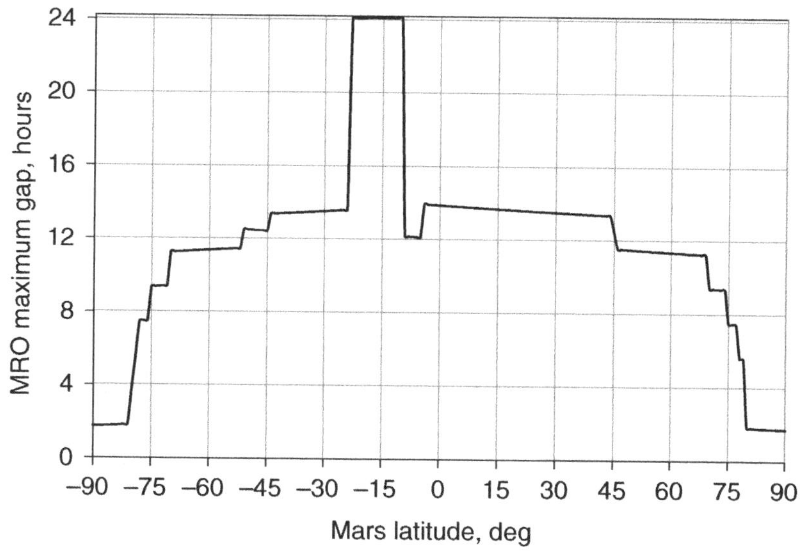

Fig. 6-6. Maximum gap between potential MRO contacts versus Mars latitude.

timing of the orbiter forward or backward in its orbit so that when a spacecraft arrives at Mars the relay orbiter will be in a good orbit position to provide telecom and navigation support for critical events surrounding arrival. Communications during EDL would normally be one-way (return link to MRO only).

The antenna placement on an arriving/descending vehicle and that vehicle's attitudes relative to the orbiter are critical to maintaining communication during EDL. It may be possible to coordinate roll steering of up to ±30 deg to point MRO's antenna to improve EDL coverage.

Plasma outages on the lander–MRO return link during atmospheric entry may occur depending on the entering spacecraft's approach angle and velocity.

6.3 Telecommunications Subsystem Overview

6.3.1 X-Band: Cruise and Orbital Operations

Uplinks to MRO and downlinks from MRO at X-band are the primary means of communication between the MRO and the DSN antennas in California, Spain, and Australia.

The X-band communication system on the orbiter uses a 3-meter-diameter (10-foot) HGA and a 100-watt (W) X-band TWTA to transmit signals to Earth. Each of these devices is more than twice as capable as those used by previous Mars missions. As a result, MRO has been sending data back to Earth more than 10 times faster than previous missions.

At a maximum distance from Earth (400 million km [250 million miles]), the orbiter is designed to send data at a rate of at least 500 kbps. At closer ranges, the signal strength can be greater, so higher data rates are possible. When the orbiter is at its closest ranges (about 100 million km [60 million miles]), for several months the orbiter was planned to be able to send data to Earth at 3 to 4 megabits per second (Mbps).

The MRO project scheduled two 34-m Deep Space Stations (DSSs) daily for an average of 16 hours per day during the science phase. Twice a week, the 70-m antennas were also requested.

With its large antenna, high-powered TWTA, and fast computer, the orbiter has transmitted data to Earth at rates as high as 6 Mbps. This rate is quite high because MRO achieves it while 100 million kilometers from Earth. Over its 2-year primary mission (2006–2008), the spacecraft transmitted more than 73 terabits of science data, about twice what was originally expected. This is about 20 times as much data as previous Mars missions and more data than all previous planetary missions combined. During the first extended science mission (2008–2010), MRO sent down another 53 Tb of science data [5].

From the viewpoint of a DSN antenna on Earth, the orbiter spends about one-third of its time in every orbit behind Mars. During these times, the orbiter is occulted (has no line-of-sight communications path with the Earth) and cannot communicate with the DSN. Out of 16 hours daily that DSN tracking could potentially be scheduled during the orbital mission, MRO actually has sent data to Earth for 10 to 11 hours for more than 700 days. The data rate has averaged between 0.5 and 4 Mbps depending on Earth-Mars distance.

Figure 6-7 is a block diagram of the MRO telecom subsystem. Of the redundant active elements (EUTs, USOs, SDSTs, and X-band TWTAs), only one is powered on at a time.

The subsystem mass and spacecraft power input are summarized in Table 6-3.

The mass values are the totals for both redundant units for the SDSTs, X-band TWTAs, and UHF transceivers. The mass of microwave components, cabling, and waveguides (WGs) not individually called out is summed for the major telecom functional elements.

The project book keeps the HGA gimbals and their drive motors in a different subsystem. However, they are included in Table 6-3 as they would not be on the spacecraft except to direct the HGA to Earth.

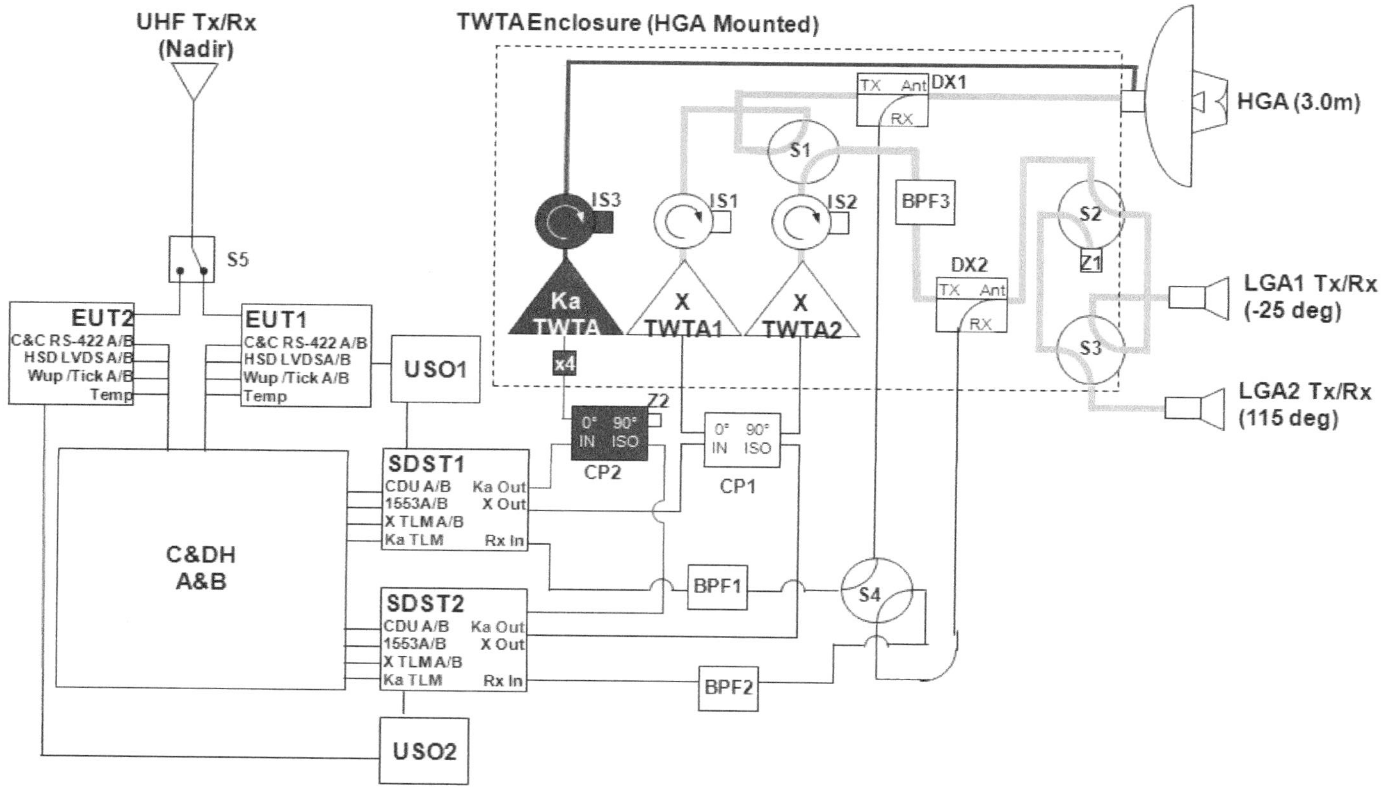

Fig. 6-7. MRO Telecom Subsystem block diagram. Please visit www.wiley.com/go/deepspace communications to access the color version of this figure.

Table 6-3. MRO telecom mass and power summary.

Assembly	Subtotal, kg	Total mass, kg	Spacecraft power input, W	RF power output, W	Note
X-band transponder		6.4	16		Orbit average power
SDSTs (2)	5.8				
×4 frequency multiplier bracket	0.1				
Other microwave components	0.5				
Traveling-wave tube amplifiers		12.1			
X-band TWTAs (2)	1.9		172	102	100 W nominal
Ka-band TWTA	0.8		81	34	35 W nominal
X-band electronic power converters	3.0				
Ka-band electronic power converter	1.5				
Diplexers and brackets	1.8				
Waveguide transfer switches	1.5				
Other microwave components	1.4				
Miscellaneous TWTA hardware	0.2				
X-band and Ka-band antennas		22.6			
HGA prime reflector	19.1				
Antenna feed assembly	1.6				
LGAs and polarizers	0.8				
Miscellaneous antenna hardware	1.1				
HGA gimbals and drive motors		45.0	14		Orbit average power

Table 6-3. (Continued)

Assembly	Subtotal, kg	Total mass, kg	Spacecraft power input, W	RF power output, W	Note
Waveguides and coax		8.3			
USOs (2)		1.7	5		Orbit average power
UHF subsystem		11.5			
Electra transceivers (2) (each transceiver has an integral solid-state RF power amplifier)	10.1		71	5	On, full duplex (17.4 W standby)
UHF antenna and radome	1.4				
String switch (S)	0.1				
Telecom total		**107.7**	**359**		

The X-band system was designed to have no single point of failure (with the exception of the HGA, couplers, and diplexers), and to minimize circuit loss. The coupler (CP) and diplexer (DX) are waived because the probability of failure of these components is very low. Both are passive radio frequency (RF) components with no moving parts and no electronics.

X-Band Microwave Elements: In Fig. 6-7, S1, S2, and S3 are waveguide transfer switches. S1 allows for the output of either TWTA to be sent either to the HGA or to either LGA. S2 and S3 allow for the selection between LGA1 and LGA2. S4 is a coaxial (coax) transfer switch that routes the uplink to either SDST1 or SDST2.

The RF switches are designed such that the switches will fail in either of two switch positions. The probability that the switch will fail in between positions is remote.

The bandpass filters (BPFs) BPF1 and BPF2 are coaxial bandpass filters centered at the X-band receive frequency (7.183 GHz). They are used to filter out interference from the X-band TWTA output that could leak from the transmit port of the diplexer to the receiver port.

BPF3 is a waveguide bandpass filter that is centered at the transmit frequency (8.439 GHz) and is used to filter out the harmonics of the transmit frequency. This is needed to prevent interference to ground receivers operating in frequency bands that are the second, third, or fourth harmonics of the X-band output (that is, 16.9 GHz, 25.3 GHz, and 33.8 GHz), in particular during the first few days after launch when the power flux density of the downlink signal is high. BPF3 has no effect on transmissions through the HGA.

The isolators (ISs) IS1 and IS2 are X-band isolators to protect the X-band TWTA in case of a temporary short in the transmit path to the antenna. IS3 is the Ka-band isolator. The couplers in between the SDSTs and the TWTAs allow either SDST to drive either TWTA.

The USOs are cross-strapped (cross-strapping not shown) so that, if one fails, the other can be used by either SDST.

Ka-Band Elements: The Ka-band telemetry streams are cross-strapped. SDST1 gets its input data for Ka-band from command and data handling side A (C&DH-A) only, and SDST2 gets its input for Ka-band from command and data handling side B (C&DH-B) only. The Ka-band transmit chain is part of an operational demonstration experiment and therefore does not have to be single-fault tolerant.

6.3.1.1 High-Gain Antenna. The HGA consists of three main components—the feed, an ellipsoidal subreflector, and a 3-m offset parabolic main reflector. The HGA subreflector is 0.45 m in diameter and is located near the focal point of the main reflector. The X-band feed is a corrugated horn design, while the Ka-band feed is a disc-on-rod design. There is no uplink reception at Ka-band, only downlink transmission. The feeds contain polarizers at X-band and at Ka-band to generate right circularly polarized (RCP) microwaves.

Figure 6-8 shows the HGA pointing loss (the antenna gain relative to a reference 0 decibel (dB) value at boresight) at X-band transmit and receive frequencies.

Figure 6-9 shows the HGA pointing loss at the Ka-band transmit frequency.

The pre-launch HGA patterns are representative and were confirmed by in-flight calibrations as described in section 6.5.1.

The HGA, deployed shortly after launch, has since served as the primary means of communication to and from the orbiter.

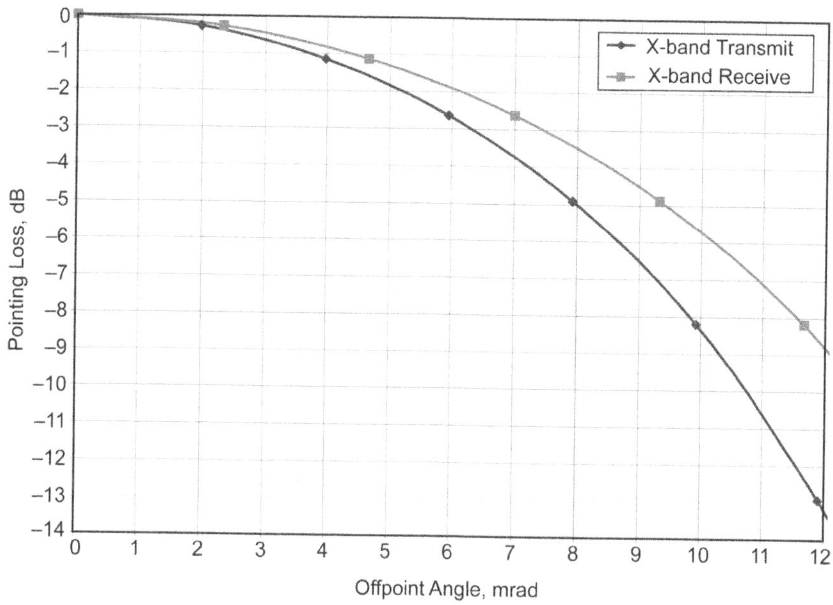

Fig. 6-8. HGA X-band transmit and receive pointing loss relative to boresight.

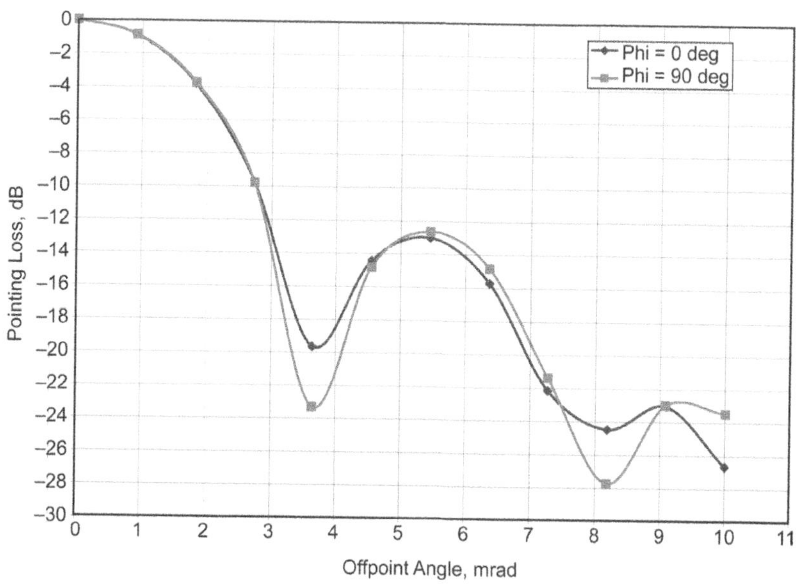

Fig. 6-9. HGA Ka-band transmit pointing loss relative to boresight.

The HGA must be pointed accurately and therefore is steered using the gimbal mechanism. The requirement for HGA pointing accuracy is 2.08 milliradians (mrad) at 99.7 percent circular error probability (CEP). This is a requirement on the mechanical system, in particular the gimbal motor that affects the link performance.

There are three gimbal mechanisms onboard Mars Reconnaissance Orbiter:

- One that allows the HGA to move in order to point at Earth
- Two that allow the solar arrays to move to point at the Sun.

Each of the gimbals can move about two axes. As the spacecraft travels around Mars each orbit, these gimbals allow both solar arrays always to be pointed toward the Sun, while the high-gain antenna can simultaneously always be pointed at Earth.

6.3.1.2 Low-Gain Antenna. Two LGAs are present for lower-rate communication during emergencies and special events, such as launch, MOI, or safe mode. The data-rate capability when using these antennas is lower because they focus the radio beam much more broadly than does the HGA. Figure 6-10 shows the pointing loss of the LGA at X-band transmit and receive frequencies. The LGA does not provide Ka-band capability.

The LGA is a horn design. It is essentially an open waveguide with RF choke rings at the end for pattern uniformity and sidelobe control. A septum polarizer placed before the waveguide horn provides RCP.

The two LGAs are mounted on the HGA dish—one on the front side and one on the back—and are moved with it. In that placement, the two LGAs make communication with the DSN possible at all times, no matter what the position of the spacecraft might be at a given time.

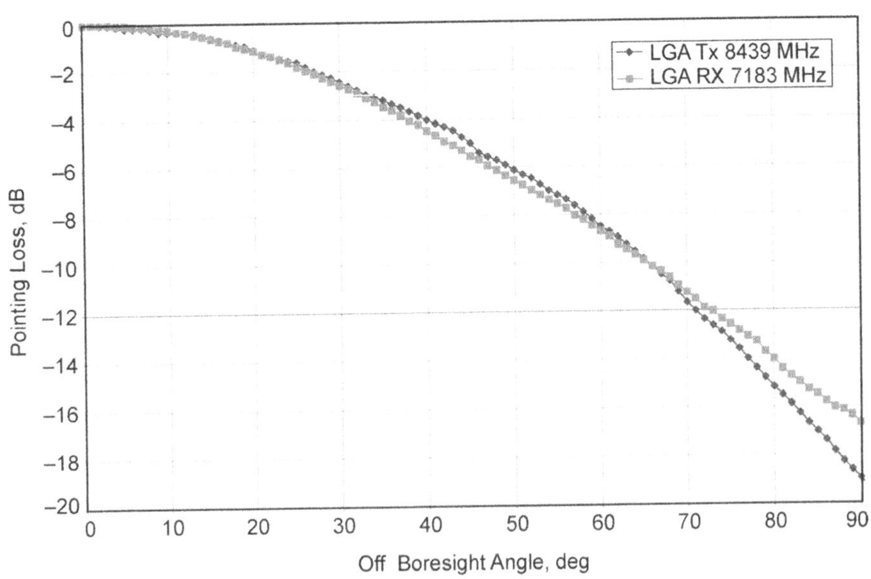

Fig. 6-10. LGA X-band transmit and receive pointing loss relative to boresight.

Table 6-4. LGA and HGA antenna link parameters.

Parameter	LGA X-band transmit	LGA X-band receive	HGA X-band transmit	HGA X-band receive	HGA Ka-band transmit
Boresight gain	8.8 dBi	8.4 dBi	46.7 dBi	45.2 dBi	56.4 dBi
Gain tolerance	±0.5 dB	±0.5 dB	±0.5 dB	±0.5 dB	±1.0 dB
Axial ratio (max)	2 dB	2 dB	1.1 dB	2.2 dB	2.3 dB
Polarization	RCP	RCP	RCP	RCP	RCP
Antenna return loss (max)	–18 dB	–18 dB	–19 dB	–23 dB	–19 dB
Half-power beamwidth			0.69 deg		0.18 deg
Pointing error budget (3-sigma)			2.08 mrad	2.08 mrad	2.08 mrad

dBi = decibels with respect to isotropic gain

The forward-facing LGA1 is mounted near the rim of the HGA and is canted 25 deg from the HGA boresight. The cant angle was selected based on the off-point angle at critical spacecraft events, such as during TCMs and MOI, when the HGA is locked in position and not tracking Earth. The aft-facing LGA2 is mounted on the TWTA panel and is canted at –115 deg from the HGA boresight.

Table 6-4 summarizes key HGA and LGA link parameters as determined before launch.

6.3.1.3 Transponders. MRO carries two small deep-space transponders (SDSTs). The SDSTs provide identical functions, and only one is powered on at a time. The SDST is a proven transponder with heritage from previous missions, such as Deep Space 1, Mars Odyssey, and MER A change from the MER (Group 1 buy) SDST to the MRO SDST was the addition of Ka-band [8].

The SDST is responsible for tracking the uplink carrier, demodulating commands from the carrier, generating the downlink carrier (coherent or non-coherent with the uplink frequency), performing convolutional coding, producing different subcarrier frequencies, modulating telemetry on the subcarrier or directly on the downlink carrier, demodulating and modulating turnaround ranging signals, and generating differential one-way ranging (DOR) tones.

The SDST is composed of four different modules: the digital processing module (DPM), the downconverter module, the power module, and the exciter module. The MRO SDST has several features differing from previous SDST designs:

- The MRO × 4 (times-four) multiplier that is used to generate the 32.2-GHz Ka-band signal from the 840f1 frequency output[7] (8052 MHz) is external to the SDST and placed on the TWTA panel (whereas the SDST is located middeck); this is done to minimize coaxial cable loss at Ka-band. In Deep Space 1 (DS1), the ×4 multiplier was internal to the SDST.
- The line receivers in the DPM are now low-voltage differential signaling (LVDS) receivers to support high-rate transmission over the compact peripheral component interconnect (cPCI) bus.
- A field programmable gate array (FPGA) with 72 thousand gates was added to the MRO SDST to support quadrature phase-shift keying (QPSK). The FPGA also performs (7,1/2) convolutional coding[8] for QPSK.
- Wideband DOR (8f1 DOR) capability was added at Ka-band.

The SDST has an internal, five-pole, 5.8-MHz low-pass filter (LPF) that filters input voltage to the phase modulator. Nominally, the MRO SDST will be configured to operate in the filtered mode. The filter reduces the amplitude of high-frequency components in the telemetry downlink to avoid interference to other missions. Use of the unfiltered mode is permitted only when the telemetry spectrum would not interfere with another mission.

Table 6-5 lists some of the parameter values that determine link configuration and performance for the MRO SDST.

6.3.1.4 RF Amplifiers. Located on the back side of the HGA is the enclosure for the TWTAs and associated microwave components. The enclosure is called the TWTA panel in the Fig. 6-1 sketch of external MRO components.

---

[7] In SDST nomenclature, f1 is the fundamental frequency from which the uplink and downlink frequencies are derived. For example, the X-band downlink is 880f1, and the X-band uplink is 749f1. The Ka-band downlink carrier is 3360f1, which is 4× the SDST's Ka-band output at 840f1. The MRO SDST operates on DSN channel 32. For this channel, f1 is approximately 9.59 MHz.

[8] Note that telemetry can be convolutionally coded in the SDST as on previous missions, but only with the (7,1/2) rate planned for use on MRO. For a turbo-coded telemetry downlink, the input to the SDST has been turbo coded in the C&DH upstream of the SDST. In this case, the stream of turbo symbols at the SDST telemetry input are treated by the SDST as bits, with the SDST's convolutional coder bypassed.

Table 6-5. SDST link configuration and performance parameters.

Parameter	Value
Receiver input levels	−156 dBm (threshold) to −70 dBm
Receiver 2-sided carrier loop bandwidth	20 Hz (threshold)
Command data rates (bps, uncoded)	7.8125, 15.625, 31.25, 62.5, 125, 250, 500, 1000, 2000 bps
Command subcarrier modulation index	0.5 to 1.5 radians, peak
Minimum telemetry symbol rate	0 bps on subcarrier, 2000 symbols per second (sps) on carrier
Maximum symbol rate	Specified to 4.4 megasymbols per second (Msps) in normal (filtered) mode, tested to 6 Msps
Telemetry modulation index range	64 equal steps of modulation voltage from 0 to 135 deg
Turnaround ranging modulation index	4.375, 8.75, 17.5, 35, 70 deg peak (accuracy ±10%, stability ±20%)
DOR modulation index, peak	28 deg peak (accuracy ±10%, stability ±25%)
Ka-band output modulation bandwidth	Normal mode 5.5 ± 1.5 MHz, wideband mode 10 MHz minimum

Figure 6-11 shows the layout of the bottom side of the TWTA panel, showing two of the TWTAs, the three power converters, and most microwave elements (diplexers, X-band bandpass filter, and isolator). The Ka-band TWTA and isolator are on the top side of the TWTA panel and are not visible in Fig. 6-11.

There are three amplifiers on board, two at X-band (only one powered at a time) and one at Ka-band. The nominal TWTA RF output power is 100 W at X-band (102 W measured pre-launch) and 35 W at Ka-band (34 W measured).

Each TWTA consists of two main components, the high-voltage power supply (HVPS), also called the electronic power converter (EPC), and the traveling-wave tube (TWT).

The diplexer is a passive device that allows for routing of X-band transmit and receive frequency signals that are present simultaneously at the antenna. The diplexer has three ports: the antenna port, the receive port, and the transmit port. The isolation between transmit and receive ports is essential to avoid self-interference within the subsystem. The diplexer also provides significant attenuation of transmit frequency harmonics.

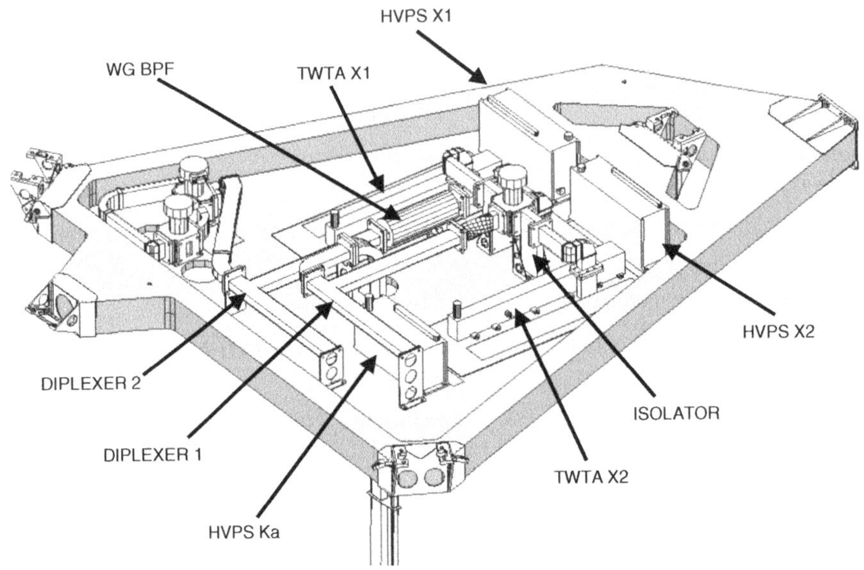

Fig. 6-11. Layout of microwave components in the TWTA panel. Please visit www.wiley.com/go/deepspacecommunications to access the color version of this figure.

The passband at the receive port is centered at 7.183 GHz to allow for the uplink signal from the antenna port to pass through to the receive port. The passband at the transmit port is centered at 8.439 GHz to allow the output of the X-band TWTA to pass to the antenna port.

Additional attenuation of transmit frequency harmonics occurs in the waveguide bandpass filter in the LGA transmit path. Each isolator (one is called out in Fig. 6-11) protects its TWTA against RF power reflected back by a momentary short at the output.

Each TWTA provides three kinds of protection for itself and the spacecraft power supply:

- **Helix Overcurrent Trip.** If helix current exceeds 5 milliamps (mA), the power converter, responding within 2 ms, goes into an automatic restart mode involving removal and reapplication of the high voltage to the TWTA.
- **Power Converter Overcurrent Trip.** If the input current exceeds a maximum value, the switching transistor is protected by cycle peak current limitation. Also, after about 2 ms, the converter goes into the automatic restart mode.
- **Bus Undervoltage Trip.** If the bus voltage at the converter input drops below 20.5 V, the high voltage switches off, and an undervoltage trip status flag is set. When the bus voltage rises above 21.5 V again, the TWTA startup sequence is initiated and preheating begins. The preheating lasts about 210 seconds. The nominal bus voltage is 28 V.

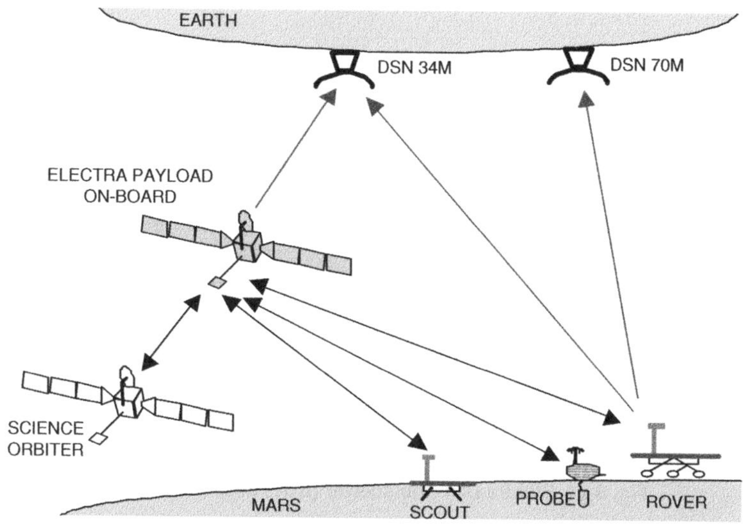

Fig. 6-12. MRO Electra payload operations concept.

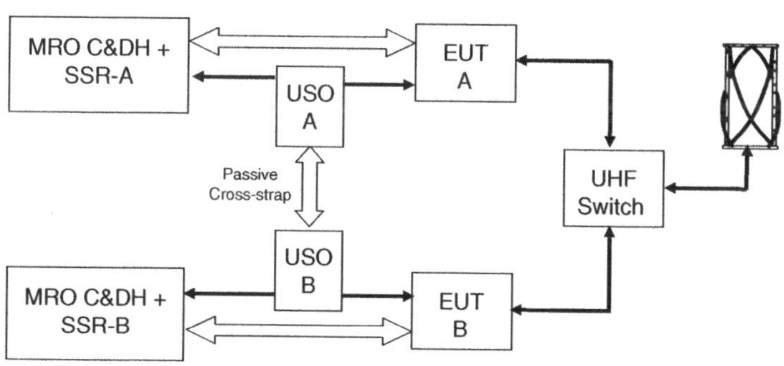

Fig. 6-13. MRO/Electra UHF block diagram and interfaces with C&DH and SSR.

6.3.2 UHF: Proximity Relay Communications

As shown in Fig. 6-12, the Electra payload in MRO becomes a network node in the Mars network constellation that provides efficient relay of high-rate in-situ mission science and engineering data. The first landing vehicles that used MRO/Electra operationally were MER, Phoenix, and MSL.

Figure 6-13 is a block diagram of the MRO UHF system and its interfaces (I/Fs) with the command and data handling (C&DH) and SSR systems. The EUTs and the USOs (which also support the X-band and Ka-band systems) are redundant. The diagram shows the allowable combinations of redundant USOs and EUTs with the C&DH sides and the redundant SSRs.

Figure 6-14 is a sketch of the EUT.

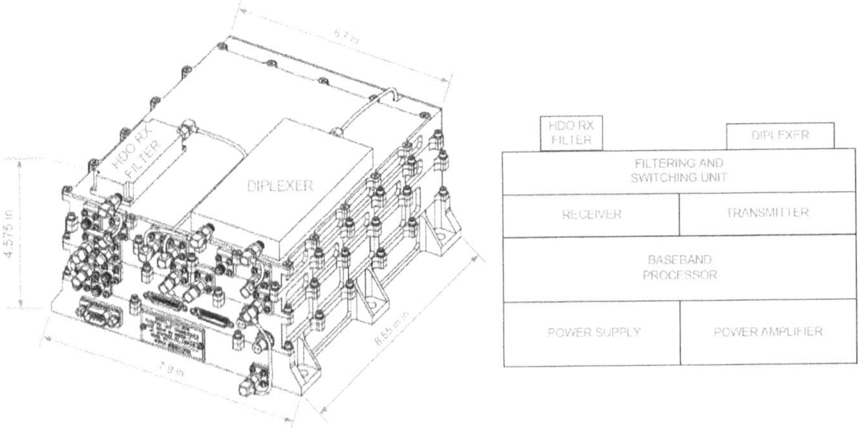

Fig. 6-14. Electra UHF transceiver (EUT) assembly.

The EUT assembly consists of five modular slices. From top to bottom, the slices are

- Half-duplex overlay (HDO) receiver filter and UHF diplexer
- Filtering and switch unit (FSU)
- UHF radio frequency module (RFM, the receiver and transmitter)
- Baseband processor module (BPM)
- Power supply module (PSM) with integral power amplifier module.

The FSU slice in the MRO EUT consists of a high-isolation diplexer, the HDO receive/transmit (R/T) switch, and the coaxial transfer switch. The BPM slice interfaces directly with MRO C&DH, the MRO SSR, the USO, and the modules that comprise the EUT.

The RFM slice consists of a single-channel UHF transmitter and receiver.

The PSM slice consists of the power supply and the driver/power amplifier. The PSM provides power to the BPM and, under BPM control, to the elements of the RFM. The PSM slice also includes a power amplifier that amplifies the modulated signal to the appropriate RF output level.

The BPM performs all signal processing, provides overall EUT control, and services the external spacecraft interfaces.

Figure 6-15 summarizes the functionality of the modem processor (MP) portion of the BPM in block diagram form, which was updated from Ref. [9].

The BPM consists of a 32-bit microprocessor, two radiation-hardened program-once field programmable gate arrays (FPGAs), and a large

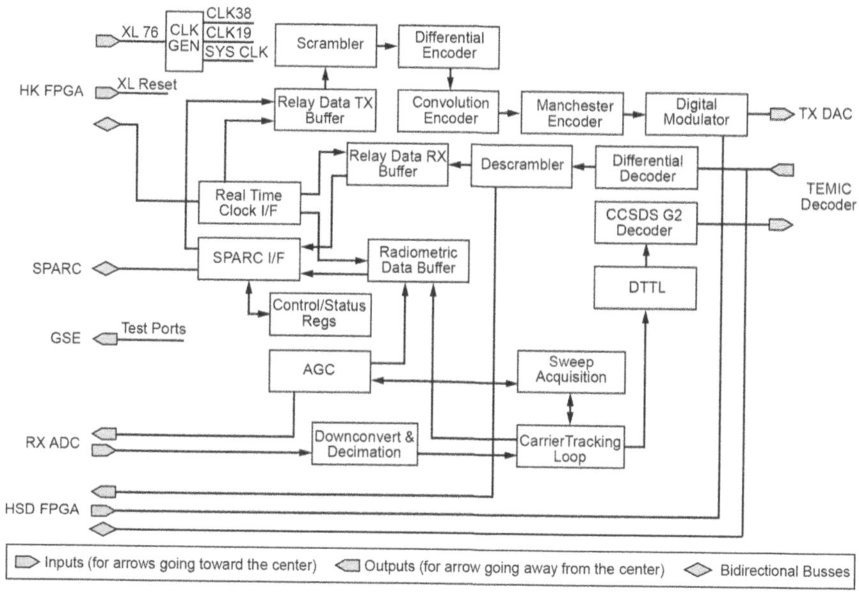

Fig. 6-15. Electra transceiver block diagram. ADC = analog-to-digital converter, AGC = automatic gain control, CLK GEN = clock generator, DTTL = data transition tracking loop, GSE = ground support equipment, SPARC = scalable processor architecture.

(~1 million gates (Mgate)) reprogrammable FPGA, along with a substantial amount of dynamic and static memory. The reprogrammable FPGA contains the modem functions and is reprogrammable post-launch. The 32-bit microprocessor manages the EUT and the relay Prox-1 protocol.[9]

In concept, one side of the BPM handles the spacecraft interfaces. A dedicated 1553 transceiver chip supports the command and telemetry interface to the host C&DH. An LVDS interface supports high-rate relay and radiometric data transfers through the high-speed data (HSD) FPGA. The other side of the BPM handles the EUT, with the housekeeper (HK) FPGA managing control and telemetry signals to and from the EUT front end, and the MP FPGA.

The main functions of the MP FPGA include

- Coding and decoding
- Modulation and demodulation

---

[9] The EUT complies with the Proximity-1 protocol defined by the Consultative Committee for Space Data Standards (CCSDS) in Ref. 11. In this chapter, the protocol is abbreviated Prox-1.

- Carrier, symbol, and decoder synchronization
- Prox-1 frame synchronization detection
- Prox-1 transmit (Tx) and receive (Rx) user data and control data buffering
- Receive signal level management, automatic gain control (AGC)
- Radiometric Doppler and open-loop record functions
- Clock (CLK) and timestamp functions
- Implementation of the physical layer of the communication link from baseband to an intermediate frequency (IF)

The MRO Electra does not have an internal clock. The clocks for the BPM FPGAs, including bit, symbol, and sample rate clocks, are derived from the external USO.

Table 6-6 [6] defines the major operating modes, functions, and constraints for the MRO EUT.

MRO Electra implements frequency agility and swappable transmit and receive bands. The EUT complies with the CCSDS Prox-1 channel definitions for eight frequency pairs. In all, Electra supports 16 preset frequency pairs, as defined in Table 6-7 [10].

In addition to the 16 preset pairs, the MRO Electra radio has the capability to tune its Tx and Rx frequencies across the entire 390-MHz-to-450-MHz band; thus, any frequency pair combination within this band is possible. For half-duplex operation, any pair of frequencies will work as an operational pair. For full-duplex operation, the Tx frequency must be chosen in the range of 435 MHz to 450 MHz, and the Rx frequency must be chosen in the range of 390 to 405 MHz.

The MRO Electra payload provides a single nadir-looking (vertical down to Mars) UHF LGA. The antenna shares the nadir deck with science payloads. Some parts of these nearby payloads that are responsive at UHF frequencies couple with the antenna and distort its nominal gain pattern. To compensate for this, the MRO mission plan allows for spacecraft roll steering of up to 30 deg to point the better parts of the UHF antenna pattern toward the surface user. The orbiter sets up for this pass by roll steering to a fixed roll angle. The Electra payload performs the pass, and then the orbiter rolls back to the standard nadir pointing position.

Figures 6-16 (437.1 MHz) and 6-17 (401.6 MHz) show antenna gain in dBi versus angle from boresight. These angles from boresight are called cone or theta angles. In each figure, the solid curve is the average gain over cuts made in the axis orthogonal to cone. Angles in the orthogonal axis are called clock

Table 6-6. MRO/Electra modes, functions, and performance.

Capability	Values
Protocol	Prox-1 (reliable and expedited link layer protocols)
Frequencies	See next section (including Table 6-7)
Modes of operation	Half-duplex[10] Rx and Tx (no Prox-1 protocol in half duplex) Full-duplex transceiver
Full-duplex carrier modes	Coherent, noncoherent
Transceiver RF output power	5.0 W full duplex, 7.0 W half duplex
Circuit loss, EUT to antenna	–0.42 dB
Receiver thresholds, at antenna	–130.8 dBm (1 kbps) to –99.6 dBm (1024 kbps) coded –126.0 dBm (1 kbps) to –91.1 dBm (2048 kbps) uncoded
Carrier modulation modes	Suppressed carrier, residual carrier (60 deg mod index)
Modulation types	Residual carrier binary phase-shift keying (BPSK) with bi-phase-L (Manchester). Suppressed-carrier BPSK
Frequency reference	Ultra stable oscillator
Rx and Tx symbol rates	1, 2, 4, 8, 16, 32, 64, 128, 256, 512, 1024, 2048 thousand symbols per second (ksps). Also, adaptive data rate mode
Received signal power range	–140 to –70 dBm
Encoding	Uncoded, (k = 7, r = 1/2) convolutional, differential symbol coding
Decoding	Uncoded, (k = 7, r = 1/2) convolutional (3-bit soft decode)
Scrambling/ descrambling	V.38
Acquisition and tracking loop	Second-order phase locked loop (PLL), with loop bandwidth 10 Hz to 10 kHz (for received signal from –140 dBm to –70 dBm)
Tracking range and rate	±20 kHz, ±200 Hz/s

[10] The term "full duplex" is used by MRO in the conventional sense of simultaneous forward and return link capability at separate frequencies. The term "half duplex" means that Electra's transmitter and receiver are not on simultaneously, even though the forward and return links may be on separate frequencies.

Table 6-7. CCSDS Prox-1 "Blue Book" channel numbers and "preset" Electra frequencies.

Channel Number	CCSDS Forward Frequency (MHz)	MRO Preset Forward Frequency (MHz)	CCSDS Return Frequency (MHz)	MRO Preset Return Frequency (MHz)
0	437.1	437.1	401.585625	401.585625
1	435.6	435.6	404.4	404.4
2	439.2	439.2	397.5	397.5
3	444.6	444.6	393.9	393.9
4	435 to 450	436	390 to 405	401.4
5	435 to 450	438	390 to 405	402
6	435 to 450	440	390 to 405	402.6
7	435 to 450	441	390 to 405	403.2
8		442		391
9		442.5		392
10		443		393
11		445		395
12		446		395.5
13		447		396
14		448		399
15		449		400

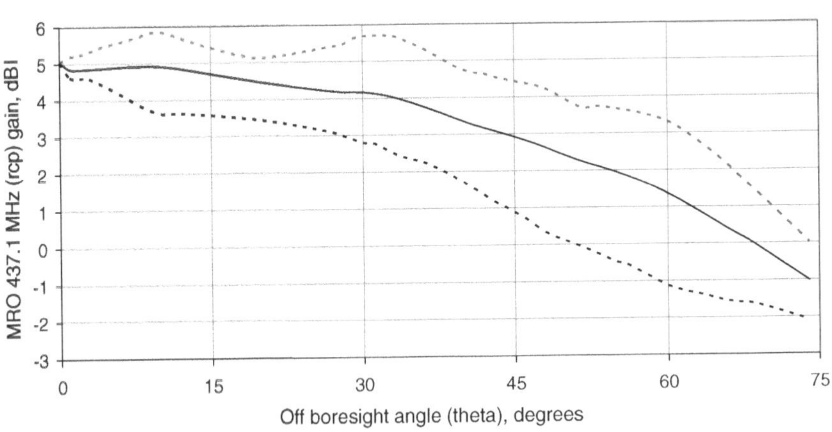

Fig. 6-16. MRO 437.1-MHz gain pattern.

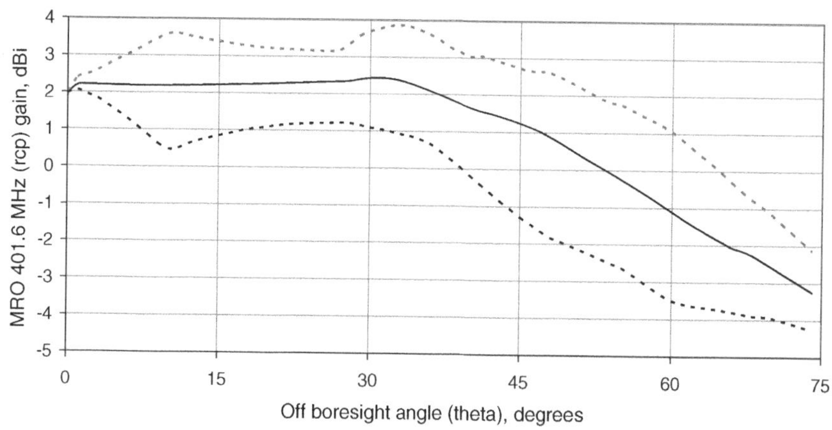

Fig. 6-17. MRO 401.6-MHz gain pattern.

or phi angles. The dotted curves above and below the solid curve are the gains for the best-case and worst-case clock cut, respectively. The antenna is RCP for both the forward and return links.

The MRO Electra transceiver is compatible with the CCSDS Proximity-1 Space Link Protocol [10, 11]. Prox-1 transfer frames are sent on both the forward link (from the orbiter to the surface vehicle) and the return link (surface back to the orbiter) using the Prox-1 protocol link management in either reliable (retransmission) or expedited (no retransmission) mode. In retransmission mode, an automatic repeat queuing (ARQ) protocol is utilized to request retransmission of any proximity frames that are not received error-free. MRO also provides a relay service (called "raw data") not utilizing the Prox-1 protocol. The orbiter also provides a form of Doppler data and a form of open-loop data. These data types or services are defined in the following.

6.3.2.1 Proximity-1 Data. Typically the MRO Electra will initiate a Prox-1 session by sending a string of "hail" data packets while looking for a response from the specific lander identified in the hail packet. This standard operating procedure can be reversed—that is, lander-initiated relay sessions are possible. The hail includes information describing the session operating mode for both the forward and return link directions. This includes, among other things, operating frequency, data rate, and channel-coding mode.

6.3.2.2 Time-Stamp Packets. Time-stamp data consist of snapshots of the local Electra clock corresponding to the ingress or egress times of Prox-1 frame-synchronization markers. Thus, time stamp data is only collected in conjunction with Prox-1 mode operations. The time stamps are paired with corresponding Prox-1 frame sequence numbers and noted as arriving or departing frames. If the other end of the relay link is also capable of collecting Prox-1 frame time stamps, the collection of these time stamps at both ends of the link can be used as a form of dual 1-way ranging and used to correlate clocks on the lander and orbiter.

6.3.2.3 Raw Data. In raw data mode, there is no hailing or link establishment protocol, nor is there any session data management or accounting protocol. A link is established by time sequence transmissions and reception at both ends of the link. In addition to coordinated sequence timing, both sides of the link must agree beforehand to the same data link mode settings—for example, frequencies, data rates, and coding.

6.3.2.4 Phase and Power Data. MRO's Electra transceiver can sample and record the phase and power level of a phase-locked received carrier signal. This radiometric information is highly accurate (being based on the MRO USO signal and with successive samples tied directly to the USO-based local clock). Each sample contains phase, AGC power, in-phase (I) amplitude, quadrature (Q) amplitude, and a USO-based time. These data form the basis for a Doppler metric. The 160 bit data format is shown below.

Phase and Power sample format

Header	Phase	AGC	Iamp	Qamp	Spare	Time Code	Trailer
bit[159:156]	bit[155:107]	bit[106:101]	bit[100:88]	bit[87:75]	bit[74:60]	bit[59:4]	bit[3:0]

6.3.2.5 I/Q (Open Loop) Record Data. MRO's Electra transceiver can capture complex in-phase/quadrature (I/Q) samples of the received signal, down-converted to baseband, at sample rates as fast as 150,000 samples/second. This mode is known as open-loop record and is primarily used in support of events like EDL where there is a concern that high signal dynamics or low signal levels will prevent real-time demodulation of the signal in the Electra transceiver. These open loop record samples are passed to the MRO solid state recorder (SSR) and later forwarded to Earth where a software receiver and spectrum analysis tools can be used to demodulate the signal and to decipher the signal dynamics. In addition to the complex I/Q samples, the AGC level is also captured for each sample. This allows open loop record to work over the full receive signal level range of the Electra radio and still allow for faithful reconstruction of the received signal later on Earth.

There are two data modes for open loop record. The first mode captures the USO clock as part of the sample data. This clock-stamped format is used for the first data sample and occasional later samples to establish a clock reference to the sample data. Most of the open loop record samples do not include a time stamp, and the time of each sample is inferred from the sample count and the sample frequency that is phase locked to the USO clock tick. The two data formats are shown below.

I/Q (open loop) sample time-coded format

Header	I	Q	Spare	AGC	Spare	Time Code	Trailer
bit[95:92]	bit[91:84]	bit[83:76]	bit[75:74]	bit[73:68]	bit[67:60]	bit[59:4]	bit[3:0]

I/Q (open loop) sample non-time-coded format

Header	I	Q	Spare	AGC	Trailer
bit[31:28]	bit[27:20]	bit[19:12]	bit[11:10]	bit[9:4]	bit[3:0]

6.3.3 Ka-Band: Operational Demonstration

The MRO spacecraft has a fully functioning Ka-band downlink equipment suite, comparable to that for the X-band downlink, including

- A one-way carrier (USO or auxiliary oscillator driven) or a two-way coherent carrier (using the X-band uplink carrier frequency reference)
- Modulation of telemetry with any of the available data rates, encoding types, and modulation index values
- Modulation of turnaround ranging from the X-band uplink, with a settable modulation index
- Modulation of differential one-way ranging tones, more widely spaced than at X band; Ka-band tones are 76 MHz from the carrier, as compared with X-band tones at 19 MHz.

The Ka-band components of the subsystem include a ×4 (times-four) multiplier, a Ka-band TWTA and its power converter, a Ka-band feed element in the HGA, and other microwave parts as defined in Section 5.3.1.

Deep Space Network 34-m antennas capable of receiving Ka-band are requested twice per week during the prime science mission as part of the demonstration.

6.4 Ground Data System

6.4.1 Deep Space Network

The three primary DSN ground complexes are located near Goldstone (California), Madrid (Spain), and Canberra (Australia). The DSN antennas are categorized according to their diameter and performance. During cruise and orbit operations, MRO was allocated use of the 70-m antenna subnet, the 34-m beam-waveguide (BWG) antenna subnet, and the 34-m high-efficiency (HEF) antenna subnet.

MRO used the 70-m antennas to support MOI and may require them for emergency mode communications (safe mode operations on the LGA).

MRO depends on the 34-m BWG antennas for the vast majority of the mission telemetry and commanding. The BWG antennas differ from the HEF antennas in that beam-waveguide optics (mainly consisting of a series of small mirrors) are used to direct microwave energy from the region above the main reflector to a location at the base of the antenna (typically the pedestal room). This allows for easier access to the microwave equipment, and the positional stability allows for use of state-of-the-art ultra-low noise amplifier and feed designs.

MRO may alternatively be allocated 34-m HEF stations (one at each complex) for passes that do not require Ka-band downlink capability. Because the low-noise amplifier (LNA) is located near the HEF antenna feed, the gain-to-noise temperature ratio, G/T, is about 1 dB better than in the BWG antennas.

With the new X-/X-/Ka-band (X-band up, X-band down, Ka-band down) feed and LNA upgrades to the 34-m BWG antennas, the upgraded 34-m BWG stations have a slightly higher gain over temperature (G/T).

The gain, noise temperature, and pointing characteristics of the antennas are listed in the *DSN Telecommunications Link Design Handbook* [12].

6.4.2 Ka-Band Demonstration Requirements

The Ka-band demonstration includes an assessment of the DSN's readiness to track Ka-band signals from deep-space missions. One operational station (DSS-25) tracked the "new technology" Ka-band downlink from Deep Space 1 in 1998–1999, and the DSN has tracked Ka-band sporadically for Cassini radio science activities. Several of the 34-m stations have Ka-band downlink capability to support the MRO Ka-band operational demonstration. These are DSS-25 and DSS-26 at Goldstone in California, DSS-34 near Canberra in Australia, and DSS-55 near Madrid in Spain.

The 34-m BWG Ka-band beam width is less than 18 millidegrees (mdeg). The basic antenna pointing capabilities required for the Ka-band demonstration include

- "Blind-pointing" of the antenna (computer driven, without input from the received downlink) must be better than 10 mdeg [13] so that the monopulse system (active pointing) will be able to operate.
- The monopulse must be operational (without it, pointing errors may cause link degradation of 4–5 dB).

Besides the normal functions of telemetry demodulation and decoding, and measurements of Doppler, two-way ranging, and delta-DOR, the Ka-band demonstration requires the following monitor data generation capabilities:

- Accurate measurement by the operational receiver of signal-to-noise ratio (SNR), particularly symbol SNR
- Accurate measurement by the operational receiver of system noise temperature (SNT)
- Sampling of receiver monitor data at the specified 5-second interval, with prompt delivery of the data to the MRO database.

These additional requirements will enable the demonstration to identify data outages caused by weather events and to separate them from outages caused by other phenomena.

6.4.3 Ground Data Network Flow for Relay Data through Electra

Figure 6-18 shows the MRO science data flow, processing, and accountability mechanisms. In the context of the five Electra relay data types (Prox-1 data, raw data, time stamps, phase and power data, and open-loop data), all are "science data." The features highlighted in red identify the flow of Electra relay data to the ground.

Electra relay return-link data are routed over a shared LVDS interface to a dedicated SSR hard partition. When collecting data from Electra, the SSR's Electra interface software limits the amount of data to the space in the Electra partition minus the amount of not-yet-read data in the partition. The software will issue a telemetry event record with the difference between actual and planned sizes of the data collection.

In Fig. 6-18, telemetry processing extracts from the MRO telemetry stream the CCSDS advanced orbiting systems (AOS) frames containing MRO CCSDS source packets. AOS frames containing telemetry from Electra are processed to extract MRO science protocol packets.

These source packets are provided to the CCSDS File Delivery Protocol (CFDP) process running on the ground, which finds the MRO CFDP protocol data units (PDUs) and reconstructs the Electra pass product. Given that there may be gaps in the telemetry stream, retransmissions may be requested from the spacecraft on the AOS frame level. Associated with the Electra pass product will be a detached Planetary Data System (PDS) label, which will contain metadata that describe the circumstances of the collection, for example, MRO identifier, orbit, and so forth. There also will be a CFDP transaction report on the holes, if any, in the Electra pass product.

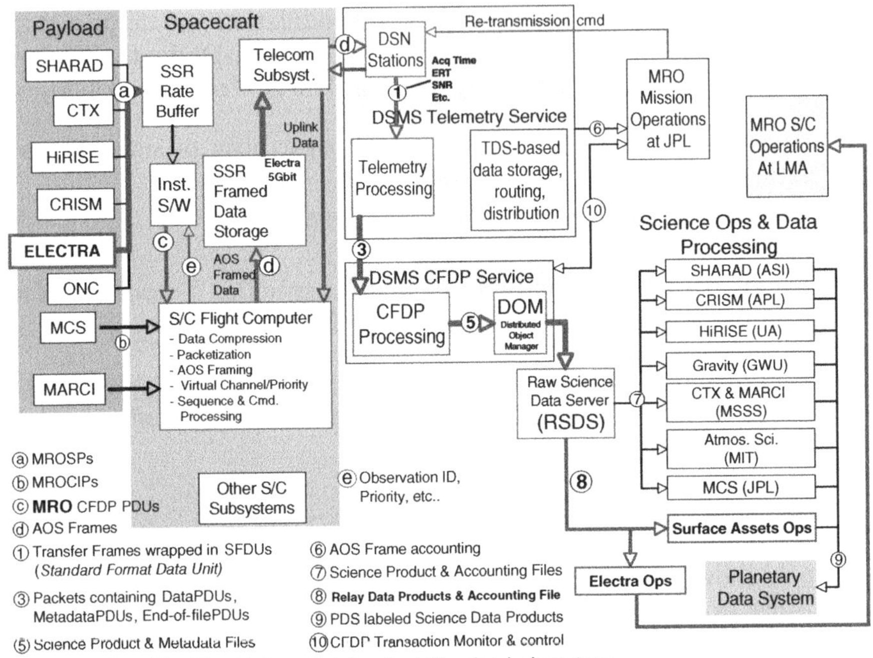

APL = (Johns Hopkins University) Applied Physics Laboratory
CIP and SP = instrument output data
DOM = Distributed Object Manager
GWU = George Washington University
HiRISE = High Resolution Imaging Science Experiment
LMA = Lockheed Martin Aerospace
MSSS = Malin Space Science Systems
UA = University of Arizona

Fig. 6-18. MRO Electra science data flow. Please visit www.wiley.com/go/deepspace communications to access the color version of this figure.

A second run of the CFDP process will take the Electra pass product and search for Electra CFDP PDUs to extract the PDUs for each Electra subproduct [relay (Prox-1 data), time stamp, raw data, phase and power, and open-loop I and Q data].

Each Electra relay telemetry product consists of a binary product file, which varies for each product type, as well as a CFDP transaction log file and a detached American Standard Code for Information Interchange (ASCII) label.

Figure 6-19 shows the relay pass product from the Ground Data System (GDS), and Fig. 6-20 shows the relay data product delivered. The pass product is a binary file with the CFDP PDUs, while the data product is a binary file with the Prox-1 transfer frames.

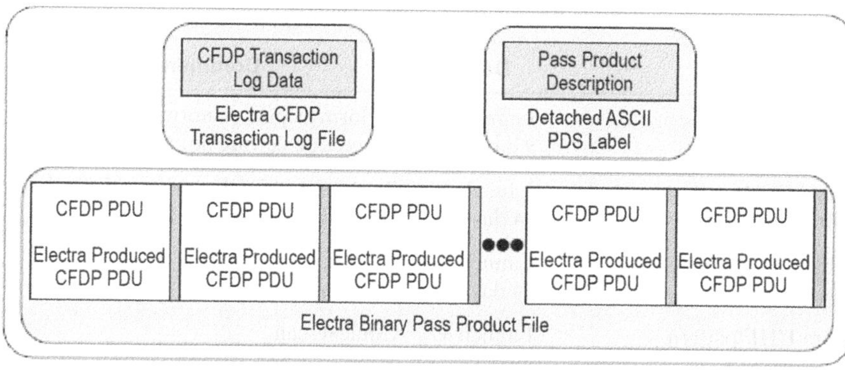

Fig. 6-19. Electra relay pass product output from GDS.

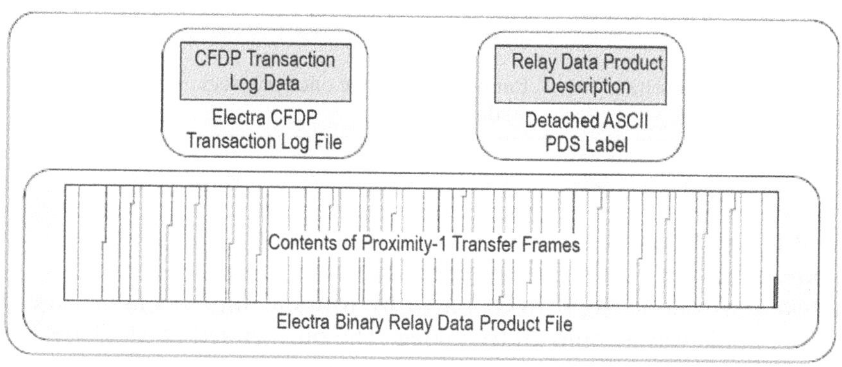

Fig. 6-20. Electra relay data product from GDS.

6.5 X-Band Telecom Operations

6.5.1 Cruise Calibrations

Planned telecom cruise calibrations are summarized in Table 6-8. During the X-band and Ka-band HGA calibration, the spacecraft articulates the HGA through a grid-like pattern (raster scan) about the pre-launch antenna boresight. By monitoring the received signal strength, the calibration determines if the HGA phase center has shifted during launch.

6.5.2 MOI Telecom Configurations

The main engine burn for MOI began at 21:24 universal time coordinated (UTC) referenced to Earth received time (ERT), the time this event was seen on the Earth. Ninety minutes prior to this, the telecom path was switched to LGA1, which then was used throughout the orbit insertion process. Thirteen minutes prior to engine start, the spacecraft began a slew to MOI attitude. It ended the slew 5 minutes before engine start, and its inertial attitude remained

Table 6-8. Telecom cruise calibrations.

Calibration type	Date	Comments
X-band LGA performance	Launch + 9 days	Perform as part of normal LGA ops. Spacecraft slewing not required.
X-band HGA pattern calibration	Launch + 24 days	Raster scan; simultaneous with Ka-band HGA calibration
Ka-band HGA pattern calibration	Launch + 24 days	
Electra UHF pattern characterization	Launch + 40 days	Conical scan
Electra UHF performance	Launch + 40 days	
X-band Delta DOR checkout	Launch + 40 days	Done once per week, starting at L + 40 d
Ka-band wideband delta-DOR checkout	Launch + 40 days	Done once per week, starting at L + 40 d

fixed at this position throughout MOI. At the beginning of the burn, LGA1 was boresighted at Earth. By the time MRO entered solar eclipse (start + 21 minutes) and was occulted from Earth by Mars (2 minutes later), the off-boresight angle was around 20 to 25 deg. The burn ended at 21:51 ERT; the turn back to Earth ended at 22:13; and occultation ended (downlink reached the Earth) at 22:16 ERT.

During the MOI itself, the DSN supported downlink telemetry with two 70-m antenna stations simultaneously (DSS 14 and DSS 63 had overlapping coverage). Figure 6-21 shows the approximate elevation angles for the 70-m antennas on March 10, 2006.

Following re-acquisition with LGA1, the X-band system was transitioned back to the HGA for spacecraft checkout. Navigation began to collect two-way radiometric data and to perform orbit determination of the capture orbit.

6.5.3 Aerobraking Telecom Configurations

During MRO aerobraking, the two main activities were the drag passes and the aerobraking maneuvers (ABMs). The drag passes occurred at the capture orbit perigee, with the spacecraft oriented with the velocity vector in order to maximize the drag coefficient. The ABMs typically were conducted at the apogee of the capture orbit and were scheduled to adjust the orbit after the drag pass if needed.

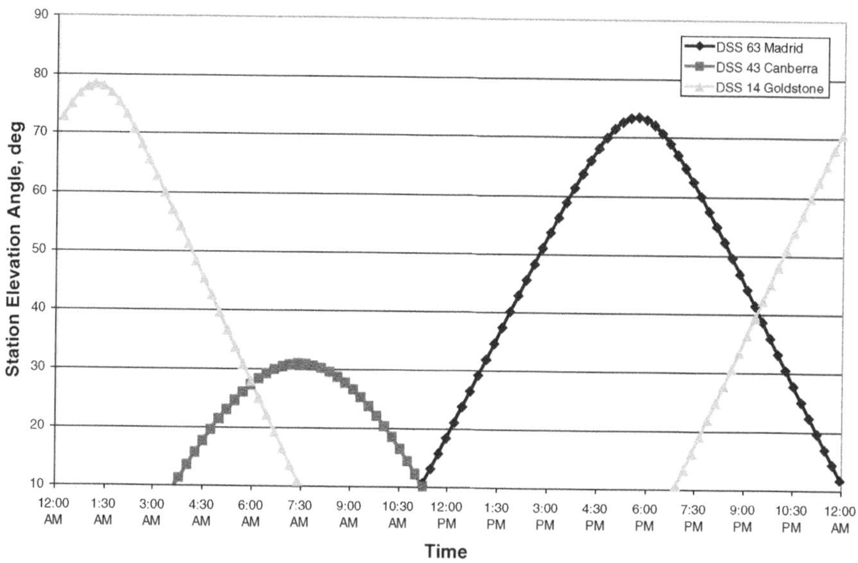

Fig. 6-21. Station elevation angles on day of Mars Orbit Insertion.

At 16 minutes prior to the start of each drag pass, an onboard sequence configured the telecom system to transmit a carrier-only downlink over LGA1. In addition, the uplink bit rate was switched to 7.8125 bps, the minimum available rate. After 1 minute for the DSN to lock up to the carrier, the HGA was locked into position and communications were through LGA1 throughout the duration of the drag pass. Ten minutes after the end of the drag pass, the sequence restored nominal downlink through the HGA.

Likewise, 16 minutes before the beginning of the ABM, the sequence configured the telecom system to transmit a carrier-only downlink through LGA1. This remained the telecom configuration until 15 minutes after the conclusion of the ABM, at which time the nominal downlink through the HGA was re-established.

6.5.4 Downlink Telemetry Modulation and Coding

The MRO project data volume goal for full mission success was to return more than 26 terabits of science data from Mars during its primary science phase, which exceeded any previous deep-space mission by more than an order of magnitude. The achieved return was 73 terabits.

The following kinds of modulation are used on MRO [14]:

- BPSK on a subcarrier, with the subcarrier modulating the carrier
- BPSK directly on the carrier
- QPSK directly on the carrier

Table 6-9. Emergency mode (7,1/2) + RS (short frame) concatenated code.

Code Type	Framed Bit Rate, bps	Symbol Rate, sps	Modulation Type	Subcarr Freq (kHz)	Tlm Mod Index (deg)	Carrier Loop Type	TLM only Threshold Pt/No (dB-Hz)	TLM + RNG Lo Threshold Pt/No (dB-Hz)	TLM + RNG Hi Threshold Pt/No (dB-Hz)
(7,1/2)+RS (I=1)	34.4	80	Squarewave Subcar	25	58	Residual	21.4	21.7	22.5
(7,1/2)+RS (I=1)	137.5	320	Squarewave Subcar	25	71	Residual	26.2	26.5	27.3

QPSK modulation capability by the SDST allows for twice the data rate to be transmitted through the same bandwidth as compared with the BPSK used in previous missions. Sequential tone ranging is not possible with QPSK because of the fully suppressed carrier.

Error-correcting codes as defined in Tables 6-9 through 6-14 are used on the downlink to the DSN. The table referenced in the title of each subsection summarizes the main MRO configuration items (bit rate and symbol rate, modulation type, modulation index, and station receiver loop type) and receiver thresholds. In the following subsections, the symbol rate is defined as the output of the SDST (that is, channel symbols), and the information bit rate is defined as the frame bit rate coming into the C&DH (at point "d" in Fig. 6-18).

6.5.4.1 Short Frame Concatenated (Table 6-9). The [(7,1/2) convolutional + Reed–Solomon (RS) (short frame)] concatenated code will be used only for the emergency mode, 34.38 bps and a MOI data rate of 139 bps. The 34.38 bit rate is chosen for heritage reasons, with the coded bit rate out of the C&DH uplink–downlink (ULDL) card at 40 bps and the SDST output symbol rate at 80 symbols per second (sps).

6.5.4.2 Long-Frame Concatenated (Table 6-10). The [(7,1/2) convolutional + RS (long-frame)] concatenated code has been proven in many prior missions and is used to cover the largest span of bit rates. Because of bandwidth limitations, the maximum rate for the concatenated code downlink is a bit rate of 3.3 Mbps at the SDST input and a symbol rate of 6.6 megasymbols per second (Msps) at the SDST output. If interference with another project is an issue, maximum rates are 2 Mbps and 4 Msps.

6.5.4.3 Turbo Code (Tables 6-11 Through 6-13). Turbo codes are to be used for bit rates above 32 kbps. This capability, implemented in C&DH hardware, provides more link margin as compared with convolutional codes of the same code rate. Currently the maximum decode rate of the ground turbo decoder limits use of turbo codes to 1.6 Mbps and below. If interference to another project could occur, the limit is 4 Msps (SDST output channel rate).

Table 6-10. (7,1/2) + RS (long frame) concatenated code.

Code Type	Framed Bit Rate, bps	Symbol Rate, sps	Modulation Type	Subcarr Freq (kHz)	Tlm Mod Index (deg)	Carrier Loop Type	TLM only Threshold Pt/No (dB-Hz)	TLM + RNG Lo Threshold Pt/No (dB-Hz)	TLM + RNG Hi Threshold Pt/No (dB-Hz)
(7,1/2)+RS (I=5)	556.9	1280	Squarewave Subcar	25	72	Residual	31.2	31.5	32.3
(7,1/2)+RS (I=5)	1740.4	4000	Squarewave Subcar	25	72	Residual	36.0	36.3	37.1
(7,1/2)+RS (I=5)	27846.8	64000	Squarewave Subcar	375	72	Residual	47.9	48.2	49.0
(7,1/2)+RS (I=5)	87021.1	200000	Squarewave Subcar	375	72	Residual	52.8	53.1	53.9
(7,1/2)+RS (I=5)	139233.8	320000	BPSK Direct Mod	None	72	Residual	54.9	55.2	55.9
(7,1/2)+RS (I=5)	208850.7	480000	BPSK Direct Mod	None	72	Residual	56.6	56.9	57.7
(7,1/2)+RS (I=5)	348084.4	800000	BPSK Direct Mod	None	72	Residual	58.8	59.1	59.9
(7,1/2)+RS (I=5)	696168.9	1600000	BPSK Direct Mod	None	72	Residual	61.8	62.2	63.6
(7,1/2)+RS (I=5)	1044253.3	2400000	QPSK	None	82	Suppressed	63.4	-	-
(7,1/2)+RS (I=5)	1305316.7	3000000	QPSK	None	82	Suppressed	64.3	-	-
(7,1/2)+RS (I=5)	478616.1	1100000	BPSK Direct Mod	None	72	Residual	60.2	60.6	62.0
(7,1/2)+RS (I=5)	1740422.2	4000000	QPSK	None	82	Suppressed	65.6	-	-
(7,1/2)+RS (I=5)	2610633.3	6000000	QPSK	None	82	Suppressed	67.3	-	-

Table 6-11. Turbo code, rate 1/2.

Code Type	Framed Bit Rate, bps	Symbol Rate, sps	Modulation Type	Subcarr Freq (kHz)	Tlm Mod Index (deg)	Carrier Loop Type	TLM only Threshold Pt/No (dB-Hz)	TLM + RNG Lo Threshold Pt/No (dB-Hz)	TLM + RNG Hi Threshold Pt/No (dB-Hz)
Turbo 1/2	745645.4	1500000	BPSK Direct Mod	None	72	Residual	60.7	61.1	62.5
Turbo 1/2	1491290.8	3000000	QPSK	None	82	Suppressed	63.5	-	-

Table 6-12. Turbo code, rate 1/3.

Code Type	Framed Bit Rate, bps	Symbol Rate, sps	Modulation Type	Subcarr Freq (kHz)	Tlm Mod Index (deg)	Carrier Loop Type	TLM only Threshold Pt/No (dB-Hz)	TLM + RNG Lo Threshold Pt/No (dB-Hz)	TLM + RNG Hi Threshold Pt/No (dB-Hz)
Turbo 1/3	66279.6	200000	Squarewave Subcar	375	72	Residual	49.6	49.9	50.7
Turbo 1/3	132559.2	400000	BPSK Direct Mod	None	72	Residual	52.6	52.9	53.7
Turbo 1/3	198838.8	600000	BPSK Direct Mod	None	72	Residual	54.4	54.7	55.5
Turbo 1/3	497096.9	1500000	BPSK Direct Mod	None	72	Residual	58.3	58.7	60.1
Turbo 1/3	662795.9	2000000	BPSK Direct Mod	None	72	Residual	59.6	60.0	61.4
Turbo 1/3	795355.1	2400000	QPSK	None	82	Suppressed	60.1	-	-
Turbo 1/3	994193.8	3000000	QPSK	None	82	Suppressed	61.1	-	-
Turbo 1/3	1325591.8	4000000	QPSK	None	82	Suppressed	62.4	-	-

Table 6-13. Turbo code, rate 1/6.

Code Type	Framed Bit Rate, bps	Symbol Rate, sps	Modulation Type	Subcarr Freq (kHz)	Tlm Mod Index (deg)	Carrier Loop Type	TLM only Threshold Pt/No (dB-Hz)	TLM + RNG Lo Threshold Pt/No (dB-Hz)	TLM + RNG Hi Threshold Pt/No (dB-Hz)
Turbo 1/6	331397.9	2000000	BPSK Direct Mod	None	72	Residual	56.1	56.5	57.8
Turbo 1/6	497096.9	3000000	QPSK	None	82	Suppressed	57.6	-	-
Turbo 1/6	662795.9	4000000	QPSK	None	82	Suppressed	58.9	-	-
Turbo 1/6	994193.8	6000000	QPSK	None	82	Suppressed	60.6	-	-

Table 6-14. RS coding only (long frame).

Code Type	Framed Bit Rate, bps	Symbol Rate, sps	Modulation Type	Subcarr Freq (kHz)	Tlm Mod Index (deg)	Carrier Loop Type	TLM only Threshold Pt/No (dB-Hz)	TLM + RNG Lo Threshold Pt/No (dB-Hz)	TLM + RNG Hi Threshold Pt/No (dB-Hz)
RS only (I=5)	130531.7	150000	Squarewave Subcar	375	72	Residual	58.2	58.5	59.3
RS only (I=5)	1740422.2	2000000	BPSK Direct Mod	None	72	Residual	69.5	69.9	71.2
RS only (I=5)	2088506.6	2400000	QPSK	None	82	Suppressed	70.0	-	-
RS only (I=5)	2393080.5	2750000	QPSK	None	82	Suppressed	70.6	-	-
RS only (I=5)	2610633.3	3000000	QPSK	None	82	Suppressed	71.0	-	-
RS only (I=5)	2871696.6	3300000	QPSK	None	82	Suppressed	71.4	-	-
RS only (I=5)	3480844.4	4000000	QPSK	None	82	Suppressed	72.3	-	-
RS only (I=5)	5221266.6	6000000	QPSK	None	82	Suppressed	74.0	-	-

6.5.4.4 RS-Only (Table 6-14). The RS-only coding is used primarily for very high data rates in situations where MRO is close enough to Earth that coding gain is less important (such as in early cruise and during Mars–Earth closest approach). The maximum rate for RS-only data is 6.6 Mbps (6.6 Msps). If interference to another project could occur, the limit is 4 Mbps (4 Msps).

6.5.5 Coordinating MRO and MER X-Band Operations

The MRO SDST operates on DSN channel 32. When it became apparent that MER-A (also on channel 32) and MER-B (on channel 29) were likely to still be active on Mars at MRO arrival, a coordination plan was agreed to between the projects. The agreement, excerpted below, focused on the MRO MOI and aerobraking phases. It continued in use through mid-2011, the end of the MER-A mission.

According to the general agreement, as shown in Table 6-15, MER-A ("Spirit") would forego use of the DSN during the MOI and aerobraking critical event periods. This agreement, to ensure MRO safety, documents the dates chosen for MER-A to use (and not use) DSN during this period.

Table 6-15. MRO–MER agreement on channel 32 X-band uplink use.

Schedule	Agreement
Before February 28	MER-A operates normally, X-band downlink (direct to Earth) and uplink (direct from Earth) as needed
Starting February 28	2006-059T00:00 - 2006-060T00:00
	No MER-A X-band operations on channel 32
March 1–5	MER-A X-band uplink allowed
March 6–12	2006-065T00:00 - 2006-072T00:00
(MOI was March 10)	No MER-A X-band operations on channel 32
March 13 through aerobraking exit (ABX) – 2 weeks	MER-A X-band uplink coordinated weekly with Mission Planning and Sequencing Team (MPST) to avoid overlaps with MRO uplink windows. MER-A will not use X-band uplinks if in conflict with MRO uplink windows
ABX – 2 weeks through ABX	No MER X-band operations (UHF with Odyssey)

Prior to the start of the critical aerobraking period (March–September, 2006), MRO and MER jointly developed a coordination plan to reduce the chances of an inadvertent MRO channel 32 uplink interfering with MER channel 32 commanding, or vice versa. The plan called for MER to define a one-hour period for each MER-A sol when the project would do any required X-band commanding for that sol. This period is outlined in Fig. 6-22. Times in the figure go from left to right on two rows. The top row is spacecraft event time at Mars. The bottom row is Earth transmit time at the station.

Figure 6-22 and the acronym MUKOW (MRO uplink keep out window) define the goal of this coordination.[11] The term "keep out" referred to scheduling the DSN uplink to MRO to be turned off at an agreed-to time, to keep it out of the MER SDST receiver when MER uplinking was required, and to scheduling the uplink to MER to be turned off to keep it out of the MRO SDST receiver when MRO uplinking was required. The MUKOW required coordinated scheduling between the projects. With that done, each project could complete its individual planning process with its group of allocated DSN stations operating independently. See chapter 7 Refs. 19 and 20.

[11] The MUKOW coordination process continued through the extended science phase (ending September 2010) and into subsequent mission phases of the MRO and MER projects, as long as DSN support of MRO spacecraft and the Spirit Rover X-band activity existed.

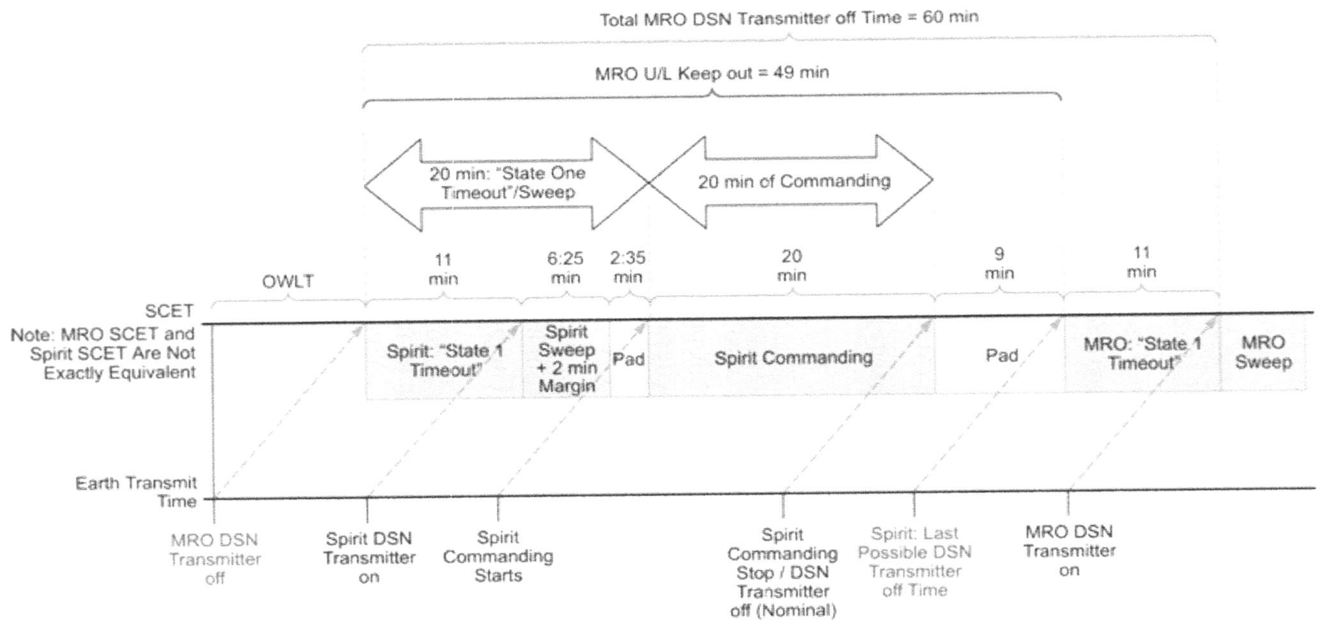

Fig. 6-22. MRO uplink keep out window (MUKOW) timing.

Every two weeks, MER delivered a spreadsheet file with two entries in each row: (1) the time the DSN transmitter supporting MRO was to be turned off and (2) the time the DSN transmitter subsequently supporting MER-A was to be turned off.

The MER and MRO SDSTs each have a "State 1 timeout" duration of 10 minutes between when the SDST receiver goes one way (in response to the station transmitter turn off) and when it forces its phase-locked loop (PLL) to best-lock frequency. Each project uses a "sweep acquisition" uplink frequency profile by the station to ensure the station's uplink carrier goes through the actual (temperature-dependent) best-lock frequency.

The coordination process, as developed on a basic weekly schedule, allowed for negotiated updates (including cancellations or additions) of keep-out times to meet significant needs of either project. These could include changes in MRO aerobraking times or anomalies that might occur with either project.

Although the agreement primarily affected uplink operations, both MER spacecraft occasionally required a direct-to-Earth (DTE) X-band downlink for onboard spacecraft clock correlation with UTC. Also, both MER spacecraft relied on the detection by the DSN station of an unmodulated X-band downlink carrier ("beep") to verify the success of the "in the blind" direct-from-Earth (DFE) command session and the consequent hand off by flight software to the new sol's master sequence [15].[12]

Successful lock-up of MER DTE passes required coordination with MRO because the MRO downlink signal level is much greater than the MER HGA DTE downlink. When MER-A needed to do a DTE, it would use a specific communications mode (504 bps or lower on a 25 kHz subcarrier, (7,1/2) convolutional code). During the time of the MER-A downlink, MRO needed to reduce its telemetry rate and be on the USO (achieved by the MUKOW ground transmitter coordination).

For a MER-B ("Opportunity") DFE uplink (three channels away), MRO can continue its normal uplink. During the time of a MER-B downlink (three channels away), it was recommended that the coordination include the absence of uplink ranging modulation to MRO. This was achieved by defining the station configuration for the MRO pass to not include ranging.

---

[12] MER beep detection does not require any special MRO configuration or action. It does require (now only for MER-B) that the station tracking MER narrow the receiver's fast Fourier transform (FFT) in both bandwidth and signal level range to reduce the effects of MRO spectral components near the beep frequency. The beep frequency is precisely known because the beep is two-way coherent with the uplink.

Though the MUKOW process continued with the MER project to mid-2011, MRO aerobraking exit (ABX) was completed in September 2006, after which the instruments were given a check and configured for solar conjunction that began in October 2006.

6.6 Ka-Band Cruise Verification

6.6.1 Ka-Band Operations Overview

The MRO Ka-band mission activity was planned to have two components: an engineering verification of basic spacecraft and station Ka-band functionality during the MRO cruise to Mars, and a communications operational demonstration during the orbital mission. Before orbital operations began, an onboard Ka-band exciter anomaly occurred on May 26, 2006 [16], and an X-band waveguide transfer switch anomaly occurred on August 16, 2006 [17]. The project has elected not accept the risk of conducting the Ka-band operational demonstration during either the primary or extended orbital mission.

This section describes primarily the activities planned, executed and documented during cruise. At the end is a summary of the objectives of the operational demonstration relative to what had been started during the Deep Space 1 mission (Chapter 5).

6.6.2 Ka-Band Link Prediction and Performance during Cruise

MRO and four 34-m BWG stations participated in a total of 10 passes dedicated to Ka band during cruise. All three sites participated. DSS 25, DSS 34, and DSS 55 each had three passes, and DSS 26 had one.

The maximum Ka-band data rate achievable throughout the entire mission, with a 3-dB margin, would have been 331 kbps. In fact, a wide variety of data rates and modes was used during cruise. These were changed by the background sequence, with modulation index also changed by real-time command, as described in the next section. To simulate the occultation of MRO by Mars and to verify the ability of the stations to reacquire the Ka-band downlink, the Ka-band TWTA was turned off and back on.

During the dedicated pass on October 7, 2005, DSS-25 decoded turbo-coded data from the Ka-band downlink for the first time. During the pass on October 31, DSS-55 received a total of 133 Gbits from Ka-band, at rates as high as 6 Mbps; these represented the largest data volume and the highest data rate from deep space to that time.

In the critical area of station antenna pointing, the best performance to date was with DSS-34, which consistently achieved 4- to 5-mdeg blind-pointing

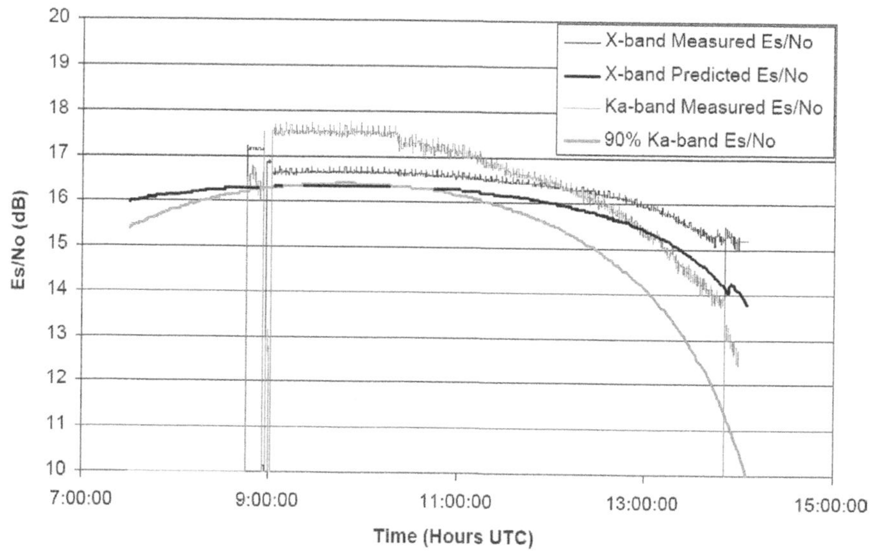

Fig. 6-23. Comparison of Ka-band and X-band telemetry (DSS 34, December 26, 2005). Please visit www.wiley.com/go/deepspacecommunications to access the color version of this figure.

error according to the monopulse system's correction offsets. It appears that DSS-34 has the best sky models for blind-pointing in the regions for MRO during cruise.

The MRO Ka-band team demonstrated functionality and characterized at both bands the performance of the radiometric data types used for navigation and radio science. Two-way Doppler and ranging performance (both downlink bands using the X-band uplink) were comparable at X-band and Ka-band and met or exceeded project requirements.

The two sets of shadow passes operated with both X-band and Ka-band set for 550 kbps RS and (7,1/2) convolutional concatenated coding. Table 6-10 shows the information rate was ~480 kbps in this mode, and the SDST output symbol rate was 1.1 Msps. Ranging was off for Ka-band in the first set of shadow passes and at 17.5-deg modulation index for the second set. The station monopulse system (not required for X-band) was not used in the first set of shadow passes, but was tried for some of the passes in the second set. Significant findings from the cruise tests included the following:

- Ka-band SNT measurements are sufficiently accurate, when the monopulse works, at signal levels corresponding to the shortest Earth–Mars distance.
- The MRO Ka-band system (35-W RF) can outperform the X-band system (100-W RF) in good weather, as shown in Fig. 6-23 [13].

The lessons learned section describes some of the other findings from the cruise tests [18].

6.6.3 Ka-Band Communications Demonstration Plans

The motivation for verifying the operational use of Ka-band was to build upon the results from Deep Space 1 and the MRO cruise experiment to achieve increased available bandwidth and, therefore, a higher available data rate. The deep-space allocation at X-band (8.4–8.45 GHz) is 50 MHz, and that at Ka-band (31.8–32.3 GHz) is 500 MHz. However, weather effects cause much larger fluctuations on Ka-band than on X-band. This characteristic makes the traditional link design for data return power inefficient for Ka-band. The traditional method involves a single downlink rate per pass and assigns a margin sufficient to provide a required data availability and to overcome a defined weather-effects severity. The margin is larger than required most of the time. By testing the operational use of Ka-band, MRO could have demonstrated the potential for greater average data rate using a concept of operations requiring significantly less power for the same total data volume. Several variations of this concept [19] use optimization techniques involving multiple data rates during a tracking pass. The rate-selection criteria account for the station elevation-angle profile as well as the distance of the spacecraft to Earth during each pass.

The MRO test was planned in the form of a telecommunications technology demonstration. The purpose of the demonstration was to develop operational procedures specific for Ka-band that account for the weather variations and are still compatible with the way the MRO flight team sequences the spacecraft. The demonstration would involve the use of data rate (and coding and modulation index) selection algorithms with input from time-variable weather models (and possibly forecasts). On the ground, the MRO cruise mission phase had already shown how the ability to point the station antenna accurately enough and to monitor signal-to-noise ratio and system noise temperature are factors to be resolved in order to determine the operational feasibility of using Ka-band for future missions.

The Ka-band experiment team planned to characterize and create or update models of Ka-band link performance. The operational demonstration planned to determine:

- The accuracy of existing Ka-band models
- Effects of weather forecasting/predicting on Ka-band telemetry
- Benefits of data-rate optimization during Ka-band passes
- Ka-band link performance during solar conjunction
- Differences in ranging and Doppler performance between X-band and Ka-band.

None of these plans were accomplished because the MRO Ka-band was not switched on during orbital operations.

6.6.4 Spacecraft X-Band and Ka-Band Constraints and Operational Factors

Considering one link (X-band or Ka-band) at a time, the codes available in the C&DH are

- Turbo codes with block length 8920 bits and rates 1/2, 1/3, and 1/6
- (255,223) RS block code.

The SDST has the capability to concatenate a (7,1/2) convolutional code on the RS, making a third coding type available.

The C&DH imposes the following limits on the coding and data rates available. The channel symbol rate is defined at the SDST output, so these limits refer to convolutional code symbols for the concatenated code.

- If the X-band and Ka-band downlinks have different data types (carry non-identical data streams), one has to use turbo coding and the other has to use RS coding.
- If X-band and Ka-band carry different data types, the combined channel symbol rate of the two bands should not exceed 6 Msps.
- If both bands carry identical data, the symbol rate on each band cannot exceed 6 Msps.

The ground turbo decoder limits the rate 1/2 code to a maximum bit rate of 1.5 Mbps.

At Mars, to minimize interference to missions in the same station antenna beamwidth and operating on nearby channels, X-band uses QPSK modulation for symbol rates higher than 2 Msps. For Ka-band, BPSK modulation is always used.

During the prime science mission, the MRO mission plan was to allocate the Ka-band demonstration two passes per week and one delta-DOR pass a month. The SDST allows for independently configurable telemetry subcarrier frequencies and modulation index values, independent control of DOR (on/off), turnaround ranging (on/off), and ranging modulation index.

The operations demonstration concept was based on maximizing the average data return subject to minimum availability. Three different scenarios were planned [19]:

1) Nominal link operations using link designs (predictions) based on long-term monthly or weekly statistics

2) Link operations using short-term forecasts

3) Link operations during superior solar conjunction.

The demonstration was planned to take advantage of the fact the spacecraft can be sequenced through two different procedures: background sequencing and mini-sequencing [13]. The approved sequence represents the onboard programming that controls subsystem configuration and operation. In addition, real-time commands can be used to change simple functions such as Ka-band modulation index. Use of real-time commands is very limited because of possible interaction with the planned sequences that must be validated before being uploaded.

The lead time for normal MRO sequencing meant that it would have been a challenge to incorporate very short term (1–2 days) weather forecasting into any Ka-band operational concept. Normal background sequencing programs the spacecraft for 28 days, and a background sequence goes through a 28-day cycle to design, test, and upload. During the cruise experiment, the Ka-band link's data rate and modulation index were changed according to the background sequence.

Mini-sequencing programs the spacecraft for specific events such as instrument calibrations or trajectory correction maneuvers. A second important use of the mini-sequence is to modify an already executing background sequence according to later information available to the project. The development cycle for mini-sequences typically takes a week. Using mini-sequencing, MRO Ka-band telecom parameters, such as data rate profile and modulation index, could be changed weekly – at best – rather than monthly [18].

6.6.5 Delta-DOR X-Band and Ka-Band Operations and Performance

Data at both X-band and Ka-band were acquired for seven delta-DOR passes during MRO cruise.

At X-band, the technique of delta differential one-way ranging (delta-DOR) has proved to be valuable for supporting deep-space cruise navigation, especially for missions with tight targeting requirements at Mars [13]. To make a delta-DOR measurement, a very long baseline interferometry (VLBI) system at each of two stations makes high rate recordings of signals from the spacecraft and angularly nearby radio sources. Driven by a sequence of events called the DSN Keywords File, the antennas at both stations point alternately, every few minutes, at the spacecraft and at the radio source in synchronism as the recordings are made. The radio source observations calibrate the system. For each source, the difference in signal arrival time between stations is determined and delivered to the navigation team, constituting the measurement. Operational X-band measurements are specified to provide an angular position accuracy of 2.5 nanoradians (nrad) [20].

The wider spectrum allocation at Ka-band can enable an advance in delta-DOR accuracy. To achieve a substantial improvement in accuracy at Ka-band, it was necessary to improve antenna-pointing performance at the ground stations, increase the frequency of the DOR tone at Ka-band, increase the sample rate for the VLBI data recordings, and continue work on surveying radio sources at Ka-band. These were accomplished as follows for the cruise experiment:

- Radio source flux surveys were made by the National Radio Astronomy Observatory (NRAO) at 24 GHz and 43 GHz.
- The VLBI receiver's front-end bandwidth was widened to accommodate the ±76 MHz spanned by the MRO DOR tones.
- Initial models for station antenna "blind-pointing" were developed for pointing to the radio sources where monopulse could not be used.

Except for a few radio source observations that were degraded due to known ground station pointing problems, the measurement accuracy at Ka-band was comparable to the accuracy at X band and within expectations. The factor-of-four increase in the DOR tone frequency for MRO at Ka-band relative to X-band was just enough to offset the lower SNR for sources at Ka band relative to X-band.

6.6.6 Planned Solar Conjunction Experiments

Communications experiments were planned to compare X-band with Ka-band during the two solar conjunctions that bracketed the primary science mission. These conjunctions were in October–November 2006 and 04 2008. The minimum Sun–Earth–probe (SEP) angles were 0.39 deg on October 23, 2006, and 0.46 deg on December 5, 2008. Because the Ka-band exciter anomaly and the waveguide transfer switch anomaly occurred prior to the first solar conjunction, no formal experiments were done. Similar to other projects, MRO invoked for each solar conjunction an uplink command moratorium and planned X-band data rate reductions based on the profile of SEP angles.

The solar conjunction communications experiment would have primarily been at DSS-25 at Goldstone, where Earth weather (tropospheric) effects are usually minimal. The experiment plan built upon data from previous missions by characterizing the solar charged-particle effects on the Ka-band and X-band carrier and telemetry links. At both frequency bands, the effect of the solar charged particles increases as the Sun-Earth-probe (SEP) angle decreases. The primary objectives were

- To evaluate Ka-band performance as a function of SEP angle against the concurrent X-band performance
- To measure any degradation to the link that occurred during solar coronal transient activity, such as coronal mass ejections.

Solar effects are smaller at Ka-band than at X-band, and this advantage of using Ka-band was known from previous in-flight experiments. One goal was to determine how low the SEP angle could possibly go (for each band) while maintaining carrier lock and achieving reasonably reliable telemetry (with care given to telemetry modulation index and station receiver loop bandwidth parameters). An X-band link using BPSK begins to degrade near a 2-deg SEP angle. Based on comparable solar effects, it is believed that a Ka-band link would begin to degrade somewhere near 1 deg.

To isolate the downlink effects from the uplink, the experiment planned to use the USO as the downlink frequency reference for passes when the SEP angle was 5 deg or less.

Below 1-deg SEP angle, the experiment plan included simulated frequency shift-keying (FSK) modulation using the carrier to demonstrate information flow at the equivalent of 1 bps.

6.7 Lessons Learned

The initial MRO X-band and Ka-band lessons learned were documented in the MRO Post-Launch Assessment Review (PLAR) [21], which was held in October 2005. Some of these problems have subsequently been resolved or worked around. As the project documents additional lessons, documentation updates will continue.

The major after-launch telecom hardware issues, discussed on previous pages, include

- The Ka-band exciter anomaly
- The X-band waveguide transfer switch anomaly
- The use of an SDST operating on the same X-band channel as the SDST on another active Mars mission and the consequent need for inter-project coordination between MRO and MER.

In addition to these, other issues have arisen during flight. Some issues have subsequently been resolved or worked around. The following paragraphs briefly summarize the MRO issues in terms of lessons-learned by frequency band: X-band, Ka-band, and UHF.

6.7.1 X-Band

Earth Network Bandwidth (from DSN stations to JPL): The term "bandwidth" refers both to the maximum bit rate a channel can carry and the longest time delays from one end of the channel to the other.

The MRO requirements are as follows:

- The Deep Space Mission System (DSMS) shall provide the capability to ensure that a version of a payload product set containing any data can be made available within 24 hours of receipt of the Earth-receive time of that data.
- For the aerobraking phase, the project requests a guarantee of real-time data at a rate of at least 220 kbps.

The MRO project documented their concerns and worked with the DSN to resolve them prior to the critical aerobraking mission activity. These concerns regarded the timely transfer of spacecraft data from the stations to the mission operations and the computational horsepower required to process the data for subsequent aerobraking sequences. Generalized to a lesson-learned as missions become more and more bandwidth intensive, these concerns are:

- The station-to-JPL link may at times have too much data transfer delay to support critical engineering and primary science rates. The data links are provided by the NASA Integrated Services Network (NISN).
- Computing power may not be sufficient to unwrap real-time engineering packets during planning activities with very short turnaround times.
- Reliable Network Service (RNS) may not be able to handle the full downlink rate from station to project, particularly when one site (for example Madrid) might be tracking several spacecraft, each with high rate downlinks.

6.7.2 Ka-Band

In-Flight HGA Calibration: The first HGA calibration was performed over DSS-55 on September 9, 2005. The calibration point spacing (1 deg by 1 deg) was too large to resolve the Ka-band antenna pattern. The calculated Ka-band boresight had an error of 0.1 deg. This is equivalent to an uncertainty (loss) in effective isotropic radiated power (EIRP) of 5 dB.

The received downlink Ka-band signal level was too high for a normally configured station to measure the antenna pattern accurately.

Station antenna pointing errors and atmospheric effects could significantly affect the calibration. Active measures to reduce or account for their impact are recommended.

6.7.3 UHF

Electromagnetic interference (EMI) to the Electra receiver: The EMI problem was documented as a result of pre-launch testing, it was subsequently confirmed in cruise phase testing, it was a continuing concern during the primary science phase, and it remains a concern in the relay phase.

Pre-flight EMI tests showed that almost all unwanted MRO payload and MRO spacecraft subsystem UHF output appeared as tones. Box level testing of various science payloads during MRO development revealed interferers that produced tones exceeding a specification threshold signal level of –140 dBm, with perhaps hundreds of tones in total. One tone from the CRISM instrument, inside the nominal Electra receive band, was measured at –70 dBm centered near 400 MHz. The power in this single EMI tone is higher than any signal level we expect to receive from any Mars lander. The tones have been identified as harmonic overtones of switching power supplies, data buses, or clock mechanisms.

For return-link communications at rates of less than 256 kbps, the Electra radio team reprogrammed the MRO FGPA modem post launch to include a digital filter that partially suppresses this large interference tone and allows relay support at the standard 401.6 MHz return-link frequency. For data rates greater than 256 kbps, the prescribed approach is to move the return-link center frequency more than 3 MHz away from the 400-MHz interference tone. Doing this allows the surface acoustic wave (SAW) filter in Electra to eliminate this tone completely. This approach is used for MSL support using a return link center frequency of 391 MHz.

References

[1] J. Taylor, D. K. Lee, S. Shambayati, *Mars Reconnaissance Orbiter Telecommunications, DESCANSO Design and Performance Summary Series*, Article 12, Jet Propulsion Laboratory, California Institute of Technology, Pasadena, California, September 2006. http://descanso.jpl.nasa.gov/DPSummary/summary.html (accessed October 30, 2014)

[2] "Mars Reconnaissance Orbiter public homepage," Jet Propulsion Laboratory, California Institute of Technology, Pasadena, California. ttp://mars.jpl.nasa.gov/mro/ (accessed December 29, 2010)

[3] R. Lock and R. Sharrow, *MRO Mission Plan*, Rev. C, JPL D-22239 (internal document), Jet Propulsion Laboratory, California Institute of Technology, Pasadena, California, July 2005.

[4] D. Morabito, "Communicating with Mars During Periods of Solar Conjunction," Jet Propulsion Laboratory, California Institute of Technology, Pasadena, California, March 11, 2002, http://trs-new.jpl.nasa.gov/dspace/bitstream/2014/11771/1/02-0400.pdf (accessed December 29, 2010)

[5] P. Varghese, *MRO Mission*, JPL D-68093 (internal document), Jet Propulsion Laboratory, California Institute of Technology, Pasadena, California, December 2010.

[6] J. Taylor and D. Bell, *Mars Relay Description for Scout 2007 Proposals*, JPL D-93013 (internal document), Jet Propulsion Laboratory, California Institute of Technology, Pasadena, California, April 2006.

[7] C. H. Lee, K.-M. Cheung, C. Edwards, S. J. Kerridge, G. K. Noreen, and A. Vaisnys, "Orbit Design Based on Global Maps of Telecom Metrics," Aerospace 2005 IEEE Conference, Big Sky, Montana, March 5–12, 2005.

[8] G. Glass, Small Deep Space Transponder (SDST) Evolution, D-689094 (internal document), Jet Propulsion Laboratory, California Institute of Technology, Pasadena, California, July 18, 2005.

[9] E. Satorius, T. Jedrey, D. Bell, A. Devereaux, T. Ely, E. Grigorian, I. Kuperman, and A. Lee, "The Electra Radio," Chapter 2, *Autonomous Software-Defined Radio Receivers for Deep Space Applications*, J. Hamkins and M. K. Simon, eds., Jet Propulsion Laboratory, California Institute of Technology, Pasadena, California, http://descanso.jpl.nasa.gov/monograph/series9/Descanso9_02.pdf (accessed October 30, 2014), also John Wiley & Sons, Hoboken, New Jersey, 2006.

[10] *Proximity-1 Space Link Protocol—Physical Layer, Recommendation for Space Data System Standards*, CCSDS 211.1-B-4 (Blue Book), Consultative Committee for Space Data Systems, December 2013. http://public.ccsds.org/publications/archive/211x1b4.pdf (accessed October 30, 2014)

[11] *Proximity-1 Space Link Protocol—Data Link Layer, Recommendation for Space Data System Standards*, CCSDS 211.0-B-5 (Blue Book) Consultative Committee for Space Data Systems, December 2013, http://public.ccsds.org/publications/archive/211x0b5.pdf (accessed October 30, 2014)

[12] *DSN Telecommunications Link Design Handbook*, Rev. E, JPL 810-005 (D-19379), Jet Propulsion Laboratory, California Institute of Technology, Pasadena, California, [Note: the document has the organizational name of DSN (Deep Space Network); in the early 2000s, it was DSMS for Deep Space Mission Systems.], August 25, 2006. http://deepspace.jpl.nasa.gov/dsndocs/810-005/ (accessed December 29, 2010.)

[13] S. Shambayati, D. Morabito, J. S. Border, F. Davarian, D. Lee, R. Mendoza, M. Britcliffe, and S. Weinreb, "Mars Reconnaissance Orbiter Ka-band (32 GHz) Demonstration: Cruise Phase Operations," AIAA 2006-5786, AIAA SpaceOps 2006, Rome, Italy, June 18–23, 2006, American Institute of Aeronautics and Astronautics, 2006.

[14] D. K. Lee, MRO Telecom Design Control Document, Rev. A, JPL D-22724 (internal document), Jet Propulsion Laboratory, California Institute of Technology, Pasadena, California, July 18, 2005.

[15] J. Taylor, A. Makovsky, A. Barbieri, R. Tung, P. Estabrook, and A. G. Thomas, *Mars Exploration Rover Telecommunications, DESCANSO Design and Performance Summary Series*, Article 10, Jet Propulsion Laboratory, Pasadena, California, October 2005, http://descanso.jpl.nasa.gov/DPSummary/summary.html (accessed October 30, 2014)

[16] E. D Archer, *MRO Small Deep Space Transponder Ka-band Exciter Anomaly Final Report*, JPL D-31195 (internal document), Jet Propulsion Laboratory, California Institute of Technology, Pasadena, California, May 15, 2007.

[17] T. Bayer and L. Epp, *Mars Reconnaissance Orbiter Waveguide Transfer Switch Anomaly Final Report*, JPL D-31194 (internal document), Jet Propulsion Laboratory, California Institute of Technology, Pasadena, California, March 30, 2007

[18] S. Shambayati, J. Boder, D. Morabito, and R. Mendoza, "MRO Ka-and Demonstration: Cruise Phase Lessons Learned," IEEE Aerospace Conference 2007, Big Sky, Montana, June 18, 2007.

[19] S. Shambayati, F. Davarian, and D. Morabito, "Link Design and Planning for Mars Reconnaissance Orbiter (MRO) Ka-band (32 GHz) Telecom Demonstration," IEEEAC paper # 1383, *IEEE Aerospace Conference*, Big Sky, Montana, March 5–12, 2005.

[20] J. S. Border, "Innovations in Delta Differential One-way Range: From Viking to Mars Science Laboratory," *Proceedings of 21st ISSFD (International Symposium on Space Flight Dynamics)*, Toulouse, France, September 28–October 2, 2009. http://issfd.org/ISSFD_2009/OrbitDeterminationI/Border.pdf (accessed April 19, 2011)

[21] *MRO Post-Launch Assessment Review* (internal document), Jet Propulsion Laboratory, California Institute of Technology, Pasadena, California, April 24, 2006.

[22] T.-H. You, E. Graat, A. Halsell, D. Highsmith, S. Long, R. Bhat, S. Demcak, E. Higa, N. Mottinger, and M. Jah, "Mars Reconnaissance Orbiter Interplanetary Cruise Navigation," *20th International Symposium on Space Flight Dynamics*, Annapolis, Maryland, 2007. http://issfd.org/ISSFD_2007/3-4.pdf (accessed July 24, 2014)

[23] B. Arnold and D. Bell, *MRO/Stanford UHF Test Report*, JPL D-31161 (internal document), Jet Propulsion Laboratory, California Institute of Technology, Pasadena, California, February 23, 2006.

Chapter 7
Mars Exploration Rover Telecommunications

Jim Taylor, Andre Makovsky, Andrea Barbieri, Ramona Tung,
Polly Estabrook, and A. Gail Thomas

This chapter describes and assesses telecommunications of the two rovers launched in 2003 and named Spirit and Opportunity [1]. Throughout this chapter, the names MER-A and Spirit are used interchangeably, and likewise MER-B and Opportunity. Generally, the term "spacecraft" refers to the vehicle before landing, and the term "rover" refers to the vehicle after landing.

For each spacecraft (rover), there were three phases of the Mars Exploration Rover (MER) primary flight mission:

- As a cruise spacecraft, MER communicated with the tracking stations of the DSN via an X-band uplink and downlink.
- During entry, descent, and landing (EDL), the cruise stage had been jettisoned; the MER lander continued to communicate via an X-band downlink to the Deep Space Network (DSN), and it initiated an ultrahigh frequency (UHF) return link to the Mars Global Surveyor (MGS) orbiter.
- On the surface, the lander opened up to reveal the rover, which stood up and completed egress by driving off from the lander after several sols. The rover communicates with the DSN and with MGS as well as with the 2001 Mars Odyssey (ODY) orbiter and the European Space Agency's Mars Express (MEX) orbiter.

Some of the grey scale figures in this book have color versions in the companion website www.wiley.com/go/deepspacecommunications. Please refer to this website where the text refers to colors.

Deep Space Communications, First Edition. Edited by Jim Taylor.

The primary surface missions for the Spirit and Opportunity rovers ended as planned in April 2004, after 90 sols, with extended missions continuing for both rovers. As of the end of 2010 each rover had accumulated more than 5 Earth years of surface operations. Opportunity remains healthy and continues to drive and collect and transmit science data back to Earth, primarily through its UHF links to both Odyssey and the Mars Reconnaissance Orbiter (MRO). Spirit remains silent at her location on the west side of the plateau area known as Home Plate. No communication has been received from Spirit since Sol 2210 (March 22, 2010), as the fourth Martian winter of surface operations was beginning [2].

This chapter provides, mainly in Section 7.3, a description of the MER X-band and UHF telecommunication subsystems, with emphasis on both their development and operational challenges and lessons learned.

The MER spacecraft were designed, built, and tested at the Jet Propulsion Laboratory (JPL) in Pasadena, California. The MER Flight Team is located at JPL.

Much of the telecommunication (telecom) subsystem design information in this chapter was obtained from original primary mission design documentation: the X-band Operations Handbook [3] and the UHF Operations Handbook [4]. "MER Reports" [5] is an on-line compilation of detailed sol-by-sol science and engineering reports in the form of downlink reports from each operational area, including the telecom flight team. Reference 6 is a DocuShare library containing project reports and operational documents. (References 5 and 6 are only accessible from within JPL.)

7.1 Mission and Spacecraft Summary

7.1.1 Mission Objectives

The MER project had an initial primary objective of placing two mobile science laboratories on the surface of Mars to remotely conduct geologic investigations, including characterization of a diversity of rocks and soils that might hold clues to past water activity. The project intended to conduct fundamentally new observations of Mars geology, including the first micro-scale studies of rock samples, and a detailed study of surface environments for the purpose of calibrating and validating orbital spectroscopic remote sensing. The project aimed to achieve these objectives in a manner that would offer the excitement and wonder of space exploration to the public.

The Mission Plan [7] quantifies the objectives of a 90-sol surface mission in terms of minimum and full mission success. The project required that minimum mission success be achievable through use of X-band only or UHF only.

The rovers achieved more than full mission success. One example of the success criteria relates to the requirement to drive and use the instruments:

Full success: Drive the rovers to a total of at least eight separate locations and use the instrument suite to investigate the context and diversity of the Mars geologic environment. Every reasonable effort shall be made to maximize the separation between investigation locations to increase site diversity, without compromising overall mission safety or probability of success.

Minimum success: Drive the rovers to at least four separate locations and use the instrument suite to investigate the context and diversity of the Mars geologic environment.

With drives of nearly 8 km for Spirit and more than 25 km for Opportunity, and a total surface campaign lasting nearly 6 years through 2010, each rover has completed the "full success" objectives multiple times. In fact, there have been spirited debates in science planning about where stops can be made, and for how long, balancing the science that can be done at any given stop against achieving the long term driving objectives.

7.1.2 Mission Description

MER-A and MER-B are identical. Each had a launch mass of 1,063 kilograms (kg). MER-A was launched using a Delta II 7925 launch vehicle from Space Launch Complex 17A (SLC-17A) at the Cape Canaveral Air Force Station (CCAFS) in Florida. MER-B was launched using a Delta II 7925H launch vehicle from SLC-17B at the Cape Canaveral facility. The launch period and arrival dates were as follows:

Mission	Open Window	Close Window	Actual Date	Arrival
MER-A	May 30, 2003	June 16, 2003	June 10, 2003	January 4, 2004
MER-B	June 25, 2003	July 12, 2003	July 7, 2003	January 25, 2004

The two 18-day launch periods were separated by a minimum of 8 days. The launch vehicle provider required 10 days to turn around launch operations, and if MER-A had not launched until the last day or two of its launch period, MER-B would have been delayed so that 10 days would have separated the launches. Each launch day had two instantaneous daily launch opportunities, providing a high probability of liftoff within the back-to-back MER launch periods. A fixed arrival date was used to make the planning for each of the MER-A and MER-B missions tractable.

Table 7-1. Mission phases and planned dates for MER-A and MER-B (detailed to 2005).

Phase	Definition	MER-A Open Phase Start	MER-B Open Phase Start
Launch	Launch to thermally stable, positive energy balance state, launch telemetry played back	May 30, 2003	June 25, 2003
Cruise	End of Launch phase to Entry -45 days	May 31, 2003	June 26, 2003
Approach	Entry -45 days to Entry	November 20, 2003	December 11, 2003
EDL	Entry to end of critical deployments on sol 1	January 4, 2004	January 25, 2004
Postlanding through Egress*	End of EDL to receipt of DTE following successful placement of rover wheels on the Martian surface	January 4, 2004***	January 25, 2004***
Surface Operations**	End of Egress to end of Primary Mission	January 8, 2004	January 28,2004
Primary Mission End	Successful receipt of last scheduled UHF data return the night of sol 91	April 6, 2004	April 27, 2004
Extended mission		*May 2004*	*May 2004*
First Continuations of extended mission		*October 2004 (start FY 2005)*	*October 2004 (start FY 2005)*

* Sometimes referred to as "egress" for short, or as "impact through egress" (ITE).

** Sometimes referred to as "surface" for short.

*** The planned minimum duration of ITE (for Spirit) was 4 sols, establishing the planned start date of surface operations.

"Extended missions" refers to MER surface operations beginning in May 2004 and continuing into 2016.

Most of Table 7-1, from the Mission Plan, summarizes the planned phases of the primary mission. The last two rows (italicized) define the first two extensions of the mission. The initial extended mission was approved to the end of FY2004 (September 28, 2004). A 6-month extended-extended mission began the next day and concluded March 27, 2005. Since then, NASA has extended the mission several times, and it is currently into 2016 for the still active Opportunity rover. The extensions have been granted (funded) based on detailed project proposals for the kinds and value of the science that each extension would make possible.

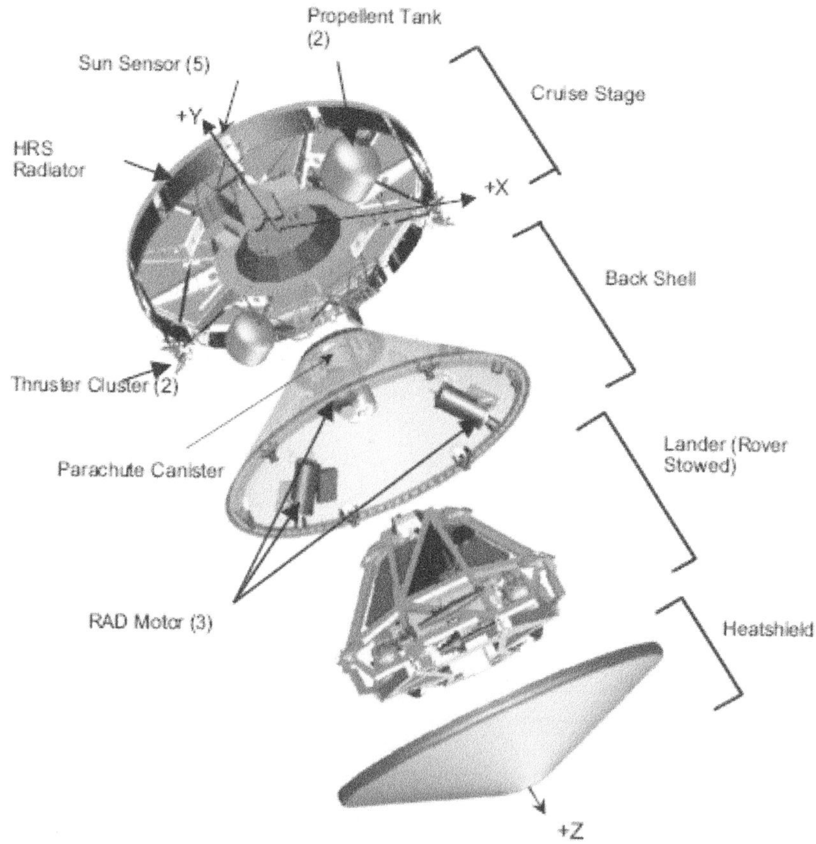

Fig. 7-1. MER Flight System, "Exploded" View. Please visit www.wiley.com/go/deepspacecommunications to access the color version of this figure.

7.1.3 The Spacecraft

The MER Flight System [7],[1] which is based on the Mars Pathfinder (MPF) cruise and EDL systems, delivered a large (185-kg) rover to the surface of Mars. The rover design is based on the Athena rover (carrying the Athena science payload), which began development under the Mars 2001 and Mars Sample Return (MSR) projects. An exploded view of the MER Flight System is shown in Fig. 7-1.

The Flight System consists of four major assemblies: 1) cruise stage, 2) aeroshell (heat shield and backshell), 3) lander, and 4) rover. The following description, table, and diagrams are from Ref. [7]. Table 7-2 summarizes the assembly masses.

---

[1] See Fig. 7-11 for a block diagram of the telecom subsystem elements discussed in the following paragraphs.

Table 7-2. Flight System mass breakdown.

Component	Allocated Mass (kg)	Cumulative Mass (kg)
Rover	185	185
Lander	348	533
Backshell/Parachute	209	742
Heat Shield	78	820
Cruise Stage	193	1013
Propellant	50	1063

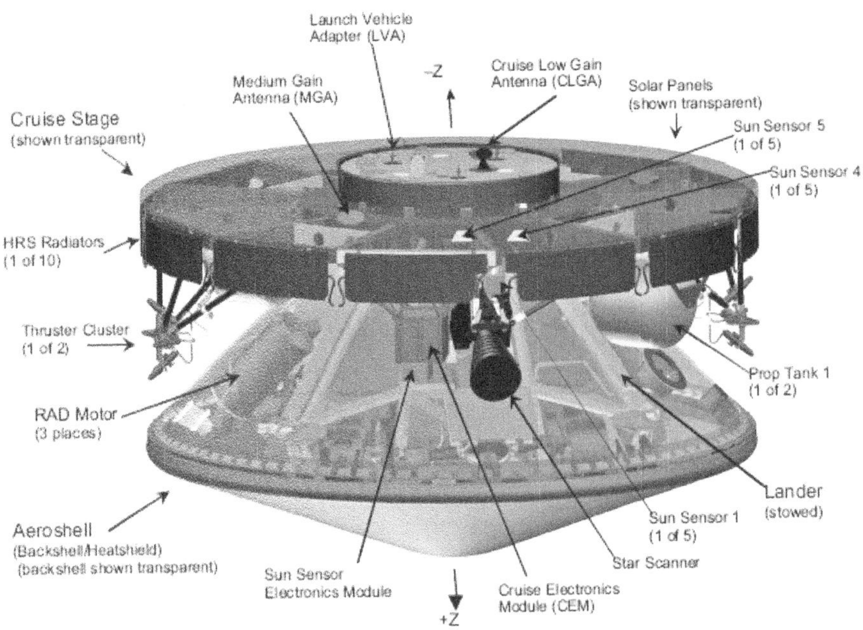

Fig. 7-2. MER Spacecraft in cruise configuration. Please visit www.wiley.com/go/deepspacecommunications to access the color version of this figure.

7.1.3.1 Cruise Stage. The spacecraft in its cruise configuration is shown in Fig. 7-2.

The cruise stage is very similar to the MPF design and is approximately 2.65 m in diameter and 1.6 m tall (attached to aeroshell) with a launch mass of 1063 kg. During flight, MER is a spin-stabilized spacecraft with a nominal spin rate of 2 revolutions per minute (rpm). Six trajectory correction maneuvers (TCMs) were planned during the flight to Mars, as well as payload and engineering health checks.

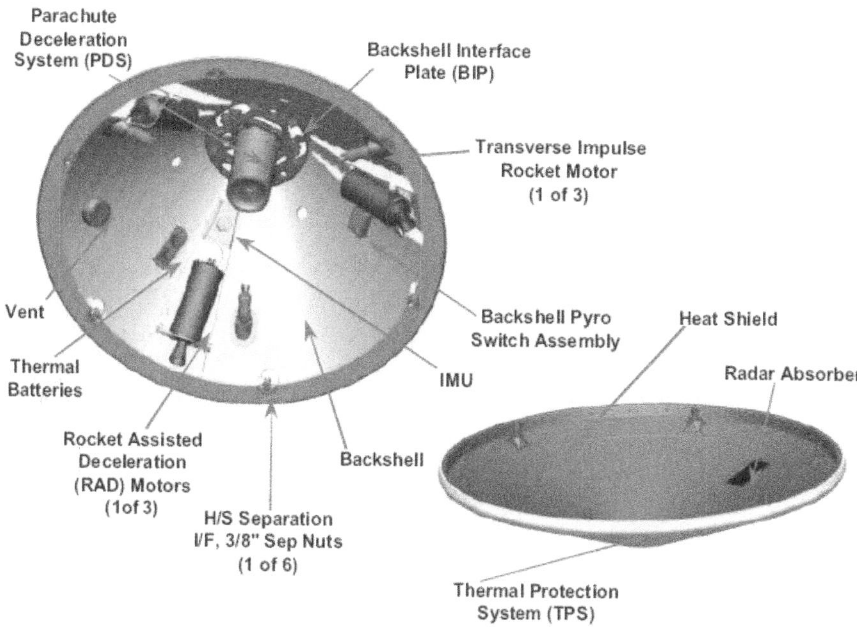

Fig. 7-3. Aeroshell configuration. Please visit www.wiley.com/go/deepspace communications to access the color version of this figure.

7.1.3.2 Entry, Descent, and Landing Systems (Aeroshell and Lander). Approximately 15 minutes (min) prior to entering the Martian atmosphere, the cruise stage was separated from the aeroshell containing the lander and rover. The aeroshell, shown in Fig. 7-3, is based on the MPF design, utilizing a Viking-heritage heat shield and thermal protection system. Stowed at the top of the backshell was an MPF/Viking-heritage parachute that was scaled up to approximately 15 meters (m) in diameter to accommodate MER's heavier entry mass of 825 kg.

Several other components used during EDL were mounted on the backshell. These included the backshell pyrotechnic device (pyro) switch assembly with relays controlling EDL pyro events, as well as redundant thermal batteries to power the pyros. A Litton model LN-200 Inertial Measurement Unit (IMU) mounted on the backshell propagated spacecraft attitude during entry and was also used to determine parachute deploy time based on deceleration in the atmosphere. Three small solid rockets mounted radially around the backshell constituted the Transverse Impulse Rocket System (TIRS); they provided horizontal impulse. The three large solid rockets of the Rocket-Assisted Deceleration (RAD) system nulled vertical velocity just before landing.

After ~4 min of atmospheric deceleration, at an altitude of ~10 kilometers (km) and an atmospheric relative velocity of ~450 meters per second (m/s),

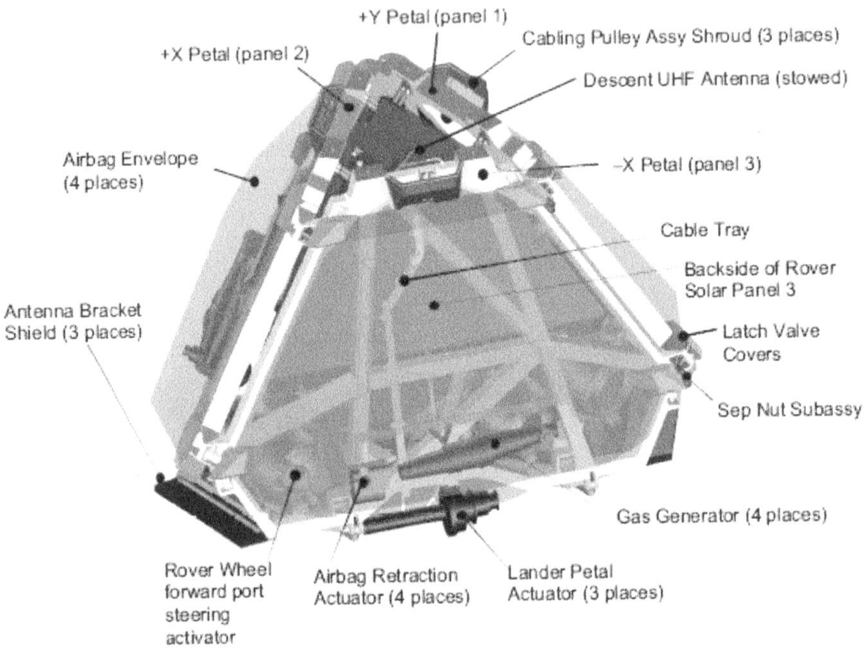

Fig. 7-4. Lander in stowed configuration. Please visit www.wiley.com/go/deepspace communications to access the color version of this figure.

the parachute was deployed. The heat shield was released using six separation nuts and push-off springs. The lander was lowered from the backshell on a Zylon[2] bridle, ~20-m-long, which was stowed in one of the lander side petals. The separation rate was controlled by a descent-rate limiter, which consisted of a friction brake and steel tape and was deployed with the bridle. The bridle incorporated an electrical harness that allowed the firing of the solid rockets from the lander/rover as well as providing data from the backshell IMU to the flight computer in the rover.

Figure 7-4 shows the lander in its stowed configuration and Fig. 7-5 in the extended position, ready for rover egress.

A radar altimeter unit, whose antenna is mounted at one of the lower corners of the lander tetrahedron, was used to determine distance to the Martian surface. Radar acquisition occurred within 2.4 km (~7900 ft) of the surface, ~5 min after entry, with the descent system traveling ~75 m/s. The radar data

---

[2] Zylon is a trademarked name for a range of thermoset polyurethane materials manufactured by the Zylon Corporation. These materials are members of the synthetic polymer family. Somewhat related to Kevlar and nylon, Zylon is used in applications that require very high strength with excellent thermal stability (from Wikipedia).

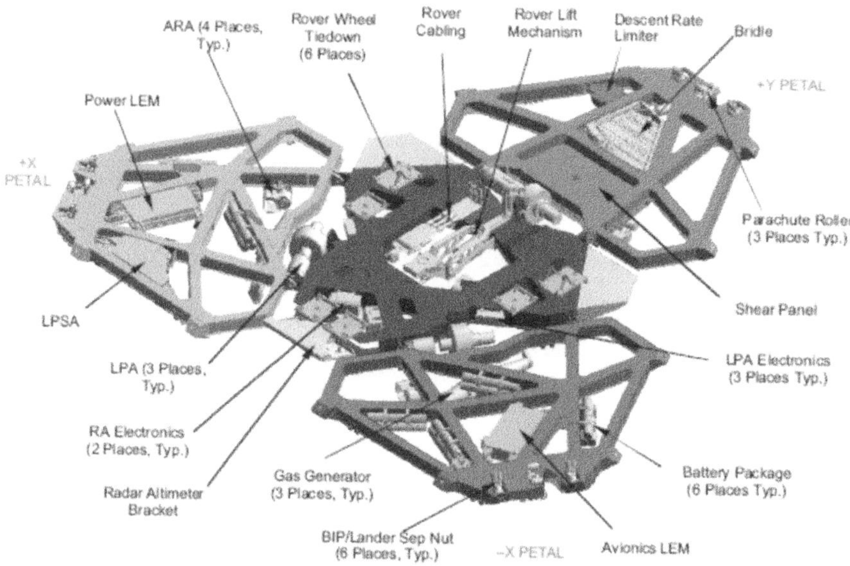

Fig. 7-5. Lander in deployed configuration (for clarity, egress aids are not shown). Please visit www.wiley.com/go/deepspacecommunications to access the color version of this figure.

was used to determine a firing solution for the RAD solid rockets on the backshell.

A soft landing was achieved by using the RAD to slow the lander to zero vertical velocity 10–15 m from the surface. A major concern during RAD firing was any backshell tilt that might have been introduced by winds in the lower atmosphere. The TIRS, an addition over MPF, could be fired in any combination to reduce a tilt effect.

The Pathfinder-heritage airbag system was used to cushion the impact of the lander on the surface. The radar provided data to determine (on-board) the RAD firing solution. Then, before RAD ignition or TIRS firing, the airbags were inflated to ~1.0 psig (as for MPF) via three pyro-initiated gas generators. The system was (correctly) expected to bounce many times and roll before coming to rest on the surface several minutes after initial contact.

The lander's primary structure was four composite petals with titanium fittings. The base petal connected to the three side petals through the high-torque lander petal actuators (LPAs), which could independently adjust the petals from the stowed position. The Flight Team could then command adjustment of the petals up or down to potentially improve the conditions for egress of the rover. Egress aids, or "ramplets," were connected between the side petals and were passively deployed when the petals opened.

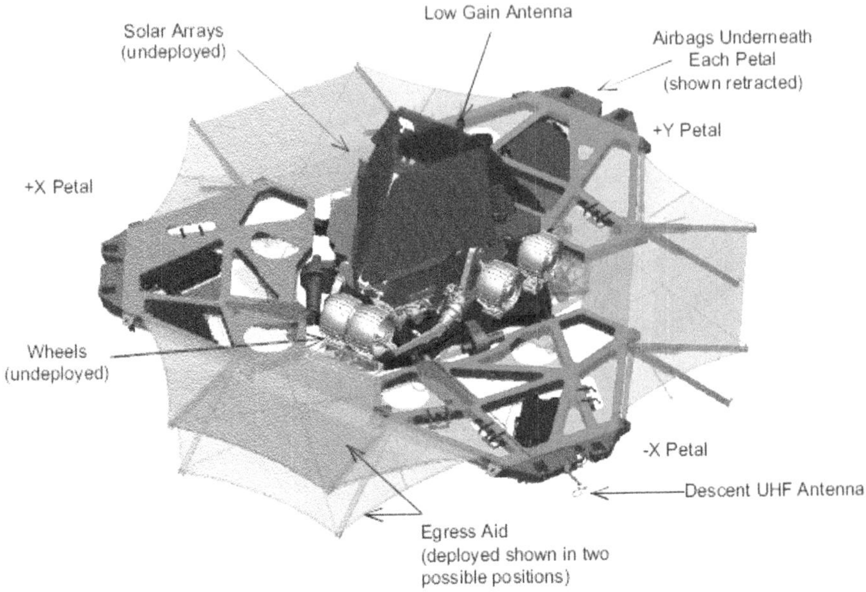

Fig. 7-6. Rover stowed on lander after petal opening. Please visit www.wiley.com/go/deepspacecommunications to access the color version of this figure.

7.1.3.3 Rover. At the heart of the MER spacecraft is the rover, shown in Fig. 7-6 in its stowed configuration, as it looked just after the lander had opened its petals.

Figure 7-7 shows the rover deployed. At its wheelbase, the rover is approximately 1.4 m long and 1.2 m wide. At its solar panel, the rover is 1.8 m wide and 1.7 m long. In its deployed configuration, with the Pancam Mast Assembly (PMA) deployed, the rover is just over 1.5 m tall and has ground clearance of at least 0.3 m. The rover body and primary structure, called the Warm Electronics Box (WEB), is an exoskeleton of composite honeycomb lined with aerogel[3] for insulation. The top face of the box, a triangular panel called the Rover Equipment Deck (RED) completes the WEB enclosure.

---

[3] Aerogel is a highly porous solid formed from a gel, such as silica gel, in which the liquid is replaced with a gas. Often called frozen smoke or blue smoke, it is composed of 99.8 percent air and is a stiff foam with a density of 3 milligrams per cubic centimeter (mg per cm^3), which makes it the world's lowest-density solid. The substance has extremely low thermal conductivity, which gives it its insulative properties. (from Wikipedia)

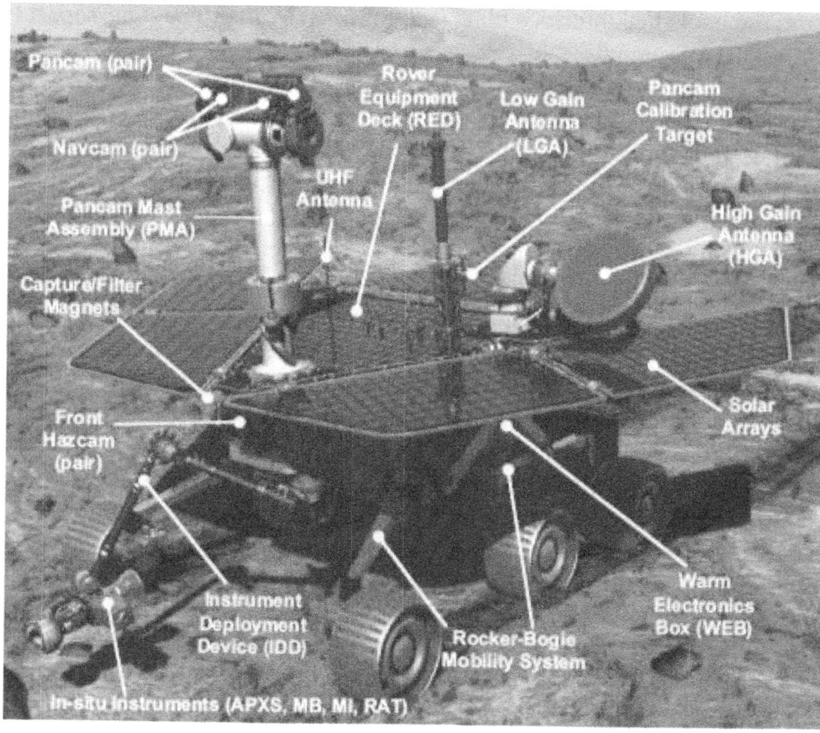

Fig. 7-7. Rover in deployed configuration. Please visit www.wiley.com/go/deep spacecommunications to access the color version of this figure.

7.2 Telecommunications Subsystem Overview

7.2.1 X-Band: Cruise, EDL, Surface

Communication functions on the rover are provided by an X-band transponder (the Small Deep-Space Transponder [SDST]), a solid-state power amplifier (SSPA), and a UHF transceiver located in the rover WEB. The SDST and SSPA operate in all mission phases. During cruise, the SDST received and transmitted via the Cruise LGA (CLGA) or the Medium-Gain Antenna (MGA). The CLGA served for the first few weeks after launch and for some TCMs. The MGA provided added capability as the Earth-to-Mars distance increased.

Communication during EDL was required to provide information to help reconstruct a fault should one occur. The LGAs available during EDL accommodated wide variations in orientation. During EDL, the X-band system transmitted multiple-frequency shift-keying (M-FSK) tones or semaphores, indicating the spacecraft state and completion of major EDL phases; the tones could be received at the expected orientations. (A similar, but simpler concept was used for MPF.) The Backshell LGA (BLGA) was used to radiate out from the backshell interface plate (see Fig. 7-3) from cruise stage separation until lander deployment.

Once the lander was separated from the backshell, the Rover LGA (RLGA) was then used to radiate from the top of the lander. In addition, a small patch antenna, mounted on the base petal (petal LGA [PLGA]), was used once the lander reached the surface. The rover cycled between the RLGA and PLGA once per minute to increase the probability that the signal would be received on Earth independent of which petal the lander came to rest on.

During the primary and extended surface missions, the X-band transponder has been supported by either an HGA or the RLGA mounted on the RED (Fig. 7-7). The RLGA has provided near omnidirectional coverage for both command and low rate telemetry data. Throughout the surface missions, the rover has been able to receive commands at a minimum rate of 7.8125 bps and transmit telemetry at a minimum rate of 10 bps on the RLGA. The HGA is a steerable, flat-panel, phased array, providing high-rate reception of command and transmission of telemetry data. During the surface missions, the uplink and downlink rate-capability via the HGA has depended on the Mars-Earth distance. At smaller ranges, command rates up to the 2-kilobits per second (kbps) maximum and telemetry up to the 28.8-kbps maximum have been used.

7.2.2 UHF: EDL, Surface

In addition to the X-band system, the UHF system was also used for the portion of EDL where the lander was suspended on the bridle. Following lander separation, a Descent UHF Antenna (DUHF, a small monopole antenna mounted at the top of the petals) was deployed to communicate with Mars Global Surveyor (MGS) at 8 kbps, providing engineering telemetry that was later relayed to Earth.

On the surface, the UHF system operated in a relay mode using both the Odyssey orbiter and the MGS orbiter's Mars balloon relay system. A relay/command demonstration with the MEX orbiter was also conducted. The rover's UHF system is implemented using a Cincinnati Electronics transceiver (Model CE-505) and was designed to be especially compatible with a like transceiver on Odyssey. The system uses a rover UHF antenna (RUHF, a 19-cm monopole antenna) mounted on the RED. This radio is capable of rates of 8, 32, 128, or 256 kbps for either transmission (rover to orbiter) or reception (orbiter to rover). The rover Flight System design limited the forward link to a single rate, 8 kbps. After some checkouts in the primary mission, the MER project coordinated with Odyssey to use either 128 kbps or 256 kbps on the return link for each pass, depending on which rate would give the greater data return. See Section 7.5.2.4.

7.2.3 Direct-to-Earth Downlink Capability

Figure 7-8, from Ref. [7], shows the prelaunch predicted direct-to-Earth (DTE) data-rate capability from MER-A landing to MER-B end-of-primary-mission. Each capability is a series of decreasing rates caused by the increasing Earth–Mars range over the time span. The least capability is RLGA to the 34-m stations (bottom curve), with the RLGA to the 70-m stations the second least. The greatest capability is the Earth-pointed HGA to 70-m stations, with the HGA to 34-m stations the second greatest. For a given combination of rover antenna and station type, on average the (15,1/6) code provides slightly greater capability than the (7,1/2) code.

7.2.4 UHF Relay Capability

UHF downlink data relays were planned through both the Odyssey and MGS orbiters. As defined for the primary mission, this link is used for the return of noncritical science and engineering telemetry.

More than 60 percent of the total mission data return[4] was planned to come through the UHF relay channel. An average of 1.8 potential communications passes above 20 deg elevation (with respect to the landing site) per sol per orbiter are available with a minimum of three passes every two sols and a maximum of four passes. These passes range in duration from 2 to 8 minutes, and the return-link data rate from the rover to both orbiters was planned to be as great as 128 kbps.[5] Maximization of the data downlink volume necessitates the use of as many of these UHF passes as possible.

Each rover had the potential for UHF relay passes with each of the two orbiters in the local morning and the local afternoon, providing as many as four UHF passes per rover per day. The orbiter morning passes are distributed between midnight and sunrise (local solar time), and the afternoon passes from midday to late afternoon. Figure 7-9 shows a typical distribution of passes for the Spirit rover with both Odyssey and MGS in local solar time units and the corresponding maximum elevation of each pass. The figure shows Spirit could be planned to communicate with MGS at about 01:30 and

---

[4] Data-return statistics for the Spirit and Opportunity primary missions through September 2005 are in Section **7.5.2**. In summary, about 92 percent of the total data return was to Odyssey, 5 percent to MGS, and 3 percent over the X-band DTE link.

[5] The specific plan was to return data from the first few post-landing passes at the lowest rate, 8 kbps, then to jump to 128 bps if the link performed as expected and could support that rate. This plan was achieved. In fact, the 256-kbps rate was used in the extended missions for many Spirit and Opportunity relay passes.

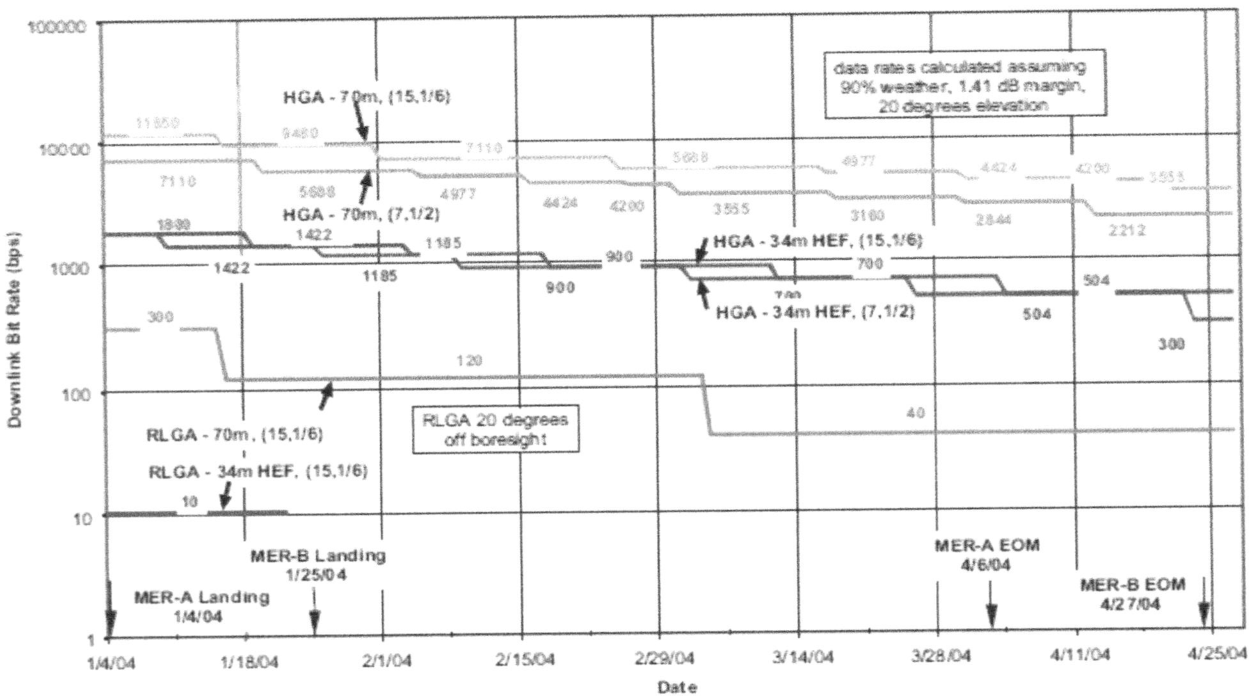

Fig. 7-8. X-band predicted downlink capability for mission planning. Please visit www.wiley.com/go/deepspacecommunications to access the color version of this figure.

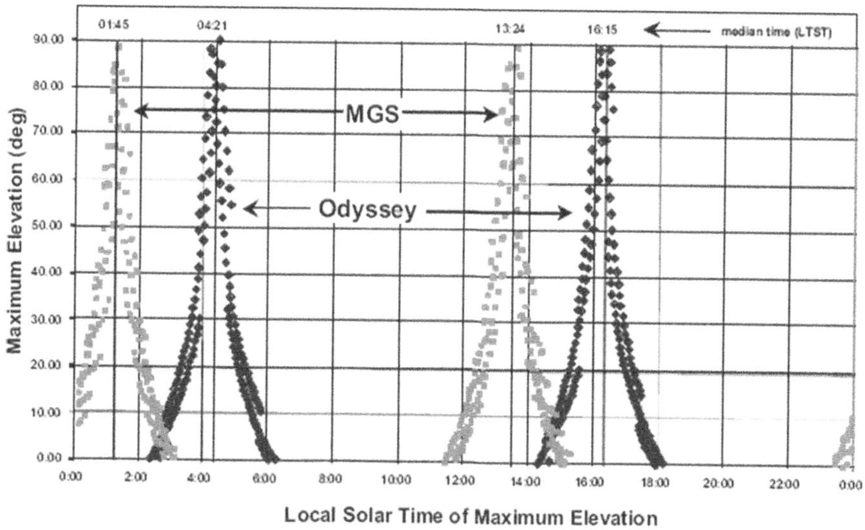

Fig. 7-9. Distribution of Odyssey and MGS overflight times and maximum elevations (MER-A site).

13:30 local solar time, and with Odyssey at about 04:30 and 16:30. As the MER missions continued, the MGS orbiter mission ended, to be replaced for UHF relay by the MRO mission.

Rover tilt was expected to be a minor factor in link performance, as rover-orbiter distance dominates the tilt as a factor in link performance.[6] Rover azimuth, however, strongly affects link performance due to the asymmetry in the antenna-gain pattern. In addition, the same pass that returns 50 megabits (Mb) in a favorable azimuth, could see that return cut in half if the HGA assembly blocks the view. The average data-return volume is estimated to be about 56 Mb/sol per rover for Odyssey and about 49 Mb/sol per rover for MGS.

---

[6] The first postlanding relay planning predicts were based on the average of those made for every 10 deg in azimuth since data-return volume was not initially a factor in planning rover orientation. Before too long in the primary mission, the rovers were sometimes deliberately oriented in azimuth after a sol's science activity to increase the data return. Still later, the relay link-prediction program was augmented with a capability to predict for tilt as well as azimuth. In one case, on sol 278 (November 4, 2004), Opportunity was driving through steep and rocky terrain and was tilted as much as 31.04° during the Odyssey afternoon pass. The difference between no-tilt and 31-deg tilt predicts was 57.4 Mb versus 41.5 Mbs.

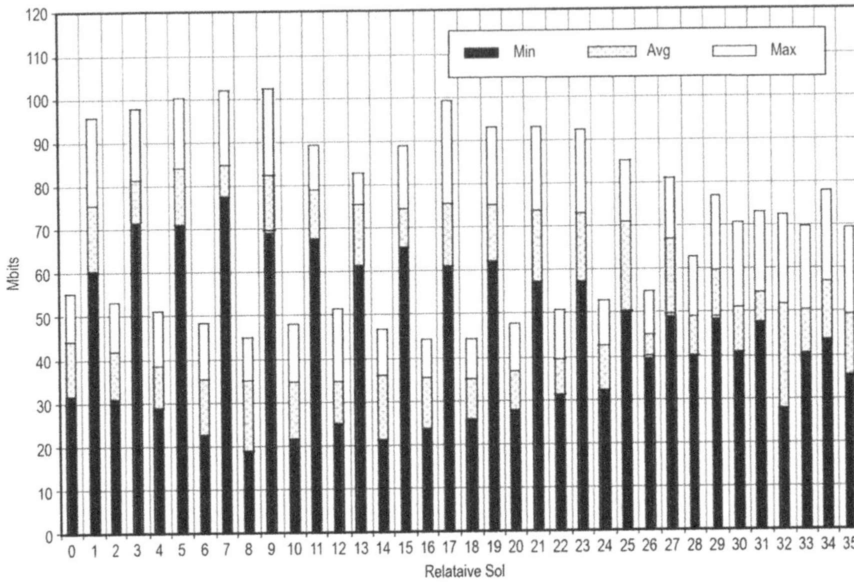

Fig. 7-10. Odyssey 128-kbps data volume from MER per sol from a 0°-latitude landing site (averaged over all azimuths, no tilt).

It must be noted, however, that the actual volume of data that can be returned via the UHF link varies from pass to pass, and depends on both the highly variable maximum elevation angle and rover orientation. Higher maximum elevation angle results in both a longer pass time and more time at a shorter slant range. The project chooses higher elevation passes that can support a higher data rate and thus usually a larger total data volume for the pass. Figure 7-10 provides an example of the sol-to-sol variability of the data volume returned via Odyssey showing both the effects of variable pass durations and various rover azimuths. Similar results have been obtained for the MGS relay.

The potential data-return volume was further constrained by the availability of Odyssey onboard memory. The Odyssey UHF Relay Operations Plan made prior to MER surface operations allocated a total of 100 megabits 12.5 megabytes) of Odyssey onboard memory to both MER rovers (and to Beagle II, which did not operate). The allocation was later increased to 120 Mb per rover for the primary mission. Thus, the volume of data that may be relayed through Odyssey is constrained by data that may remain in the Odyssey buffer from the previous relay pass. How quickly the buffer can be emptied is a function of the DSN coverage allocated to Odyssey for downlinking this data.

7.3 Telecom Subsystem Hardware and Software

7.3.1 X-Band Flight Subsystem Description

7.3.1.1 X-Band Functions. The telecommunications subsystem was designed to perform the following functions:

- Receive an X-band uplink carrier from the DSN.[7] This carrier may be unmodulated or modulated by command data or by a ranging signal or both.
- Demodulate the command data and the ranging signal.
- Generate an X-band downlink carrier either by coherently multiplying the frequency of the uplink carrier by the turn-around ratio 880/749, or by utilizing an auxiliary crystal oscillator (aux osc).
- Phase-modulate the downlink carrier with either of two signals (or both):
  - A composite telemetry signal, which consists of a square-wave subcarrier (25 kilohertz [kHz] or 375 kHz) that is binary-phase-shift-keying (BPSK)–modulated by telemetry data provided by the avionics subsystem.
- As modulation for navigation, either
  - A ranging signal that was demodulated from the uplink during cruise (this is referred to as two-way or turn-around ranging), or
  - A set of unmodulated tones, used for delta differential one-way ranging (delta-DOR) during cruise. The SDST DOR module generated these tones.
- Permit control of the subsystem through commands to select signal routing (for example, which antenna should be used) and the operational mode of the subsystem (that is, the configuration of the elements of the subsystem). Examples are command data rate, telemetry subcarrier, convolutional code, downlink ranging modulation index). This commanding can be done either directly from the ground (with real-time commands) or through sequences of commands that were previously loaded on the spacecraft.
- Provide status telemetry for monitoring the operating conditions of the subsystem. Examples are aux osc temperature, SDST current, subcarrier frequency, ranging channel state (on or off) coherent/noncoherent operation, and receiver lock state (uplink carrier in or out of lock).

---

[7] The DSN is a global network of antennas and related support facilities, managed by JPL for NASA. The DSN provides both command uplink and navigation to deep-space probes and downlink telemetry to the Space Flight Operations Facility and the end-users it serves.

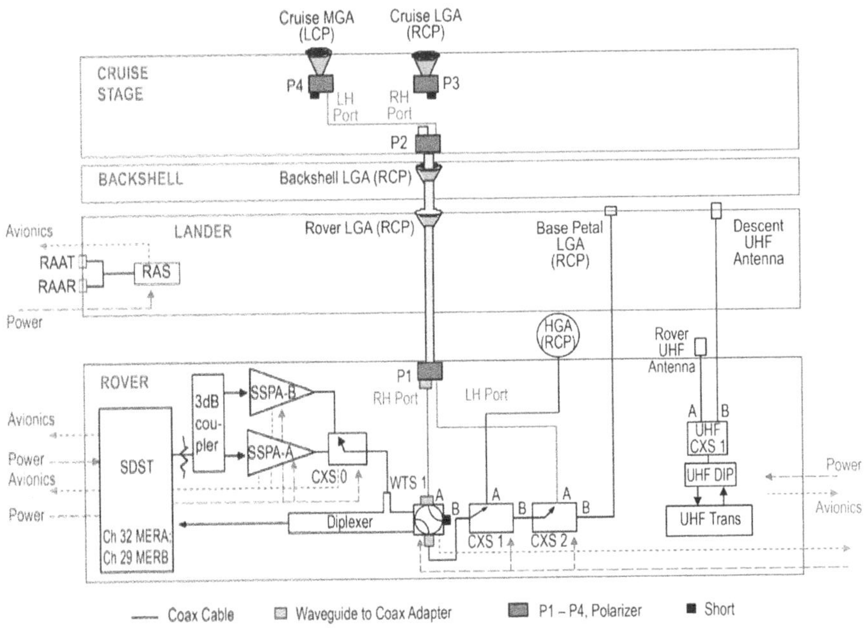

Fig. 7-11. X-band Telecom Subsystem block diagram. (RAAT and RAAR are Relay Antenna Assembly Transmit and Relay Antenna Assembly Receive.) Please visit www.wiley.com/go/deepspacecommunications to access the color version of this figure.

- For the radio frequency (RF) transmitter, provide on/off power control to permit the conservation of power.
- Upon a power-on-reset (POR), the system is placed into a single, well-defined operating mode. This provides a known subsystem state from which the ground can command the telecom subsystem during safe-mode (emergency) operations.

In addition, as planned for the EDL phase, the SDST could generate and transmit the so-called M-FSK tone described in Section 7.2.1 above. In this alternative to telemetry, a unique subcarrier frequency is used to signal (as a semaphore) that a particular spacecraft event has occurred. The M-FSK tones were used during the EDL portion of the mission, where the expected signal level was too low and the Doppler environment too dynamic to provide telemetry via a conventional phase-coherent receiver.

7.3.1.2 Functional Block Diagram. Figure 7-11 is a block diagram of the X-band telecom subsystem, with the functional elements as described in the four major assemblies of Fig. 7-1.

7.3.1.3 Interfaces with Other Subsystems. The telecom subsystem interfaces with the spacecraft are illustrated in Fig. 7-12.

The interfaces with the avionics subsystem and the power subsystems are as follows:

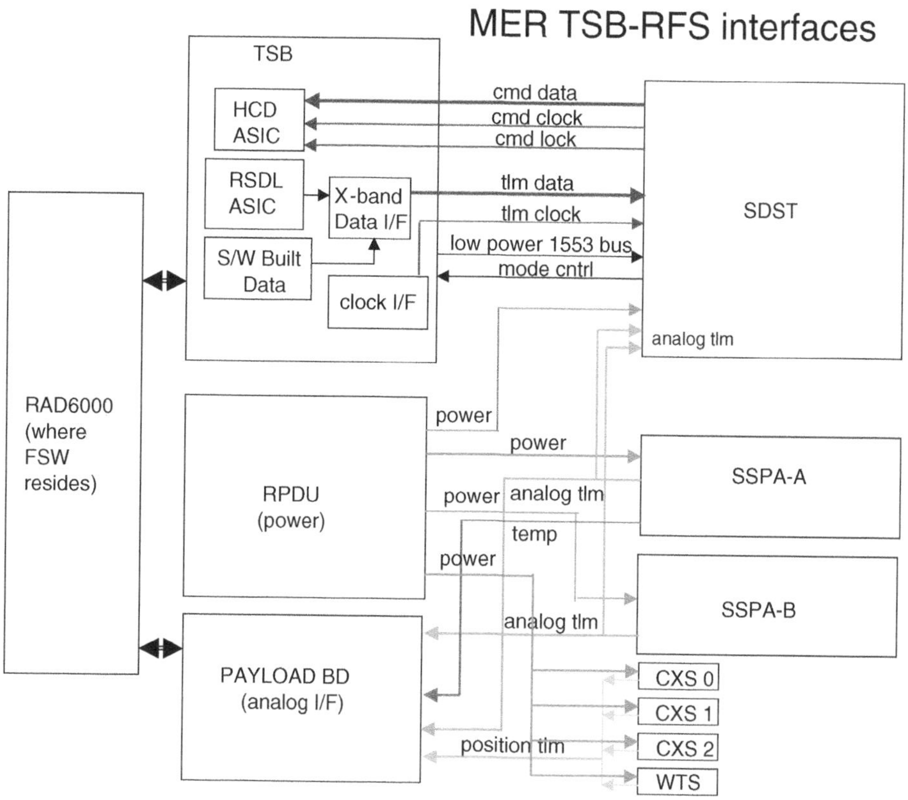

Fig. 7-12. Telecom Subsystem interfaces. Please visit www.wiley.com/go/deepspace communications to access the color version of this figure.

Avionics includes hardware and the flight software. The telecom subsystem relies on avionics to control its operating mode. This control can be done via

- A real-time command from the ground, demodulated from the X-band uplink carrier and provided to avionics, or
- A sequence of commands stored on board and issued by the sequence engine, or
- Communications behavior, where the change of state occurs as the result of opening of a communications window[8] or the closing of the window (that is, return to the current default or background state), or

[8] A communications window (comm window) delivers a set of communications parameters to the rover using a single command. The parameters include start time, duration, choice of rover antenna (which determines whether the window is X-band or UHF), durations for real-time and recorded data-priority tables (DPTs), uplink (or forward) and downlink (or return) data rate, hardware configuration table to invoke, and an optional sequence for the window to initiate at its start time. Comm windows operate within a "communications behavior" portion of the flight software. A comm window does not rely on the rover's sequence engine.

- Fault protection, where the change of state occurs as the result of a response algorithm that activates when the fault-protection software detects a defined fault.

In each case, it is the avionics subsystem that issues the commands that control how the telecom subsystem is configured. The only exception is the POR state. If a POR is triggered, the SDST will enter its POR state.

The avionics subsystem provides the telecom subsystem with the telemetry data to be downlinked, as well as a data clock to drive the convolutional encoding done by telecom. The clock is to be either data clock × 2 for (7,1/2) encoding or data clock × 6 for (15,1/6) encoding. Avionics does the frame and packet formatting and the Reed-Solomon (RS) encoding of the telemetry data that is to be transmitted by telecom.

Avionics selects the frame size (either long or short) and whether the data sent to telecom is RS-encoded or checksum (CS)–encoded. The CS mode was not used for X-band during mission operations. Data to be RS-encoded is produced by the RS downlink (RSDL) application-specific integrated circuit (ASIC) on the telecom support board (TSB).

For the uplink, telecom provides avionics, specifically, the hardware command decoder (HCD), with

- The detected command bits it has demodulated from the uplink signal sent by the DSN station,
- The bit clock, and
- The command detection in-lock status.

Telecom relies on avionics to do error-control of the uplink data stream. That is, avionics determines what is a valid command and what is not a valid command.[9]

SDST mode control commands (such as: telemetry mod index, ranging on/off, coding, coherency) are done via the 1553 bus; they are issued by avionics.

RS422 interfaces exist between the SDST and the avionics TSB card, for (a) telemetry data and clock and (b) command data, clock and lock status (to the HCD.)

---

[9] We discovered one instance in the MER extended mission where the HCD and the flight software failed to handle gracefully a command containing multiple-bit errors. The error-filled "command" that went to flight software wrote to an incorrect location and caused rover entry to safemode. ISA Z84599 [8].

Telecom relies on the power subsystem to drive the waveguide transfer switch (WTS) and coaxial switches (CXSs), which select the X-band SSPA and the X-band and UHF antenna.

7.3.1.4 Description of X-Band Components

7.3.1.4.1 Antennas. As described in Section 7.3 and shown in the block diagram Fig. 7-11, each MER had several antennas, used during different phases of the mission:

- Cruise communications were through the MGA and the CLGA, both located on top of the cruise stage;
- During EDL, as the cruise stage and then the backshell were jettisoned, the spacecraft used the BLGA; and for the first day of deployment on the surface, the PLGA.
- For surface operations, the X-band antennas were the RLGA and the HGA.

Table 7-3 summarizes the major RF characteristics of the antennas and, at the bottom, their size and mass. The rover X-band antennas (RLGA and HGA) and the rover UHF antenna are mounted on the RED as shown in Fig. 7-13.

The CLGA, the BLGA, and the RLGA are RF horns mounted on the same circular waveguide "stack" that is designed to break off in sections as described in Section 7.3. The RLGA is the shortest section of waveguide; hence, the RLGA circuit losses are the smallest while those of the CLGA are largest.

The HGA is mounted on a two-axis gimbal located on top of the RED, so it became available only after deployment of the rover.

7.3.1.4.2 Radio Frequency Subsystem. The Radio Frequency Subsystem (RFS) is a general name for the three active X-band elements of the telecom subsystem and the passive elements that connect them[10]. The active elements are the SDST and the two SSPAs. The other active telecom subsystem element is the UHF transceiver, along with its diplexer. Figure 7-14 shows the locations of the SDST and SSPAs on one side of the rover electronics module (REM) along with the X band switches and diplexer. The UHF

[10] Spacecraft power into the RFS and the UHF transceiver that is not radiated as RF is converted to heat that must be managed. During cruise when the RFS was powered on continuously, MER thermal control was accomplished by the Heat-Rejection Subsystem (HRS). Figure 7-14 shows a heat pipe, part of the HRS, between the SDST and the SSPAs.

Table 7-3. MER X-band antenna characteristics.

Mission Phase	Cruise		EDL		Surface	
Antenna	CLGA	MGA	BLGA	PLGA	RLGA	HGA
Receive frequency, MHz	7183.118057 MER-A 7179.650464 channel 29 (MER-B) 7183.118057 channel 32 (MER-A)	Same	Same	N/A	Same	same
Transmit frequency, MHz	8435.370372 MHz channel 29 (MER-B) 8439.444446 MHz channel 32 (MER-A)	Same	Same	Same	Same	same
Gain, boresight, RX, dB	7.68	18.1	N/A	N/A	5.73	20.5
Gain, boresight, TX, dB	7.18	19.2	7.71	6.0	6.89	24.8
Polarization*	RHCP	LHCP	RHCP		RHCP	RHCP
Beamwidth, deg	±40 RX ±42 TX	±10.3 RX ±9.3 TX	N/A RX ±35 TX	N/A RX ±52 TX	±46 RX ±37 TX	±5.0 RX ±4.2 TX
Axial ratio, on b/s, dB	0.49 RX 0.85 TX	1.01 RX 0.27 TX				6.34 RX 4.47 TX
Axial ratio, off b/s, dB	85° off boresight: 7.70 dB RX 6.00 dB TX	20° off boresight: 6.29 dB RX 7.53 dB TX				
Design	Open-ended waveguide with choke	RF conical horn	Open-ended wave-guide with choke	Microstrip array 1.5 × 1.5 in. (3.8 × 3.8 cm)		0.28–m-dia. Printed dipole array
Mass, kg	0.431	0.499	0.235	0.020	0.775	1.1

* The polarization of the RLGA (and BLGA) is normally right-hand circular polarization (RHCP or RCP). It could be set to left-hand circular polarization

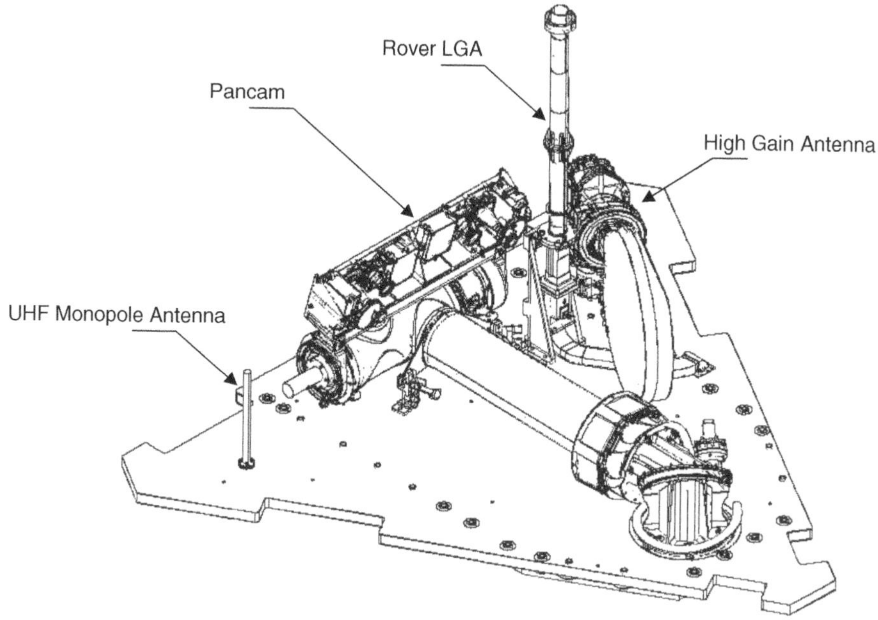

Fig. 7-13. X-band and UHF antennas on the RED.

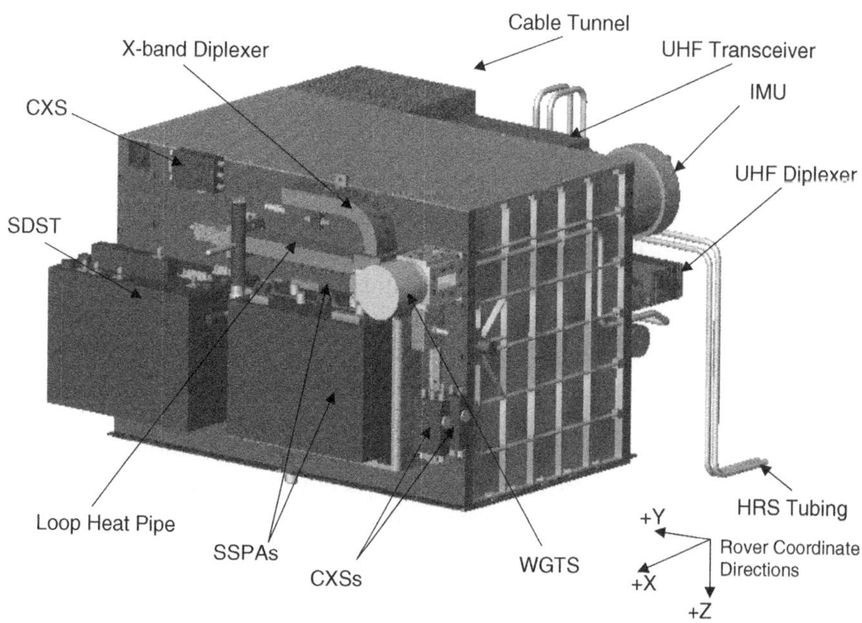

Fig. 7-14. RFS mounted on the sides of the REM. Please visit www.wiley.com/go/deepspacecommunications to access the color version of this figure.

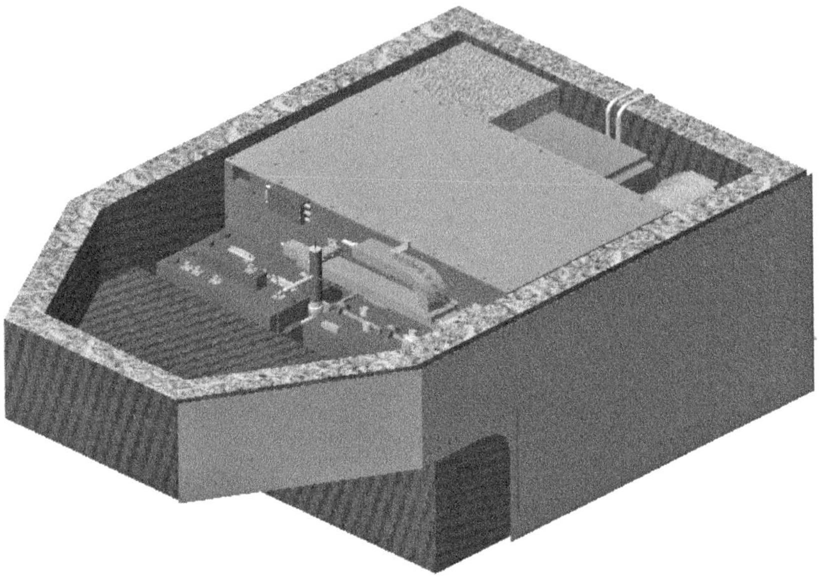

Fig. 7-15. MER warm electronics box. Please visit www.wiley.com/go/deepspacecommunications to access the color version of this figure.

transceiver is on the other side of the REM. The REM is inside the WEB, as Fig. 7-15 shows.

During cruise, the cruise stage's HRS evacuated unwanted heat generated by the SSPA. Upon arrival at Mars, the HRS tubing was severed, as designed, at the interface with the aeroshell. Subsequently, excess heat was evacuated from the rover by passive thermal control.

On the surface, the WEB kept the rover warm at night, when no heaters could be left on. During the day, when X-band and UHF transmitters operated successively three or four times, the rover temperature would rise toward the hot temperature limits[11] because X-band and UHF heat-generating elements were so near each other on the REM. The amount of heat generated by operating X-band and UHF elements limited the durations and intervals between successive X-band and UHF transmitter operations.

7.3.1.4.3 X-Band Diplexer. The diplexer is a device that allows signals to be simultaneously transmitted at one frequency and received at another frequency. It provides sufficient receive side rejection of the SSPA generated

[11] The upper (hot) temperature limits were 50°C allowable flight temperature (AFT) and 60°C protoflight qualification limit for SDST; 50°C AFT and 70°C protoflight qualification limit for SSPA; and 55°C AFT and 70°C protoflight qualification limit for UHF transceiver.

Table 7-4. X-band diplexer functional parameters.

Parameter	Diplexer Port	Parameter Value
Passband	TX	8.29–8.545 GHz
	RX	7.1–7.23 GHz
Insertion Loss	TX	26 dB max
	RX	9 dB max
Isolation	TX/RX	95 dB min
		100 dB nominal

Table 7-5. Transfer switch functional parameters.

Parameter	WTS Value	CXS Value
Frequency, GHz	7.1–8.5	7.1–8.5
Insertion Loss, dB	0.05	0.15
Return Loss, dB	23	20
Power Handling Capability, watts (W)	1000	70
Isolation, dB	>60	>60
Switching Time, ms	50	5

transmitter signal preventing damage of the SDST receiver front end or interference with the uplink signal from Earth. It allows simultaneous transmit and receive signals to use the same antenna. X-band diplexer functional parameters are shown in Table 7-4.

7.3.1.4.4 Transfer Switches (WTS and CXS). Refer to the block diagram in Fig. 7-11. There are two types of transfer switches, coaxial and waveguide (CXS and WTS). The subsystem has three CXSs and one WTS. Transfer switch functional parameters are shown in Table 7-5.

- CXS 0 allows us to select either SSPA-A or SSPA-B for the downlink. Since launch, CXS 0 has been set to SSPA-A.
- CXS 1 selects between the HGA and the input to CXS 2.
- CXS 2 selects between the LCP port of polarizer P1 and the base petal LGA (PLGA) with left-hand circular polarization (LHCP or LCP).

The WTS (also known as a "baseball switch") is mounted on the output of the diplexer port 2. The WTS is commanded to select between the LGA stack, and the input to CXS 1.

Table 7-6. 3-dB coupler functional parameters.

Frequency Range	7.1–8.5 GHz
Insertion Loss	0.5 dB
Isolation	20 dB
Coupling	3 dB
Power Handling	5 W

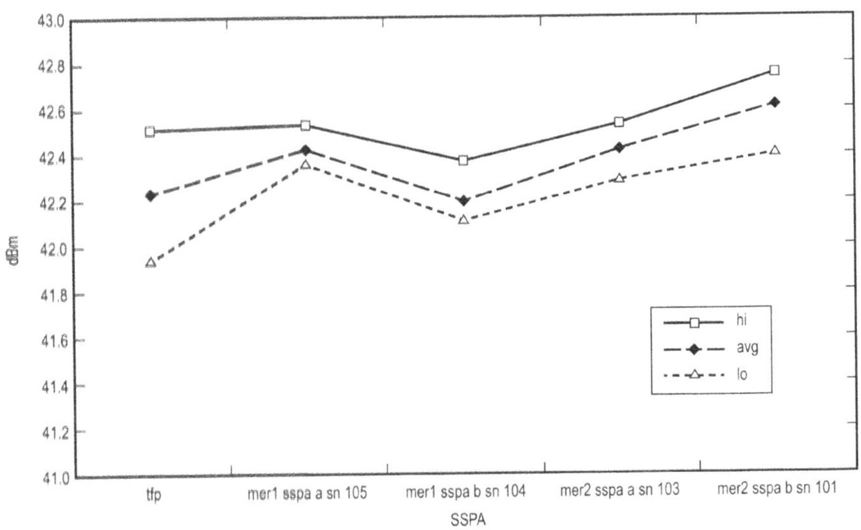

Fig. 7-16. RF output of the two MER-B (MER-1) and two MER-A (MER-2) SSPAs.

A WTS is heavier than a CXS. Because it has lower insertion loss, the WTS is used for the most important low-gain transmit path. A CXS is used on other paths where a higher insertion loss can be tolerated. These include the paths leading to the MGA, the HGA, and the PLGA. Though an LGA, the PLGA was used only on the first day of Mars surface operations.

To select a particular antenna for X-band receive and transmit may require commanding the WTS, CX1, and CX2. The connections between switches also enable use of the HGA and RLGA in surface operations even if the WTS should get stuck in the CXS1 position.

7.3.1.4.5 Solid-State Power Amplifier. Each of the two redundant SSPAs receives its RF input from SDST exciter via a 3-dB coupler, as shown in Fig. 7-11. Table 7-6 defines the major functional parameters of the 3-dB coupler.

The active SSPA provides about 16.8 W (42.25 decibels referenced to milliwatts [dBm]) of RF output power, as shown by Fig. 7-16, a graph taken from

Fig. 7-17. MER SDST. Please visit www.wiley.com/go/deepspacecommunications to access the color version of this figure.

test data. The first point (mean and tolerances) is the prediction program model, and the four points to the right of the model point represent prelaunch measurements of the four MER SSPAs.

The direct current (DC) power input for each SSPA is about 58 W. The DC input varies a little with temperature.

7.3.1.4.6 Small Deep-Space Transponder. The MER SDST is based on the proven design first flown on Deep Space 1 (DS1) in 1998, but its phase modulator was improved so as to be more linear (it is now a dual-stage modulator). Figure 7-17 is a photograph of the SDST. The SDST consists of four slices (boards): the power-converter module, the digital-processor module (where the signal processing is done), the down-converter module (where the analog part of the receiver phase-locked loop is) and the exciter module (where the telemetry and or ranging or DOR is modulated onto the downlink RF carrier). Receiver carrier-loop parameters are shown in Table 7-7.

Table 7-7. Receiver carrier loop.

Parameter	Parameter Value			
Noise Figure, dB	Temp	60°C	25°C	–40°C
	Channel 29 SDST (S/N 203)	2.59	2.15	1.27
	Channel 32 SDST (S/N 201)	2.58	2.12	1.91
Tracking Threshold	–155 dBm			
Tracking Rates	200 Hz/s for uplink $P_t \leq -120$ dBm			
Capture Range	±1.3 kHz			
Tracking Range	Greater than ±30 kHz at 200 Hz/s for uplink P_t down to –140 dBm			
Loop Noise Bandwidth at Threshold ($2Bl_0$)	20 Hz			
Loop Noise Bandwidth for Strong Signals	231.3 Hz two-sided, at P_c/N_0 = 100 dB-Hz			

Having a POR state is very desirable. It ensures that the SDST comes up in a known state, for example every morning at rover wake-up. The flight software then has only to enter a limited set of well-defined commands to place the SDST into its desired operating state. Table 7-8 shows the POR state for the SDST.

Ranging Performance: Ranging is a means to determine the position of the spacecraft by measuring how long radio signals (ranging codes) take to travel from Earth, to the spacecraft, and back to Earth. Accuracy of the measurement depends on knowing how much of the total delay is produced in the transponder, the spacecraft antenna cabling, and the station ranging equipment.

Table 7-9 shows the delay the ranging signal experiences as it goes through the SDST. See Table 7-10 for total delay through the spacecraft.

In Table 7-9, one range unit (ru) = 1478/221 * 1/Ftx = 0.931 nanoseconds (ns) for MER-A and B.

7.3.1.4.7 Range Delay after Integration on Spacecraft. The total range delay through the spacecraft (Table 7-10) will vary depending on which antenna path is used. This is because the cable lengths are significantly different. The table does not include values for the RLGA or HGA because ranging was not used for surface operations.

Table 7-8. Power-on-reset state table.

Controlled Parameter or Mode	Value at POR
Auto Coherent/Noncoherent Transfer	Enabled
VCXO*/aux osc Transfer	Enabled
Command Data Rate	7.8125 bps
Normal TLM Encoding Mode	(7,1/2)
Normal TLM Mod. Index	50°
Normal TLM Mode	Subcarrier
Ranging Mod. Index (Gain)	17.5°
Ranging Mode	Baseband
Ranging	Off
Remote Terminal Time-out	Disabled
Remote Terminal (RT) Event Counter	0
SDST Event Counter	0
State 1 Time-out	Enabled
Subcarrier Frequency	25,000 Hz
Transponder Mode	Normal Operation
Wideband TLM	Off
X-band DOR	Off
X-band Exciter	On

* VCXO = voltage-controlled crystal oscillator.

Table 7-9. SDST range delay (in range units).

Parameter	S/N 203—Channel 29	S/N 201—Channel 32
Range delay, average	1388.66 ru	1386.75 ru
Range delay variation at one temperature	±2.5 ru	±2.5 ru
Carrier suppression, dB	0.3 (17.5° nom)	0.3 (17.5° nom)
	1.2 (35° nom)	1.2 (35° nom)
Ranging channel noise equivalent bandwidth	1.96 MHz	2.24 MHz

Table 7-10. SDST range delay after spacecraft integration (in nanoseconds).

Antenna Path	SDST (S/N203)—Channel 29	SDST(S/N201)—Channel 32
CLGA up/CLGA down	1383.9 ns	1384.0 ns
MGA up/MGA down	1393.5 ns	1394.5 ns

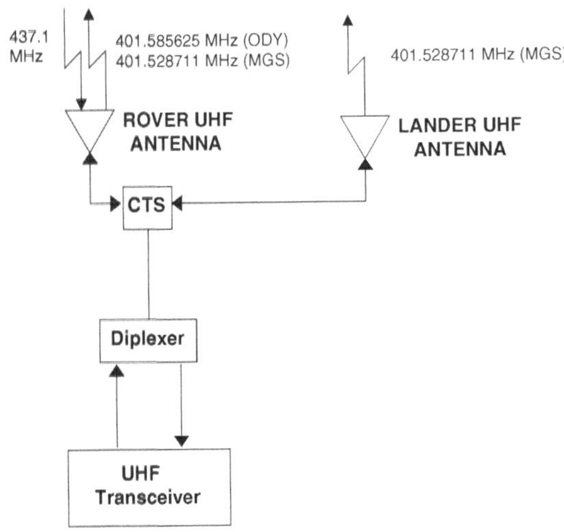

Fig. 7-18. UHF subsystem block diagram.

7.3.2 UHF

The MER UHF subsystem, a block diagram of which appears in Fig. 7-18, consists of the following components:

- Transceiver, which performs transmission and reception of UHF communications. It is also the interface with the avionics subsystem.
- Two UHF antennas: the DUHF (on the lander), used to transmit to MGS during EDL, and the RUHF, used to transmit and receive with orbiters during surface operations.
- Diplexer and coaxial switch to connect the transceiver to one of the two antennas.

7.3.2.1 UHF Antennas. The descent and rover UHF antennas are quarter-wavelength monopoles. Figure 7-19 shows photographs of the rover UHF antenna.

The DUHF has an additional mechanism that deploys the antenna parallel to the bridle after backshell separation. While the monopoles are nominally linearly polarized with a toroidally shaped gain pattern, parasitic coupling of the UHF transmit and receive signals with structures on the spacecraft create significant distortions to both gain and polarization. This is especially true for the RUHF, due to vertically oriented structures (mainly the LGA and PMA) on the deck that act like passive parasitic antenna elements.

A right-hand polarization pattern, as measured on a rover mock-up in the JPL antenna range, is shown in Fig. 7-20. The figure shows the RUHF

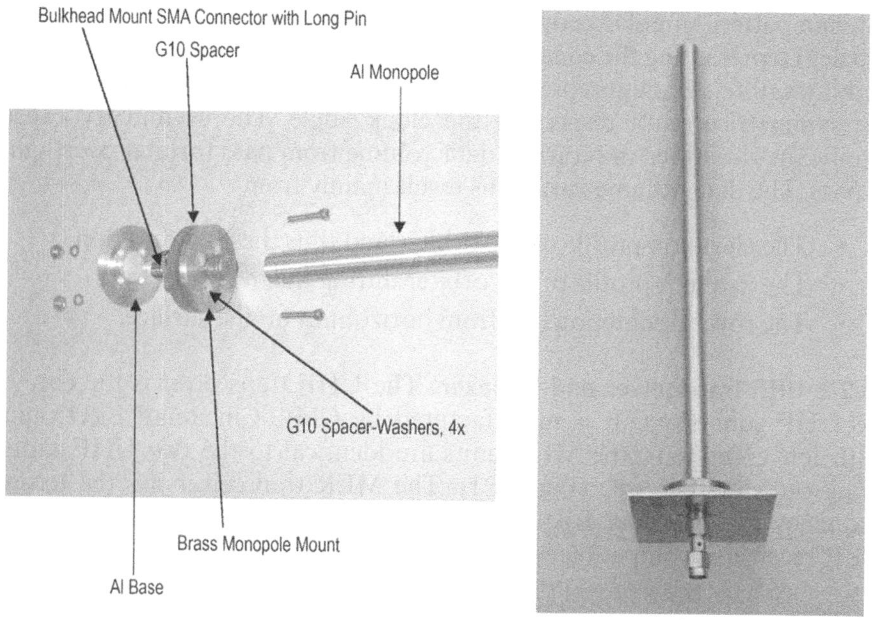

Fig. 7-19. Rover UHF antenna. Please visit www.wiley.com/go/deepspace communications to access the color version of this figure.

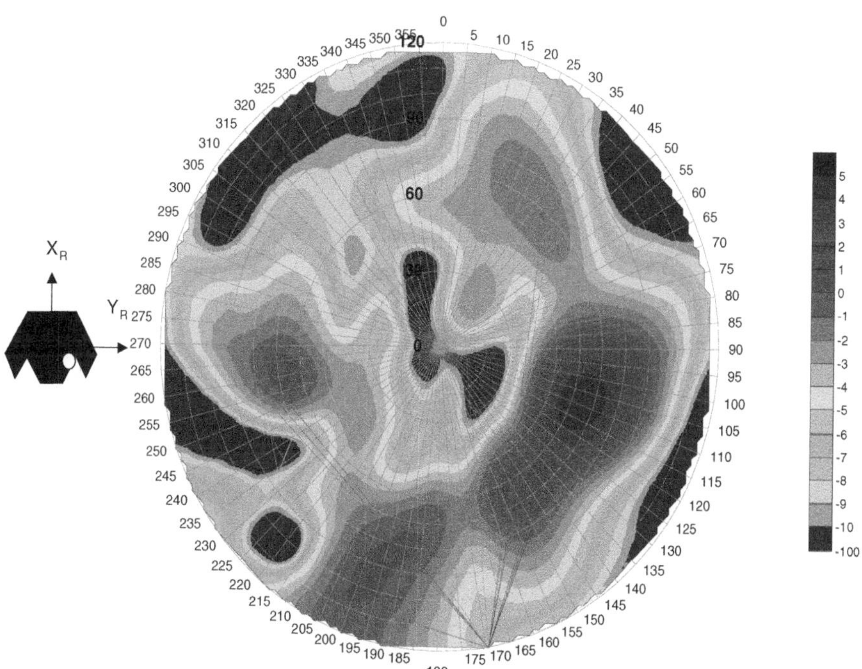

Fig. 7-20. Rover UHF antenna pattern as measured on a mock-up at 402 MHz. Please visit www.wiley.com/go/deepspacecommunications to access the color version of this figure.

antenna pattern in polar coordinates, with the concentric grid markers (0 to 120 deg) representing the cone angle (angle from the boresight) and the radial grid lines (0 to 360 deg) representing the clock angle. The RUHF pattern is not symmetrical with respect to the clock angle. The asymmetry causes significant variations in returned data volume from pass (orbiter overflight) to pass. The data-volume variations result mainly from

- The elevation profile of the orbiter and thus the pass duration,
- The azimuth profile of the orbiter during the overflight, and
- The rover orientation (tilt from horizontal) on the surface.

7.3.2.2 UHF Transceiver and Diplexer. The UHF transceiver is the core of the UHF subsystem. It is manufactured by CMC Cincinnati Electronics. With few exceptions, the MER units are identical to the two UHF radios flying on Mars Odyssey (Fig. 7-21). The MER transceiver has the receive frequency and transmit frequency swapped relative to Odyssey's, and the MER receiver is compatible with MGS as well as with Odyssey.

CMC also manufactured the MER UHF diplexer used to isolate transmit and receive frequencies for simultaneous operation. The transceiver and diplexer were thoroughly tested as a single subsystem.

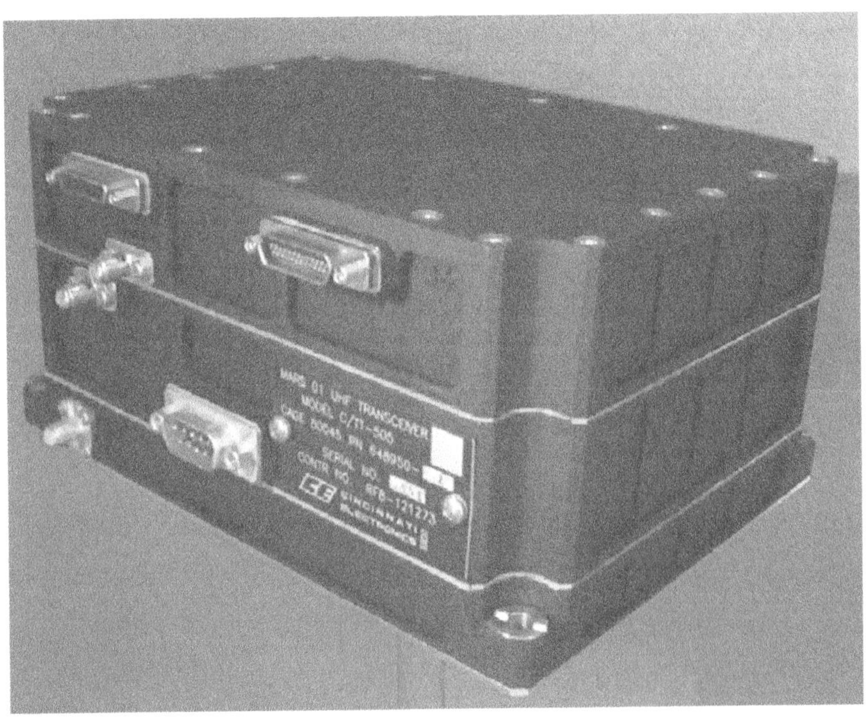

Fig. 7-21. Odyssey UHF transceiver. Please visit www.wiley.com/go/deepspace communications to access the color version of this figure.

7.3.2.3 UHF System Operation

7.3.2.3.1 Physical Layer. At the physical layer [23], the following are the main characteristics of the MER UHF system:

- Power (measured)
  - Power consumption 6 W (receiving only), 43 W (transmitting/receiving)
  - RF output 12 W (typical, transmitting)
- Frequency
  - One forward frequency (orbiter to rover) of 437.1 MHz
  - Two return frequencies (rover to orbiter):
    - 401.585625 MHz (Odyssey and MEX)
    - 401.528711 MHz (MGS)
- Modulation
  - PCM/Bi-Phase-L/PM modulation with residual carrier, with a modulation index of 1.05 radians (60 deg)
- Data Rates
  - Forward link: 8, 32, 128, 256 kbps[12]
  - Return link: 8, 32, 128, 256 kbps[13]
- Encoding
  - Forward link: none
  - Return link: convolutional with rate 1/2 and constraint length 7
- Carrier Acquisition at ±8 kHz off center frequency (forward link)
- Receiver threshold, typical, forward link, for bit error rate of 1×10^{-6}
  - 8 kbps phase-shift-keyed, uncoded: –117 dBm

7.3.2.3.2 Data Frame Layer (Odyssey and Mars Express). At the data frame layer, MER implements the Consultative Committee for Space Data Systems (CCSDS) Proximity-1 Space Link protocol (UHF1) [8], which is the standard used for relay communications by all the missions currently at Mars, except MGS, launched in 1996.

The data layer of the Proximity-1 protocol provides the structure (frame sequence number and forward error coding) that allows the establishment of a compatible link and the exchange of error-free information between the

---

[12] The UHF radio was implemented to support these four rates. However, MER required, tested, and operated the forward link with only the 8-kbps rate. The command path to the rover has a low data-volume requirement.

[13] Operationally, the highest return rate to MGS is 128 kbps. Initially, the highest rate to Odyssey was also 128 kbps. Later in the primary mission, the 256-kbps rate was also used. See Section **7.4** of this chapter.

orbiter and a surface vehicle such as the rover. It also allows verification that the orbiter is communicating with the intended surface vehicle.

The link with a surface vehicle is always initiated by the orbiter at 8 kbps, sending a Proximity-1 transfer frame (17 bytes long) with Set Transmit and Set Receive directives in order to configure the transceivers at both ends in a compatible mode. Information about communications mode, data rates, coding, and modulation to be used are all contained in this frame.

The nominal mode of communications with a surface vehicle is the sequence-controlled service defined in the Proximity-1 protocol. This mode ensures the error-free transmission of the input bit-stream to the receiving end. The serial data from the transceiver transmit buffer is formatted in the data field of the Proximity-1 transfer frame.

The following are the most important fields of the transfer frame header:

- Attached Synchronization Marker to allow identification of the start of the frame
- Spacecraft ID of the surface vehicle
- Frame Sequence Number to allow the receiving end to verify that data is being received in the proper order
- 32-bit cyclic redundancy check (CRC) appended after the frame to allow the receiving end to detect if any bit of the packet suffered an error during transmission.

In the sequence-controlled mode, MER implements a Go-Back-2 [frames] Automatic Repeat Request (ARQ) protocol. This protocol permits transmission of the next sequenced frame while waiting for the acknowledgment (ACK) for the one previously sent. In this way, the throughput is increased relative to a Stop-and-Wait protocol. In the case where an ACK is not received before the end of the transmission of the second frame, the orbiter will continue sending the same two transfer frames still to be acknowledged. MER can receive and send Proximity-1 frames up to 1024 bytes long.

To transfer data, the sequence-controlled service needs both a forward link and a return link to be active. If an anomaly (such as a failure of a transmitter) has occurred in one of the two links, data can still be sent on the remaining functional link by operating in the so-called unreliable bit-stream mode. In this mode, the Proximity-1 protocol is bypassed, and delivery is not guaranteed to be error-free or in order.

All forward- and return-link equipment are operational on the Odyssey and MRO orbiters and on Opportunity (though Spirit's condition has been unknown since March 2010). The unreliable bit-stream mode has not been known to be required since EDL. However, the unreliable mode was verified on Opportunity/orbiter return link tests, and has been routine in post March 2010 Spirit return link planning for both Odyssey and MRO.

7.3.2.3.3 MGS Operations. The MER UHF transceiver is also backward-compatible with the Mars Balloon Relay protocol (MBR, also called UHF2) implemented on MGS (originally designed in support of Russian and U.S. missions consisting of small landers, balloons, and penetrators).

The UHF2 protocol has no data-layer protocol. During a 16-s cycle, the forward link is used to send two types of tones:

- One of three request commands (RCs) that allow MGS to address any one of three surface vehicles at the same time.[14] After detection of the RC tone, the surface vehicle will send a pseudonoise (PN) code while waiting for the transmit command (TC).
- The TC is sent by MGS when its receiver achieves bit-sync-lock on the initial return link. After detection of the TC tone, the surface vehicle starts sending its science and engineering data.

If the return power-to-noise ratio drops below threshold, MGS begins transmitting a carrier only. Upon receiving the carrier, the surface vehicle radio will stop transmitting. Due to timing issues and the fact that no data layer is present, the quality of the UHF link to MGS is less than what is possible in the link to Odyssey or MEX.

7.3.3 MER Telecom Hardware Mass and Power Summary

The mass and input power of the elements of the telecom subsystem are summarized in Table 7-11.

7.4 Ground Systems

7.4.1 Deep Space Network

7.4.1.1 Background. Communication between the MER spacecraft and the DSN has been at X-band for all mission phases (cruise, EDL, and surface operations, and continuing into the extended missions). Furthermore, even though the MGS and Odyssey orbiters have received surface data from rovers via a UHF link, the data from the orbiters was transmitted to the DSN via X-band. Specific station operating modes and configurations to support MER are in the Network Operations Plan [9].

---

[14] Both Spirit and Opportunity respond to the same tone RC1, since it was required that the two UHF radios be swappable between rovers during ATLO. Because Spirit and Opportunity landed on opposite sides of Mars, there is no possibility of overlap during an overflight.

Table 7-11. MER X-band and UHF mass and input power summary.

Assembly	Input Power, W	RF Power out, W	Mass, kg	Quantity	Mass Total, kg	Dimensions, cm
			X-Band			
SDST each			2.682	1	2.682	18.1 × 11.4 × 16.6
Receiver (R) only	11.0					
R+exciter, two-way (coherent)	13.3					
R+exciter, one-way (aux osc)	13.8					
SSPA	58	16.8	1.300	2	2.600	4.4 × 17.2 × 13.4
Hybrid			0.017	1	0.017	2.5 × 1.0
WTS			0.378	1	0.378	4.1 × 9.65 × 10.9
CTS			0.062	3	0.187	5.3 × 3.0 × 4.0
Coax			0.057	4	0.228	
Diplexer			0.483	1	0.483	27.7 × 5.6 × 7.9
Attenuator			0.004	1	0.004	0.79 × 2.18
HGA			1.100	1	1.100	28.0 dia.
CLGA			0.431	1	0.431	10.0 × 2.3
BLGA			0.235	1	0.431	10.3 × 3.5
RLGA			0.775	1	0.431	60.2 × 3.1
PLGA			0.020	1	0.020	1.5 × 1.5
MGA			0.499	1	0.499	23.4 × 13.4 at rim
Terminations, dummy loads, etc.			0.006	4	0.026	
X-band totals	71.8 max	16.8	5.367		6.835	

Table 7-11 (Continued)

Assembly	Input Power, W	RF Power out, W	Mass, kg	Quantity	Mass Total, kg	Dimensions, cm
			UHF			
UHF transceiver	6 rx only 43 rx/tx	12 *	1.900	1	1.900	5.1 × 6.8 × 3.7
Diplexer			0.400	1	0.400	2.9 × 3.7 × 1.3
CTS			0.083	1	0.083	5.3 × 3.0 × 4.0
RUHF			0.100	1	0.100	16.9 × 1.9 × 1.9
DUHF			0.100	1	0.100	16.9 × 1.9 × 1.9
Coax			0.300	1	0.300	

* UHF RF power out is measured at diplexer output.

CTS = coaxial transfer switch, WTS = waveguide transfer switch.

Cruise passes were conventional, most of them 6–10 hours long with both uplink and downlink. Ranging or delta-DOR navigation signals shared the carriers with command-and-telemetry modulation. Cruise commanding could be initiated any time after MER's mission controller (call sign ACE, the real-time interface with the DSN) verified that the uplink sweep was successful by seeing the downlink frequency transition from one-way noncoherent to two-way coherent. This transition confirmed that the spacecraft receiver was in lock with the uplink carrier and ready to receive commands. During cruise and again beginning in May 2005, the one-way light time (OWLT) was less than 10 min, and the tracking passes were long, so it was feasible to wait for confirmation of sweep success before commanding.

Surface operations during the first portion of the primary mission used two-way DTE passes 30–60 min in duration, with both uplink and downlink. Later surface operations relied on uplink receive-only passes called direct-from-Earth (DFE). These were 20–30 min in duration and had no downlink. DFE passes were used to reduce spacecraft power use. Neither delta-DOR nor ranging was used during surface operations, since other means of determining rover position were accurate enough.

The OWLT began to exceed 15 min shortly before the end of the primary mission and did not again fall below 15 min for nearly a year. Fifteen minutes is significant compared to the duration of the communications pass. To avoid tying up rover operations for an extra round-trip light time (RTLT), extended-mission commands were radiated prior to receipt of confirmation of uplink sweep success. The normal downlink mode was coherency-enabled, not only to obtain two-way Doppler data, but also because SDST temperature varied continually during a sol. Temperature changes caused frequency variations in the SDST aux osc output that made one-way downlink difficult or impossible to acquire and track.

7.4.1.2 Stations Used by MER (34-m and 70-m, All Complexes) For cruise and surface operations phases, all three 70-m stations, all three 34-m high-efficiency (HEF) stations, and all of the operational 34-m beam waveguide (BWG) stations tracked MER. During launch, a 26-m station's X-band acquisition aid antenna was used to initially detect the downlink and to help with station pointing correction in case of deviations from the nominal trajectory. During cruise, a DSN array of stations successfully tracked MER as a demonstration.

7.4.1.3 DSN Changes Instituted during the MER Mission

7.4.1.3.1 34-m BWG 20-kW Transmitter and X/X/Ka-Feed Upgrades. Station transmitter power has generally been less of a concern to MER than is using a standard uplink (command) bit rate consistently to avoid confusion and errors over the rate. However, MER mission planning became simpler when all of the 34-m BWG transmitters were upgraded from 4 kW to 20 kW. This meant that the X-band uplink performance of all DSN 34-m antennas could be treated as essentially the same, and a single uplink rate could be used for long periods of time. Two of the 34-m BWG stations (DSS-26 and DSS-55) also received new feeds that allowed them to transmit at X-band and receive at both X-band and Ka-band, with a lower X-band system noise temperature than with the previous feed. Though MER transmits and receives X-band only, the X/X/Ka feeds improved X-band downlink performance for these stations, making them comparable to (or slightly better than) 34-m HEF antennas.

The nominal cruise uplink rate was 125 bps. Because of the shorter communications periods (comm windows, defined in Section 7.3.1.3) during surface operations, the uplink rate via the HGA was initially 1000 bps until increasing Earth–Mars distance reduced this to 500 bps. Similarly, the uplink rate via the RLGA was initially 31.25 bps, and later was made 15.625 bps.

On launch day, the first three passes were with 34-m stations operating at a reduced uplink power (200 W). If the received power at the spacecraft had been too high, risks would have included digital-to-analog converter (DAC) rollover glitches[15] or even damage to the SDST hardware.[16]

For the cruise and surface flight software loads involving large uplink file loads, the 20-kW transmitters supported 2000 bps (highest uplink rate available) on the cruise MGA and the rover HGA during the primary mission. In the extended mission, the flight software update was uplinked at 1000 bps over many passes (~30 min each). A flight software patch was uploaded at 2000 bps in February 2005.

7.4.1.3.2 Network Simplification Project Changes. The Network Simplification Project (NSP) changes were largely transparent to MER.

The project had to change station monitor channels to reference newly defined Monitor-0158 channels in the data monitor and display (DMD) and query processes. However, MER incorporated a set of multimission monitor DMD pages that were already developed and tested by the Lockheed Martin Aerospace (LMA) Mars operations team. Not having to develop these from scratch saved MER flight operations considerable time.

Twice during cruise, as documented in Incident, Surprise, Anomaly report (ISA) Z82482 [10], the new ability of a DSN station to transmit and receive on different polarizations was accidentally invoked, despite the fact that the spacecraft antenna in use always transmitted and received with the same polarization at any given time. Because of less-than-perfect isolation in the spacecraft polarizers, imperfect termination of an unused port on the WTS, and coalignment of the boresights of the MGA (connected to the left-hand [LH] port) and the CLGA (connected to the right-hand [RH] port), there were leakage paths that allowed uplinks sent with the wrong polarization to get into the SDST.

---

[15] The SDST's receiver has a DAC. The DAC rollover glitch is a known idiosyncrasy. When the receiver static phase error (SPE) crosses binary rollover points (for example, 8, 16, and 32 DN) as the frequency to the in-lock SDST receiver is increasing, the DAC generates a current spike that can knock the receiver out of lock. The SDST is most susceptible to this glitch at strong signal levels and cold temperatures.

[16] Use of 200-W uplink power ensured that the maximum uplink power would not exceed –60 dBm on the first pass after launch, taking into account station-to-spacecraft range, and angle to the spacecraft LGA. The specified SDST damage threshold is +10 dBm.

One occurrence was during a critical spacecraft cold-reboot activity when the CLGA was selected, but a left-hand-circular-polarized (LCHP, or LCP) uplink (and commands) got in through the MGA via a leakage path. The opposite situation occurred later in cruise when the MGA was selected, but a right-hand-circular-polarized (RHCP, or RCP) uplink sweep got in through the CLGA (no commands were sent). In the first case, the off-boresight angle from the MGA to the Earth was only about 2.5 deg; in the second, the angle from the CLGA to Earth was about 8 deg.

The polarizers (septum design) have inherent port-to-port isolation of better than -20 dB. However, in the stack configuration, there are significant mismatches at several interfaces that contribute to degrading the isolation. The use of a dead short on the unused port of the WTS (to save spacecraft mass) allows oppositely polarized signals to leak into the other port of the polarizer. A secondary leakage path results from the imperfect polarization generation of the polarizers.

Since surface operations began, only the RH port has been used (for either the HGA or RLGA), so it is unlikely that any LHCP uplink from the DSN would affect the spacecraft.

7.4.1.3.3 Multiple Spacecraft per Aperture. In late cruise, MER began regularly participating in Multiple Spacecraft per Aperture (MSPA) sessions with the Odyssey and MGS orbiters once the MER spacecraft came close enough to Mars to be in the same station antenna beamwidth as these orbiters. For surface operations, MSPA has in fact become a valuable capability for MER, in addition to the inherent ground-system efficiency improvement of being able to track two or three simultaneous downlinks.

Because MER surface operations at X-band used 20- to 60-min communications sessions of the same order of magnitude as the OWLT (10–20 min), or without a downlink at all, stations could not Conscan[17] on the MER downlink signal in time for it to improve uplink pointing. Furthermore, when MER was downlinking via the RLGA, Conscan was generally not used. Ripples in the RLGA

---

[17] Conscan (from "conical scanning") is an antenna-pointing technique that relies on the antenna system using its received signal to minimize the angle between the antenna's boresight and the direction of the received signal. To begin, the boresight is intentionally moved a small angle away from the predicted pointing direction, then continuously scanned in a cone around the predicted position at that small angle. The Conscan algorithm estimates the position around the cone where signal strength is the highest and moves the boresight in that direction. In contrast with the predict-driven pointing that sometimes caused significant (3- to 5-dB) pointing errors with MER surface downlinks, Conscan is not dependent on modeled Earth atmospheric refraction.

pattern (several decibels from peak to peak) would be misinterpreted by Conscan as pointing errors, causing the DSN antenna to change its pointing (adversely) in an attempt to compensate. Enabling Conscan on an orbiter X-band downlink (via the HGA) improved 70-m pointing for the MER uplink by 3 to 5 dB for many uplink passes, as later determined from recorded spacecraft telemetry sent back over the UHF relay link. MSPA was also useful for troubleshooting anomalous signal characteristics in the MER uplink and downlink. Comparing the signatures with those of the orbiter uplink and/or downlink (when available) helped determine whether the cause was the DSN, weather, or the spacecraft.

7.4.2 Entry, Descent, and Landing Communications

Figure 7-22, from the Mission Plan [7], summarizes the events and representative relative times for MER-A and MER-B during the EDL mission phase.

EDL was divided somewhat arbitrarily into the segments listed below. Together they took about 6 min, hence the nickname for this period, “six minutes of terror.”

- Cruise (prior to atmospheric entry [E])
- Entry (from E to E + 230 s)
- Parachute deployment (from E + 230 s through E + 270 s)
- Bridle deployment (E + 270 s through E + 360 s)
- Landed (beyond E + 360 s)

The most challenging period of the MER-to-ground communications was during EDL. As each vehicle entered the Martian atmosphere, it slowed dramatically. The extreme acceleration and jerk caused extreme Doppler dynamics on the 8.4-GHz (X-band) signal received on Earth. After the vehicle slowed sufficiently, the parachute was deployed, causing almost a step in deceleration. After parachute deployment, the lander was lowered beneath the parachute on a bridle. The swinging motion of the lander imparted high Doppler dynamics on the signal and caused the received signal strength to vary widely due to changing antenna pointing angles. All during this time, the vehicle was transmitting important health and status information that would have been especially critical for future missions if the landing had not been successful.

Even using the largest station antennas, the weak signal and high dynamics rendered it impossible to conduct reliable phase-coherent communications. Therefore, a specialized form of M-FSK was used. The signal processing that was required to demodulate the X-band DTE data tones used, as a point of departure, the methods of the Mars Pathfinder mission. However, the process for MER extended these to allow carrier tracking in conjunction with tone demodulation. The M-FSK scheme used 256 different signal

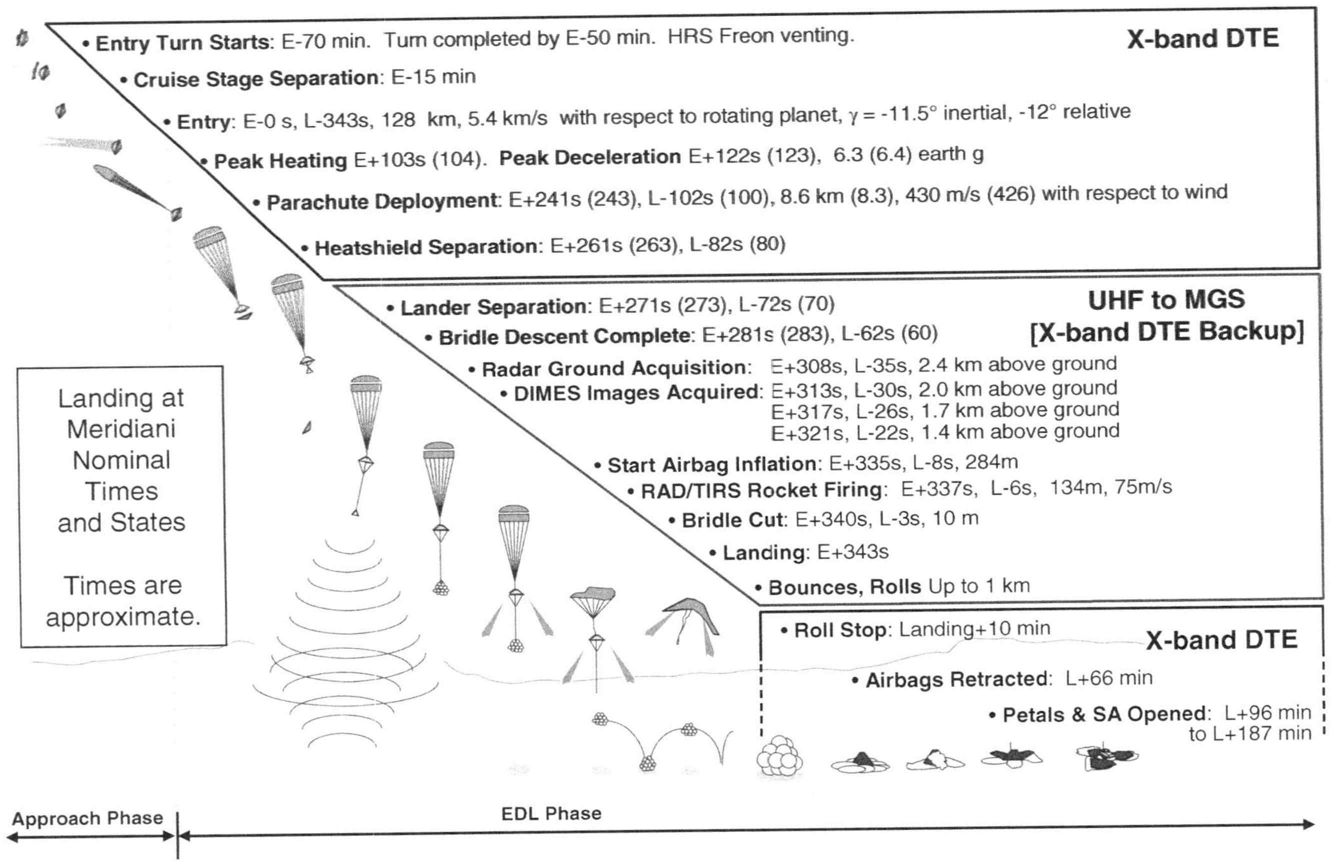

Fig. 7-22. MER-A EDL representative timeline (MER-B times in parentheses).

frequencies, each a semaphore to indicate the completion of a particular EDL event or the status of the flight software and fault protection at a particular time.

The following summary of carrier-frequency and signal-level variations that occurred during EDL has been adapted from the plans and expected variations described in [11]. The signal frequencies were modulated on the carrier, one at a time, as a subcarrier, using the SDST's capability to produce many distinct subcarrier frequencies. During hypersonic entry, the signal frequency could be switched every 10 s, resulting in the communication of 8 bits of information each 10 s. When the lander was suspended from the bridle, and the UHF link was prime, the duration of the modulation frequencies was extended to 20 s to better facilitate detection during this period of highly varying signal-to-noise ratio (SNR). This would result in fewer messages, but each would be of higher reliability than would be possible with the use of a 10-s duration.

The expected MER-B dynamics profile, magnitude, and uncertainty are illustrated in Fig. 7-23. The profiles are shown for one of the candidate landing sites. Three different profiles are shown—the nominal entry path angle (centered) and two other path angles (to the left and right) that correspond to the estimated maximum deviations from the nominal profile. For each entry angle, the spacecraft-to-Earth Doppler shift at the X-band frequency is shown in Fig. 7-23 (a). The range of Doppler shift is approximately 90 kHz, and the (two-sided) range of Doppler uncertainty is approximately 50 kHz. Figure 7-23 (b) shows the expected Doppler rate, or first derivative of Doppler frequency, due to acceleration.

The first maximum occurred due to atmospheric drag during hypersonic entry, at 150 s to 220 s past entry. The maximum varied from 700 Hz/s to 1200 Hz/s, depending on entry angle. The second maximum was a spike in Doppler rate due to parachute deployment. During the hypersonic entry, the range of uncertainty in Doppler rate was roughly the same as the maximum possible Doppler rate. For example, at approximately 150 s past entry, the acceleration could be anywhere from approximately 0 Hz/s to 1200 Hz/s. The same is more obviously true for the parachute release. Figure 7-23 (c) shows the second derivative of Doppler frequency due to jerk. During hypersonic entry, the value ranged from approximately −25 Hz/s^2 to 40 Hz/s^2. The exact values shown at parachute deployment are not precise due to the inaccuracy in the numerical differentiation used to obtain them.

The predicted SNR for the MER-B downlink signal during EDL is shown in Fig. 7-24. It is the ratio of total power-to-noise spectral density of the X-band signal received at a 70-m DSN antenna. The total power received at Earth from the spacecraft depends on the angle of the spacecraft with respect

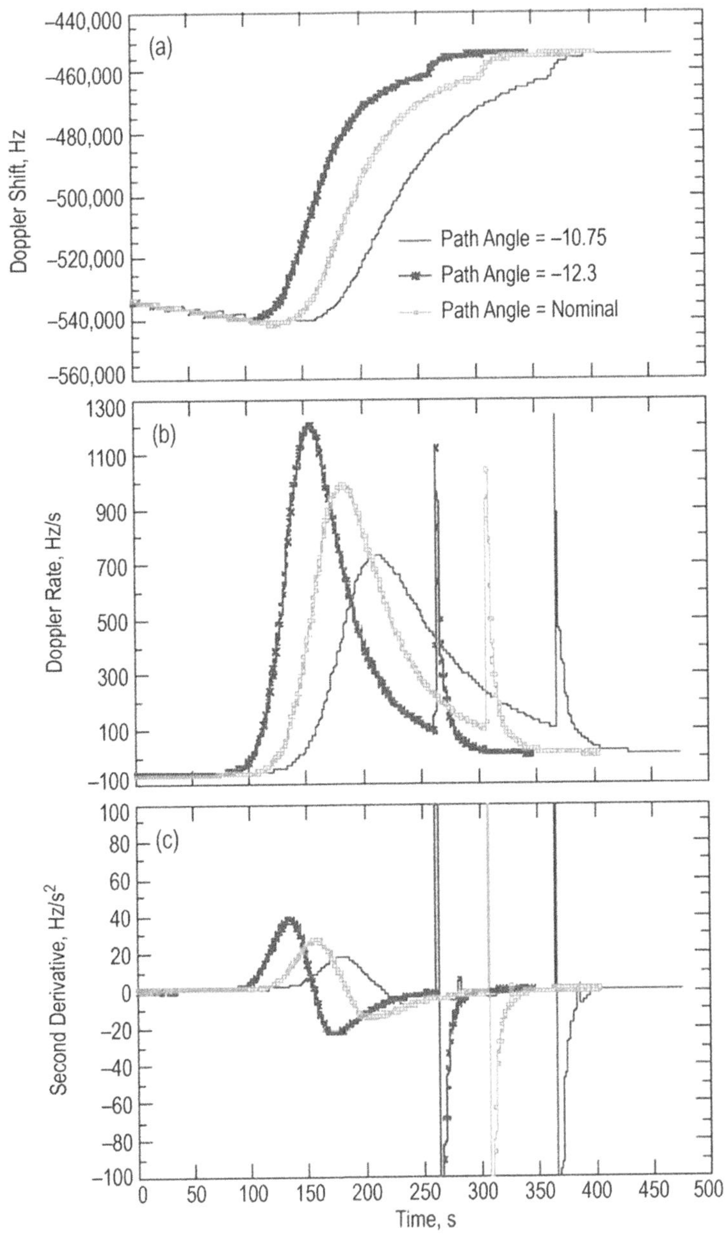

Fig. 7-23. MER-B EDL dynamic properties of (a) Doppler, (b) Doppler rate, and (c) Doppler acceleration. (The nominal path is at the center and the other two path angles to the left and right are the estimated maximum deviations from nominal.) Please visit www.wiley.com/go/deepspacecommunications to access the color version of this figure.

to the Earth and on the antenna-gain pattern. The antenna gain depends both on the angle off the axis of rotation of the spacecraft and on the rotation angle. The center curve in Fig. 7-24 is the nominal expected total power SNR

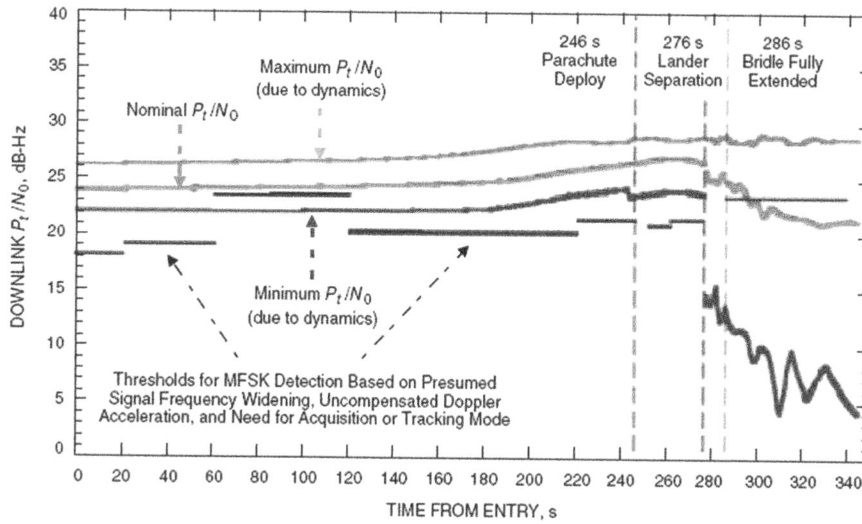

Fig. 7-24. Predicted X-band downlink signal levels during MER-B EDL. Please visit www.wiley.com/go/deepspacecommunications to access the color version of this figure.

versus time. This nominal SNR is based on the spacecraft axis orientation being the nominal angle, and on the nominal antenna gain with respect to rotation angle. The upper curve is the maximum SNR that might be achieved and is based on the most favorable orientation angle, and the lower curve is the minimum expected SNR. The three vertical dashed lines indicate the nominal times of the key events of parachute deployment at 246 s past entry, lander separation from the backshell at 276 s past entry, and full extension of the bridle with the lander at its end at 286 s past entry.

Figure 7-25 (a) shows the block diagram of the EDL data analysis (EDA) processor[18] and Fig. 7-25 (b) the EDL tracking process.

[18] A NASA Tech Brief [12] documents the EDA, described as a system of signal-processing software and computer hardware for acquiring status data conveyed by M-FSK tone signals transmitted by a spacecraft during descent to the surface of a remote planet. The design of the EDA meets the challenge of processing weak, fluctuating signals that are Doppler-shifted by amounts that are only partly predictable. The software supports both real-time processing and post processing. The software performs fast-Fourier-transform integration, parallel frequency tracking with prediction, and mapping of detected tones to specific events. The use of backtrack and refinement parallel-processing threads helps to minimize data gaps. The design affords flexibility to enable division of a descent track into segments, within each of which the EDA is configured optimally for processing in the face of signal conditions and uncertainties. A dynamic-lock-state feature enables the detection of signals using minimum required computing power—less when signals are steadily detected, more when signals fluctuate. At present, the hardware comprises eight dual-processor personal-computer modules and a server. The hardware is modular, making it possible to increase computing power by adding computers.

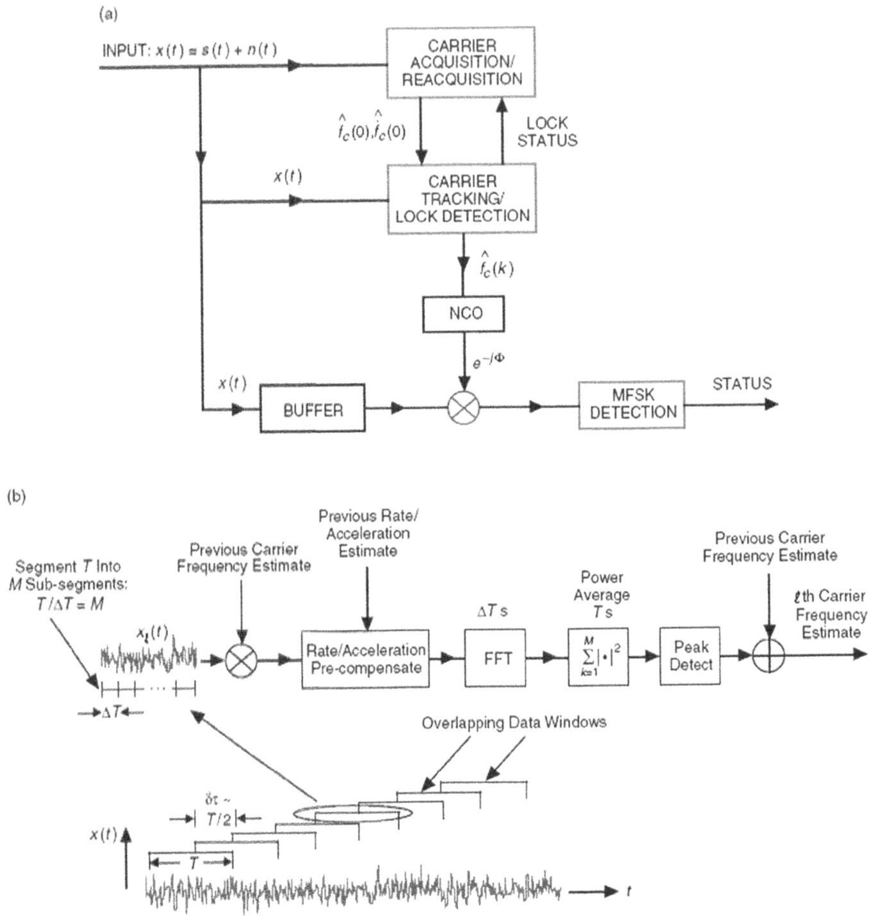

Fig. 7-25. Entry, descent, and landing (EDL) (a) signal processor and (b) tracking process.

During the higher-dynamics portions of EDL (pre-entry cruise, entry, parachute deployment, and bridle deployment), the detection interval, T, used for carrier tracking and acquisition was made 1 s (2 s in the lower-dynamics cruise portion). However, in the final phase of EDL, once the lander came to rest, the dynamics remained very low. A much longer interval (T ~15 s) could be used and in fact was desirable due to the lower SNR conditions. On the other hand, the tone-detection interval throughout was matched to the symbol duration (10 s) since the effects of carrier dynamics had been removed to a large extent by the carrier tracker.

7.4.3 Relay Data Flow

Active orbiters (Odyssey and MGS in the early days of MER and Odyssey and MRO in the latter part of the surface mission) have a relay package on board that allows the reception of data from vehicles (landers, rovers, etc.) on or near the surface of Mars. This surface-to-orbiter link can be referred to as the return link or, by analogy to DTE, downlink.

7.4.3.1 Odyssey. The total allocation in the Odyssey memory for surface vehicle data is approximately 260 Mb.

At the beginning of the primary mission, each of the two rovers was allocated 120 Mb.[19] Data received in the relay was divided into fixed length packets with a distinct application process identifier (APID) for each rover. These packets have fairly high priority on the Odyssey downlink with data rates to the DSN of up to 110 kbps at the beginning of the mission. As the Mars-Earth distance increased, Odyssey rates dropped to approximately 40 kbps into a 70-m DSN station, 14 kbps into a 34-m antenna. Odyssey can also operate in bent-pipe mode, that is, downlink to Earth while at the same time receiving data from landers at UHF (for the passes where Odyssey does not need to transmit data to the rover at UHF).

Like any other data source on board Odyssey, MER relay data can overflow its buffer allocation; if this occurs, the oldest data in the buffer is deleted by the new data.

When the RUHF return rate was increased to 256 kbps for some Odyssey passes, it was recognized that the MER buffer allocation might be exceeded. Since Odyssey downlink rates were also decreasing due to increasing Mars–Earth distance, it was decided to combine the allocation of the two rovers into a single buffer. This arrangement worked well initially because it is practically impossible for a single overflight to overflow the allocation at 256 kbps (the best pass recorded returned 170 Mb); the likelihood of having two consecutive passes with very high data volume is also very small.

Later in the extended mission, at near maximum Earth range, not overwriting relay data became more problematic. Two consecutive rover passes to Odyssey might be only 2 hr apart. With a minimum Odyssey X-band downlink rate of 14 kbps to the DSN, Odyssey could downlink approximately 50 Mb per hour, including Odyssey data with higher priority than the stored MER relay data.

---

[19] This allocation provided ability to store on board up to 15 min of data received at 128 kbps. Maximum Odyssey overflight time—horizon to horizon—can be as long as 17 min, but due to antenna pattern and other link considerations, the best UHF pass at 128 kbps was on the order of 110 megabits. The remainder of the memory was allocated to the Beagle 2 lander, but unfortunately no Beagle data was ever received.

MER has automatic tools to query the Odyssey ground system after each pass for the packets with the APIDs assigned to MER. The packet header is then stripped off, and the data is sent to the MER ground data system for frame synchronization; at that point the data looks as if it came directly via the MER X-band downlink.

7.4.3.2 Mars Global Surveyor (operated until November 2006). On the MGS spacecraft, the interface with the UHF radio was the Mars Orbiter Camera (MOC). Relay data from the rovers and the MOC images shared the same buffer allocation. The total data volume available for MER relay during the primary surface mission was approximately 77 Mb.[20] This allocation was routinely overflowed during MER operations at 128 kbps. In contrast to Odyssey relay storage, if the MGS MOC buffer had no additional space available, any new MER data is not recorded, and the old data is preserved.

The relay data in MOC packets reached the principal investigator for the instrument (at Malin Space Science Systems in San Diego), where the relay data was extracted from the MGS-to-Earth downlink and sent at JPL for frame synchronization.

7.4.3.3 Commanding the Rover via UHF. The UHF link from an orbiter to the rover is called the forward link. A forward link is comparable in function to an X-band DFE link and in general can provide commanding of the rover. However, MGS could not send data at UHF to a lander. The Odyssey and MRO forward links can provide a UHF backup to the X-band that is normally used to command the rovers.

Commands destined for the Odyssey–rover UHF link are sent from the MER ground system to the Odyssey ground system, where they are bundled in files. Each of these files is uniquely identified by a number, the spacecraft identifier (SCID) of the destination (Spirit or Opportunity), the pass number, and the day of the year. These files are then wrapped into Odyssey telecommand frames and uploaded to Odyssey memory. At the time of the specified overflight, these files are pushed into the Odyssey UHF transceiver buffer for transmission. While the Odyssey forward link is being used for commanding, return-link data cannot be simultaneously transmitted to Earth via X-band. That is, bent-pipe rover-to-Odyssey-to-DSN immediate relay is not possible. Odyssey

[20] The MGS project defined storage volume in the MOC buffer in terms of "frags" of 240 kilobytes (kB) (1.92 megabits) each. The maximum data volume allocation was 40 frags or 76.8 megabits. However, by mutual agreement between the MER and MGS projects, the relay allocation was nominally between 30 and 37 frags (51 to 71 megabits). Occasional passes were allocated only 15 to 20 frags (29 to 38 Mb) if MGS was performing compensated Pitch-and-Roll Targeted Observation (cPROTO) imaging activities or if MGS DSN coverage was limited.

stores the rover data on board and waits until the forward pass to the rover is finished before relaying the stored data back to the ground.

7.5 Telecom Subsystem and Link Performance

7.5.1 X-Band: Cruise, EDL, and Surface

In cruise, the MER spacecraft received an X-band uplink from the DSN and transmitted an X-band downlink back to the DSN. On the Mars surface, the X-band uplink is often referred to as a DFE link, to distinguish it from a UHF link received by the rover via relay from the Odyssey or MGS orbiter. The X-band downlink is often referred to as a DTE link.

Refer to Chapter 5 (Deep Space 1) for standard uplink and downlink spacecraft–DSN design control tables (DCTs). MER has uplink and downlink DCTS (not shown) that are similar to these tables, though modeled with MER parameters (transmitter power, receiver system noise temperature, circuit loss, antenna gain, antenna pattern).

This section begins with the performance of the on-board telecom subsystem and the Deep Space Network at the other end of the links during two critical mission phases: initial acquisition after launch, and EDL.

7.5.1.1 X-Band Performance during Initial Acquisition. Link performance during the first pass after launch was different for MER-A and MER-B. The MER-B trajectory produced higher required tracking antenna angular rates. The station antenna pointing on MER-B was also hampered because the launch vehicle performance (and hence its trajectory) were slightly different from predicted values. Consequently, station antenna pointing was off, resulting in lower than expected signal strengths. The suspicion that the MER-B trajectory was off-predict was substantiated when the DSN tried adding various time offsets (as great as ±50 s) to the pointing predicts on the backup station's antenna, and got a significant increase in signal strength using one of the nonzero time offsets. Link performance improved substantially when MER Navigation delivered their first post-launch trajectory update less than 24 hours later.

The downlink signal, especially for MER-A, was so strong as to produce unexpected signatures in the station receiver monitor data, as shown in Figs. 7-26 and 7-27. The symbol SNR (SSNR) estimator and Maximum Likelihood Convolutional Decoder (MCD) bit SNR estimator saturate at approximately 40 dB bit SNR. The actual received values were much higher (about 70 dB), but the Block V Receiver (BVR) and the MCD would read values higher than 40 dB as still only 40 dB. Fortunately, the uplink received power level (the SDST wideband automatic gain control [AGC]) was telemetered with reasonable accuracy. By tying together the saturated downlink measurement (reading an expected 30 dB too low) and the more accurate uplink measurement that

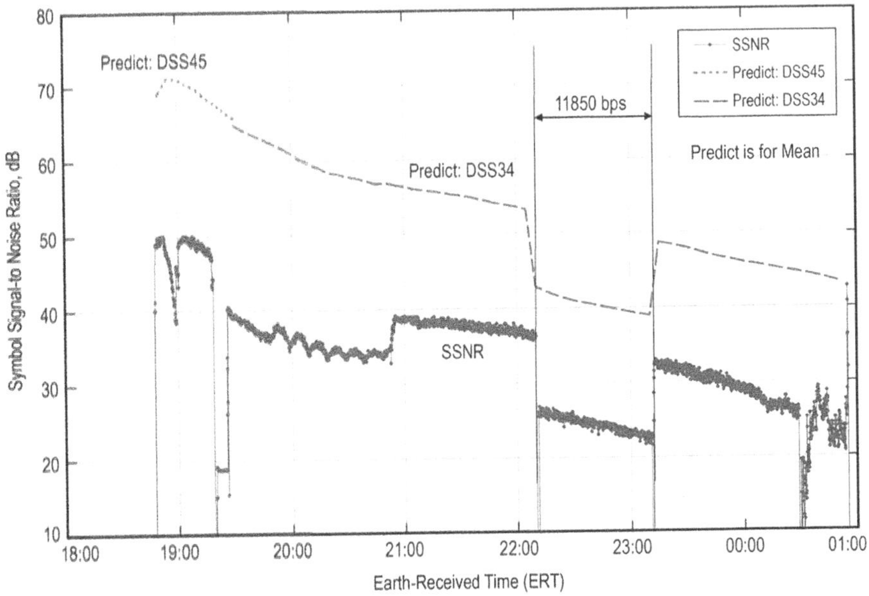

Fig. 7-26. MER-A initial acquisition symbol signal-to-noise ratio. Please visit www.wiley.com/go/deepspacecommunications to access the color version of this figure.

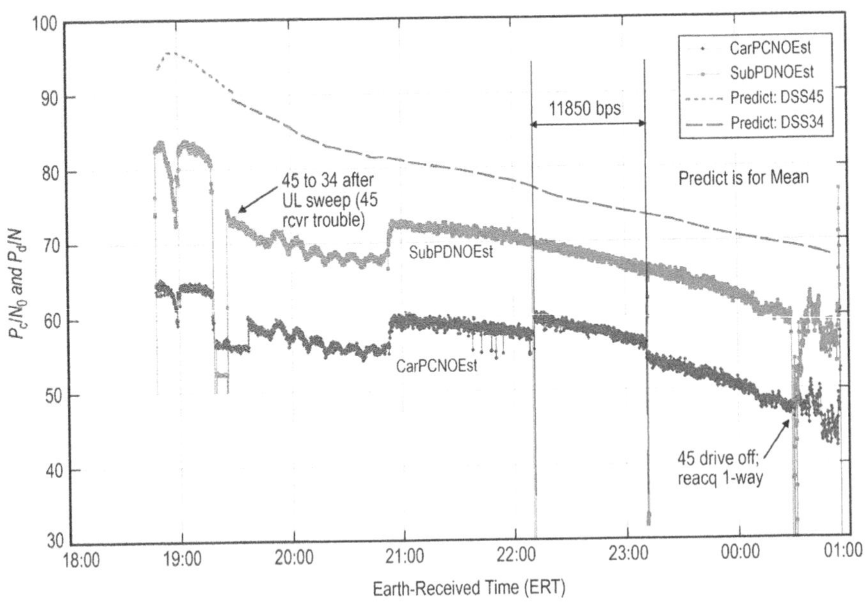

Fig. 7-27. MER-A initial acquisition, P_c/N_0, and P_d/N_0. Please visit www.wiley.com/go/deepspacecommunications to access the color version of this figure.

matched predicts, the telecom analyst on the MER Flight Team was able to assert that the MER telecom subsystem and the DSN were both performing normally. The cause of the discrepancy has been documented for use on launch day by future missions that will face similarly strong uplinks and downlinks.

Similarly, the carrier SNR (ratio of carrier power to noise spectral density, P_c/N_0) estimator saturates as P_c/N_0 increases from below 80 to above 90 dB-Hz. Also, P_c/N_0 decreases due to bleed-through of (strong) P_d harmonics into carrier noise estimation bandwidth, raising the noise floor. This bleed-through effect persists until range increases to the point that P_d harmonics are below the equipment floor noise. The amount of this P_d bleed-through differs for the 375 kHz subcarrier used for the 11850 bps playback telemetry rate and the 25 kHz subcarrier used for the 1185 kbps real time telemetry rates, causing the jumps at the transition to and back from 11850 bps.

However, another effect was noticed by the operations analysts: when the telemetry subcarrier frequency was 25 kHz (for the initial acquisition data rate of 1,185 bps), the BVR estimate of P_c/N_0 was less than later in the same pass when the subcarrier frequency was switched to 375 kHz (for the launch data playback data rate of 11,850 bps), even though the telemetry modulation index for both rates was set to the same value. The reason for this phenomenon is that the telemetry data is modulated onto the subcarrier. With the strong downlink signal levels, harmonics from the data spectrum contributed significant "noise" power into the BVR carrier-loop bandwidth. In fact, this noise source dominated over thermal noise in the channel, increasing the apparent noise spectral density, P_c/N_0. Increasing the subcarrier frequency moved the data spectrum further away from the carrier-loop bandwidth, decreasing the noise power, and thereby resulting in a larger P_c/N_0.

For both MER-A and MER-B, the saturated SNR and P_c/N_0 estimator idiosyncrasies were effectively gone within 48 hours after launch.

Another effect commonly seen on missions shortly after launch is the ranging "pedestal effect" (where turned-around uplink noise is the dominant downlink noise source, and raises the effective noise floor), but this was not noticeable on MER due to weaker downlink-signal levels.

7.5.1.2 X-Band Performance during EDL. Section 7.4.1 describes the special ground-system elements required to process the downlink modulation during EDL, and Fig. 7-22 describes the spacecraft configurations and the on-board telecom hardware and communications link transitions through the EDL sequence. The following summary of X-band carrier frequency changes is taken from Ref. 13; it applies to both Spirit and Opportunity EDL, except as noted.

Before the onboard EDL sequence started, the spacecraft was in the nominal cruise configuration, transmitting a two-way coherent signal from the MGA. The first telecom subsystem and link configuration change from the onboard sequence occurred one hour and forty-five minutes before Mars atmospheric entry interface (henceforth entry), when the spacecraft transitioned from using the MGA to the CLGA and to a telemetry rate of

10 bps. The standard DSN closed-loop receivers were reconfigured to look for the one-way signal using RH polarization.

When the aux osc came on, the downlink carrier underwent a warm-up frequency transient that was observed in the Radio Science Receiver (RSR). In-flight measurements confirmed preflight testing, showing a frequency increase of approximately 300 Hz over the first 15 s of aux osc operation, then a slow decay to steady state. The temperature of the aux osc was stabilized by the HRS, which was used during the interplanetary cruise to keep the temperature of the spacecraft environment stable. The cyclic behavior of the HRS caused the aux osc frequency to oscillate. HRS cycles tended to last approximately 6 min. Figure 7-28 (a) shows a 2-rpm spin rate of the cruise stage superimposed as ripples on the aux osc drift and HRS cycling signatures.

The CLGA is not on the spin axis, so the spin signature in Doppler frequency became more prominent at the turn to entry attitude. The first effect of the aux osc drift was movement of the mean frequency from 170 Hz at 03:20 to a peak

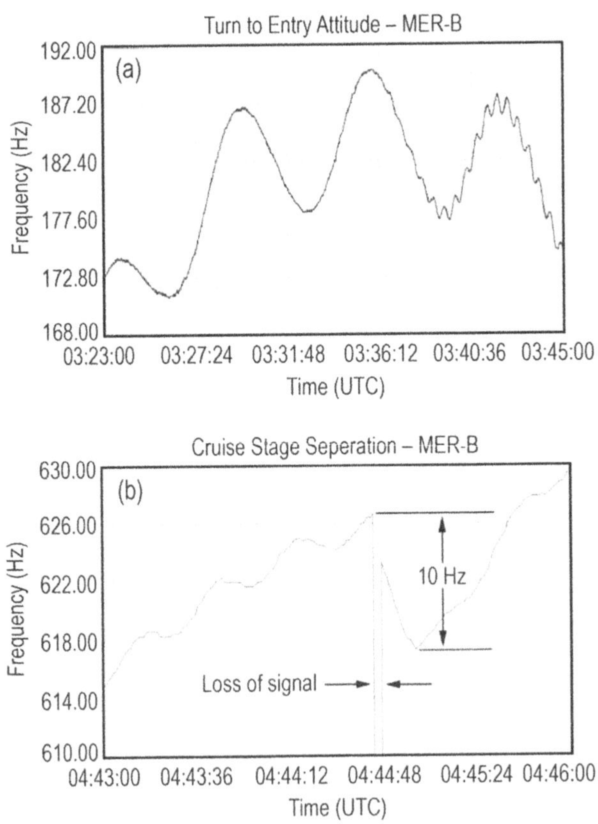

Fig. 7-28. MER-B signature of (a) CLGA spin and HRS cycling and (b) cruise stage separation.

of 185 Hz at 03:36, then back down to 180 Hz by 03:45. The second effect was the cycling of the HRS, evidenced by the 6-min, 12-Hz peak-to-peak oscillations. The third effect was the spacecraft spin Doppler. After the turn to entry, the peak-to-peak one-way frequency variation was 3.3 Hz at 2 rpm.

Before the lander entered the Martian atmosphere, the HRS was disabled. Without the temperature control provided by the HRS, the temperature of the aux osc increased from the cruise temperature of approximately 0°C. A 2-kHz rise was expected from 0°C up to 8°C, followed by a 7-kHz drop before landing as the temperature reached 25°C.

The cruise stage was jettisoned from the landing package 15 min before entry. The firing of the pyros imparted a force on the lander. Seen in Fig. 7-28(b), a 10-Hz Doppler shift occurred in the received signal. The discarded cruise stage blocked the downlink signal path on its way to burning up in the Mars atmosphere, causing the 2-s signal outage.

The beginning of the entry segment of EDL, the lander hitting the top of the Martian atmosphere, was rather benign so far as Doppler frequency effects were concerned. Soon, however, the friction caused the velocity to drop dramatically, and the spacecraft transitioned from speeding up towards Mars to slowing down, as seen in Fig. 7-29 for MER-A. Then, the deployment of the parachute caused an almost instantaneous 7-kHz jump in the received signal. This event caused the closed-loop DSN tracking receivers to go out of lock. Good closed-loop lock was not regained until the landers were stationary on the surface. The RSR and EDA were able to identify the signal.

All of the Doppler shifts came from changes in the acceleration of the craft. When the frequencies and the accelerometer data recorded on board MER-A are overlaid on one another, as in Figs. 7-29 and 7-30, the correspondence between the two can be seen. RAD firing is –4 s in Fig. 7-30. The data in that figure, collected during the bouncing, shows that the downlink signal was maintained until the beginning of the seventh bounce, at which time the signal was lost for a period. Review of the accelerometer data in non-real-time shows that the magnitude of the impacts decreased at precisely this time. A portion of the energy had been converted into rotational energy, and the higher spin rate caused a larger Doppler shift that was not tracked. When the lander's spin began to slow again on about the 23rd impact, the signal was identified once again.

The following summary of signal-level changes and the operation of the EDA is synthesized from Refs. 13, 14, and 15. Transmission of the M-FSK signal (described in Section 7.4.2) directly to Earth via X-band continued until RAD firing. For Spirit, ~4 s prior to landing, the RAD system on the backshell decelerated the lander from 240 to 0 km/h. Three seconds later, the lander cut its bridle and fell freely to the surface. It hit the soil at an expected

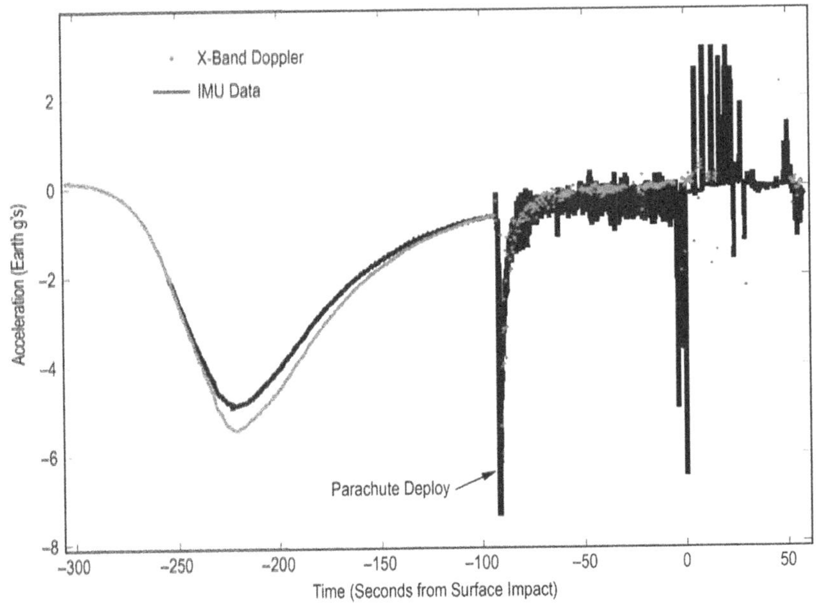

Fig. 7-29. Spirit EDL Doppler frequency and accelerometer data: entry compared to landing. Please visit www.wiley.com/go/deepspacecommunications to access the color version of this figure.

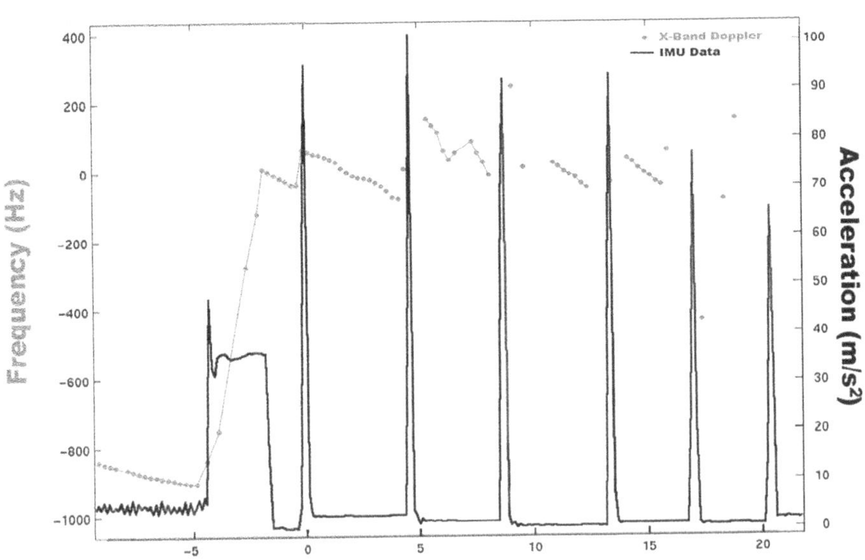

Fig. 7-30. Spirit Doppler frequency and acceleration due to RAD firing and bounces. Please visit www.wiley.com/go/deepspacecommunications to access the color version of this figure.

speed of over 80 km/h with a force of 40 g. The X-band carrier-only signal and a UHF 8-kbps signal were transmitted through bridle-cut, touchdown, and the subsequent bouncing on the surface of Mars.

The ability of the DSN or the MGS spacecraft to receive these signals could not be guaranteed as it depended primarily on lander orientation. Spirit and Opportunity each bounced for about 90 s after touchdown until they came to rest on Mars. When the lander flight software transitioned into the critical deploy state, the UHF transmitter was sequenced off and the lander was sequenced to transmit a set of five subcarriers—each 30 s long—via the RLGA, then to switch to the PLGA for transmission of the carrier-only signal for 3 min prior to repeating the original five subcarriers. These subcarriers signaled the lander state prior to the critical mechanical deployments.

Figure 7-31 shows the received carrier SNR per Hz (P_c/N_0) and the data SNR per Hz (P_d/N_0) in dB-Hz for the Opportunity EDL.

This Opportunity plot spans from about 44 min before cruise stage separation until after the second set of landed tones were transmitted. Each division in the plot represents 10 min. The entry point occurred at 4:59:46 Universal Time, Coordinated (UTC). Landing occurred at 5:05:28 UTC. The Opportunity lander bounced until 5:07:15 UTC. From that time until 05:22, six peaks separated by deep nulls can be seen in the received X-band carrier (orange points in Fig. 7-31). These variations were caused by multipath between the direct and reflected-from-Mars X-band signals as the Earth set at the landing site.[21] The carrier and all of the entry and descent tones were received in real

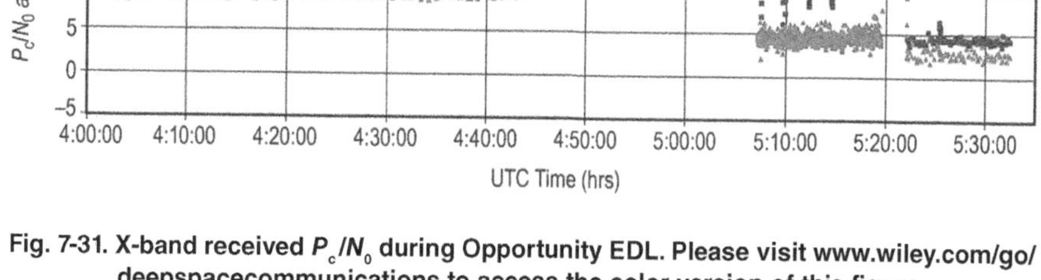

Fig. 7-31. X-band received P_c/N_0 during Opportunity EDL. Please visit www.wiley.com/go/deepspacecommunications to access the color version of this figure.

[21] Multipath means the receiver sees two (or more) radio waves, one of them coming directly from the transmitter and the other reflected from something (such as the surface of Mars) and so arriving via a different path. Because the reflected path is longer than the direct path, the two waves may arrive in phase (constructive interference, stronger signal) or out of phase (destructive interference, weaker signal). As the Earth sets and the path lengths change, the signal level versus time shows a characteristic variation called fading.

time for both Spirit and Opportunity. For Spirit, there was a loss of signal for more than 15 min after landing. Most of the data from this outage was recovered in post-EDL signal processing.

Figures 7-32 and 7-33 and the following description from [15] summarize the results from the EDA real-time and non-real-time processing during Spirit's EDL in terms of the X-band carrier and tone SNR levels through EDL.

For Spirit's EDL, Fig. 7-32 (tone power to noise spectral density ratio) and Fig. 7-33 (carrier power to noise spectral density ratio) and the following description come from Ref. 15. Figure 7-32 provides P_d/N_0 values from the DSS-14 EDA in real-time. Figure 7-33 shows (a) real-time P_c/N_0 values and (b) post-processed P_c/N_0 values from DSS-43. Note the start/stop times in Figs. 7-32 and 7-33 are not the same. Blue in Fig. 7-32 indicates low-quality

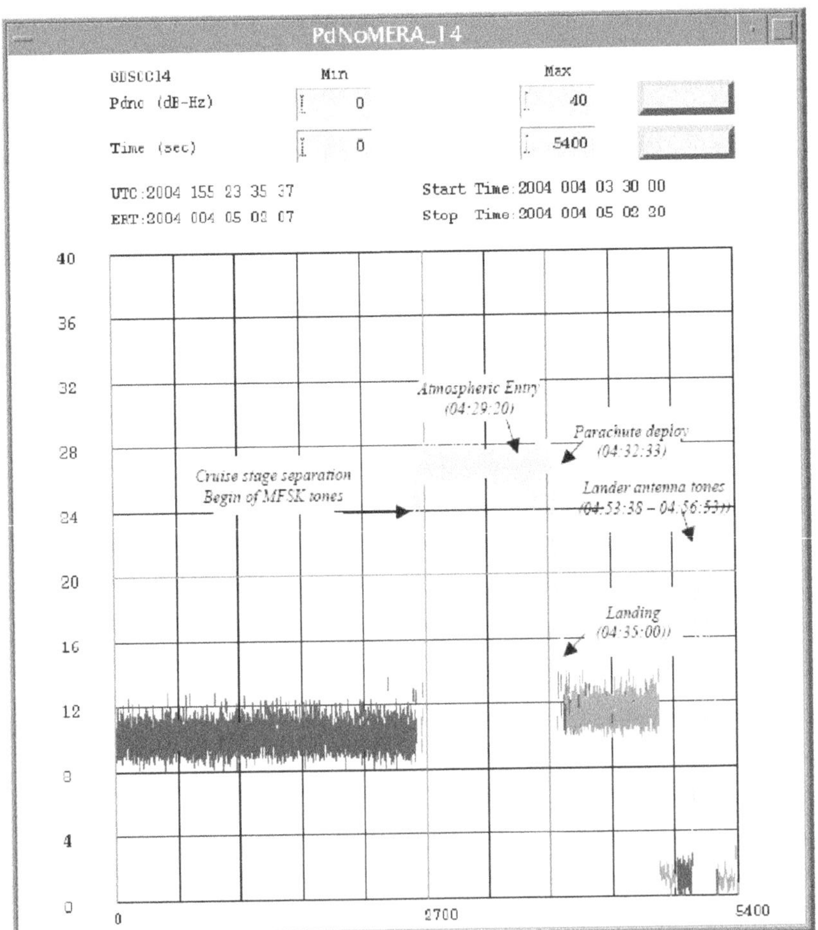

Fig. 7-32. M-FSK tone power during Spirit's EDL. Please visit www.wiley.com/go/deepspacecommunications to access the color version of this figure.

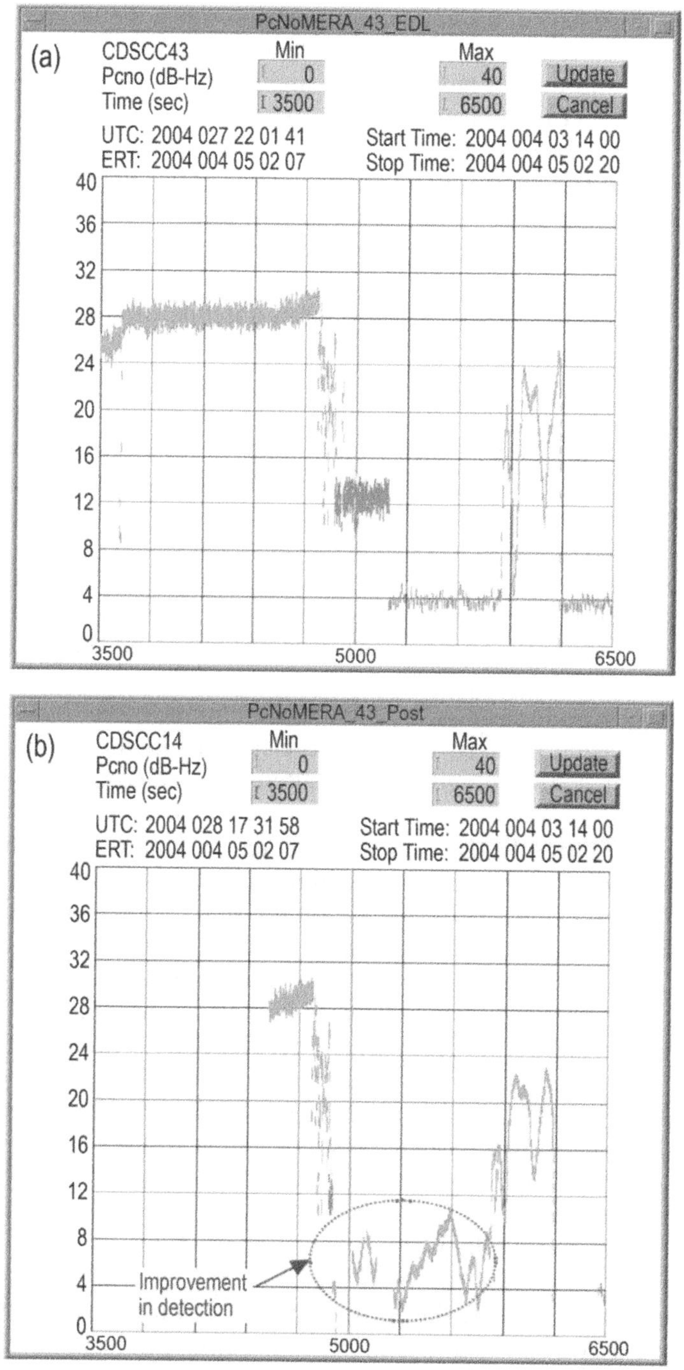

Fig. 7-33. Comparison between (a) real-time and (b) postpass carrier detection for Spirit EDL. Please visit www.wiley.com/go/deepspacecommunications to access the color version of this figure.

data (mostly prior to cruise stage separation) that was not analyzed. Green in both figures indicates periods of analyzed valid data, and violet shows periods of analyzed invalid data. Figure 7-33(b) highlights an interval of improved post-detection data (now green).

In Fig. 7-32, the tones do not begin until after cruise stage separation, at about the time marked 2700 s on the time scale. In this figure, the signal level indicated prior to 2700 s is for a subcarrier modulated with telemetry. After separation, the output tone power is constant, so the different levels shown come from spacecraft antenna selection and antenna orientation relative to the Earth.

Overall, the results from the actual EDLs were better than originally anticipated. All tones marking critical events (such as cruise stage separation, parachute deployment, and PLGA deployment) were detected during real-time operation.

Early in mission operations planning, there was a concern about the ability to maintain contact with the spacecraft during the parachute-deployment and bridle-descent segments. This concern was prompted by the known possibility that the RLGA would point away from the Earth during the swinging motion. Also of concern was a potential communications blackout upon entry due to plasma induced by hyperdeceleration. In the 1997 landing of Mars Pathfinder, a 30-s outage was attributed to this factor.

In the case of Spirit's EDL, constant contact was maintained during this whole period, all the way until touchdown. The suspense came between touchdown and the first received post-landing signal. The project expected a few minutes of communications outage during this period. However, the actual outage lasted more than 15 min. Later reprocessing with longer integration time and wider frequency-rate search recovered 11 min of this gap. Figure 7-33 shows the comparative results of real-time and post-pass processing, with data in the top half of the figure from DSS-43 and data from DSS-14 in the bottom half. Green indicates valid data processed in real time, and violet shows periods of receiver noise (lost data). At DSS-14 (Fig. 7-33(b)), the green segment over the period 5000 s to 5800 s after start includes the post-pass recovered data. At DSS-43 (Fig. 7-33(a)), the corresponding period shows violet unrecovered data.

During the 15-min outage, the Radio Science Team reported unexpected detection of LH-polarized signal. Post-landing analysis by the MER project confirmed the possibility of LH reception. It was attributed to the orientation of the lander relative to the Earth as it came to a full stop, and to the antenna polarization ellipticity in the direction away from the main beam axis.

Because of this experience with Spirit, additional EDA equipment was deployed to process any LH signal during the landing of Opportunity. Figure 7-34 shows the EDA carrier detection of Opportunity from DSS-14 in oppositely polarized (RH and LH) channels. As in Figs. 7-32 and 7-33, blue indicates low-quality data, green indicates valid data in real time, and violet is lost data. LH signal power was 4–8 dB lower than its RH counterpart, as expected; however, at 4930 s after start, the LH channel remained detected for an additional 1.5 min while the RH channel experienced outage.

Detection of Opportunity's landing was even better than that of Spirit's. Again, all critical tones were detected. The post-touchdown outage was only ~1 min, compared to 15 min for Spirit. That outage occurred 3950 s after start, as shown in Fig. 7-34.

7.5.1.3 Performance versus Predicts: Cruise. As with other recent JPL deep-space missions, MER predictions for the X-band links are made using the Telecom Forecaster Predictor (TFP) ground software. TFP details are included in the user's guide [16]. Project-specific models (antenna gains and patterns, SDST receiver and transmitter parameters, etc.) in TFP are initially based on pre-launch subsystem tests that then are updated as required during flight.

During cruise, with stable performance day after day, the standard downlink criterion of (mean minus 2-sigma) worked well for both the CLGA during the early days and the MGA later. Telecom analysts became familiar with the usual link signatures. The time between sequence approval/command uplink and sequence execution was at least several days, allowing for an orderly process between link evaluation and data-rate planning.

X-band performance compared well with the TFP models. Because of an excellent prelaunch telecom test program, no spacecraft-specific models needed to be updated during the mission.

Despite the excellent telecom performance, midway through cruise it became apparent that generating the number of Attitude Control Subsystem (ACS) turns originally planned to maintain fairly good MGA pointing toward Earth was creating excessive workload for the systems and ACS members of the Flight Team. As part of a cruise "workload simplification" strategy, the project cut this number approximately in half, trading off larger off-Earth angle (thus a lower data rate) for a smaller off-Sun angle. In retrospect, it did not really simplify the overall workload because telecom, thermal, and power flight team analysts had to rerun studies to see if the changes in the plan were okay (even though we were deleting turns). The rerun studies had to be run yet again if the planned turn dates had to be changed to accommodate other constraints. Nonetheless, creative use of TFP and Excel tools alleviated the telecom analyst workload somewhat.

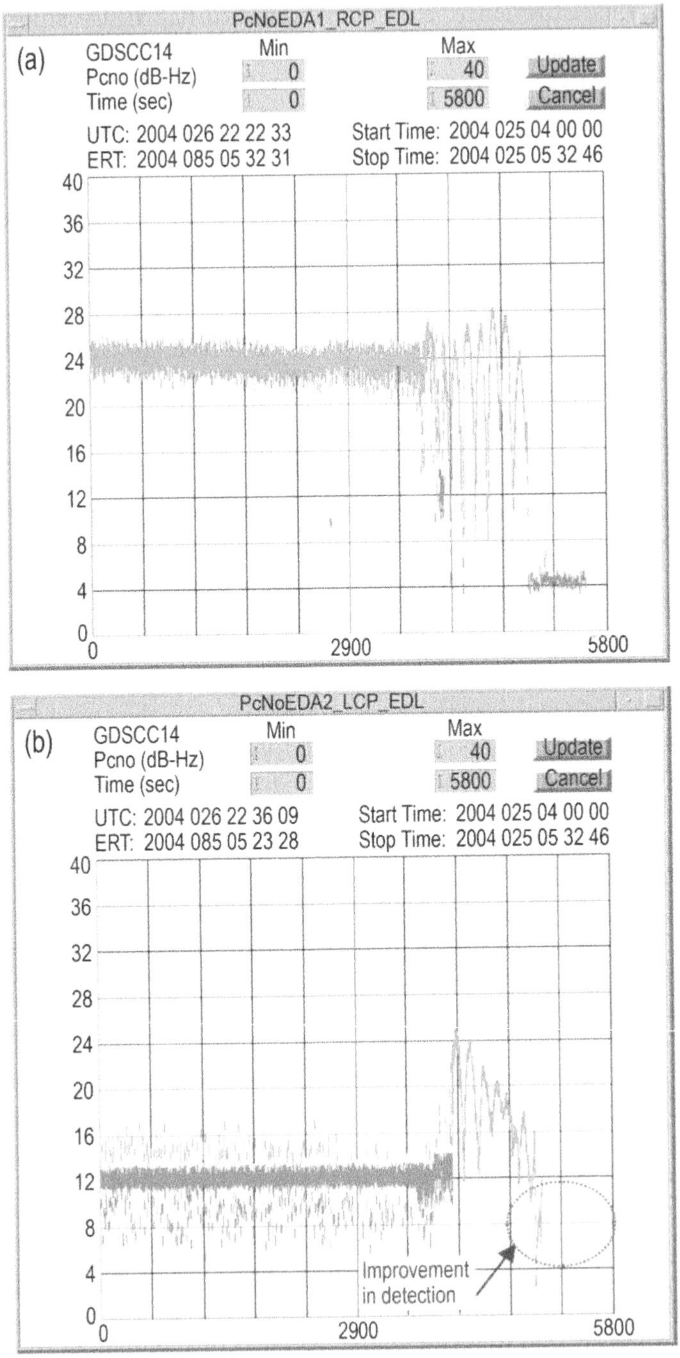

Fig. 7-34. Comparison of RCP and LCP carrier detection during Opportunity EDL. Please visit www.wiley.com/go/deepspacecommunications to access the color version of this figure.

The 10-bps telemetry rate was not used during cruise because there were alternatives to accepting its long lockup time and slow transfer of data: (a) request a 70-m station in case of a spacecraft emergency that would otherwise require 10 bps, or (b) wait until the elevation angle at the 34-m station was high enough (about 30 deg) to support the planned data rate of 40 bps.[22]

7.5.1.4 Performance vs. Predicts: Surface. On the surface of Mars, the mean-minus-2-sigma downlink criterion proved to be too optimistic because there was a shorter time between link evaluation and planning and because little or no data loss could be tolerated. A 2-sigma criterion meant there was only about a 1- to 1.2-dB margin to absorb performance variation. Any weather- or pointing-related problems larger than that would cause data loss. In retrospect, allowing more margin (3 sigma) would have been preferable. Instead, the work-around often was scrambling to reduce the data rate to the next available lower one (by building a real-time command and radiating it to the spacecraft just prior to a communications session) when weather threatened to make the downlink unsupportable.

Later-than-planned lockup by the DSN of the MER telemetry was a continuing challenge. The time of lockup is when the station has locked the carrier, subcarrier, and symbols so that after lockup the telemetry data will be valid. The project had always planned for transmission of only lower value "real-time" data until planned lockup. Lockup time varies with bit rate, and it is typically about 1 min for HGA data rates. Because the time to lock up varies somewhat from one pass to the next, the project had to decide how long a lockup time to plan for before starting the higher-value playback data. Plan too short, and some valuable data that has been sent will be lost. Plan too long, and the amount of valuable data that can be sent is reduced. The consequence of specifying a too-short lockup time becomes even greater because data organization on board places the highest-priority playback data (such as fault and warning event reports [EVRs] and the reports of spacecraft

---

[22] Besides 10 bps, 20 bps was available with short coding. The short RS code, while allowing for faster acquisition time, had a high coding-overhead penalty. Because of this, the MER fault-protection engineer wanted to avoid using the short code in cruise, which made 20 bps also unattractive. The P_t/N_0 thresholds for the low data rates are more closely spaced than 3 dB (due to higher station-receiver-system losses for operating at low SNRs), so it does not cost as many decibels (about 3 dB from 10 bps to 40 bps, instead of the 6 dB expected for a 4:1 ratio) to increase the data rate. Conversely, you do not get as much "bang for the buck" by lowering the data rate.

If DTE telemetry had been required over the RLGA at the larger Earth–Mars distances during the extended mission, 10 bps would have been the only RLGA-supportable downlink rate, even with a 70-m station.

health) earliest in the DTE pass. Considering these factors, and using experience from previous DTEs, the project changed the comm window parameter for lockup time from 1 min to 3 min (for downlink rates of 3160 bps and higher). Three minutes use up 10% of the data-return capacity of a 30-min comm window. The parameter has been set to 2 min for some windows in the later extended missions.

During diagnosis of and initial recovery from an anomaly in the Sprit flash memory file on sol 18, the RLGA supported a 40-bps or 300-bps downlink rate, as controlled by a high-priority comm window that overrode the 10-bps default mode. This downlink rate provided the repeating EVRs that led us to suspect the flash memory as the source of the problem in the first place.

7.5.1.5 Pancam Mast Assembly Occlusion. In addition to the general performance issues just described, a specific surface DTE and DFE problem that was difficult to characterize was "PMA occlusion." The problem so named comes about when the Pancam Mast Assembly (PMA) is directly or nearly in the field of view of the HGA. For use during surface operations, the ACS engineer modeled in the Tball attitude geometry visualization tool[23] the expected timing of possible PMA occlusions, but the model did not adequately account for the variability in effect of the PMA shape and size for different camera bar positions; and therefore, it was not adequate to capture the magnitude of the problem.

During prelaunch development, both telecom and ACS analysts recommended characterizing and modeling for PMA occlusion, knowing that it would occur during surface operations. Analyzing and modeling occlusion was not made a priority by the project. As a result, ACS is able to predict probable occlusion in DTE passes, but there are no models to predict accurately the severity or duration of an occlusion. Unfortunately, during the Spirit flash-memory-file anomaly, the rover happened to be parked such that PMA occlusion degraded most of the HGA sessions from 12:00 local solar time (LST) and later (usually as great as –8 dB; once up to –14 dB), resulting in substantially compromised downlink-rate capability.[24] Because the rover's instrument deployment device (IDD) was deployed, the attitude could not be changed to move the PMA out of the HGA field of view. PMA occlusion effects on the next sol had to be estimated based on the empirical data from the current sol. In retrospect, good prelaunch characterization would have

[23] Tball is a JPL-developed 3D visualization program that depicts the celestial sphere with the spacecraft at the center. It permits computation of Sun-to-spacecraft position/velocity vectors at the desired epoch using the latest ground-based spacecraft ephemeris.

[24] During the extended-extended mission, Spirit suffered another period of PMA occlusion to the LGA during sols 557 through 570 (July 27–August 10, 2005).

helped resolve the problem more quickly, by removing the PMA occlusion factor in link-performance variability.

7.5.1.6 X-Band Carrier-Only Beeps. For the surface mission to date, the sequences for most sols have included 5-min carrier-only downlinks called "beeps" (when transmitted with the rover stationary) or "honks" (when transmitted while driving) to convey information when scheduling a DTE is not practical (due to power, thermal, or activity constraints). The beeps or honks are most often used to indicate successful upload and execution of the new sol's master sequence. The new master sequence is uploaded each morning "in the blind" (without downlink confirmation of uplink sweep and command success). The beeps are first detected by the station operator visually by using the open-loop carrier fast Fourier transform (FFT). After the carrier has been detected in the FFT, the station will try to lock up the signal using the closed-loop receiver, and this is nearly always successful. The timing of the received beep conveys whether the upload and initiation of the new master were successful or not. In the absence of onboard faults, the beep will occur at one of two deterministic times: a nominal beep time or an off-nominal beep time. Each sol's master has two beeps sequenced: one at the nominal beep time for the new sol, and one at the off-nominal beep time for the following sol (new sol + 1). If the upload succeeds, the new master terminates the old master sequence (before the time the old off-nominal beep would be sent), and sends its beep at the nominal time. If the morning load fails, the old master remains alive and performs two actions. It executes a "run out" (canned science sequence), and it sends its beep at the off-nominal time.

From very shortly after EDL on both rovers, beeps have been sequenced, with SDST coherency enabled. The certainty of beep detection and lockup is increased when coherency is enabled and the SDST receiver is in lock on an uplink. At weak SNR levels, the stable two-way downlink frequency is far easier to detect than is a noncoherent one-way downlink, which may drift by several kHz in a few minutes. For about the first 20 sols for each rover, the project requested Radio Science to provide beep detection using the RSR, in parallel with the station using the FFT and the closed-loop receiver. The stations soon became proficient in detecting, locking up, and reporting the times of beeps, so RSR support is no longer routinely used.

In the primary mission, nearly 100 percent of the beeps sent were detected by the DSN. In the first extended mission, the DSN missed all but one of the inadvertent one-way beeps (no uplink in lock) and one coherent two-way beep. Not unexpectedly, several two-way beeps were missed during solar conjunction due to solar scintillation effects and the weakness of the beeps. Carrier-only beeps are reliably detectable in the standard receiver down to a P_t/N_0 of 12 dB-Hz. Occasionally the DSN has been able to detect and lock up on slightly weaker beeps. Figure 7-35 shows the predicted P_t/N_0 of the 11 a.m.

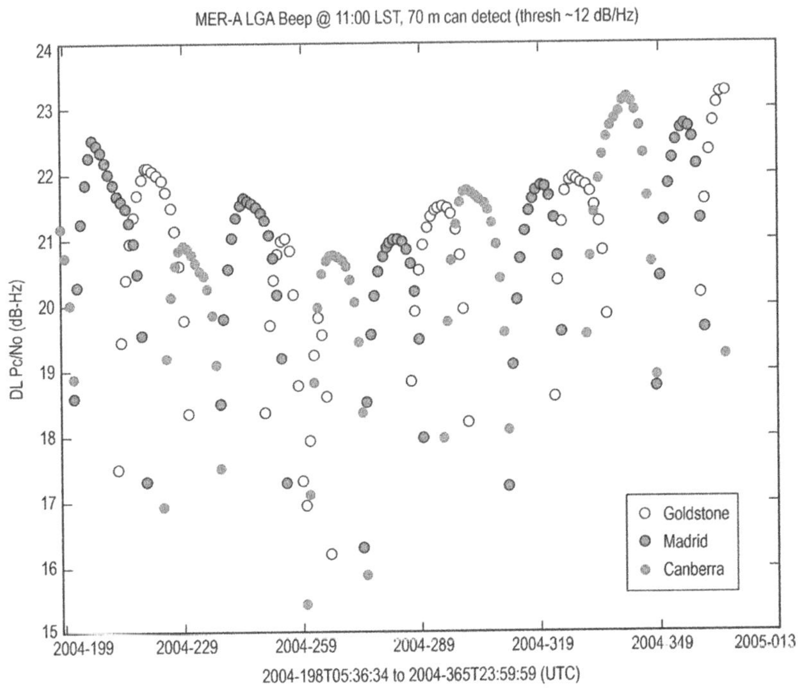

Fig. 7-35. Predicted MER-A 11 a.m. beep received P_t/N_0 at 34-m stations. Please visit www.wiley.com/go/deepspacecommunications to access the color version of this figure.

(local solar time) beeps for MER-A through the end of 2004. The colors indicate the DSN sites: gold for Goldstone, red for Canberra, and green for Madrid. The lower the station elevation angle at beep time, the lower the predicted P_t/N_0.

7.5.1.7 Antenna Pointing

7.5.1.7.1 Station. A periodic problem with DFE passes has been a type of excessive uplink pointing loss called aberration (see Section 8.4.1.3). It is caused by a geometric effect Earth and Mars in their orbits. The resulting degradation in SDST received power was as high as 8 DB for 70-m passes in July 2004 at larger Earth-Mars distances. It occurred with both MER rovers and at all 70-m stations. The cause of this uplink pointing loss is a combination of the angular motion of the spacecraft during a round-trip light time (RTLT)[25] and the DSN predict-driven (blind-pointing) error. With nonzero values of RTLT and angular motion, a station antenna cannot point perfectly for an uplink and perfectly for a downlink at the same time.

---

[25] This loss is currently not modeled in the MER adaptation of the multimission Telecom prediction tool, but it has been incorporated in the MSL adaptation.

Because the downlink is at a higher frequency than the uplink, the station antenna beam is narrower for the downlink. Also, downlinks usually have a significantly lower signal margin than do uplinks. For these reasons, the pointing algorithm favors the downlink and points the antenna toward the current spacecraft location for receiving the downlink as nearly perfectly as possible. But this pointing causes a problem for the uplink. This problem is called aberration. Think of a station antenna pointed to receive a rover downlink now. By the time an uplink that is transmitted now reaches the rover at Mars, the rover is no longer in the direction the antenna was pointing when it sent the uplink. The current radiated uplink does not arrive at the spacecraft until an OWLT later, and in the meantime, the spacecraft has moved with respect to the station antenna's pointing.

Figure 7-36 shows the uplink pointing error (between pointing positions for rover-transmitted downlink and rover-received uplink signals) caused by the angular motion of Mars during the signal travel time. This effect is called aberration. Figure 7-37 compares the uplink pointing loss for 70-m and 34-m antennas that results from the pointing errors in Fig. 7-36.

The performance loss in dB is greater for 70-m than for 34-m antennas because the former use narrower beamwidth than do the latter. The pointing error worsens with increasing Mars-Earth range (longer RTLT). It reached a maximum of about 3.25 dB near solar conjunction in September 2004, when Mars motion was perpendicular to the line-of-sight of the Earth. Because the

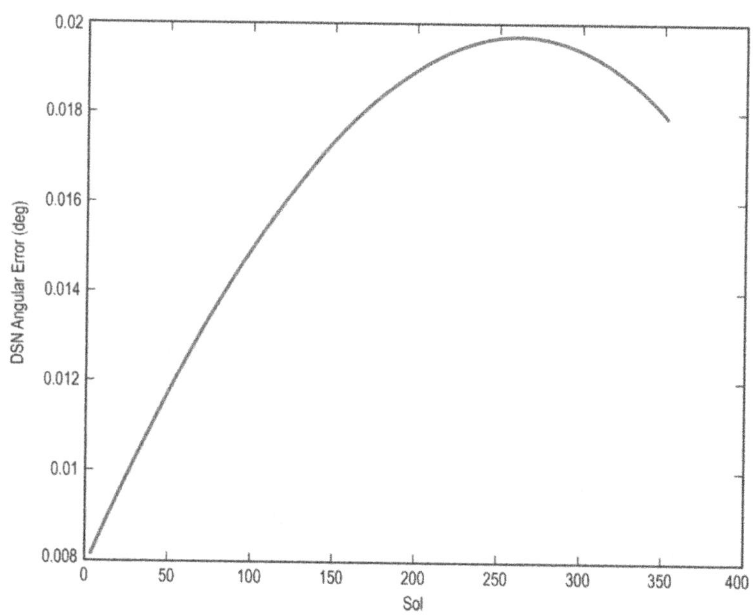

Fig. 7-36. MER-A angular motion with respect to DSN during a RTLT.

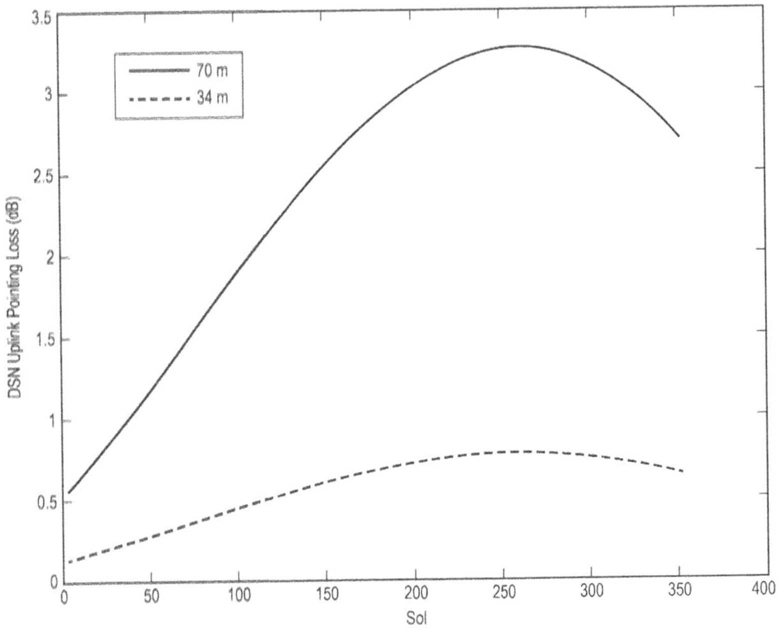

Fig. 7-37. MER-A uplink pointing loss due to angular motion in RTLT.

uplink operates nominally with an inherent pointing error up to 0.019 deg, any incremental DSN pointing changes/errors (including those of Conscan) have a magnified effect on the uplink when compared to the downlink, which operates nominally with a pointing error of 0 deg. On the occasions when an uplink pointing loss as great as 8 dB was observed from 70-m stations, analysis of signal strength telemetry suggests an overall pointing error of 0.03 deg, with 0.02 deg due to spacecraft motion and 0.01 due to 70-m blind-pointing error. The effect of a 0.01 deg blind-pointing error on the downlink is only 1 dB, which is why not much improvement is observable by the station when Conscan is enabled.

To reduce the effects of station pointing issues during the first extended mission, the project decided to uplink to the HGA rather than to the RLGA and to limit the command rate to 500 bps from all stations. The lower command rate increased the link margin and thus the pointing errors that could be tolerated. The 70-m stations were also requested to Conscan on MGS or Odyssey downlinks during MSPA tracks.[26] Conscan improves downlink pointing somewhat, but of course it does not eliminate the uplink aberration problem mentioned above. Fortunately, the excess margin at the 500-bps command rate

[26] Conscan is not requested for MEX MSPA tracks. MEX has a highly elliptical orbit, and MER did not seek an agreement with MEX to allow MER to affect that mission's uplink and downlink by requesting Conscan.

at the 70-m stations has accommodated the amount of uplink pointing loss that has occurred so far, whether or not Conscan is enabled.

7.5.1.7.2 Rover. The control of HGA pointing toward Earth is subject to ACS subsystem errors caused by such factors as temperature and bus voltage variations. The ACS analyst on the flight team periodically recommends the correction of HGA pointing through an activity called the fine-attitude update. These updates are infrequent because science activity dominates the rover resources. The telecom analyst tries to separate out rover HGA pointing errors and station errors by comparing HGA performance with RLGA performance during the same sol and by comparing HGA performance at the same station before and after an update. HGA pointing accuracy was determined to be generally within 2 deg in the primary mission, and downlink data-rate capability planning has been based on a 2-deg error assumption.

Figures 7-38, 7-39, and 7-40 show typical uplink received carrier power (from the SDST telemetry channel called carrier-lock accumulator [CLA], cla_snr, expressed as a signal-to-noise ratio) at a 34-m station and a 70-m station (with and without Conscan), respectively. Each figure includes both HGA and RLGA periods, and the TFP predicts together with the telemetry data. All the curves in each plot are labeled. A quick key to these figures: the HGA predict is the generally horizontal line near the top, and the RLGA predict is the somewhat decreasing line about 13 dB lower. In each figure, the data curve in the shape of an ascending staircase is the telemetered temperature of the SDST voltage-controlled oscillator (VCO), with the temperature scale in degrees Celsius on the right axis.

The cla_snr from the HGA begins at about 20:30 UTC in Fig. 7-38, 03:20 in Fig. 7-39, and 01:35 in Fig. 7-40. It ends 20 min later, corresponding to a 20-min comm window. The cla_snr from the RLGA is the short interval

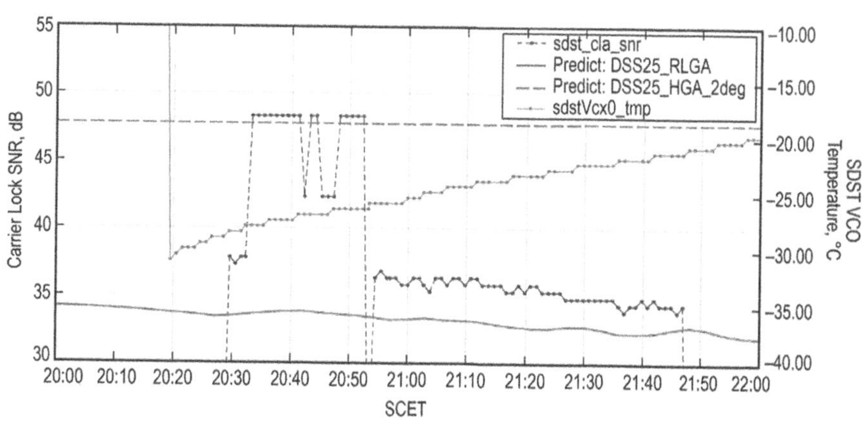

Fig. 7-38. MER-A surface uplink performance from a 34-m station (without Conscan). Please visit www.wiley.com/go/deepspacecommunications to access the color version of this figure.

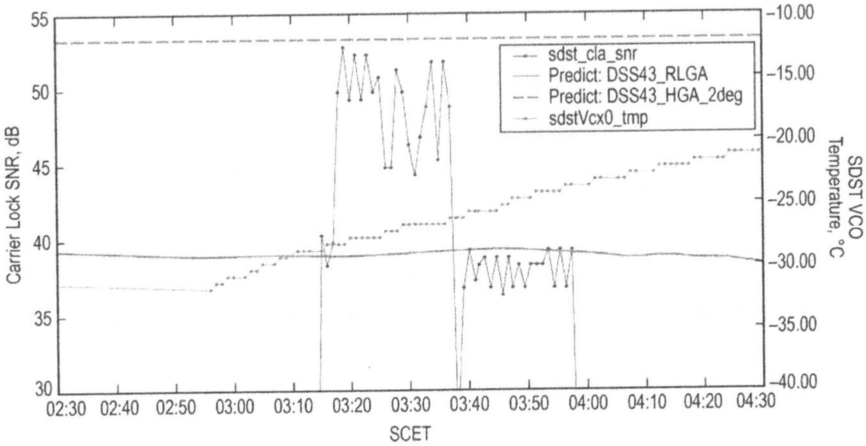

Fig. 7-39. MER-A surface uplink performance from a 70-m station (with Conscan). Please visit www.wiley.com/go/deepspacecommunications to access the color version of this figure.

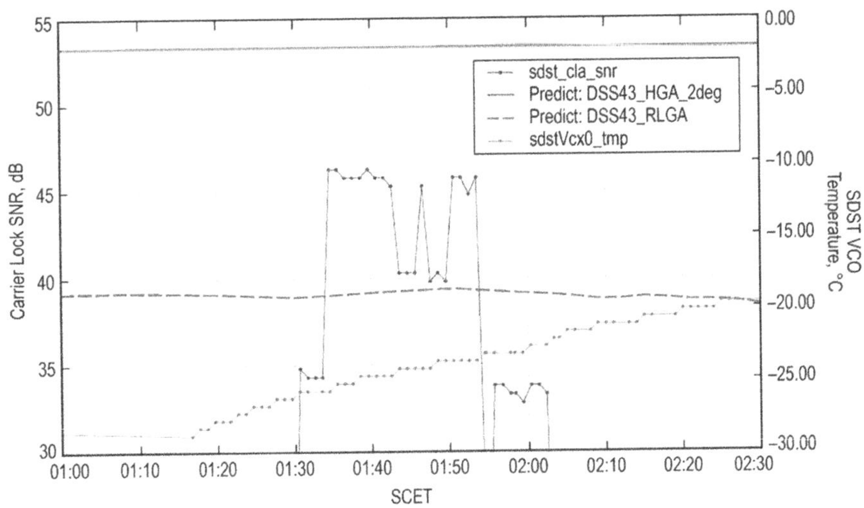

Fig. 7-40. MER-A surface uplink performance from a 70-m station (without Conscan). Please visit www.wiley.com/go/deepspacecommunications to access the color version of this figure.

preceding the HGA, and the longer period following it. During the HGA interval, command transmissions (with 5.8-dB carrier suppression) cause the deep, short-duration dips in each figure. The predictions are run without command modulation, so they are compared against the noncommand values of cla_snr. Comparing Fig. 7-39 with Fig. 7-40 shows the effect of Conscan. Conscan improves the overall 70-m uplink level by optimizing pointing on the (orbiter) downlink, thereby moving the average uplink level to 3 dB below the level for perfect uplink pointing. However, the signal level becomes more variable due to the periodic conical scanning performed by the

DSN antenna, which introduces additional pointing errors on the order of the size of the scan radius.

7.5.1.8 Uplink Acquisition Problems Caused by Rover Temperature Variations. The SDST best-lock frequency (BLF) is the uplink frequency that results in zero static phase error (SPE); hence it places no stress on the tracking loop. BLF varies with the temperature of the VCO in the tracking loop. In cruise, the VCO temperature changed very slowly (unless there was an attitude-changing TCM), generally much less than 1°C from day to day. Because of that stability, the project needed to provide BLF updates to the DSN only a few times for each MER during cruise. The DSN uses BLF as a scaling input to the uplink and downlink frequency predicts that are sent to the station for each tracking pass. Each pass requires a new set of predicts since each pass has a unique Doppler profile.

Using the uplink predicts, the station acquires the uplink using the Magellan acquisition (MAQ) tuning template, with ramped uplinks for all passes. The MAQ is a template to –130 dBm or less. To accommodate the –130-dBm constraint, the 70-m stations were requested to operate at 10-kW transmitter power during sweeps until the Earth–Mars distance increased to the point that full power (20 kW) could again be used. MER-B did not spend as much time in the coldest temperature regime as did MER-A. MER-B used a different day/night power profile called "deep sleep" that was started in July 2004 and was used during the periods of lowest power available in the Martian winters.

Characterization of the BLF during surface operations, as compared with prelaunch testing data, yielded the plots in Figs. 7-41 and 7-42.

The frequency of the VCO on MER-A could change by as much as 15 kHz over a period of 10 min due to temperature-dependent coherent leakage, as discussed in Section 7.5.1.9.1. To account for VCO temperatures down to –20°C, the MER-A BLF and sweep range (SR) about that center were both changed on some sols. Eventually as rover wake-up temperatures decreased toward –30°C and subsequently to lower than –40°C, it became necessary to use an uplink frequency reference offset (FRO) for the morning acquisitions.[27] Uplink FRO values as large as –13 kHz had been requested for MER-A.

---

[27] An uplink FRO is a constant frequency adjustment added at the time of the pass to the pregenerated BLF-based uplink (ramped) frequency from the DSN Predicts Group. Use of an FRO allows for morning and midday passes and simplifies operations, allowing a single TSF despite greatly different VCO temperatures at the two times. The Telecom analyst can verify that the proper FRO has been used by verifying the presence of a constant bias in the downlink Doppler residual. The bias is the FRO multiplied by the SDST X-band "turn-around" ratio (880/749). This offset occurs because downlink frequency predicts are not adjusted for the FRO. The downlink receiver can still acquire the biased carrier frequency, provided the carrier frequency remains within the acquisition FFT range.

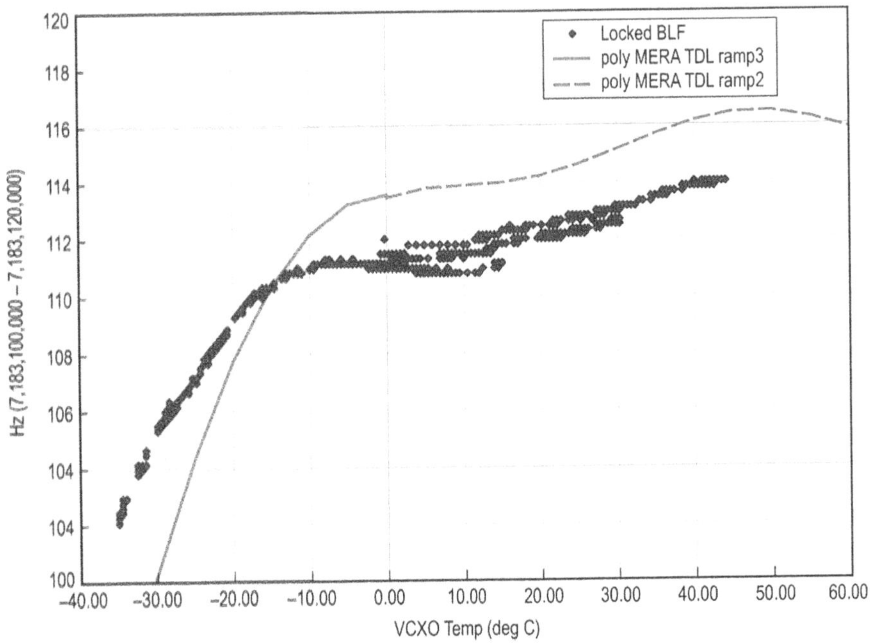

Fig. 7-41. MER-A surface best-lock frequency (in flight vs. prelaunch test).

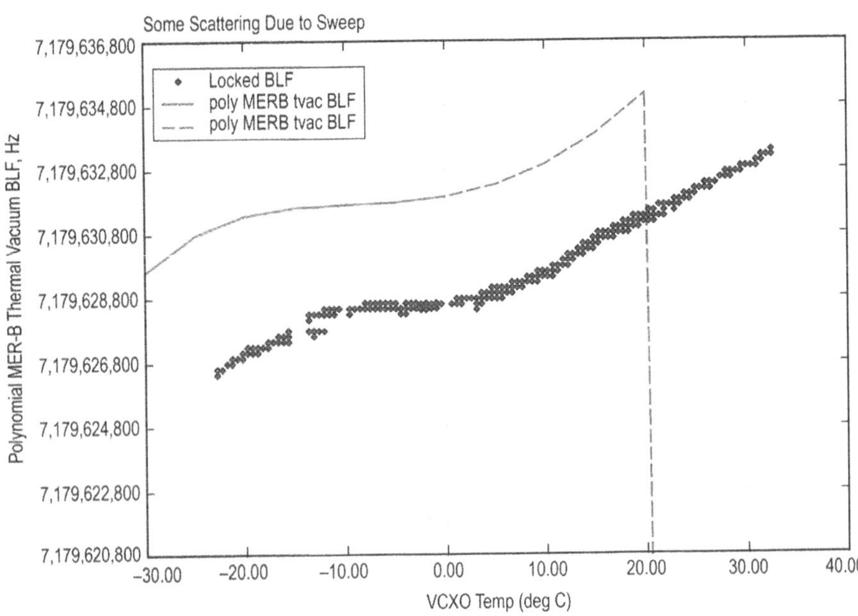

Fig. 7-42. MER-B surface best-lock frequency (in flight vs. prelaunch test).

To accommodate even colder temperatures, the SR for MER-A was increased to 8 kHz. For initial surface operations, SR for MER-A was made 3 kHz (the cruise value), then increased to 5 kHz, then 8 kHz, with even wider values eventually being required. To keep the sweep duration the same, some in-flight tests were made with an SR twice as large (16 kHz), and a sweep rate twice as fast (200 Hz/s).[28] For the MER-B SDST, which does not have coherent leakage or minimum temperatures as low as those of MER-A, FROs have been limited to –5 KHz. The MER-B BLF has stayed within the nominal ±5-kHz SR for all VCO temperatures so far.

7.5.1.9 Other Key X-Band Technical Issues

7.5.1.9.1 Coherent Leakage in MER-A SDST. This leakage, present only in the MER-A SDST, causes buildups in the receiver static phase error of as large as 15 kHz during periods between the SDST's state 1 (S1) time-outs. The S1 time-out resets the carrier loop to its BLF (SPE = 0). After an S1 time-out, the receiver will lock to an uplink sweep centered at BLF. Depending on the SPE drift rate, during some portion of the 10 min between time-outs, when the SPE becomes large enough, the receiver will not lock to an uplink centered at BLF.

The effect of the leakage is most severe at cold temperatures, the maximum drift magnitude increasing sharply below –25°C. The direction of the drift between time-outs may be positive or negative depending on the specific temperature. The operational mitigations include

- Increasing the SR to 8 kHz,
- Using an FRO to center the actual sweep around the predicted BLF, and
- Trying to time the acquisition to get an S1 time-out in the middle of the sweep. The S1 time-outs can be predicted from the rover wake-up time.

Figure 7-43 is a plot of MER-A static phase error in the SDST receiver loop. The relatively smooth variation between 14:50 and 16:10 occurred with the receiver in two-way lock with a station, and the 10-minute cycling between 0 and several kHz after that occurred with the receiver out of lock.

---

[28] To conserve power and maximize the time available for science, the project minimized the period between wake-up and start of the DFE comm window and also the duration of the comm window. These intervals have been designed to work with a sweep-duration of about 4 min maximum, whether the sweep is to the RLGA in the morning prior to the window or to the HGA at midday within the window.

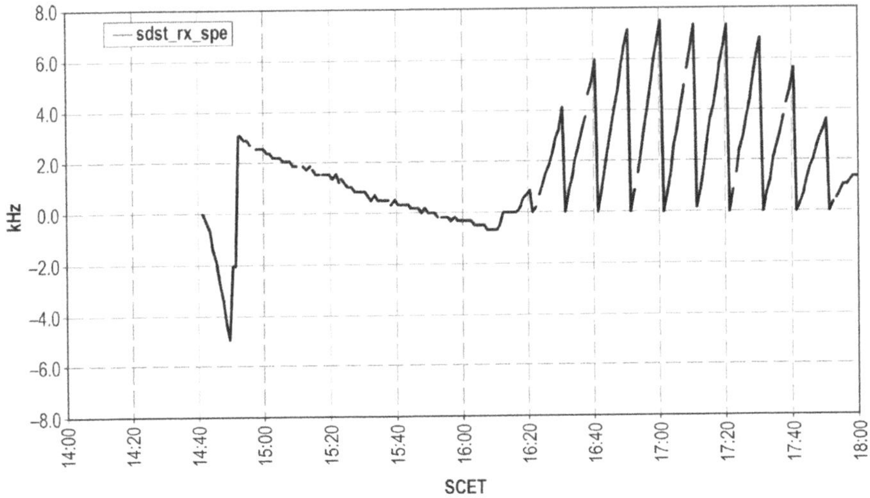

Fig. 7-43. MER-A SDST receiver static phase error showing effect of coherent leakage.

The sawtooths before and after this time are one-way, with the S1 time-outs at 10-min intervals resetting the SPE to 0. An SPE of 0 indicates the SDST will acquire carrier lock at BLF.

7.5.1.9.2 SDST Receiver DAC Rollover Glitch. A DAC in the receiver induces a voltage spike in the VCO when the digital value of the SPE rolls over from all ones to mostly zeroes. The voltage spike could cause a loss of already-achieved uplink lock. The receiver is most susceptible to this problem for positive-going sweeps, at temperatures lower than –15°C, and for strong uplinks, greater than –130 dBm. This is a problem seen during prelaunch testing and during rover operations in the extended missions.

The operational mitigations that have been used to avoid DAC rollover include

- Sweeping into the RLGA rather than the HGA to reduce received power,
- Using the MAQ template, to sweep downward from above to below BLF, sweeping positive only while returning to BLF,
- Reducing 70-m transmitter power to 10 kW to limit the received power to –130 dBm for HGA acquisitions at smaller Mars–Earth distances, and
- Keeping the SR narrow to avoid most rollovers.

7.5.2 UHF: EDL and Primary Mission Surface Operations

In this chapter, the UHF link from the orbiter to the rover is called the forward link, analogous to the X-band uplink from the DSN to the rover. The UHF link from the rover to the orbiter is called the return link, analogous to the rover X-band downlink to the DSN.

7.5.2.1 EDL UHF Link Predictions and Performance. After parachute deployment, the MER lander transmitted a UHF return link at 8 kbps to MGS, whose orbit had been phased so that the orbiter would be in view of the descending MER. Both the X-band DTE carrier-only signal and the UHF 8-kbps signal were transmitted throughout bridle-cut, touchdown, and subsequent bouncing on the surface of Mars. During the EDLs for both Spirit and Opportunity, the UHF 8-kbps signal was received by MGS from the time when the lander separated from the backshell until the time when MGS set at the landing site. During this interval, the UHF link returned about 3.5 Mb.

For mission design purposes, a link margin of 10 dB was kept for the UHF EDL phase. Such a high margin was justified by the challenges of getting a good antenna measurement on the lander mock-up at UHF; in addition, there were great uncertainties in the geometry for this mission phase (for example, MGS position, and angles between the antenna and MGS due to swinging on the bridle). Due to the possibility of the antenna's breaking off during airbag deployment and the challenges of guaranteeing a good signal while the lander was bouncing on the surface, no requirement was specified in UHF performance after RAD rocket firing.

Figure 7-44 shows the received UHF power at MGS and the lock status of the carrier, bit synchronizer, and Viterbi decoder during the MER-B EDL. In this timeline, bridle-cut was at 04:54:21 UTC, roll-stop at 04:56:08 UTC, and MGS-set at 05:02:38 UTC (0-deg horizon for the planned rover landing site).

For both MERs, UHF performance during EDL exceeded predictions. The lander UHF antenna was not damaged during the inflation of the airbags, and the MGS receiver was able to stay in lock even while the lander was bouncing.

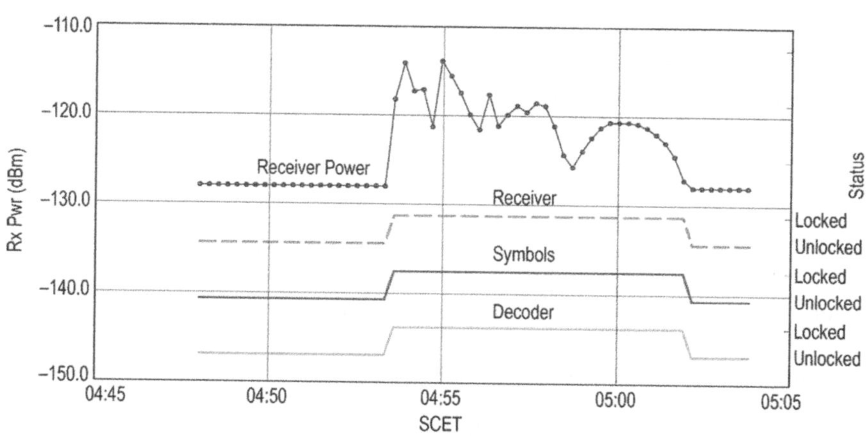

Fig. 7-44. MER-B EDL (January 25, 2004), UHF received power and lock status.

7.5.2.2 Primary-Mission Surface UHF Link Predictions. At X-band, predictions are often generated in the form of data-rate capability, and the results are used to set the downlink rate to the DSN from among more than a dozen possibilities, depending on Earth–spacecraft range, spacecraft antenna pointing angle, and station-elevation angle. For MER UHF, data-volume predictions have proved to be especially useful because of the limited variation in range between the rover and the orbiter, the lack of modeled elevation-angle effects, and the small number of data rates from which to choose.

The Generalized Telecom Predictor (GTP) tool, together with a series of scripts, generates a data-volume capability file (DVCF) for all overflights for the specified rover-orbiter pair in a given time period. The output is in the form of tabular summaries displaying volume (in megabits) for each view period (potential relay pass) and for all rover yaw angles in steps of 10 deg. The DVCF predictions assume that the rover is not tilted. Not tilted means that at the landing site, the RED is horizontal and the RUHF is vertical.[29] The output can be displayed in several forms. Figure 7-45 shows an example of tactical use of Generalized Telecom Predictor/data-volume capability file (GTP/DVCF) volume predictions. Part (a), top, shows the predicted volume in megabits at every 10 deg in azimuth on a polar plot, for each of two potential low-elevation passes. Part (b), bottom, shows the geometry of each of these passes superimposed on the rover UHF antenna pattern (oranges and reds indicate higher gains) for the actual yaw of 297 deg. Based on the yaw and the predictions, the pass shown to the right was selected, and it returned 75 Mb.

Based on these kinds of predictions, and after verification of normal UHF link performance on the surface, the return-link rate was raised to 128 kbps for all MGS and Odyssey passes shortly after landing.

DVCF predicts showed that the return link to Odyssey could often support a rate of 256 kbps, but this rate had very limited testing before launch, and was initially restricted from use. However, in February 2004,

---

[29] For much of the primary and early extended mission, "no tilt" was a good approximation, with the actual tilt generally less than 4 deg. As the extended missions have continued, with the rovers climbing hills or descending slopes into craters, Odyssey-MER communications have occurred with the relevant rover at a tilt as great as 31 deg. Though DVCFs are still run with "no tilt," the exactness of predictions has now increased with scripted use of GTP, taking into account telemetered rover roll, pitch, and yaw angles. These predictions have also sometimes accounted for line-of-sight blockage from the local terrain rather than using a simple, fixed, minimum elevation angle.

(a)

(b)

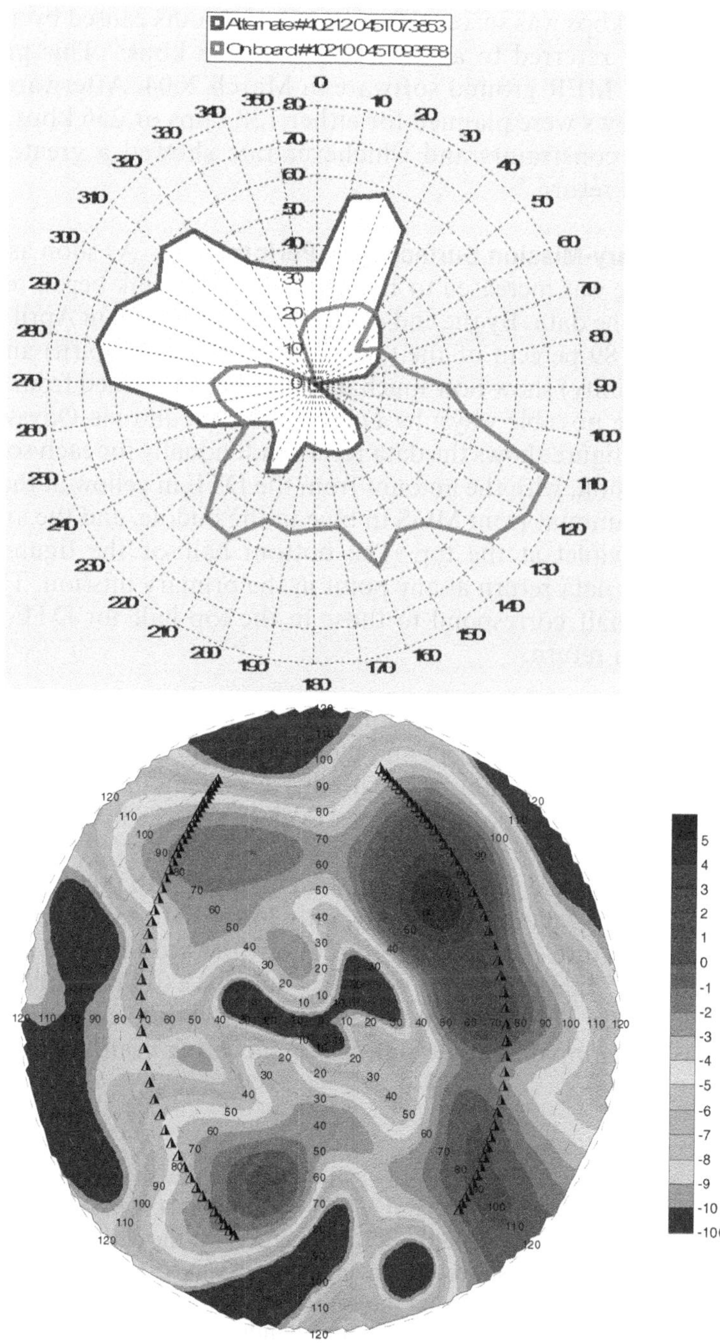

Fig. 7-45. Example of tactical use of GTP/DVCF volume predictions with (a) showing predicted volume and (b) showing the geometry of each of these passes superimposed on the rover UHF antenna pattern. Please visit www.wiley.com/go/deepspace communications to access the color version of this figure.

a test of 256 kbps was successful except for dropouts caused by a transceiver idiosyncrasy referred to as "extra byte at 256 kbps." This problem was corrected by MER ground software in March 2004. Afterwards, Odyssey comm windows were planned for either 128 kbps or 256 kbps, depending on Odyssey constraints and whichever rate showed a greater predicted data-volume return.[30]

7.5.2.3 Primary-Mission Surface UHF Performance. As soon as the return-link data rate was increased to 128 kbps, the UHF link began returning the majority of the data. By the end of the primary mission in April 2004, UHF data totaled 89 percent of the total. Figure 7-46 (for Spirit) and Fig. 7-47 (for Opportunity) show how much data has been returned from the rover in each of three possible ways: by DTE, via MGS, and via Odyssey. The top half of each figure shows the data return individually for each sol during the primary mission, with the amount from the DTE in yellow at the bottom of the bar, the amount from MGS in blue in the middle, and the amount from Odyssey in violet at the top. The bottom half of the figure shows the accumulated data return at any point in the primary mission. The colors in the bottom half correspond to those in the top half for DTE, MGS, and Odyssey data return.

During the first extended mission, the portion of total data returned via UHF increased to 95 percent and by October 2005 to 97 percent. Additionally, relay communications compatibility was demonstrated with the MEX orbiter during a few coordinated passes at 32 and 128 kbps.

The UHF links were generally compared in detail with predictions only when the actual performance was significantly lower than expected or showed an unusual signature. To get high-resolution forward-link telemetry data on received UHF power, a "UHF report" would have to be included in that sol's rover sequence. This report provides telemetry with a sampling resolution as small as 1 s, compared to the standard engineering health (or housekeeping)

---

[30] Once normal surface operations began, UHF window planning was always more fully automated than was X-band window planning. However, plans for a week's worth of UHF windows could not fully account for the changing yaw angle that could result from sol-to-sol driving plans. As the extended mission went on, UHF windows were changed in the sol-to-sol "tactical" process to optimize for 256 kbps or 128 kbps, and some planned windows were cancelled if their data volumes were predicted to be significantly lower than other possible windows 2 hr before or after. Eventually, some changes also took into account the rover tilt and horizon obstructions. As described in the next section, MER developed a process for notifying the Odyssey and MGS ACEs via e-mail of cancelled windows or ones with changed data rates.

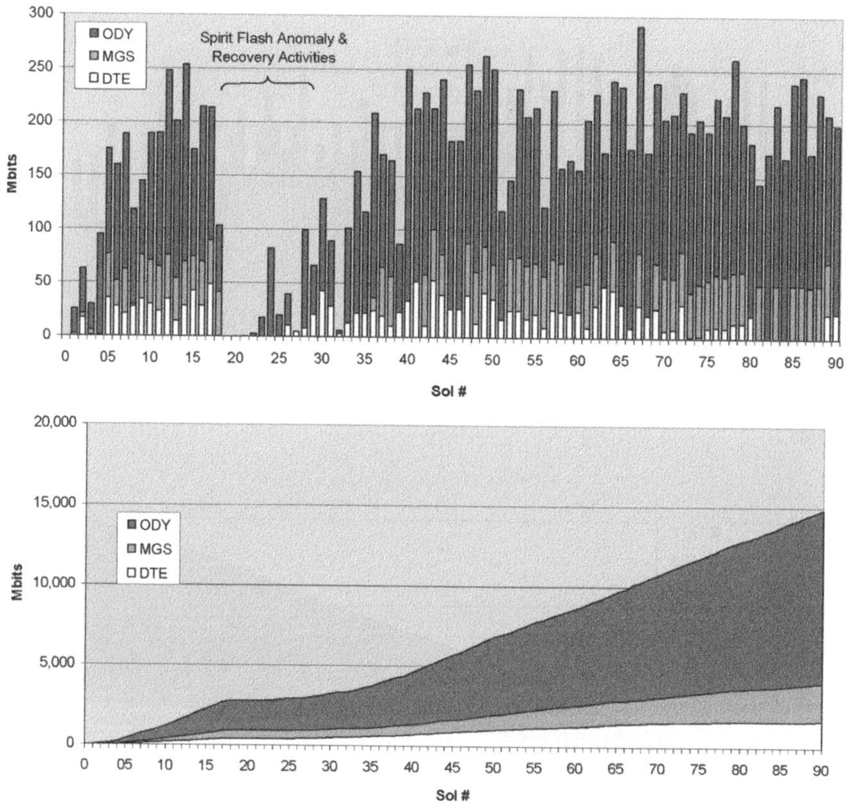

Fig. 7-46. MER-A primary-mission data sources: volumes per sol (top) and accumulated (bottom). Please visit www.wiley.com/go/deepspace communications to access the color version of this figure.

and accountability (EH&A) rate of 30 to 60 s.[31] In the primary mission there were as many as four passes per rover per sol, most often one by MGS and the rest by Odyssey.

---

[31] In the primary mission, the UHF telecom analyst requested UHF reports in order to analyze specific relay passes and to characterize relay operations for planning. During the extended mission, UHF reports, initially assigned a low priority, were routinely generated for all Odyssey passes by attaching a "generate UHF report" sequence that began when the UHF comm window started preparation. To mark a selected UHF report for transmission, MER data management would generate a command to raise its priority. Data management routinely reprioritized the UHF reports generated for all passes every seventh sol so that Telecom could spot-check UHF performance. Telecom could also request reprioritization of UHF reports that were of "interesting" passes. Data management marked for automatic deletion any UHF reports older than 7 sols.

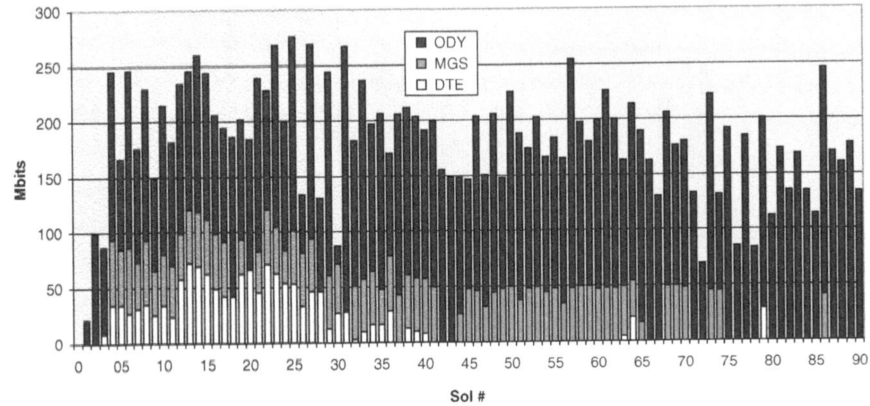

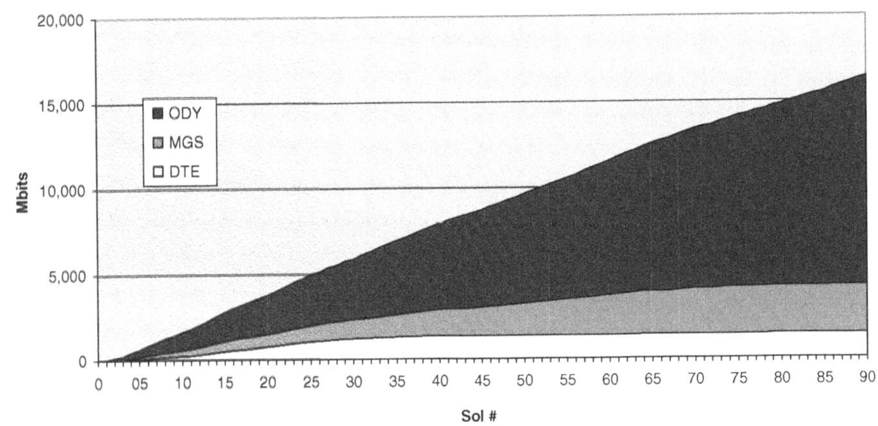

Fig. 7-47. MER-B primary-mission data sources: volumes per sol (top) and accumulated (bottom). Please visit www.wiley.com/go/deepspace communications to access the color version of this figure.

7.5.2.4 UHF Pass Planning and Optimization. Primary mission UHF pass selection was a coordinated effort involving both long-term (strategic) and short-term (tactical) planning.

Strategic planning was conducted several weeks in advance by a multi-mission team of representatives from MER, Odyssey, and MGS.[32] Geometrically, there are four overflights per sol per orbiter. The LST of the passes is typically 2–3 a.m. and 2–3 p.m. for MGS, and 4–5 a.m. and 4–5 p.m. for Odyssey. The orbiters' view periods (time above the horizon) are short; as a result, a typical UHF comm window is about 15 min long. During the primary mission, all passes with geometric view periods greater than 5 min were sequenced. MER Mission Planning designated a subset of these overflights as "requested" passes, and chose the return-link rates based on DVCF results (128 kbps for MGS;

[32] Coordination for the Mars Express interoperability demonstration (experiment) was done separately.

128 or 256 kbps for Odyssey). The remaining "unrequested" passes were sequenced on the orbiters with a default link configuration. Among the criteria considered for pass selection were geometry, data volume, and the potential to minimize orbiter buffer overflow. Sequences were built to generate comm windows to support the requested passes. Depending on how many passes were planned per sol, the sequences were uplinked every one or two weeks.

Tactical planning was conducted on a sol-by-sol basis, and it considered passes occurring in the next sol or two. Attention was focused on optimizing UHF data return, subject to various constraints, such as rover attitude, available energy, and expected time of data receipt on the ground. DVCFs identified desirable yaw angles for parking the rover. If the rover was significantly tilted (particularly in the east-west direction), data volume could vary significantly from zero-tilt DVCF predicts, especially for low-elevation passes or 256-kbps passes. Sometimes the tilt was so large (~20°) that the orbiter was occluded by the rover deck for most or all of a pass. In these cases, GTP was run using the estimated rover attitude for more accurate link assessment. Results from tactical planning included identification of passes to keep, modify, or delete, yaw(s) to park the rover for maximum data return (for mobility planning), and predicted data volume for those passes (for science planning). Maximizing UHF data return was so important that on some occasions the rover was commanded to turn (change its yaw direction) in the time between two afternoon overflights. This was to maximize the total data return. In the example of Fig. 7-37, it can be seen that if the rover were turned by 180 deg, the east pass would be on the high-gain region of the antenna pattern.

Odyssey uses the CCSDS Proximity-1 Space Link protocol (UHF1) [8], which is designed to ensure error-free delivery of data by using a Go-Back-N (frames) protocol. Idiosyncrasies in the design of the radio do cause the Odyssey return link to have a few (0 to 15) data gaps per pass, each starting with loss of lock and ending with reestablishment of the link. Despite the gaps, the Odyssey link can achieve a throughput of 97 percent when the SNR in the link is high.

MGS, which was launched several years before Odyssey, implements the Mars Balloon Relay protocol (MBR or UHF2) [17]. At 128 kbps, typically two rover transfer frames every 16 s were lost while the MGS radio changed modes (and meanwhile stopped accepting data), and the MER radio, not detecting the change, continued to send data. Because of the large number of gaps, only lower-priority data was sent during MGS passes. In 2006, after the end of the MGS mission and after MRO had arrived at Mars, UHF passes were via Odyssey or MRO. The MBR protocol is less efficient than Proximity-1. Even when the bit error rate at MGS is low (indicating a solid link), data is transmitted for only 13.3–13.8 s out of every 16 s.

In addition to data protocol issues, MGS passes were data-volume limited, and they ran a high risk of buffer overflow, as described in Section 7.4.3. Because UHF data was recorded in the MOC buffer, MGS limited the amount of UHF data it would collect per pass (typically 30 to 60 Mb). Once the buffer allocation was reached, MGS stopped collecting data even though the UHF link might still be active. As a result, any rover data sent after the MOC buffer was full was lost and had to be retransmitted by MER during another pass. On some sols, MGS passes were used in place of afternoon HGA passes to get higher data volume (especially when the rover was energy-limited).

MER UHF tactical plans are communicated to Odyssey and MGS via the "uhf-tactical" e-mail list. Receipt of messages and actions taken by the orbiters are also confirmed via this list. Normally after confirmation of successful receipt of the daily command load, the MER ACE sends an e-mail identifying the passes to be kept and deleted by MER. Notification of pass deletions are a courtesy that allows MER to avoid unnecessary troubleshooting for missing data. In addition, the MER tactical team uses the e-mail mechanism to document orbiter data collection from previously unrequested passes.[33]

Changes to UHF link parameters are handled via an orbiter relay state-change (ORSC) request since the orbiter (which does the hailing) has to be commanded to change the link configuration. The most common request by far has been to change the return-link rate (from 128 kbps to 256 kbps or vice versa). The ORSC request to change return-link rates involves sending commands to Odyssey twice: one command before the overflight changes a global variable to override all sequenced return-link rates and use the specified one, and one command after the overflight changes the global variable back to honor the previously sequenced return rates. ORSC requests must be e-mailed and received by the orbiter ACE before the drop-dead uplink time (DDUT) to allow time for the state-change command(s) to be radiated from the tracking station to Odyssey.

7.5.2.5 Commanding the Rover via Odyssey UHF Link. Forward-link verification activities were run for the first few days on the surface. These activities explored a UHF frame-duplication idiosyncrasy that can cause problems with the forward link from Odyssey to a rover. This problem can result in loss of parts of commands or repeated execution of immediate, virtual channel 1 (VC-1) commands. In order to maximize the chances for success, recommendations were developed based on the

---

[33] During the strategic planning process, unrequested passes are sequenced on the orbiters, but not on MER.

results of the UHF forward-link verification activities. Recommendations included

- Duplicating the commands within a single uplink session (in case of partial command loss),
- Padding the desired VC-1 immediate command uplink transfer frames front and back with "no operations" (no_op) commands (so that the only immediate command that can be executed twice is a no_op),
- Delaying the Odyssey forward-link start time until several minutes into the overflight, when the geometry is better, and
- Lowering the return-link rate (to reduce link dropouts, which can induce the frame-duplication problem).

Routine commanding of each rover during surface operations through September 2005 was via the DSN DFE link. Prior to the extended missions, commanding via the UHF link (which is possible only with Odyssey) was generally limited to verification tests. A significant exception occurred in June 2004, when the next available HGA DFE window was still hours away, and commanding via the LGA at 15.625 bps was either too slow or not possible. To correct a rover onboard power profile, MER-A was commanded via Odyssey during a 256-kbps UHF comm window to change the power modes. The MER project transmitted a command file (in which each of the short commands was repeated several times) to the Odyssey control center, and the Odyssey operations team sent the file to their spacecraft for relay to the rover. The rover responded properly, and later telemetry showed that all commands got in.[34]

A practical reason for the limited use of UHF commanding of the rovers during the extended missions was the LST of the morning Odyssey pass, typically between 4 and 5 a.m. This time is more than 5 hr before the typical 10 a.m. LST of the X-band command window. When working on Mars time, the science and sequencing teams would have had 5 hr less to plan and prepare activities for the next sol after the afternoon receipt of data if they commanded at UHF instead of at X-band. The commands would have had to be ready before the DDUT, and the Odyssey ACE (who was generally not living on Mars time after the MER primary mission) would have had to be available to send them to the orbiter.

UHF commanding of the rover, based on the rover team operating on Earth time and with a revision of the sequencing activities timeline, was re-evaluated for the March 2006 MRO Mars orbit insertion. This was to avoid X-band uplink interference with the MRO with MRO and the Spirit

[34] MER also conducted UHF forward-link tests in September and October 2005 in preparation for a one-week UHF-only operational demonstration in late October 2005.

Rover (both operating on DSN channel 32) were within the station antenna beamwidth. The co-channel operation had not been thought likely with the planned 90-Sol MER mission starting in January 2004. Aside from using UHF for MER, a more common technique called MUKOW (MRO uplink keep out window) in which station X-band transmitter operation is carefully timed between MER and MRO when both vehicles are in view of the station (not occulted by Mars).

7.5.2.6 UHF Link Analysis. UHF link analysis was conducted during the primary mission using several data sources. Typically, they included a combination of queried MER telemetry, queried Odyssey telemetry (e-mailed by the Odyssey team), and GTP predictions. Higher visibility was obtained by including UHF report data products (which had to be requested by the telecom analyst). The data was plotted using various Excel tools developed by MER telecom. Generally, link volume predictions have compared well with actual data return (usually within 10–20 percent). Occasional outliers are due to excessive tilt (not considered ahead of time), obstruction or occlusion by surroundings or rover deck, and operation near threshold for a significant portion of the pass.

As an example of the UHF link analysis achieved, Figs. 7-48 through 7-53 compare the performance of a high-volume and a low-volume UHF pass. In each case (high volume and low volume), there are two figures that show mean and adverse prediction curves, offset vertically from each other, as well as a plot of the forward-link received power from rover telemetry or return-link received power from Odyssey telemetry. Both forward- and return-link margins (Figs. 7-48 and 7-49) are also shown as dotted lines for reference since the Proximity-1 protocol dictates that both links must be

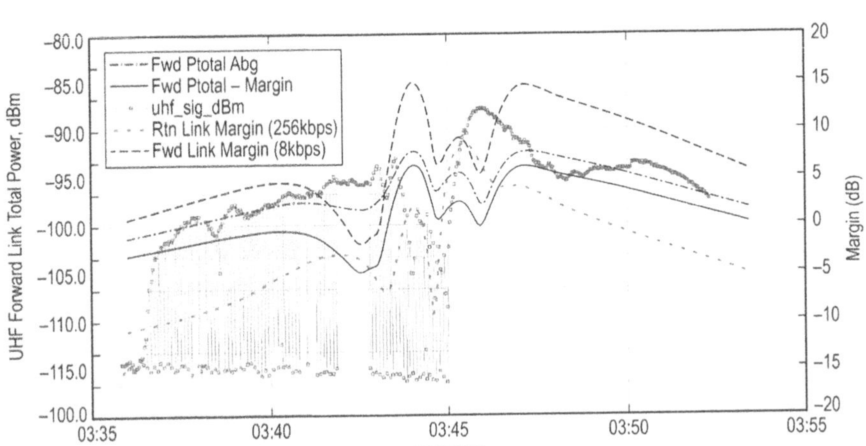

Fig. 7-48. High-volume forward link—Odyssey to MER-B, sol 104 p.m. (5/10/2004). Please visit www.wiley.com/go/deepspacecommunications to access the color version of this figure.

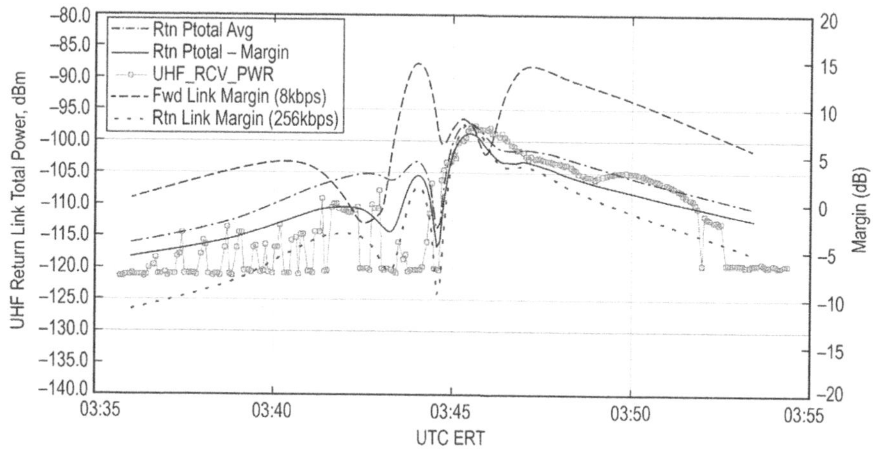

Fig. 7-49. High-volume return link—Odyssey to MER-B, sol 104 p.m. (5/10/2004). Please visit www.wiley.com/go/deepspacecommunications to access the color version of this figure.

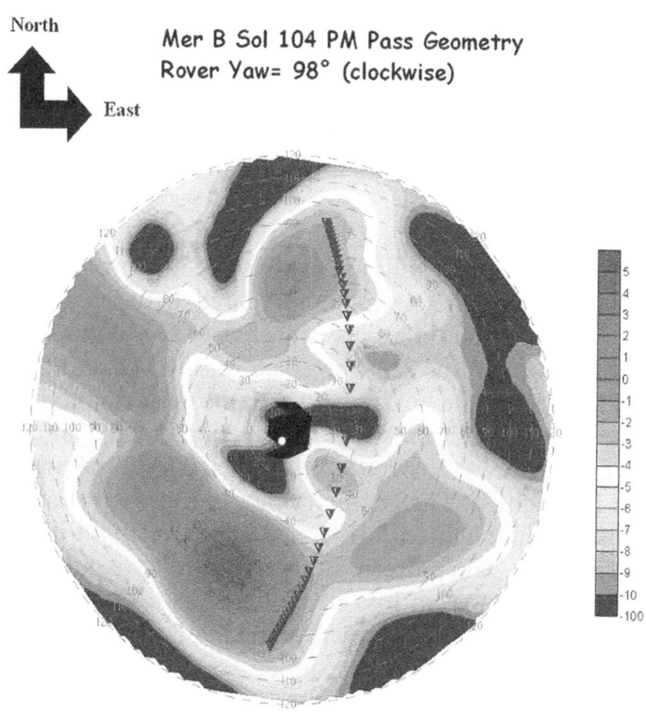

Fig. 7-50. High-volume return-link polar geometry—Odyssey to MER-B, sol 104 p.m. 5/10/2004). Please visit www.wiley.com/go/deepspacecommunications to access the color version of this figure.

above threshold for the link to be established. Figure 7-50 shows the link geometry superimposed on the MER UHF return-link pattern. The predictions account for the orientation of the rover (azimuth and tilt from horizontal). Azimuth angle is referred to as "rot" (rotation) in the two performance and prediction figures of each set and as "yaw" in the pattern figure. See figure numbers for high-volume and for low-volume links in the following two paragraphs.

In the high-volume examples (Figs. 7-48, 7-49, and 7-50), the actual received power curves generally follow the shape of the predicted total power curves, with the return link modeled much better than the forward link.[35] The large dip in the return link just before 03:45 corresponds to the orbiter passing over the null in the pattern shown in Fig. 7-50. (Figures 7-50 and 7-53 are polar antenna plots showing the angles 0 to 120 deg from boresight radially.) This 256-kbps pass was predicted to return 83 Mb, but the actual return was 125.5 Mb. The reason for the higher data volume is that the link was predicted to drop out at about 03:49, but the link performed closer to average than to marginal (a few decibels better), allowing the return link to remain above threshold for a few extra minutes. This is shown in Fig. 7-49.

There is room for improvement in our UHF prediction capability; however, it has been difficult to use the accumulated UHF reports to measure rover polarization and gain antenna patterns. We also suspect the source of some of the difference between predicted and actual performance may lie in the orbiter antenna patterns. It has proven difficult to decouple rover and orbiter quantities. Antenna measurement on good-quality spacecraft mock-ups should be made a priority for future missions.

In the low-volume example (Figs. 7-51, 7-52, and 7-53), the actual received-power curves again generally follow the shape of the predicted total power curves. However, during the period highlighted between the two vertical black lines in Fig. 7-52, the link was expected to close and it did not. The predicted data volume for this 256-kbps pass was 80 Mb, but the actual data return was 4.4 Mb. When the link is analyzed, the performance is not too surprising despite the large discrepancy between predicted and actual data volume. Figure 7-53 shows the overflight geometry, with the predicted above-margin period highlighted. The geometry plot in Fig. 7-53 shows that during the part of the pass highlighted in Fig. 7-52, the overflight was in a steeper portion of the antenna-gain pattern. Because it is not possible to separate the antenna gain and polarization loss to model each accurately, errors in the modeling or small differences between predicted and actual pointing angle could have pushed the link below threshold.

During the latter portion of the above-margin period, the predicted link margin was only slightly above zero for 2 min. A predicted 30 Mb was not relayed during that period. This highlights another weakness in the current

---

[35] The return link has been closer to predict than the forward link for the entire surface mission. However, since performance is usually limited by the higher return-link rates, this has not caused a problem in MER planning.

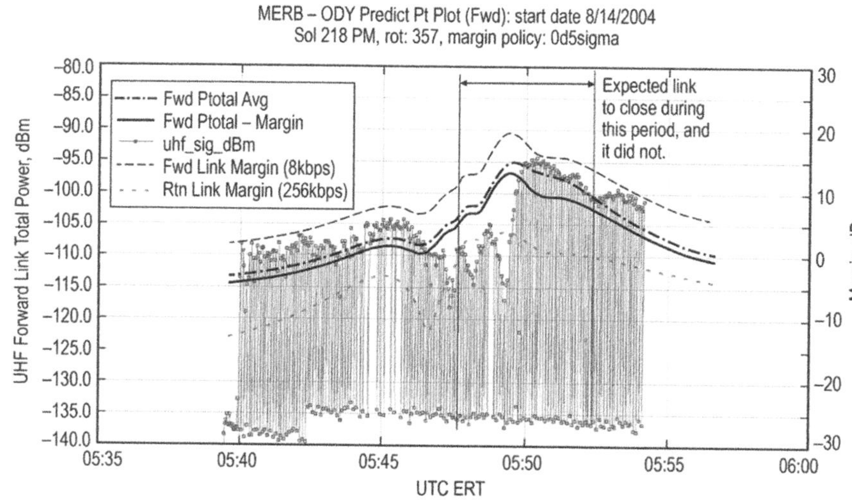

Fig. 7-51. Low-volume forward link—Odyssey to MER-A, sol 218 p.m. (08/14/2004). Please visit www.wiley.com/go/deepspacecommunications to access the color version of this figure.

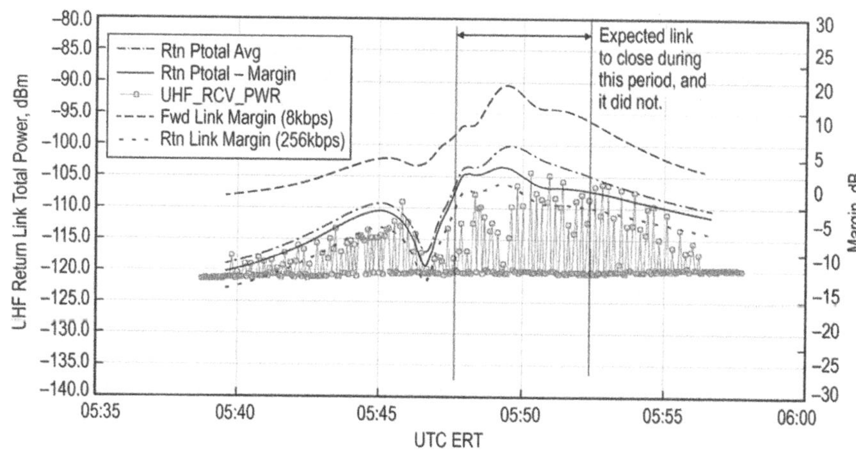

Fig. 7-52. Low-volume return link—Odyssey to MER-A, sol 218 p.m. (08/14/2004). Please visit www.wiley.com/go/deepspacecommunications to access the color version of this figure.

method of UHF data-volume estimation. When the link margin is above zero, the link is predicted to close, and when it is below zero, it is predicted not to close. It is a hard-decision algorithm, that is, there is no consideration (or weight) given to how far above threshold the link is operating. This means that links predicted to operate near threshold (either above or below) for significant portions of the pass will have much higher variability in their actual data return than those links in which the above-threshold and below-threshold portions are more distinct. This is especially true for 256-kbps passes like this one.

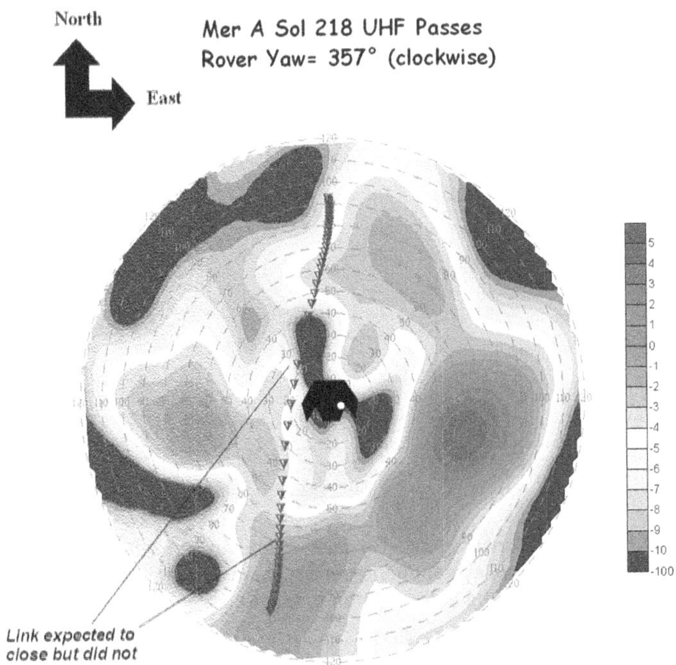

Fig. 7-53. Low-volume return link polar geometry—Odyssey to MER-A, sol 218 p.m. (08/14/2004). Please visit www.wiley.com/go/deepspace communications to access the color version of this figure.

7.6 Lessons Learned

MER has been a fantastically successful mission, with both rovers reaching Mars' surface and embarking on explorations lasting far longer than the full mission-success criterion of 90 sols each. Both the X-band and UHF parts of the telecom subsystem were well conceived, designed, tested, and operated. We would want to use these processes as models for the future. Even so, there were some problems that made it difficult to build, test, and operate the hardware. Other problems made it difficult to predict the UHF data volume and to assess why the predictions were in error, required peaks in telecom staffing, made telecom planning time-consuming, or resulted in lost data. Lessons drawn from both the good and bad experiences could smooth the operation of telecom subsystems for future Mars surface projects.

These lessons learned are grouped by major mission phases: development; assembly, test, and launch operations (ATLO); and the cruise, EDL, and surface portions of the mission operations phase. Because the DSN is an integral part of any project's telecom operations, and two Mars orbiters played a major part in rover surface operations, this section includes MER, DSN, and orbiter operations lessons learned.

In some cases, a lesson may look overly obvious in hindsight. However, the experience documented here did occur, and can be used to make future deep space telecom endeavors better.

7.6.1 What Could Serve as a Model for the Future

7.6.1.1 X-Band Development

Schedule. The biggest challenge to MER telecom subsystem implementation was the very short development schedule (about 2 years). This meant that subsystems could not wait for all the higher-level requirements to be documented; they had to order parts right away.

Tight quarters. A second challenge was the physical space allocated to telecom components. Having to fit so many hardware elements inside a tight space (the WEB) meant that the placement of some of these elements (such as cables and connectors) had to be redone several times, which meant that certain cables had to be ordered three or more times. This MER problem was not unique to telecom hardware.

Lesson: MER is widely recognized as having had an "impossible" development cycle, a low probability of both rovers successfully landing and meeting even minimum mission-success criteria, and an exhausted development team. One institutional and project lesson learned is that even development difficulties such as these do not necessarily preclude mission success. The MER project should articulate reasons (even in hindsight) why the mission could be so successful in the face of factors like an overly ambitious schedule and almost too-constrained space allocations. Future projects can use the MER information to weigh more accurately than before the risks and benefits of their own development approaches.

Communications behavior. Communications behavior was embodied in comm windows for the first time on MER. The concept differed enough from traditional sequencing of the onboard telecom hardware to make the learning curve steep.

Lesson: Test new flight software (FSW) concepts early and often. Do software-intensive tests, such as with the rover Communications Behavior Manager (CBM), as early as possible in the ATLO process, to catch and perhaps correct problems between comm windows and other parts of the FSW. The effective use of the new comm window concept on MER was significantly advanced through on-the-job training during the flight mission.

Receiver ops. One of the design principles JPL has adhered to for many years is not to turn its receiver off after launch. For this project, however, power limitations (MER runs on solar power, with batteries to get through the night) forced the project to turn the SDST off every night.

Thermal cycles. Until MER, no SDST had undergone as many temperature cycles (at least three a day) and power cycles (about two to three a day). It is a tribute to the resilience of the SDST and SSPA design that their performance has not degraded in this extreme temperature environment. A good parts program, together with assembly and subsystem testing under expected mission conditions, helps to ensure dependable operation.

Lesson: Qualify hardware for intended modes. Link the parts qualification and screening program and test program to the specific intended operating modes, especially new ones.

SDST frequencies. The Mars environment has an effect on the SDST BLF versus temperature. We found that trends in this frequency during surface operations were similar to prelaunch trends, but with some differences in terms of offset and slope. Wide temperature excursions occurred within every sol, but generally were similar from sol to sol over periods of weeks. It proved difficult for the thermal analysts to model temperature profiles in new surface modes (such as deep sleep).

Lesson: Calibration campaigns, such as the calibration of the MER X-band telemetry channels, should be continued on future projects. Calibrations include uplink received signal level, receiver frequency variation (static phase error), and power amplifier RF output, with as much data as possible collected at expected and extreme temperatures. Quantities (such as oscillator frequencies affected by pressure (should be calibrated for each distinct environment (such as vacuum of space as opposed to Mars surface atmospheric pressure).

Seeing trends. Direct measurement of link performance revealed large changes that could be attributed for the most part to certain known factors (such as DSN antenna pointing, RLGA pattern variations, and HGA occlusion or RLGA signal scattering by the PMA). Direct measurement of performance shed little light, however, on smaller trends that may have occurred (none have been observed) in other quantities (like SDST receiver sensitivity, SDST exciter RF output, or SSPA RF output changes due to aging).

Lesson: Well-calibrated and stable telemetered measurements of critical parameters like receiver sensitivity and RF output power can be more applicable to discerning slowly changing or small differences in performance than is direct measurement of link performance.

7.6.1.2 X-Band Cruise Operations

Daily operations. The telecom analysts monitored two spacecraft during cruise, each one supported by one or several DSN tracking passes per day. Planning data rates that would work with the scheduled 70-m or 34-m stations, reviewing the comm windows that implemented these data rates, and monitoring and reporting spacecraft telemetry and station monitor data for each pass became increasingly easy with repetition, but the workload was always challenging. Fortunately, there were no significant performance changes within the telecom subsystem itself during cruise. Even so, characterizing the interaction of the telecom hardware with other onboard or ground subsystems required looking at many instances of the same configuration to see if any unusual performance was repetitive and perhaps due to a particular configuration.

Lesson: Automate repetitive ground software activities. For MER, certain macros for these activities were developed during cruise and perfected during surface operations: data-query scripts were developed in UNIX and trending macros in Excel for X-band hardware (SDST and SSPA) and station monitor (MON) data. Not requiring much user input, these macros provide comprehensive display of telemetry data as "digitals" (tabulations of data numbers or state values as a function of time) or plots (graphical displays of the data numbers [DNs] or engineering units [EUs] of one or more quantities versus time) and comparison of selected quantities such as SDST signal level with predicts. The tabulations and plots were then converted to Portable Document Format (PDF) and e-mailed to all telecom team members for review.

Automating the generation of link predictions. Telecom predicts for use by the ACE for each pass were partially automated in the "dkf2pred" scripts during cruise and by the "genmer" and "pred2pdf" scripts during operations. Even so, generating predicts for every pass during the final weeks before EDL was time-consuming. Each pass had to be set up individually by the analyst for start and end times, station, and downlink rate. Automation might be to create a script that calls TFP to have the ability to read in a previously generated project file and extract from the file the information that an analyst would otherwise type in to the TFP or GTP graphical user interface (GUI). Several projects now use predict maker scripts for each station pass as defined in the station allocation file (SAF).

Lesson: Fully automate telecom predicts. Besides being faster and easier to generate, than predicts that come from manual inputs, automated predicts produce consistent output formats (the same tabulated or plotted quantities always output, in the same order). Analysts become familiar with the format and therefore make fewer errors in using the automated predicts.

7.6.1.3 X-Band Entry, Descent, and Landing

EDL planning. From eight hours before EDL until two hours after, the MER project scheduled both of the 70-m stations and many of the 34-m stations with MER in view at Goldstone and Canberra. Rehearsals with EDA and RSR components at the stations and at JPL verified configurations and procedures to route downlink signal inputs from each station antenna (front-end assembly) to the EDA and the RSR for processing. Because the EDA was equipment developed especially for MER, and the RSR configurations for EDL were unique, the rehearsals included participation by EDA and RSR experts on site and at JPL to operate and monitor the equipment to process the signal inputs.

Lesson: Rehearse complex activities. A full-up in-flight EDL rehearsal during late cruise—involving the spacecraft, elements of the Flight Team, all participating stations, and the EDA and RSR—proved invaluable in wringing out procedural and interface issues. As a result, the EDL and telecom teams found the lander performance in both real EDLs easy to assess as compared to the simulated performance during the rehearsal.

7.6.1.4 X-Band Surface Operations

Comm window changes. Similar to the bit-rate optimization of late cruise operations, comm window optimization occurred in surface operations during the "tactical" (just-in-time, sol-by-sol) sequence development process. This optimization was essential for the complex and rapid-turnaround activities on the surface. Comm windows developed by Mission Planning during the "strategic" (multisol) process used the data-rate capability file (DRCF). During MER surface operations, the telecom analyst checked the 17 parameters in each comm window. Changes to X-band comm windows, particularly to the start time or duration parameters, had to be carefully—and manually—checked against station allocations and uplink timing. Manual checking of changes in individual windows is time-consuming and error-prone.

Lesson: If only manual checking is available, minimize changes to existing comm windows and to subsystem configuration changes during windows.

Global window changes. Constraint-checking processes more automated than those of MER would facilitate changes in the timing or data rate of existing comm windows. Fortunately, an automated process allowed tactical leads to change certain parameters (downlink bit rate, duration, or start time as a function of a reference time on Mars such as 8:00 a.m.) in whole groups of windows at once. These "global" changes worked well and did not require a separate check beyond the original DRCF validation.

- The "add_seq_to_window" parameter is handy and has been used. It allows the comm window to kick off another (possibly unrelated) activity. Very often these kicked-off sequences are used to generate UHF reports (Section 7.5.2.3).
- The ability to modify comm configuration within a window's execution time using individual secondary commands to the SDST, such as downlink rate change, was used sparingly but proved useful.

Window-checking scripts. During the extended missions, the telecom team began to find the time to develop scripts to check applicable flight rules, many of which involve the interrelated modulation telemetry parameters of comm windows. Others dealt with the interaction of window start time (in spacecraft event time [SCET]) and duration with station ground times (as transmit time or receive time). These time relationships are particularly onerous to check manually because there are many of them in a sequence, and they involve both OWLT and conversions between Earth time and Mars time.

Lesson: Automate window-checks that can be defined by rules. Give particular emphasis to comparing start times and durations of windows with activity times in the station-allocation files that define MER station passes. These include beginning of track, uplink acquisition (including duration of the uplink sweep), uplink handover, and end of track. For MER, such checks have helped to ensure that commanding is not attempted before the SDST receiver is in lock, and likewise that a nominal or off-nominal beep is not scheduled with the uplink out of lock (and thus the downlink in one-way mode).

RLGA operations. Considering boresight gain alone, one would observe that the RLGA is some 13 dB less capable than the HGA. However, when planned signal levels permit, using the RLGA, with its separate location on the RED and its wide gain pattern, is sometimes a means to avoid two factors that compromise the capabilities of the HGA: signal-scattering caused by PMA occlusion (described in Section 7.5.1.5 above) and the timing constraints imposed by HGA "flop" (described in Section 7.6.2.1, below).

Lesson: Consider telecom hardware characteristics and system factors, not just gain, when planning communications. At cold temperatures, when the uplink received power at the spacecraft should be limited to avoid DAC rollover glitches, sweeps can be performed into the RLGA rather than the HGA. Carrier-only "beeps" via the RLGA instead of the HGA may result in fewer interactions and constraints. Using the RLGA for a beep does not require HGA actuator heating or interruption of science activities. The RLGA can be used for a honk while the rover is driving, whereas the HGA would require a stationary rover for pointing.

Coherent downlinks. Usually one-way noncoherent downlinks are thought to be easier to manage than are two-way coherent links because they do not rely on an uplink being in lock. However, temperature swings on the Mars surface caused very large variations on the one-way downlink carrier frequency from the aux osc. The station receiver could not lock to the rapidly changing carrier frequency at the available downlink level. Very quickly it became standard that all DTE and beep downlinks be made two-way coherent. Doing so required planning for the SDST receiver to be in lock when the rover transmitted planned DTEs and beeps. Providing for coherent downlinks eventually included configuring coherent mode for the onboard fault responses, except for the final step in the response algorithm.

Lesson: The simplest mode may not be the best. Take advantage of the greater frequency stability in the ground station to combat temperature changes in spacecraft oscillator frequencies.

Blind commanding. Usually, having telemetry in lock is considered necessary to monitor the progress of commanding a spacecraft. Commanding without immediate command confirmation is called "blind commanding." On most deep-space missions, blind commanding is done only in an emergency. MER surface operations have required it routinely, whenever the command period (uplink windows) and the light time are comparable and the next downlink pass or UHF relay is hours away. In the primary surface mission, blind commanding caused only one command error and one failed command load (both involving the same station) to the RLGA.

Lesson: MER surface operations prove that blind commanding can be very reliable. However, successfully establishing and maintaining the uplink for such commanding requires repeatable behavior of the receiver, precise use of the tuning template and its parameters, well-trained command operators (ACEs), and good command system monitoring capability.

Beeps. The beep has become an enduring marker for success (or not) of the command upload and initial operation of each new master sequence. The beep (a 5-minute X-band carrier-only downlink) is a simplified form of the semaphores (M-FSK tones) used in EDL. The timing of the detected beep designates it as either "nominal" (all okay) or "off nominal."

However, the failure to detect a beep has not invariably meant that there is a problem on the rover. Most often the telemetry sent back during a subsequent UHF pass has shown that the SDST and SSPA sent a beep at the planned time. During the extended missions, this experience sometimes gave the project confidence to press on with planned activities (such as sending the command–loss-timer command) even after a station failed to detect the planned nominal beep.

Lesson: Use of a beep (or semaphore or other simple go/no-go signal) may make it possible to proceed with planned activities in cases where standard telemetry is unavailable to support such a decision.

RSR operations. When the beep-detection process was new in the primary mission, the project requested beep detection by the RSR as a backup to the DSN's beep detection. As a result, after staffing for a few beeps, the JPL

Radio Science group made available a prototype Web page for requesting RSR support for beep detection over specified time periods without the need for intervention by station personnel or staffing by Radio Science. Given Mars-time beep scheduling, this proved a useful automation. It worked at some stations, some of the time.

Lesson: To reduce overall project staffing costs, consider cross-training non-specialists to run the RSR remotely and evaluate the output in a simple preset mode, such as to detect a beep. As an example, improve the operability of the remote (Web-based) access to the RSR setup.

Multipath. The telecom analyst could reliably predict uplink and downlink performance and operate 80 deg from the RLGA boresight (10 deg from the horizon with the rover level), and with the HGA Earth-pointed with the direct signal path close to the rover deck. Signal variations with the characteristic fading that may have been caused by multipath occurred on a few passes, but this variation never degraded planned DTEs.

Lesson: Multipath may not be a problem. The MER experience should be applied to the analysis of potential multipath in predicting the telecom link performance for other missions where it may occur.

7.6.1.5 UHF Development. The keys to success in the UHF test program (two rovers, three kinds of orbiters, a short development schedule) included

- The full-time availability of Odyssey and MGS test sets, and MER's own UHF system test equipment (STE)
- For surface operations, choosing a few out of the many available transceiver modes and a single forward-link rate
- Insisting on testing only in the most key areas, such as measuring the extent of electromagnetic compatibility (EMC) with surface subsystems and instruments
- Knowing which equipment can be operated during UHF passes (that is, which equipment is least likely to cause interference with the UHF receiver or be interfered with by the UHF transmitter), knowledge that has proved valuable in the time- and power-constrained Martian winter
- The Proximity-1 protocol, which ensures that if data comes down at all, it is error-free data.

Lesson: Ensure that similar trades are made a part of future mission implementation.

7.6.1.6 UHF Surface Operations. Communications behavior works well for UHF windows.

UHF windows. UHF comm windows are significantly less work to create and review than are X-band windows because the UHF radio has fewer "adjustment knobs" (such as modulation index and subcarrier frequency) than does the SDST.

Few window changes. Parameters of strategically delivered comm windows are not modified. Because they are of fixed duration and span all or most of the geometric overflight view period, they cannot be moved in time. This means the tactical team does not spend its limited time reviewing UHF comm windows.

Window deletions and rates. In the primary and extended missions, tactical changes have been limited to

- Deletion of some strategically planned UHF comm windows because others (also strategically planned and in the sequence) provided greater expected data volume or because of rover power constraints
- Changes between 128-kbps and 256-kbps return-link rate (or between coherency and noncoherency) on Odyssey, using the ORSC process.

Lesson: Simplify a "utility" like communications when it makes sense to do so. Relay link planning between Mars' surface and orbiters involves fewer comm issues—such as HGA pointing, station weather, or station antenna pointing—than does planning for links that originate or end on the Earth's surface.

Relay protocol. The Proximity-1 protocol means that if relay data comes down at all, it is error-free data. Analysis of performance is much less labor-intensive for UHF than for X-band.

Lesson: Consider the appropriate use of modern communications protocols in deep-space missions.

7.6.2 What Could Be Improved

7.6.2.1 X-Band Development

FSW simulator. Because there was no avionics simulator before the start of ATLO, the debugging of problems related to onboard hardware performance was rather time-intensive and required the interaction of many teams. One specific example was that resolution of a bit-timing problem at 10 bps led to using a DSN test facility (DTF-21) twice, and tied up DSN test operators and Ground Data System (GDS) personnel as well as ATLO test personnel.

Lesson: Provide for stand-alone project facilities (in this case a flight software [FSW] simulator) to test new capabilities without requiring the early and repeated involvement of multimission facilities.

Downlink rates. Uplink rates are in factors of two. A finer resolution between adjacent downlink rates between 40 bps and 120 bps possibly would have reduced the time to resolve the MER-A sol 18 flash-memory-file anomaly (see Section 7.5.1.4). The 3:1 ratio between 120 and 40 bps means it takes nearly 5 dB more link performance to support 120 bps.

Providing low downlink rates is a challenge because they take the longest to test and are affected in performance by factors that do not vary linearly with data rate.

Lesson: Thoughtfully trade the complexity of implementing, testing, and using numerous bit rates against the utility of specific rates, including in contingencies.

Surface environment. The time available to characterize the SDST in Mars-like conditions, especially cold temperature and partial vacuum, was limited because of the need to debug several serious FSW problems involving rover instruments.

In retrospect, the telecom hardware areas to focus on should have been the SDST BLF and acquisition and tracking characteristics at cold temperatures and partial vacuum (to approximate the thin atmosphere in which the rover operates on Mars). This was particularly so for MER-A, where SDST coherent leakage (see Section 7.5.1.9.1) made cold-temperature uplink acquisitions operationally demanding.

Frequency calibration. In addition to the unlocked static-phase-error (SPE) drift (coherent leakage) in the SDST on MER-A (but not on MER-B), surface operations were made more difficult by the relatively coarse calibrations of the uplink signal level (cla_snr) as a function of temperature. The SPE drift makes blind uplink acquisitions problematical at some temperatures, and the coarse calibration makes separating out the effects of rover's antenna pointing and station pointing difficult.

Lesson: Consider the environmental factors, the intended equipment use, and any specific deficiencies in particular units when designing the test and characterization program.

RF leakage. The implemented onboard X-band system had some opposite-polarization leakage paths that became apparent twice during cruise when a station inadvertently transmitted with the wrong uplink polarization to MER-B [10]. The incorrectly polarized uplink signal still made it into the SDST. The SDST telemetry data led to a quick correction of the configuration.

Refer to Fig. 7-11, which shows that when the MGA is selected (for both uplink and downlink), the CLGA is not selected, and vice versa. In the first occurrence, a cold-reboot activity, the selected antenna path was the CLGA (RH polarization). The tracking station had been wrongly configured to transmit LH polarization, though it was correctly configured to receive RH. Because of an RF leakage path through the MGA antenna (LH), the SDST received from the MGA a lower-than-predicted (for the CLGA) uplink level, but the SDST still properly decoded commands sent with the wrong polarization. This "success" caused a great deal of confusion until the alternate uplink path was identified. In the second incident, the station was incorrectly transmitting RH, and the selected antenna path was the MGA (LH polarization). The SDST acquired carrier lock via a CLGA (RH) leakage path. However, ranging modulation was below threshold. Ranging data was lost until the uplink polarization could be corrected.

Lesson: Controlling (and measuring the magnitude of) leakage paths is a necessary consideration in spacecraft microwave-component selection and configuration.

Lesson: The ability to absolutely verify uplink and downlink polarization settings has not kept pace with the ability to separately control these settings at a station.

PMA occlusion. Obstruction of the X-band downlink via the HGA by the PMA was a significant problem during communication attempts to resolve the MER-A sol-18 flash-memory-file anomaly. Also, obstruction of the X-band uplink via the RLGA has occurred in some rover orientations during the extended missions.

HGA flops. An HGA "flop" will occur when the required HGA pointing nears a singularity in a gimbal axis. Testing of surface operations in a testbed uncovered an FSW flaw that would cause a fatal software error during an HGA flop. The problem was traced to improper CBM and HGA interaction during flops. As a result of the testing, comm windows during the primary mission that were predicted to be interrupted by the ACS software autonomously performing an HGA flop were cancelled or moved. Rover attitudes were carefully chosen to avoid flops until the FSW could be patched. No planned or autonomous flops have occurred on either rover.

Lesson: Fully characterize antenna pointing and antenna interaction problems (PMA occlusion, risk of HGA flops), and develop operational workarounds before flight.

Antenna characterization. MER is not the first project that has been operated with relatively poorly characterized antenna-gain and polarization patterns.

Lesson: Spend the necessary time and resources to characterize antenna performance with a high-fidelity spacecraft model before launch. This applies to both UHF and X-band antennas, both uplink and downlink. A project that fully characterizes spacecraft antennas (including the obstructive and scattering effects of nearby portions of the spacecraft) can make solid plans to use higher downlink rates to return more downlink bits per pass, and higher uplink rates to complete commanding and get on to science activities more quickly than would otherwise be possible. If the project did not need higher downlink or uplink rates, with antenna characterization, it could elect to conduct operations with smaller and less costly ground stations.

7.6.2.2 X-Band Cruise Operations

Comm window types. Communications behavior (default states and comm windows) proved reliable during cruise with standard configurations (telemetry and ranging, or delta-DOR). Telecom subsystem configurations during cruise were controlled, for the most part, using normal comm windows, with each window having its start time defined as a parameter.

High-priority comm windows (HPCWs) could be made that would execute immediately upon receipt because the start time was in the past. However, an HPCW always reinforces amplifier and antenna switch states. This is undesirable in principle when it involves pulsing (reactuating) an existing switch position and cycling the SDST exciter and SSPA off then on to enforce SSPA selection. Cycling these units off also interrupts the downlink.

Lesson: A normal window that does not reinforce switch states but that starts as soon as it is received combines two good usable features of HPCW and regular comm window types.

Comm window usage. During cruise, there was a great deal of debate about whether to use HPCWs or regular comm windows for data-playback events. Regular comm windows won out because HPCWs cycle hardware (notably the SSPA and SDST exciter) to reinforce telecom hardware states.

With increasing range to Earth, telemetry rates could no longer support all the real-time and playback data that the team had come to expect. It proved to be a large burden on systems and telecom analysts to optimize downlink rate with individualized comm windows for each pass, because window parameter and timing checking was manual during cruise. In the extended missions, a window-checking script eased the manual workload.

Lesson: MER was the first deep-space project to use comm windows. The experience MER gained in generating and reviewing comm windows, then modifying or deleting them when necessary, points to ways that another project may wish to improve on MER's first-generation communications behavior and comm windows. Besides the immediate-upon-receipt versus defined-start-time trade, a project may wish to consider how to simplify the generation and review process when changing a single parameter such as bit rate while still working within the full power of communications behavior. Another trade may be combining comm windows with a simplification of the X-band communications modes (as has been done with MER UHF).

Downlink reports. During cruise, each subsystem used the same facility, software on the MER server named Quill, to complete a daily downlink report. Like a word processor, the software allowed the user to input a character string; then the software would search for all previous instances of that character string. This provided the analyst a quick means of finding and referring to earlier instances of recurring problems or activities.

Initially in surface operations, with different downlink report software, Quill did not have the string-search capability. Late in 2004, a new version of Quill was implemented. Its response in moving from one downlink report to another is much faster than that of the version in use during the primary and first extended surface mission; and it has restored the string-search capability.

Lesson: Consider the typical repetitive uses that a person will make of required software in an intense operations environment, and implement capability to enable or improve those uses.

7.6.2.3 X-Band Surface Operations

Link margin criteria. Standard criteria were developed and used to set telecom subsystem configurations and data rates for the cruise and surface mission phases. These criteria were intended to account for the inherent variability from one instance to the next of a comm link. These criteria included

- A margin policy: predicting was based on adverse margins defined as mean minus 3-sigma for commanding and mean minus 2-sigma for telemetry.
- A tolerance on HGA pointing: A 2-deg off-point of the HGA was included in predicts for data-rate planning during surface operations.
- Allowing time for the station to lock up the downlink. For cruise, this time was 1 min; for surface operations it was changed to 3 min, then later back (sometimes) to 2 min.

Lesson: Establish consistent link-performance margin, timing, and operability criteria. The usual conflicting objectives are to make the criteria sufficiently conservative that no data will be lost, but not so conservative that the amount of planned data falls below what the project can tolerate. There may be no way other than gaining experience using the criteria in order to change them to meet the project's specific needs. The MER experience suggests the following:

- Although mean minus 2-sigma for downlink performance is standard, there was somewhat more data loss during the primary surface mission than the project was comfortable with. A larger link margin would have reduced replanning by accommodating such factors as weather worse than the defined 90 percent, worse-than-expected ground antenna pointing, some amount of occlusion of the HGA or RLGA by the PMA, etc.
- There is about a 2-dB difference between allowing for 2 deg and 4 deg for HGA off-point. Pointing error is an input to the TFP GUI. Perhaps allowing for more HGA pointing error could have provided the necessary additional link margin described above.
- Unmodeled or insufficiently modeled effects, especially station-pointing error, on performance may make the criteria seem insufficiently conservative. Aberration effects on uplink performance became significant near maximum range in August through October 2004, with the station pointing its antenna based on the downlink currently being received, not on where Mars would be an OWLT later.
- The cruise value of 1 min for telemetry lockup proved insufficient on the surface, given the occasional longer-than-normal lockup time and consequent loss of the most valuable recorded data that comes down first. Midway through the primary mission, a comm window parameter value was changed so that only real-time data was transmitted for the first 3 min rather than 1 min (before valuable data started). Later, the data loss/opportunity balance again shifted. Particularly valuable windows now sometimes are planned to allow 2 min for telemetry lockup.

Thermal modeling. On MER-A sol 38, the HGA elevation-axis actuator stalled during the calibration portion of the morning comm window, causing the DTE to fail, with the HGA 30 deg off-pointed from Earth. Telemetry for problem evaluation and restorative commanding was via the RLGA. Subsequent analysis uncovered shading of the HGA by the PMA, which caused the motor to stall because it had not been sufficiently warmed up (see ISA Z83273 [18]). Following that incident, HGA heater tables were reconstructed to always assume worst-case shading, and HGA calibrations were removed from morning comm sessions.

Lesson: Occurrences like this dramatize the insufficiency of a prediction model that overlooks or oversimplifies certain factors. A sufficiently robust system design can withstand such surprises without permanent damage or irretrievable data loss.

Station antenna pointing. With no downlink confirmation in a blind-commanding session, the consequences of unexpectedly large station pointing error can be the loss of the commands. Except for aberration on the uplink, previously discussed, such station pointing error can largely be mitigated by Conscan if there is an orbiter downlink being received in MSPA mode (see Section 7.5.1.7.1).

The cause of station-pointing errors can sometimes be determined by project and DSN cooperative analysis of uplink signal level returned in later telemetry against the predicts, and by comparing times of large pointing errors with station logs.

Lesson: Follow up immediately on any suspected station-pointing error to minimize the impact on subsequent operations for the affected project or others being tracked by that station.

Lesson: Define a consistent Conscan strategy among the stations supporting a project. For example, MER experience has led to the following:

- Do not Conscan on the RLGA downlink. Neglecting rover tilt, the RLGA remains vertical to Mars, and there is a large signal variation resulting from Earth's going through a wide range of angles on the RLGA pattern.
- Always Conscan on an orbiter if one is available during an MSPA session.
- Conscan at all stations (of a given size). This was not the case during the MER primary mission due to DSN implementation and operational differences among the 70-m stations. At some stations there was at least the perception by operators that Conscan could at times drive the antenna off even a stable downlink.

Beep detection. At the beginning of the primary mission, the MER project negotiated beep detection by the DSN as a "best-efforts" activity, meaning that the formality of the JPL Discrepancy Report (DR) process could not be relied on to ensure timely assessment of missed beeps to reduce the chances of missing more beeps due to the same cause.

Lesson: Negotiate early with the DSN regarding the required level of support for any previously nonstandard capability. The DSN puts priority on analyzing problems that are covered by a DR. With limited problem analysis and resolution resources, problems involving best-efforts processes may also be resolved only on a best-efforts basis.

Project interaction at Mars. The group-buy of X-band SDSTs included several that operated on the same uplink/downlink DSN channel.

As each project acquired its SDSTs, the JPL frequency management organization took into consideration the locations of these missions (for example, at Mars) and the planned durations of the missions. Because MER's primary surface mission was planned to end in 2004, few anticipated that the rovers would still be operating well years after their landings (Spirit to mid-2011 and opportunity into 2016). Meanwhile, MRO, which arrived at Mars in March 2006, had been allocated channel 32 for its SDSTs, the same channel as MER-A (Spirit).[36]

Under the auspices of the Mars Program Office, a working group with representatives from the MER and MRO projects, the JPL Telecommunications Division, and the DSN developed a set of recommendations in May 2005 [19]. A plan incorporating these recommendations was based on the assumption that both rovers would still be active and with both UHF and X-band capability in March 2006 and for an indefinite period during the MRO prime mission afterwards. The plan required MER to develop and test a capability to command Spirit on UHF via Odyssey when critical MRO X-band operations (such as aerobraking) would be compromised otherwise.

Lesson: The group that developed the plan [20] also published the following lessons:

- Bandwidth is a program consumable (especially at Mars and the Moon).
- Bandwidth-efficient modulation approaches (for example, Gaussian-filtered minimum-shift keying [GMSK]) are needed at X-band.
- Continue to actively move high-bandwidth missions to Ka-band.

---

[36] The rover longevity also required a look at another possible interference case. The Deep Impact (DI) project used SDSTs operating on DSN channel 29, the same as Opportunity. The DI primary mission was from January 2005 through August 2005, overlapped a portion of Opportunity's surface mission. An extended mission (named Epoxi) using the Deep Impact spacecraft concluded at the end of 2010. In addition, the Dawn mission (launched in 2007) continues along with Opportunity's surface mission as of 2016. Epoxi, Dawn, and MER-B transponders all operate on DSN channel 29. The JPL multi-mission Spectrum Analysis Group ran predictions for these three missions to determine potential periods of interference between each pair of missions. Interference levels were a function of the relative received uplink or downlink signal power as well as the frequency offset resulting from the specific trajectories. When potential interference was identified, the projects would negotiate together to "deconflict" the interference through scheduling of tracking passes or cancellation of less critical tracking passes. In the 2005–2016 interval, this strategy has been effective, and no critical navigation, command, or telemetry data has been lost yet.

- MER command and telemetry operations should be conducted via UHF, but DTE should continue to be available as contingency and backup.
- The next-generation DSN and deep-space transponder should
  - Retain current SDST operational capabilities and flexibilities
  - Have synthesized frequency generation
  - Be fully software-defined (for update after launch) while maintaining current SDST reliability against failures. Reconfiguration includes changes in the operating channel to meet needs unforeseen at launch. The benefits of transceiver software reconfiguration after launch are being realized in the Electra UHF transceiver on board MRO.

Solar conjunction. See ISA Z84599 [21]. To generate data for a radio propagation study intended to improve future near-Sun spacecraft commandability, the communications research section at JPL requested a test to send sets of no_op commands to the rovers during the 2004 solar conjunction at SEP angles down to the minimum of 1 deg. From previous similar uplink work [24] on the Near Earth Asteroid Rendezvous (NEAR) spacecraft in 1997 and the Cassini spacecraft in 2003, it was known that solar effects would degrade the uplink and introduce bit errors on the command waveform presented to the HCD.

Those evaluating the request did not consider the effects of multiple-bit errors within a single, 64-bit (actually 63-bit + 1-fill-bit) uplink code block. A particular vulnerability in the MER HCD setup caused the HCD to see a code block with three or more bit errors as zero or one error and "correct" the one error; then the FSW might correctly use the incorrectly decoded code block to write into sections of the program that it should not touch. In the ISA incident, a writing mistake caused the software to declare a fatal error, halt the sequence, and do a warm reboot.

Lesson: Evaluation of nonstandard command activities should involve representatives from all potentially affected subsystems. The evaluation should be particularly strict for an activity likely to induce errors on a command link.

7.6.2.4 UHF Development

Schedule and mass. During development, the MER project was informed of large pattern and polarization variations in its selected UHF monopole antenna. However, a better-performing UHF antenna weighed more. By the time the significance to mission planning of the performance differences became apparent, it was too late to implement the heavier antenna on the spacecraft. In hindsight, performance testing of the antenna was not sufficient. Consequently, MER retained a monopole design that was not characterized well enough for accurate data volume planning (including return link data rate selection) during surface operations.

Lesson: The UHF antenna measurement on a high-quality spacecraft mock-up should always be a high priority. This is particularly true when the antenna system has pattern amplitude and polarization variations as significant as those of the MER UHF monopole.

Lesson: Design decisions made for valid developmental reasons may have large operations impacts. When this becomes apparent, consider mitigation such as additional testing.

7.6.2.5 UHF EDL

Return bit rate. The UHF 8-kbps return-link performance to MGS during EDL was very good; however,

- MGS was required to phase its orbit so that ideally it would be directly over the planned rover landing site at the middle of the EDL UHF relay period. The orbit-phasing plan, which accounted for the expected prelanding and postlanding MER geometries during the period, was accurate to within 30 s of the ideal.
- The MGS UHF antenna needed to be pointed toward the landing site for optimum gain, and this would have required an MGS spacecraft reorientation.

Lesson: Consider a lower rate (perhaps 2 kbps) for future EDL links to minimize operational impact on the orbiter. And, in doing so, watch out for latency and frame size.

7.6.2.6 UHF Surface Operations

Antenna pattern workarounds. The rover UHF (RUHF) antenna asymmetry greatly increased operational complexity.

- Polarization loss for the monopole-to-orbiter antenna is challenging to model.
- The accuracy of the orbiters' antenna patterns remains unknown.
- Forward-link prediction has turned out to be less accurate than return-link prediction, and the reason is not yet determined.
- Even so, the Proximity-1 protocol allows the accuracy of UHF link prediction to be less critical than that of X-band link prediction.

Despite plans to the contrary during development, the Flight Team has reoriented the rover whenever possible to maximize the data return, particularly during the power-limited Martian winter.

Lesson: The pressure to increase data return during operations is unstoppable. The system must be calibrated and configurable to make that possible.

Using inherent capability. An error in the implementation of UHF carrier-only-mode communications behavior was discovered less than two months before EDL. Refer to Problem/Failure Report (P/FR) Z82586 [18]. For contingency operation, a pure-carrier UHF return-link mode is required. In the implementation, there was apparently confusion between the UHF transceiver nomenclature of "tone beacon" (rover return-link carrier-only output) and "command beacon" (an orbiter forward-link hailing-signal output). As implemented, the CBM incorrectly enables the command beacon in intended return-link carrier-only modes. This combination makes detection of the intended carrier-only return link by the orbiter very difficult and therefore has been prohibited by a flight rule.

Using different nomenclature for the two modes would probably have made the implementation error less likely. The project rejected a FSW update to fix the problem so close in time to the beginning of surface operations.

The UHF cognizant engineer developed and tested a workaround, which is to attach a sequence to all UHF carrier-only comm windows. (Any comm window can specify a sequence filename, and the sequence is then initiated when the window opens.) One minute after the window begins the erroneous UHF configuration, the attached sequence commands the transceiver to standby mode, waits 10 s, then enforces the correct carrier-only transmit mode.

Lesson: With a robust basic design, such as the comm windows, an operational workaround may be available that carries less risk than an in-FSW modification.

UHF link prediction. The prediction tools (DVCF and GTP) work well to help select low-elevation extra or alternative passes, significantly increasing data volume. In good areas of the rover UHF antenna pattern, the link can be closed at a 5-deg elevation angle. However,

- The ability to account for rover tilt in the rover-orbiter geometry was added by the end of the first extended mission. In a few cases, the expected data volume differed by a factor of 2 with and without tilt included in the prediction.
- Variability from pass to pass makes setting a margin criterion (for example, mean minus 1-sigma) challenging. It may differ for each rover as well as between MGS and Odyssey for a given rover.

Lesson: Projects with relay links must be capable of accurate prediction of data return commensurate with the accuracy required in sol-by-sol activity planning.

Data return latency. Creation of UHF comm windows was more hands-off than was the creation of X-band windows, but telecom analysts on the flight team still spent inordinate amounts of time in the primary mission answering queries about when the UHF data from each window would flow into the Mission Support Area (MSA). The mission planners required the data from one sol to plan the next sol. The comm window start time and duration accurately defined when the data left the rover on its way to the orbiter. However, the time the data reached the MSA varied greatly on the particular conditions on MGS or Odyssey, and in the orbiters' ground systems.

After the primary mission, this situation improved somewhat when Odyssey provided a bent-pipe mode for data relay, thus defining when the first data from a UHF window would be on the ground. It improved further when LMA developed scripts for Odyssey passes, taking into account buffer management, to define the latest time that all data from a UHF window would be on the ground.

Lesson: Projects with relay links (especially those that depend on other projects) need to design into the end-to-end ground system the capability to estimate the latency at each step in the process, again commensurate with the accuracy required in sol-by-sol activity planning.

7.7 Beyond the Extended Mission

The preceding sections discussed MER telecom operations and performance for the primary mission (through April 2004) and the first extended mission (through September 2004).

The source article for this chapter is Article 10, *Mars Exploration Rover Telecommunications*, in the Design and Performance Summary section of the DESCANSO website

http://descanso.jpl.nasa.gov/DPSummary/summary.html

Section 6.7 of that article, published in October 2005, provides summaries of telecom planning and performance as of that date. It also has information on planning and analysis tools developed and used by the telecom flight team to make operations more efficient with a reduced flight team staff.

The MER project website http://marsrovers.jpl.nasa.gov/home/index.html provides the current status of each rover. As of January 2016, the reports on Spirit and Opportunity were:

7.7.1 Spirit

Spirit remains silent at Troy, as of Sol 2621 (May 24, 2011). Spirit became bogged down at the edge of a crater in the area called Troy. At the time, Spirit had traveled 7.73 kilometers (4.80 miles) from her landing site. More than 1,300 commands were radiated to Spirit as part of the recovery effort in an attempt to elicit a response from the rover. No communication has been received from Spirit since Sol 2210 (March 22, 2010). The project concluded the Spirit recovery efforts on May 25, 2011.

7.7.2 Opportunity

Opportunity remains active as of Sol 4267 (January 25, 2016). Opportunity, which landed on Mars 12 years ago, remained active through the shortest-sunshine days of the current Martian winter. The Mars environment lent a hand, providing wind that removed some dust from Opportunity's solar panels in the weeks before and after the Mars Southern Hemisphere's winter solstice on Jan 2. "Opportunity has stayed active this winter, in part because the solar arrays have been much cleaner than in the past few winters," said Mars Exploration Rover Project Manager John Callas.

Opportunity has been exploring the western rim of the 14-mile wide (22-km wide) crater Endeavour since 2011. This winter, it is examining rocks on the southern side of "Marathon Valley," which slices through Endeavour Crater's rim from west to east. MRO had mapped concentrations of clay minerals in this area that would have formed under wet, non-acidic conditions.

References

[1] "Photojournal," *Mars Exploration Rover Mission*, website (planetary photojournal, including images of and by the MER rovers), Jet Propulsion Laboratory, California Institute of Technology, Pasadena, California. http://photojournal.jpl.nasa.gov/index.html (accessed April 19, 2010)

[2] *Mars Exploration Rover Mission*, MER public home page, Jet Propulsion Laboratory, California Institute of Technology, Pasadena, California. http://marsrovers.jpl.nasa.gov/home/index.html (accessed April 19, 2010)

[3] A. Makovsky, *X-Band Telecommunications Systems Operations Handbook*, version 4.0, D-21239 (internal document), Jet Propulsion Laboratory, California Institute of Technology, Pasadena, California, July 29, 2003.

[4] M. Danos, *UHF Telecommunications System Operations Handbook*, D-21240 (internal document, Jet Propulsion Laboratory, California Institute of Technology, Pasadena, California, December 30, 2003.

[5] "MER Reports," *Quill MMO GDS Services*, internal web site, Jet Propulsion Laboratory, California Institute of Technology, Pasadena, California. https://quill.jpl.nasa.gov/ (accessed October 30, 2014)

[6] *Mars Exploration Rover Library*, internal website, Jet Propulsion Laboratory, California Institute of Technology, Pasadena, California. https://mars03-lib.jpl.nasa.gov/docushare/dsweb/HomePage (accessed October 30, 2014)

[7] J. Ludwinski, *Mars Exploration Rover Project Mission Plan*, D-19659 (internal document), Jet Propulsion Laboratory, California Institute of Technology, Pasadena, California, March 3, 2003.

[8] *Proximity-1 Space Link Protocol—Data Link Layer, Recommendation for Space Data System Standards CCSDS 211.0-B-5 (Blue Book)*, Consultative Committee for Space Data Systems Secretariat, National Aeronautics and Space Administration, Washington, District of Columbia, December 2013. http://public.ccsds.org/publications/archive/211x0b5.pdf (accessed October 30, 2014)

[9] *Mars Exploration Rover (MER-1/2) Project Network Operations Plan*, DSMS 871-050, Rev. A, D-25967 (internal document), Jet Propulsion Laboratory, California Institute of Technology, Pasadena, California, June 30, 2005.

[10] R. Tung and P. Illot, *Uplink/Commands Can Get into MERs in Cruise if Sent with the Wrong Polarization*, ISA Z82482 (internal document), Jet Propulsion Laboratory, California Institute of Technology, Pasadena, California.

[11] E. Satorius, P. Estabrook, J. Wilson, and D. Fort, "Direct-to-Earth Communications and Signal Processing for Mars Exploration Rover Entry, Descent, and Landing," *Interplanetary Progress Report IPN PR 42-153*, January–March 2003, pp. 1–35, Jet Propulsion Laboratory, California Institute of Technology, Pasadena, California, May 15, 2003. http://ipnpr.jpl.nasa.gov/progress_report/42-153/153A.pdf (accessed October 30, 2014)

[12] T. Pham, C. Chang, E. Satorius, S. Finley, L. White, P. Estabrook, and D. Fort, *Entry Descent Landing Data Analysis (EDA)*, NASA Tech Brief NPO-41220 BR, National Aeronautics and Space Administration, October 13, 2004.

[13] D. Johnston, S. Asmar, C. Chang, P. Estabrook, S. Finley, T. Pham, E. Satorius, "Radio Science Receiver Support of the Mars Exploration Rover Landings," *European Space Agency Third Tracking, Telemetry, and Command Systems for Space Applications Workshop*, Darmstadt, Germany, September 2004.

[14] P. Estabrook, A. J. Barbieri, C. D. Edwards, Jr., R. M. Manning, M. J. Danos, P. A. Ilott, and A. Makovsky, "Mars Exploration Rovers: Telecom System Design and Operation Highlights," *The Journal of the Institute of Electronics Information and Communication Engineers*, June 2005.

[15] T. Pham, C. Chang, D. Fort, E. Satorius, S. Finley, L. White, and P. Estabrook, "Tracking Capability for Entry, Descent and Landing and Its Support to NASA Mars Exploration Rovers," *European Space Agency Third Tracking, Telemetry, and Command Systems for Space Applications Workshop*, Darmstadt, Germany, September 2004.

[16] R. Tung, *Telecom Forecaster Predictor/Unified Telecom Predictor, User's Guide*, version 2.1, D-21499 (internal document), Jet Propulsion Laboratory, California Institute of Technology, Pasadena, California, February 21, 2002.

[17] R. Gladden, P. Hwang, B. Waggoner, B. McLaughlin, P. Pieseler, R. Thomas, M. Bigwood, and P. Herrera, "Mars Relay Coordination Lessons Learned," *IEEE paper* 0-7803-8870-4/05, *Proceedings of the IEEE Aerospace Conference*, Big Sky MT, pp. 170–190, March 5–12, 2005.

[18] R. Lindemann and H. Stone, *MER-A HGA Calibration Anomaly Sol 38*, ISA 30264 (internal document) Jet Propulsion Laboratory, California Institute of Technology, Pasadena, California, April 7, 2004.

[19] *MRO Uplink Coordination Plan*, viewgraphs in MER DocuShare (internal document), Jet Propulsion Laboratory, California Institute of Technology, Pasadena, California. https://mars03-lib.jpl.nasa.gov/docushare/dsweb/View/Collection-25395 (accessed February 2, 2011)

[20] *Frequency Sharing Channel 32 – Spirit and MRO working group*, viewgraphs in MER DocuShare (internal document), Jet Propulsion Laboratory, Pasadena, California. https://mars03-lib.jpl.nasa.gov/docushare/dsweb/View/Collection-24817 (accessed October 30, 2014)

[21] R. Denise and J. Snyder, *Fatal Event, Opportunity Sol 229*, ISA 30437 (internal document), Jet Propulsion Laboratory, California Institute of Technology, Pasadena, California, May 2, 2005.

[22] A. Barbieri, *UHF Carrier Only*, PFR 34452 (internal document) Jet Propulsion Laboratory, California Institute of Technology, Pasadena, California, December 11, 2003.

[23] *Proximity-1 Space Link Protocol—Physical Layer, Recommendation for Space Data System Standards*, CCSDS 211.1-B-4 (Blue Book), Consultative Committee for Space Data Systems Secretariat, National Aeronautics and Space Administration, Washington, D.C., Dec. 2013. http://public.ccsds.org/publications/archive/211x1b4.pdf (accessed Oct. 30, 2014)

[24] D. Morabito, "Solar Corona Amplitude Scintillation Modeling and Comparison to Measurements at X-Band and Ka-Band," *Interplanetary Progress Report IPN PR 42-153, January–March 2003*, Jet Propulsion Laboratory, California Institute of Technology, Pasadena, California, May 15, 2003. http://ipnpr.jpl.nasa.gov/progress_report/42-153/153E.pdf (accessed October 30, 2014)

Chapter 8
Mars Science Laboratory

Andre Makovsky, Peter Ilott, and Jim Taylor

8.1 Mars Science Laboratory Mission and Spacecraft Summary

The Mars Science Laboratory (MSL) mission has achieved its primary objective of placing and operating a mobile science laboratory on the surface of Mars to assess the biological potential of the landing site, characterize the geology of the landing region, investigate planetary processes that influence habitability, and characterize the broad spectrum of surface radiation. MSL is conducting fundamentally new observations of Mars geology using advanced micro-imagery and spectrometry, while assessing the radiation environment and studying the surface environments.

This chapter is written from the perspective that the MSL spacecraft was launched in 2011; it cruised to Mars; it went through entry, descent, and landing (EDL) in 2012; and the MSL curiosity rover has since operated on the surface of Mars.

Launched on November 26, 2011 with rover touchdown on Mars on August 6, 2012, the MSL mission aims to achieve its objectives on the surface of Mars in a manner that will offer the excitement and wonder of space exploration to the public. Fig. 8-1 is an artist's conception of the Curiosity rover on the surface with its instrument arm deployed.

Some of the grey scale figures in this book have color versions in the companion website www.wiley.com/go/deepspacecommunications. Please refer to this website where the text refers to colors.

Deep Space Communications, First Edition. Edited by Jim Taylor.

Fig. 8-1. Artist's conception of Mars Science Laboratory on Mars. Please visit www.wiley.com/go/deepspacecommunications to access the color version of this figure.

The span of planned launch dates was between mid-October and early December 2011 with possible arrival dates at Mars in August 2012. Geometries that enable communications during the EDL phase were a mission design driver.

The MSL candidate cruise trajectories [1,2] were limited to type I trajectories for the 2011 launch[1]. A type I Earth–Mars interplanetary trajectory carries the spacecraft less than 180 degrees (deg) around the Sun, and type II is greater than 180 deg [3].

Table 8-1 lists the opening and closing of the 2011 trajectory launch periods and the trajectory types that were considered. The designators Ia, Ib, and Ic were arbitrarily assigned for launch/arrival date pairs. The actual November 26 launch and August 6 arrival were on a Type Ib trajectory.

MSL's initial telecom capability after launch employed a non-directive low-gain antenna (LGA) and depended on the spacecraft orientation and the limited distance between Earth and the spacecraft. Starting three months after

[1] The MSL mission originally was designed for a 2009 launch and 2010 arrival at Mars. The telecommunications system design, in particular the antenna characteristics, accommodated both type I and type II candidate cruise trajectories for 2009 and 2011 launches. This chapter describes some of the antenna drivers for a launch in 2009 or 2011 and the resulting trajectory candidates.

Table 8-1. 2011 launch and 2012 arrival dates for type I trajectories.

Cruise Trajectory	Launch date	Arrival date
Type Ia open	11/25/11	8/06/12
Type Ia close	12/18/11	8/20/12
Type Ib open	11/25/11	8/06/12
Type Ib close	12/18/11	8/06/12
Type Ic open	11/29/11	8/08/12
Type Ic close	12/18/11	8/13/12

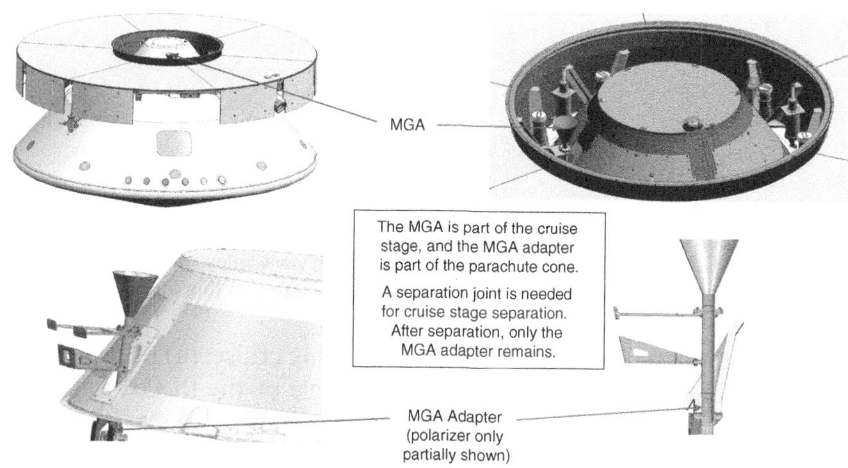

Fig. 8-2. MSL solar array and MGA locations on the cruise stage. Please visit www.wiley.com/go/deepspacecommunications to access the color version of this figure.

launch, a medium-gain antenna (MGA) was used. As shown in Fig. 8-2, the cruise stage had a solar array on a surface surrounding the horn of the MGA. The array is the flat blue surface at the top-left of the drawing. This array powered the cruise loads and charged the batteries. Therefore, the cruise-stage orientation to the Sun was driven by power and thermal subsystem constraints. The solar array normal had to be pointed near the Sun line, at a Sun-view angle that optimized the solar cell power output, without being heated too much. Conversely, the telecom link would be optimized if the antenna boresight (and, hence, the spacecraft –Z axis) were pointed towards Earth.

Figure 8-3 shows the array and the MGA edge-on in the cruise stage at the top. The –Z axis is toward the top of the figure. The competing power and telecom needs were major factors in the design trades in cruise stage orientation. The telecom and power/thermal constraints are linked by the Sun–Probe–Earth (SPE) angle [4].

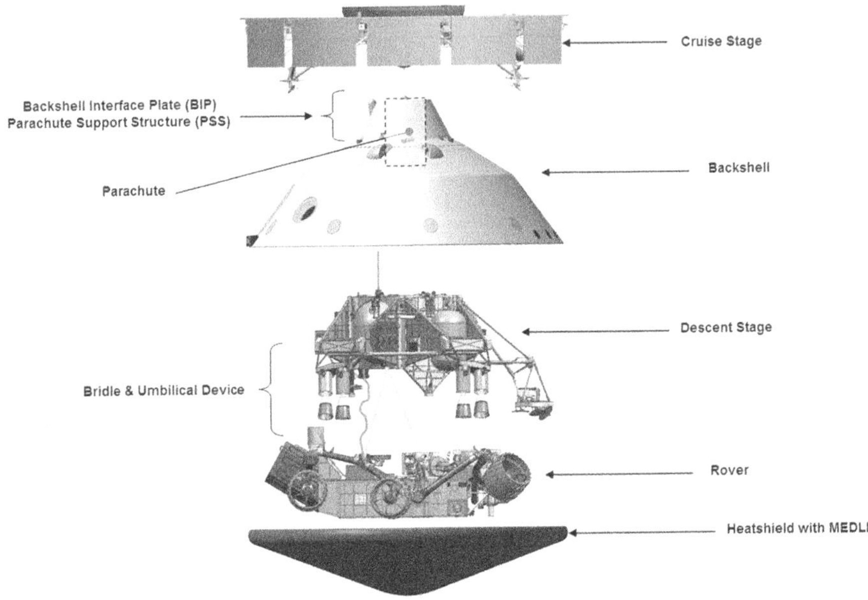

Fig. 8-3. Exploded view of the five major stages of the MSL spacecraft (flight system) (MEDLI = MSL EDL instrumentation). Please visit www.wiley.com/go/deep spacecommunications to access the color version of this figure.

Figure 8-4 shows the profiles of two angles during cruise: (a) the Sun–Craft–Earth angle, and (b) the off-Earth pointing angle of the PLGA and MGA. The broad beam of an LGA would have a poor pattern at angles greater than about 80 degrees (deg) off boresight[2] P0 due to spacecraft obstructions. The type I trajectories result in a maximum off-Earth angle of about 63 deg.

8.1.1 Mission Description

The MSL mission has completed three major phases, and is in the fourth.

1) Launch.

2) Cruise/Approach.

3) Entry, Descent, and Landing (EDL).

4) Surface Operations.

---

[2] Boresight refers to a direction in which an antenna's gain is the maximum. A fixed antenna (one not on a gimbal), such as the MSL MGA or any of the LGAs, is defined in terms of the spacecraft axis (or axes) direction along which that antenna is mounted. That direction, for example the –Z axis for both the parachute LGA (PLGA) and the MGA used during cruise, is loosely referred to as the boresight.

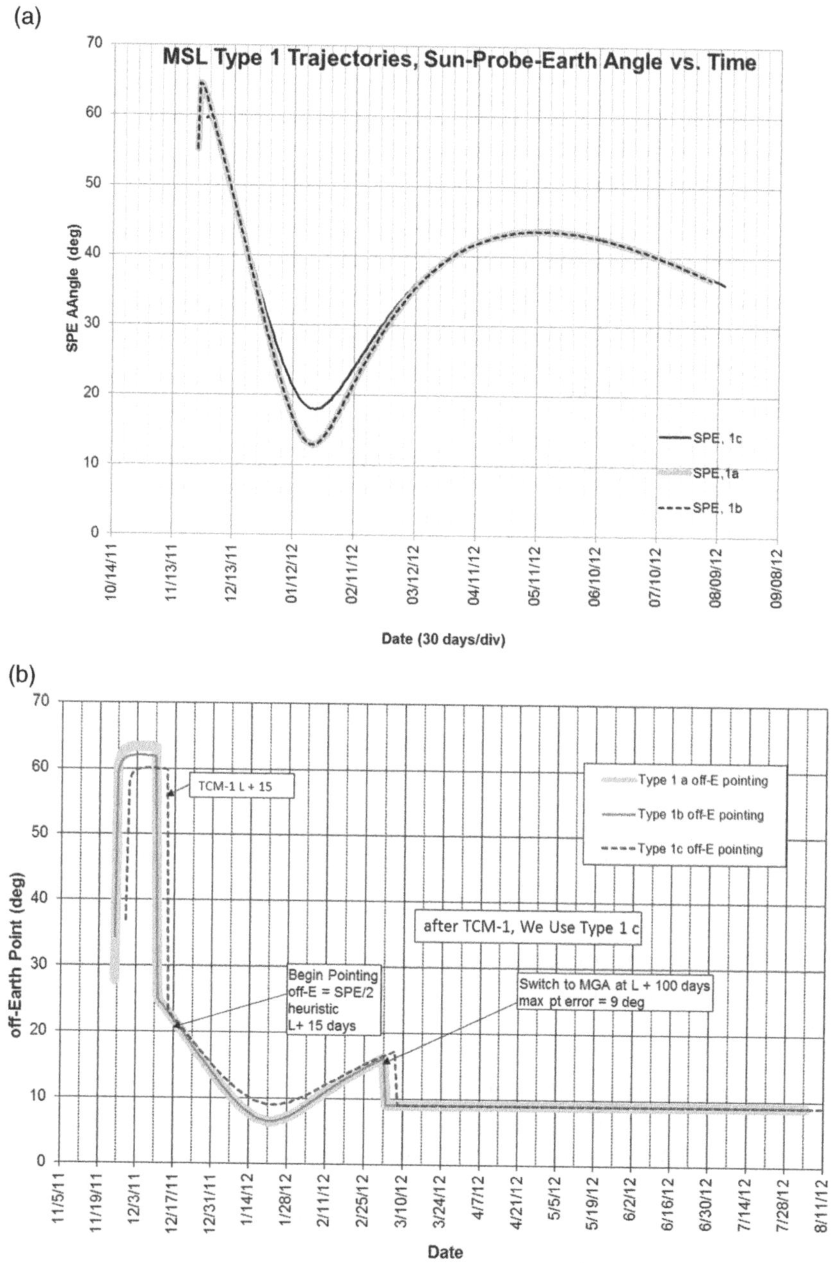

Fig. 8-4. Antenna geometry for MSL cruise trajectories: (a) Sun–Probe–Earth (SPE) angles versus time, and (b) Planned maximum off-Earth angles for PLGA and MGA (TCM is trajectory correction maneuver). Please visit www.wiley.com/go/deep spacecommunications to access the color version of this figure.

Table 8-2. MSL major mission phases.

Mission Phase	Description	Approximate Duration
Launch	The launch phase was defined to begin at the point where the spacecraft transferred to internal power prior to launch. It was complete (after spacecraft separation from the launch vehicle's upper stage) when the spacecraft reached a thermally stable, positive energy balance, commandable configuration	1 hour
Cruise	The cruise phase began when the launch phase ended, and it ended 45 days prior to atmospheric entry (E–40 days).	208 days
Approach	The approach phase was defined to begin at 45 days prior to atmospheric entry (E–45 days) and ended when the spacecraft reached the Mars atmospheric entry interface point. That point is defined at a Mars radius of 3522.2 km.	45 days
EDL	Entry, descent, and landing (EDL) began when the spacecraft reached the entry interface point (Mars radius of 3522.2 km) and ended when the rover reached a thermally stable, positive energy balance, commandable configuration on the surface.	7 minutes
Surface	The surface mission began when EDL ended, and it will end when the mission is declared complete. The design of the rover must provide for a surface mission duration of at least one Mars year (669 sols, equivalent to 687 Earth days).	Prime mission: 669 sols (with possible additional extensions)

Table 8-2 provides more detail regarding these phases. Prior to surface operations, time references are in terms of the familiar Earth hours and days.

On the surface of Mars, all planning and rover activity is in terms of sols (Martian days). One sol, in terms of Earth time, is approximately 24 hours, 39 minutes, and 35 seconds long. A sol is equivalent to about 1.027 Earth days. A sol is divided into 24 (Martian) hours, each of which is 60 (Martian) minutes, and so forth, analogous to the time units on Earth.

8.1.2 Launch/Arrival Period Selection

A year before the 2011 launch, candidate landing sites on Mars were narrowed to four (Mawrth Vallis (Valley), Gale Crater, Eberswalde Crater, and Holden Crater), spanning latitudes between 27 deg S and 25 deg N. The final selection, near Mount Sharp in Gale Crater, was made in July 2011 (4 months before launch).

With these sites, the parameter limits that drove the launch and arrival periods can be summarized as a set of constraints [5].

- Spacecraft injected mass = 4050 kilograms (kg)
- Launch-specific energy (C3) capability < 20.1 kilometers squared per second squared (km^2/s^2) (Atlas V 541, instantaneous launch window)
- Atmospheric entry velocity < 5.9 kilometers per second (km/s) (not a hard constraint due to EDL heating performance study results)
- Declination of launch asymptote < 40 deg
- Arrival no later than 30 days before the start of solar conjunction
- Launch eclipse duration ≤ 65 minutes
  - Early cruise SPE angle constraints:
    - Launch vehicle separation (SEP) attitude (SEP to SEP + 18 days):
      - Angle between –Z axis and Sun ≤ 64.0 deg and ≥ 20.0 deg
      - Angle between –Z axis and Earth ≤ 68.8 deg
  - EDL communications strategy constraints:
    - Relay:
      - Mars Reconnaissance Orbiter (MRO) Local Mean Solar Time (LMST) node as close to nominal value (3:00 p.m.) as possible
      - Odyssey LMST node at 3:00 p.m. because of propellant issues related to orbiter lifetime
      - View angle to orbiters (MRO/Odyssey) ≤135 deg
      - MRO/Odyssey elevation at landing + 1 min ≥ 10 deg
    - Direct to Earth (DTE):
      - View angle to Earth ≤ 75 deg
      - Earth elevation at landing + 1 min ≥ 10 deg
  - 20-day launch period.
  - Declination of launch asymptote < 28.5 deg
  - Atmospheric entry velocity < 5.6 km/s.
  - EDL communications strategy constraints:
    - Full ultra-high frequency (UHF) EDL coverage via MRO and Odyssey from Entry[3] to Landing + 1 minute. (Relay coverage is not possible for all of EDL due to geometric constraints from cruise stage separation (CSS until entry).
    - Full DTE EDL coverage (For type I only possible for Mawrth Vallis).

---

[3] Entry defined as the point where the entry vehicle is at a radius of 3522.2 km from nominal Mars center. Entry is also considered at a time beginning 600 seconds after nominal cruise state separation.

8.1.2.1 Cruise Mission Phase Telecom Trades. An example of the significant decisions involving telecom is the one involving cruise trajectory type (I vs. II), launch window, power subsystem constraints, and LGA pattern.

- Type I vs. type II trajectory: The type II trajectory (candidate only for the originally planned 2009 launch) had very large initial SPE angles (as shown in Fig. 8-4), greater than 120 deg at the start of the launch period. At the other end of cruise, the EDL geometry of the type II offered better UHF coverage opportunities. (One year before the 2011 MSL launch, only type I trajectories with launch dates between November 25 and December 18, 2011 were still in contention.)
- Launch period: With a 2011 launch, Juno's launch period, driven by the complex trajectory to get the Juno spacecraft into orbit around Jupiter, overlapped MSL's launch period. When MSL's launch was changed from 2009 to 2011, the MSL trajectory design (which included launch period) had to accommodate the Juno launch period of August 5–27 that was required to get that spacecraft to Jupiter. Both Juno and MSL launched on Atlas V launch vehicles, and both used the same launch pad at Kennedy Space Center (KSC). The time required to refurbish the launch pad after Juno launched and to complete MSL pre-launch activities established an earliest launch date of November 25 for the MSL Type Ib trajectory.
- Solar array[4] and antenna pointing: Power output and thermal considerations required the solar array to be pointed within an optimum range of angles from the Sun: too far from the Sun, not enough power; too close to the Sun, too much heating. This range of angles tended to force the PLGA angle to Earth to be too far off boresight in the first weeks after launch.
- Solar array pointing (thermal constraint): Until Sun–spacecraft distance increased sufficiently, the solar array was to be pointed not too close to the Sun, to avoid overheating the solar panels and losing efficiency.

Based on gain pattern measurements with a spacecraft mock-up, telecom imposed a mission design constraint of 80-deg offpoint from PLGA boresight. The measured PLGA patterns described in Section 8.2 (Figs. 8-36 and 8-37) can be compared with this 80-deg constraint.

---

[4] The solar array was on the cruise stage (Figs. 8-3 and 8-4) and augmented the radioisotope-thermoelectric generator (RTG) during cruise. Without the need for a solar array, the RTG alone, with a battery for peak loads, is sufficient for surface operations.

These trades resulted in the following telecom configurations and operating modes for launch and cruise [6].

- We launched using the PLGA. This antenna has a very broad beam, and near Earth we can cover as much as 90 deg off—afterward as much as 80 deg off.
- From a telecom downlink margin point of view, we could have launched using either the traveling-wave tube amplifier (TWTA) on the descent stage or the solid state power amplifier (SSPA) on the rover. The plan was to launch on the TWTA (to avoid having to switch to it from the SSPA early in cruise and being exposed to a switch failure).
- Operationally, being on the TWTA meant having a 210-s outage after launch vehicle (LV) separation, while the TWTA warmed up.
- For the 2011 launch, the plan was to transition from the PLGA to the MGA no later than March 2012. The driver for this is ranging, which requires a ranging power-to-noise spectral-density ratio (P_r/N_0) of at least –20 decibel-hertz [dB-Hz]. Ranging is weaker than either the command link or the telemetry link. Ranging incurs two-way space loss as well as thermal noise in the small deep space transponder (SDST) receiver's ranging channel bandwidth of ~1.5 megahertz (MHz).

8.1.2.2 EDL and Surface Mission Phase Telecom Trades. The telecom constraints during surface operations are defined in Makovsky and Danos [6], and they can be summarized as follows:

- The Mars–Earth range at arrival was greater than it was for the Mars Exploration Rover (MER) (MER was ~1.53 astronomical units [AU] [2.29×10^8 km] compared to 1.66 AU [2.48×10^8 km] for MSL). The X-band performance was correspondingly weaker than MER's, and this was reflected in lower downlink rates.
- UHF (Relay via MRO or Odyssey) is intended as primary for communications; HGA has been a low-data volume back-up and the RLGA has been used for emergency commanding)

Telemetry during EDL was transmitted via the MSL UHF subsystem and the spacecraft (sometimes called relay assets) orbiting Mars and by DTE at X-band to the Deep Space Network (DSN). The UHF relay transmitted real-time EDL data in a continuous stream at a rate of 8 kilobits per second (kbps) in bit stream mode.[5] The UHF bit stream broadcast by MSL during

[5] Bit steam mode is a non-acknowledged transmission mode that includes no Proximity-1 protocol formatting and no data retransmission mechanisms. After landing, Curiosity had multiple opportunities to transmit a superset of the real-time data that had been stored on-board during the event. The post-landing transmission used the Proximity-1 mode.

EDL was received by multiple orbiters that had their UHF relay radios set to operate in a compatible listen-only bit stream mode. The data for DTE was in the form of semaphores, the so-called multiple frequency shift keying (MFSK) tones. As with the MER landers in 2004 and the Phoenix Lander in 2008, the possibility existed to monitor the UHF carrier signal from a large Earth-located ground station that had Mars in view at the right time.

The relay assets available to MSL for EDL were the MRO and the Odyssey orbiter. Each orbiter played a unique role in capturing the UHF signal from the rover during this phase.[6] Odyssey performed realtime demodulation of the EDL relay data so UHF telemetry data returned first from Odyssey. MRO performed open loop recording for later demodulation and thus acted in a secondary role. However, if Odyssey's closed loop data capture had failed for any reason, MSL would have depended on the MRO open loop record data.

Due to the limitations of geometry during EDL for the chosen Type Ib trajectory, simultaneous coverage by both an orbiter for UHF and an Earth station for DTE was not possible for the entire EDL phase. The mission strategy ensured DTE coverage during the period from CSS until at least atmospheric entry. During this period, relay coverage began after CSS. There was substantial overlap of X-band and UHF coverage during entry, after the relay link began to be viable. During hypersonic entry, we included in our plans the expected loss of as much as 100 s of UHF coverage due to plasma blackout. The actual blackout was about 40 s. DTE coverage during the UHF blackout period was planned. The descent stage in fact continued to transmit X-band DTE MFSK tones all the way to landing.

For the 2011 launch opportunity, and to simplify the preliminary verification of EDL communications coverage, the following actions were performed in the trajectory selection process by Mission Design.

- For MRO relay coverage:
  - Check that there would be a line of sight between MSL and MRO at the time of Entry interface and at landing + 1 minute.
  - Assume that MRO could be phased anywhere in its orbit for optimal coverage of MSL EDL.
  - Check that the orbiter elevation was greater than 10 deg at landing and landing + 1 minute.
  - Check that the angle from the MSL anti-velocity vector to the orbiter at entry was less than 135 deg.

[6] It would also have been possible for MSL to plan to use the European Mars Express (MEX) orbiter in a limited capacity if either MRO or Odyssey (or both) became unavailable. However, this use would have had to be planned in advance to enable MEX to phase for the MSL EDL. MEX has subsequently performed UHF relay operations with Curiosity during surface operations.

- Similarly for DTE coverage (directly to Earth):
  - Check that there was a line of sight between MSL and Earth at the entry interface and landing times.
  - Check that Earth elevation was greater than 10 deg at landing.
  - Check that the angle from the MSL anti-velocity vector to Earth at entry was less than 75 deg.

Orbiter coverage was strongly desired since it provided return link telemetry at a single planned rate of 8 kbps versus the X-band MFSK tones at a maximum rate of a new tone every 10 seconds. Each MFSK tone could carry one of 256 messages represented as a "subcarrier frequency" (the frequency spacing between carrier and subcarrier). Each tone therefore notified the flight team of one event (for example, "parachute deploy"). In addition, use of signal processing recovered frequency characteristics of the X-band signal (such as Doppler-shift due to deceleration) could help reconstruct events during EDL. UHF, by contrast, provided a large amount of real-time engineering telemetry. The maximum planned data latency was 1 second for the UHF transmitted telemetry during EDL. Minimizing the data latency ensured that the ground received as complete a history of events as possible. Happily, a detailed forensic reconstruction was not needed.

Figure 8-5 illustrates the quality of the coverage during the post-entry phase of EDL for UHF via MRO and Odyssey and for X-band DTE, for the four

EDL Coverage Summary

Launch Period	Asset	Mawrth (24.0N)	Gale (4.5S)	Eberswalde (23.9S)	Holden (26.4S)
Type 2	MRO				
	ODY				
	DTE*				
Type 1	MRO				
	ODY				
	DTE*				

*Drop outs are possible between landing – 60 s and landing – 20 s due to parachute and powered descent dynamics.

Fig. 8-5. EDL coverage after entry for Mars Reconnaissance Orbiter and Odyssey. Please visit www.wiley.com/go/deepspacecommunications to access the color version of this figure.

candidate landing sites and the two types of trajectories. All four sites were considered viable in terms of safety and engineering considerations (including communications during EDL), allowing the final choice of Gale Crater to be on the basis of science.

The figure shows there were some differences however. Medium (green) shading indicates good coverage for the full duration of EDL for the indicated link type, light (yellow) shading indicates coverage for only part of the EDL, and dark (red) indicates little or no coverage. DTE coverage for the type I trajectory would be complete (that is, all the way until landing) only for the northern site, Mawrth Vallis. This is due to the Earth setting below the horizon for the southern sites (red shading). Conversely, UHF coverage for the type II trajectory was only partial from Odyssey for the southern sites (yellow shading), though MRO coverage was still available. Orbital phasing and survivability considerations limited Odyssey coverage. For either trajectory, only the Mawrth site offered complete EDL telecom coverage after atmospheric entry for DTE and both orbiters.

Section 8.2 of this chapter contains more detailed information regarding the EDL communications geometry and the rationale behind the 135-deg orbiter and the 75-deg Earth angles from the anti-velocity vector for relay and DTE coverage.

8.1.3 Launch Phase and Initial Acquisition

The Launch phase began when the spacecraft transferred to internal power on the launch pad. It ended when the spacecraft was declared stable, healthy, and ready to accept commands, and when the launch telemetry had been played back. The major activities in the Launch phase included the Liftoff and Boost phase of the launch vehicle; insertion into a circular parking orbit, a coast period (followed by additional launch vehicle upper stage burns necessary to inject the spacecraft onto the beginning of the planned trajectory to Mars), separation of the spacecraft from the launch vehicle, initial acquisition by the DSN, verification of the initial spacecraft health and operating conditions, and the verified execution of a minimal set of post-launch commands. Table 8-3 shows the launch window times for different phases of the launch periods (time is in Universal Time Coordinated [UTC]) [2].

In June 2006, the Atlas V 541 was selected as the launch vehicle for MSL. The Atlas 541, shown in Fig. 8-6, provides a 5-meter (m) fairing, the addition of four solid-rocket motors to the central booster, and a single-engine Centaur upper-stage. The Lockheed Martin website [7] provides more details about the launch vehicle. Figure 8-7 illustrates the launch events.

Table 8-3. Launch period / window durations.

Launch Period	Launch Date	Launch Day	Launch Window (UTC)
Type Ia	11/25/2001	Open	15:15 to 17:15
	12/06/2011	Middle	14:16 to 16:13
	12/18/2011	Close	11:48 to 13:48
Type Ib	11/25/2001	Open	15:15 to 17:15
	12/06/2011	Middle	12:35 to 14:35
	12/18/2011	Close	11:06 to 13:06
Type Ic	11/29/2011	Open	15:22 to 17:22
	12/08/2011	Middle	12:50 to 14:50
	12/18/2011	Close	11:11 to 13:11

Table 8-4 provides a legend for the acronyms used in these two figures.

Depending on the specific launch date, planning had to account for an eclipse of the Sun by the Earth after launch vehicle separation. The traveling wave tube amplifier (TWTA) was scheduled to be powered on at eclipse exit[7]. The TWTA started the X-band downlink after a 4-minute warm-up period.

The plots in Fig. 8-8 (for a type Ia first possible launch date) and Fig. 8-9 (for a type Ic last possible launch date) show the ground tracks for the first 24 hours after launch based on the optimum launch windows. Each ground track begins south and east of the launch site in Florida, goes through TIP and then SEP. Enter and exit Earth occultation times are indicated. For both trajectory types, the first DSN site to view the spacecraft after launch, called the "initial acquisition" site, was Canberra, Australia. An important mission consideration was whether the spacecraft would be in view of the initial

[7] There is nothing in terms of spacecraft power subsystem capability about associating MSL TWTA power-on with eclipse exit. However, defining it at this time allowed for a reasonable interval after the initial ascent for the microwave circuitry to vent residual gases to vacuum before subjecting it to high power radio frequency (RF) energy, thus adhering to the TWTA maker's recommendation for a period of hard vacuum before generating RF power for the first time after launch. The added time also reduced the possibility of RF breakdown or arcing in the waveguides or the TWTA itself.

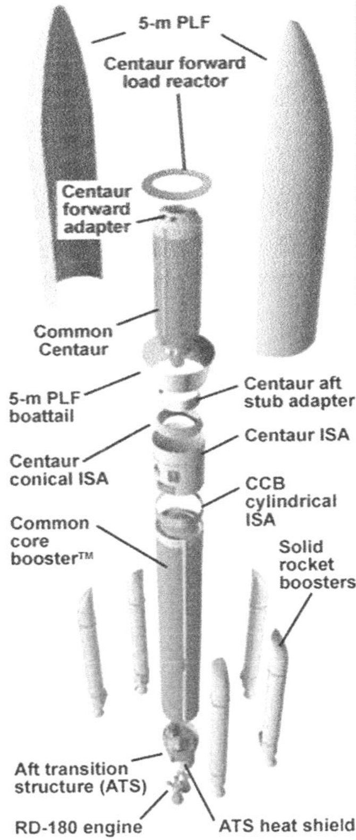

PAYLOAD FAIRING (PLF)

Features	5-m Short	5-m Medium	5-m Long
Diameter:	5.4 m	5.4 m	5.4 m
Length:	20.7 m	23.4 m	26.5 m
Mass:	3,540 kg	4,019 kg	4,394 kg

Subsystems
- Fairing: Bisector: sandwich construction with graphite epoxy face sheets & an aluminum honeycomb core
- Boattail: Fixed, composite sandwich construction
- Separation: Vertical separation by a linear piston & cylinder activated by a pyrotechnic cord; horizontal separation by an expanding tube shearing a notched frame, activated by a pyrotechnic cord

COMMON CENTAUR

Features — Common with Atlas 400 series
- Size: 3.05-m dia x 12.68-m length with extended nozzle
- Inert mass: 2,138 kg
- Propellant: 20,830-kg LH_2 & LO_2
- Guidance: Inertial

Subsystem
- Structure: Pressure stabilized stainless steel tanks separated by common ellipsoidal bulkhead
- Propulsion: One or two Pratt & Whitney restartable engine(s)
  - —Model: RL 10A-4-2
  - —Thrust: 99.2 kN (SEC) 198.4 kN (DEC)
  - —I_{SP}: 450.5 s
  - (SEC) One electromechanically actuated 51-cm columbium fixed nozzle; Four 27-N hydrazine thrusters; Eight 40-N lateral hydrazine thrusters
  - (DEC) Two hydraulically actuated 51-cm columbium extendible nozzles; Eight 40-N lateral hydrazine thrusters; Four 27-N hydrazine thrusters
- Pneumatics: Helium & hydrogen autogenous tank pressurization
- Avionics: Guidance, navigation & control vehicle sequencing, computer-controlled vent & pressurization, telemetry, tracking, range safety command, electrical power
- Insulation: Polyvinyl chloride foam (1.6-cm thick), modified adhesive bonding with optional radiation shields

SOLID ROCKET BOOSTERS (SRB)

Zero-to-Five	Ground-lit
Size:	155-cm dia x 19.5-m length
Mass:	46,559 kg (each fueled)
Thrust:	1,361 kN (each)
I_{SP}:	275 s
Nozzle cant:	3 deg

CENTAUR INTERSTAGE ADAPTER (C-ISA LARGE)

Features
- Size: 3.81-m dia x 4.46-m length
- Mass: 2,292 kg (includes ISA, aft stub adapter & boattail)

Subsystems
- Structure: Composite sandwich (aluminum core graphite epoxy face sheets)

CCB CONICAL INTERSTAGE ADAPTER

Features
- Size: 3.81-m dia x 0.32-m length
- Mass: 282 kg

Subsystems
- Structure: Aluminum machined rolled-ring forging

COMMON CORE BOOSTER™ (CCB)

Features — Common with Atlas V 400 series
- Size: 3.81-m dia x 32.46-m length
- Propellant: 284,089-kg LO_2 & RP-1
- Inert Mass: 21,336 kg for 55Z configuration
- Guidance: From upper stage

Subsystems
- Structure: Structurally stable aluminum Isogrid tanks; Integrally machined aft transition structure; composite heat shield
- Separation: Eight retro rockets
- Propulsion: Pratt & Whitney/NPO Energomash RD-180 booster engine (2 chambers)
  SL 100% thrust =3,827 kN, I_{SP}=311.3 s
  Vac 100% thrust = 4,152 kN, I_{SP}=338.4 s
- Pneumatics: Helium for tank pressurization, computer-controlled pressurization system
- Hydraulics: Integral with engine provides gimbal control
- Avionics: Flight control, flight termination, telemetry, redundant rate gyros, electrical power

Fig. 8-6. Atlas V 541 launch vehicle.

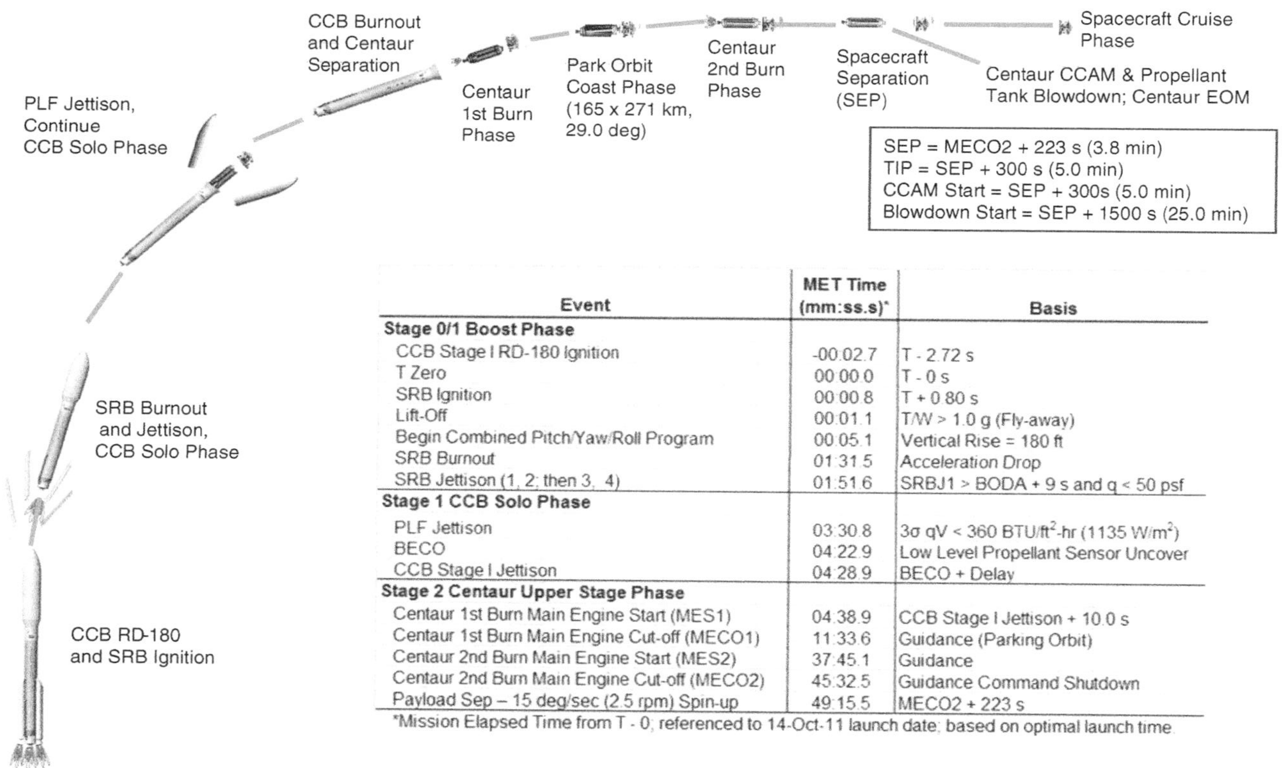

Event	MET Time (mm:ss.s)*	Basis
Stage 0/1 Boost Phase		
CCB Stage I RD-180 Ignition	-00:02.7	T - 2.72 s
T Zero	00:00.0	T - 0 s
SRB Ignition	00:00.8	T + 0.80 s
Lift-Off	00:01.1	T/W > 1.0 g (Fly-away)
Begin Combined Pitch/Yaw/Roll Program	00:05.1	Vertical Rise = 180 ft
SRB Burnout	01:31.5	Acceleration Drop
SRB Jettison (1, 2; then 3, 4)	01:51.6	SRBJ1 > BODA + 9 s and q < 50 psf
Stage 1 CCB Solo Phase		
PLF Jettison	03:30.8	3σ qV < 360 BTU/ft^2-hr (1135 W/m^2)
BECO	04:22.9	Low Level Propellant Sensor Uncover
CCB Stage I Jettison	04:28.9	BECO + Delay
Stage 2 Centaur Upper Stage Phase		
Centaur 1st Burn Main Engine Start (MES1)	04:38.9	CCB Stage I Jettison + 10.0 s
Centaur 1st Burn Main Engine Cut-off (MECO1)	11:33.6	Guidance (Parking Orbit)
Centaur 2nd Burn Main Engine Start (MES2)	37:45.1	Guidance
Centaur 2nd Burn Main Engine Cut-off (MECO2)	45:32.5	Guidance Command Shutdown
Payload Sep – 15 deg/sec (2.5 rpm) Spin-up	49:15.5	MECO2 + 223 s

*Mission Elapsed Time from T - 0; referenced to 14-Oct-11 launch date; based on optimal launch time.

Fig. 8-7. Launch Phase Illustration.

Table 8-4. Legend for LV and launch phase figures.

ATS	aft transition structure	MAD	Madrid Deep Space Center
BECO	booster engine cutoff	MECO	main engine cutoff
BODA	burnout detection algorithm	MES	main engine start
CAN	Canberra Deep Space Center	NPO	NPO Energomash is a Russian manufacturer
CCAM	collision and contamination avoidance maneuver	PLF	payload fairing
CCB	common core booster	RP	rocket propellant or refined petroleum (kerosene)
C-ISA	Centaur interstage adapter	SEC	Single-Engine Centaur
DEC	Dual-Engine Centaur	SEP	Separation
EOM	end of mission	SRB	solid rocket boosters
GDS	Goldstone Deep Space Center	SRBJ	solid rocket booster jettison
ISA	interstage adapter	TIP	target interface point
LH_2	liquid hydrogen	T/W	ratio of thrust and weight on pad
LO_2	liquid oxygen		

acquisition site when it separated from the launch vehicle. Because separation data is very important for launch-phase performance assessment, the MSL mission contracted with a non-DSN network, the Universal Space Network (USN), which would have a station with line-of-sight at separation time for any of the type I trajectories.

Figures 8-10 and 8-11 show the station elevation angles for the first 48 hours after injection for trajectories of type Ia and type Ic assuming a launch on the first possible launch date for Type Ia and the last possible launch date for type Ic. Because the initial trajectory was slightly south of the Equator for type Ia, the Canberra site, which has higher elevation angles, was favored. The opposite is true for the type Ic last possible launch date trajectory, for which Goldstone, California, and Madrid, Spain (both northern sites) would have been favored.

As was shown in Table 8-3, all launch dates (for the type Ia, type Ib, or type Ic trajectories under consideration) had windows with 2-hour durations on any launch date. The center of the 2-hour period is optimal. Figure 8-12 shows the ground track during the first ten minutes after TWTA power-on for these trajectories for launches at window open, optimal, and window close.

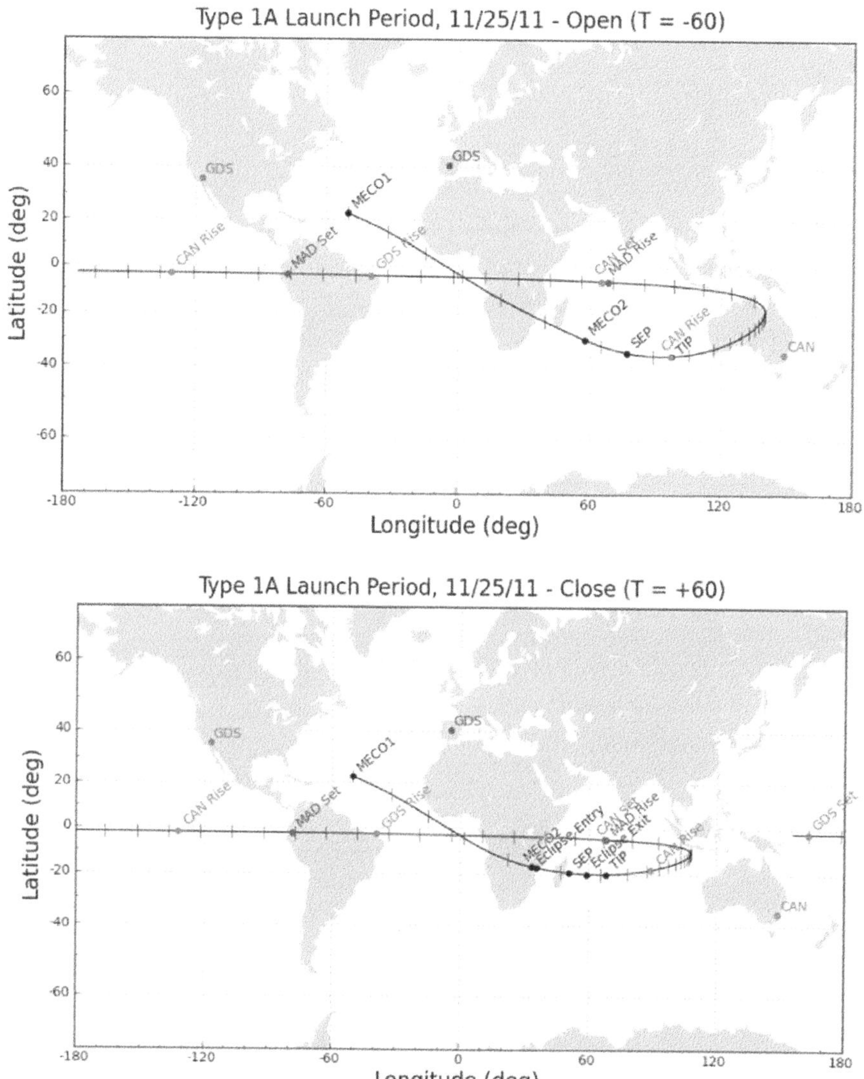

Fig. 8-8. Launch trajectory ground tracks for optimal launch windows (first possible launch day for type Ia trajectory). Please visit www.wiley.com/go/deepspacecommunications to access the color version of this figure.

A station has two minimum-elevation coverage 'masks', the higher (10 deg) for transmitting and the lower (about 6 deg) for receiving. Stations are not allowed to transmit below 10 deg (in order to limit the radio frequency (RF) power flux density that hits the Earth). Depending on the terrain in the vicinity of the station, a station does not have line-of-sight to receive below approximately 6 deg elevation; this elevation is, therefore, taken as the mask. For navigation, with both receiving and transmitting required, the minimum mask is 10 deg.

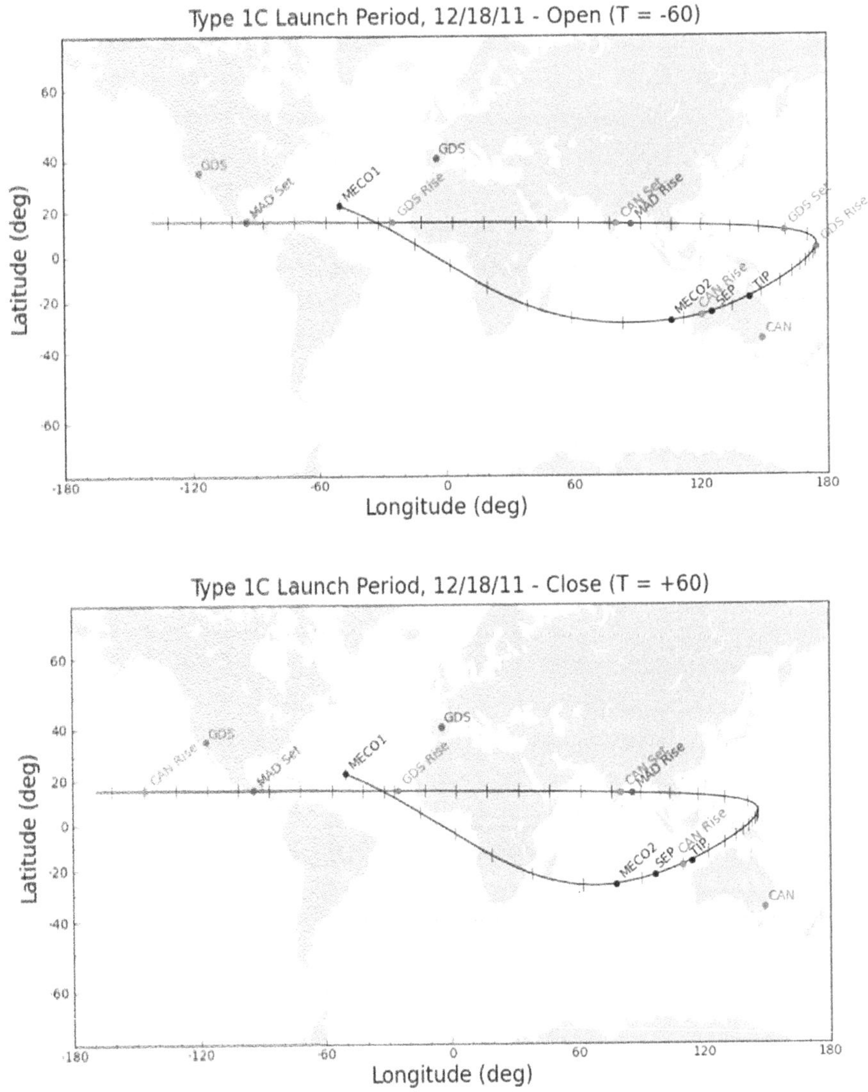

Fig. 8-9. Launch trajectory ground tracks for optimal launch windows (type Ic last possible launch day). Please visit www.wiley.com/go/deepspacecommunications to access the color version of this figure.

In addition to the DSN sites, the Universal Space Network (USN) and European Space Agency (ESA) have tracking stations at Mauritius and in Western Australia. These sites (Fig. 8-13) offered views to the spacecraft sooner after separation than the DSN sites. Figure 8-14 presents the minimum–maximum envelope of spacecraft–Earth ranges for the first 48 hours for type Ia and type Ic trajectories. The three curves are early type Ia and Ic and late type Ic trajectories (late type Ia is nearly identical to late type Ic). The purpose of this figure is to show how fast the spacecraft moves

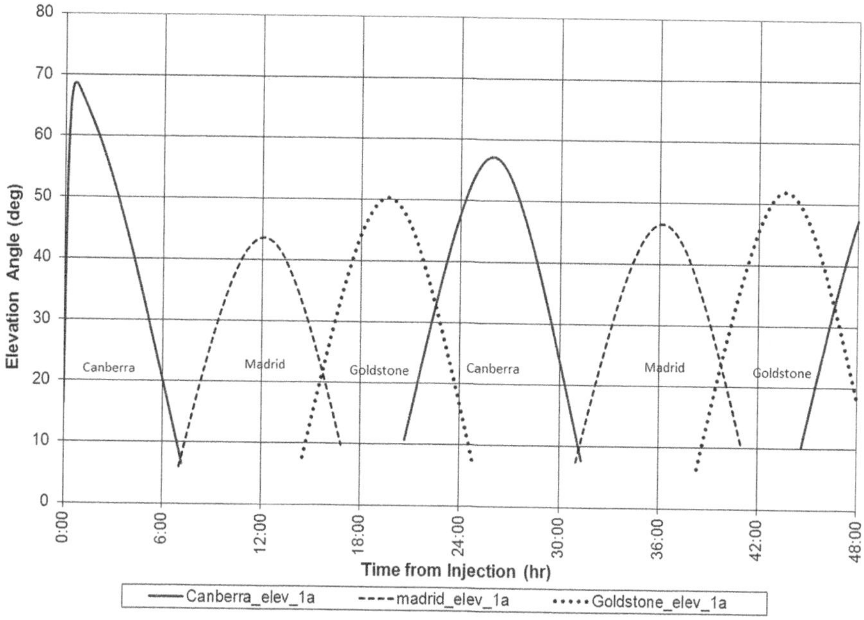

Fig. 8-10. Station elevation angle (type Ia trajectory for the first possible launch day).

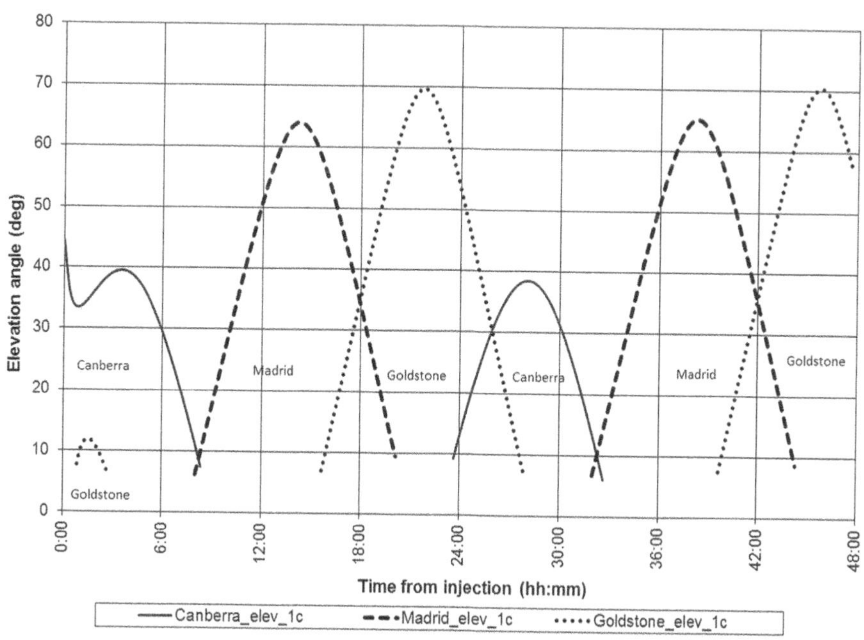

Fig. 8-11. Station elevation angle (type Ic trajectory for the last possible launch day).

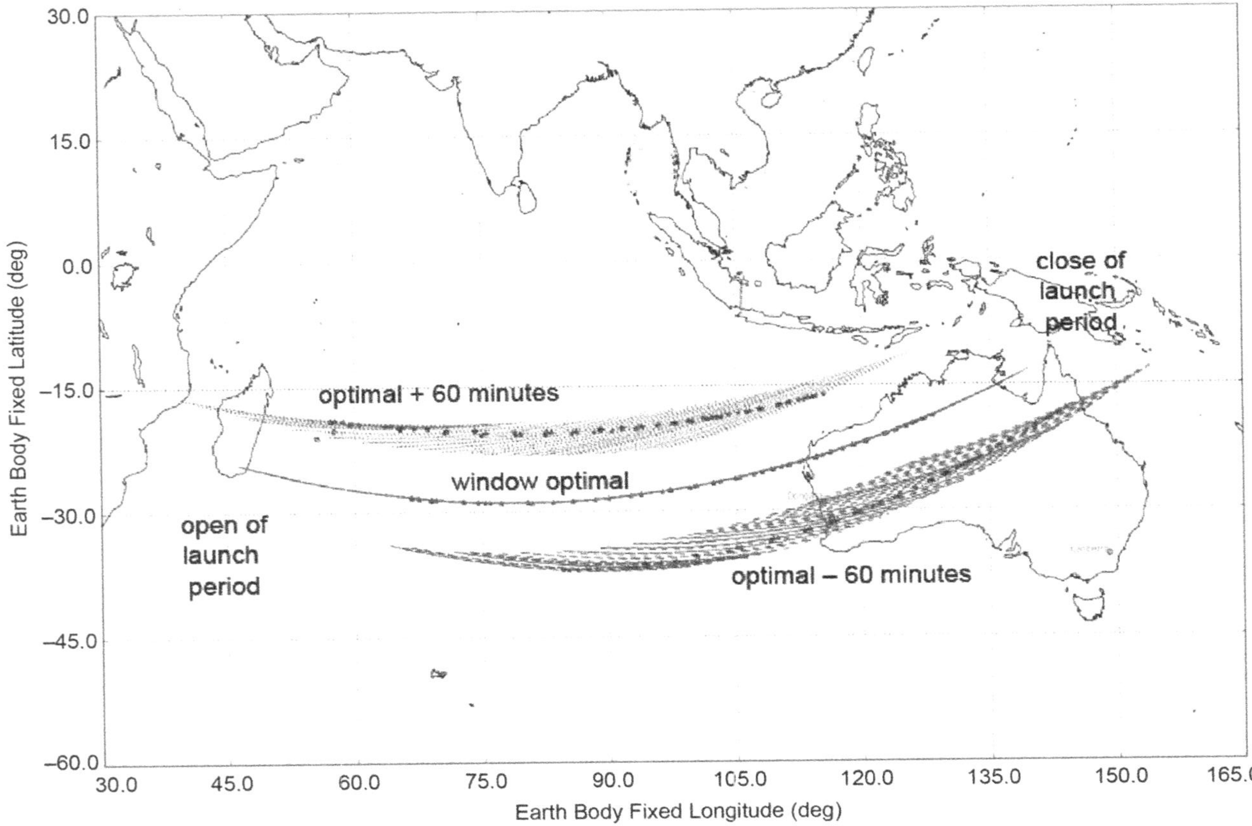

Fig. 8-12. Initial (first 10 min) ground track for type I trajectories (Earth body fixed longitude in degrees).

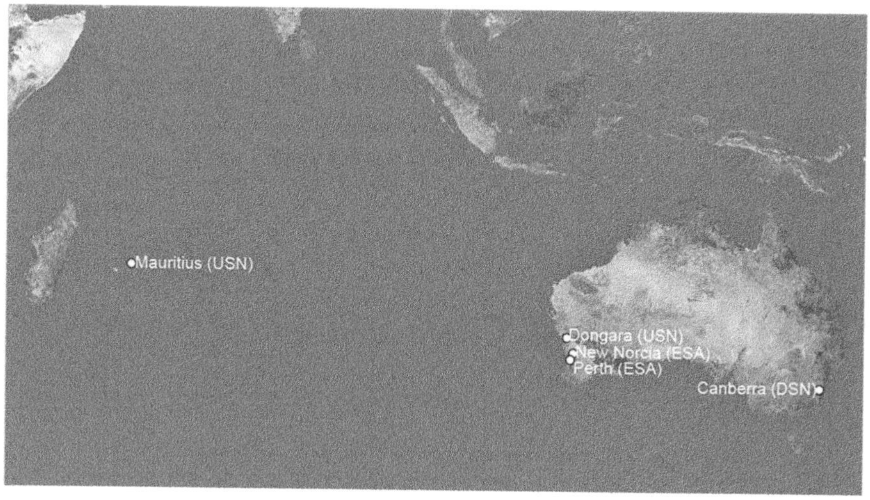

Fig. 8-13. USN and ESA tracking sites for MSL after separation.

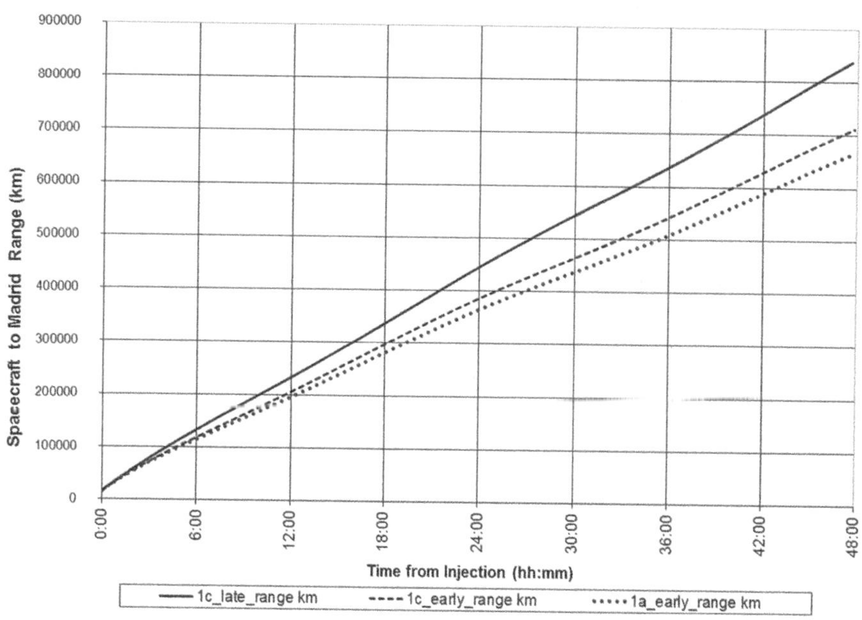

Fig. 8-14. Madrid DSS range to MSL for 48 hours after injection.

away from Earth after launch. By the third deep space station (DSS) pass (about 20 hours past injection), the spacecraft was already at Moon-distance (about 380,000 km).

With antenna pointing angles constant, the communications capability falls off with the square of the distance between transmitter and receiver. This

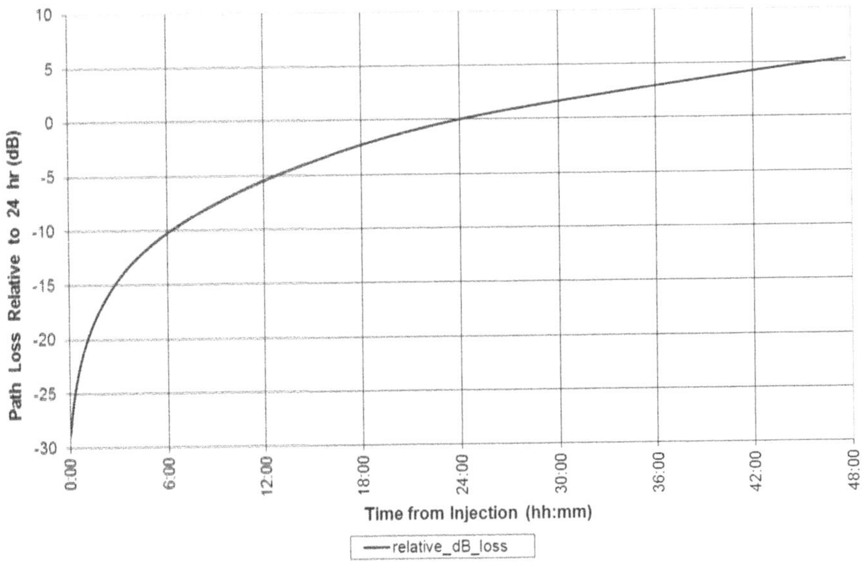

Fig. 8-15. Madrid space loss to MSL, type Ia nominal trajectory.

decrease in capability is called space loss, and it is an important factor in the link budget or design control table (DCT) that defines performance at a given point in time. As a number, the space loss changes as 1/(range squared) and in decibels (dB) as $-20 \times \log$ (range). Figure 8-15 shows how the space loss increases, as a function of time (in hours), almost 30 dB in the first 24 hours, then about another 5 dB in the next 24 hours.

After 48 hours (Fig. 8-14), depending on date within the launch window, the range would be between 660,000 and 840,000 km for a type I trajectory. Because of the logarithmic character of space loss vs. time, the span of the signal strength due to range differences for these four cases at 48 hours is approximately 2 dB, as shown in Fig. 8-15.

Like other deep-space missions, MSL began its flight under tracking station strong-signal constraints that are unique to the initial acquisition portion of the mission. To accommodate the 40-dB decrease in signal level that occurred during initial acquisition day. MSL used the following common uplink and downlink DSN configurations. (Section 8.3 includes a block diagram for a 34-m tracking station, showing the elements involved in these configurations.)

- Uplink: The station transmitter operated at 200 watts (W, normal is 18 kilowatts [kW]) for the first three passes (spanning about 24 hours total), then operated at the normal power level.
- Downlink: For the first pass only, the station microwave system was configured to receive the opposite polarization from that transmitted. For MSL, this meant the initial acquisition 34-m station received the

RCP downlink while configured for left circular polarization (LCP). Since the distance was so close and the signal so strong, polarization leakage still provided a strong enough signal to close the link.

From the project's press releases and status reports [8], the following are highlights of the launch and early cruise phases.

November 26, 2011. Liftoff of Curiosity from the Cape Canaveral Air Force Station aboard an Atlas V rocket occurred at 15:02 UTC (10:02 a.m. EST). The Atlas V initially lofted the spacecraft into Earth orbit and then, with a second burst from the vehicle's upper stage, pushed it out of Earth orbit into a 352-million-mile (567-million-kilometer) journey to Mars. Based on subsequent radio navigation data and trajectory determination, this launch produced one of the most accurate interplanetary injections ever.

December 1. The project postponed the early trajectory correction maneuver (TCM-1), to early January. That first of six planned course adjustments during the 254-day journey from Earth to Mars had originally been scheduled for 15 days after the Nov. 26 launch.

Prior to TCM-1, the spacecraft's initial trajectory had been deliberately planned and executed to miss Mars by about 35,000 miles (mi) (56,400 km). This precaution protected Mars from Earth's microbes, because the Centaur upper stage of the launch vehicle, which was not thoroughly cleaned the way the spacecraft was, left Earth on nearly the same trajectory as the spacecraft. This trajectory would miss Mars by about 38,000 mi (61,200 km).

January 11, 2012. Starting with TCM-1, trajectory correction maneuvers were planned to put the spacecraft on course and on timing to land at Mars' Gale Crater on Aug. 6, 2012, Universal Time.

Following TCM-1, if not subsequently refined by TCM-2, the trajectory would have put Curiosity about 3,000 mi (5,000 km) and 20 minutes away from entering Mars' atmosphere at the right place and time.

8.1.4 Cruise Phase

Similar to the MARS Exploration Rover[8] (MER) cruise stage described in Ref. 9 of this chapter and in chapter 7, the MSL cruise stage spun at 2 revolutions per minute (rpm) until shortly before entry into the Martian atmosphere. The cruise antennas (a medium gain antenna and a low gain antenna) were mounted with their boresights co-aligned with the

[8] The MER program included two rovers that launched in 2003. Spirit landed on Mars on January 4 and Opportunity on January 24, 2004. Spirit's end of mission was declared May 25, 2011 when the rover did not communicate with Earth after a Martian winter. Opportunity has continued to operate for more than 10 years on the Martian surface.

spacecraft –Z axis. The –Z axis was closely aligned with the spin axis of the spacecraft. In Fig. 8-3, the –Z axis is a line from bottom to top and in the plane of the drawing. Due to mass imbalances, the center of mass was slightly offset from the –Z axis, resulting in a small wobble as the spacecraft spun. Because of the spinning, the worst-case antenna gain around the axis of revolution faced Earth at least once every revolution (thus every 30 s); therefore that worst-case value was modeled for link prediction.

The major activities in the Cruise phase included checkout and maintenance of the spacecraft in its flight configuration, routine monitoring of spacecraft health and subsystem performance, characterization and calibration of the spacecraft and payload subsystems (and associated parameter updates), attitude maintenance turns; navigation activities for determining and correcting the vehicle's flight path (for example, trajectory correction maneuvers [TCMs]), and preparation for EDL and surface operations. The three forms of navigation data that involved the telecom links were:

1) Two-way Doppler, provided whenever the spacecraft receiver was in coherent mode and in lock with an uplink carrier, and the downlink carrier was in lock in the station receiver.

2) Turnaround ranging, provided with the uplink and downlink carriers in the two-way Doppler mode and when the spacecraft ranging channel was on and the uplink carrier modulated with the ranging signal.

3) Delta differential one-way ranging (DOR) provided when the spacecraft transmitted a one-way downlink and the spacecraft DOR tones were sequenced on. Two tracking stations would, in coordination, each alternately track the spacecraft and a quasar so as to fix the angular location of the spacecraft relative to the quasar.

The MSL mission planned for as many as six TCMs, with the expectation the last few would be cancelled if the trajectory remained good for entry without them (see Table 8-5 for more detailed TCM information).

The propulsion system was designed to execute axial and lateral propulsive velocity corrections in the spacecraft reference frame. A vector mode maneuver is one that combines the axial and lateral segments so that the vector sum produces the desired inertial change in velocity (the "delta V") in magnitude and direction. This is a powerful maneuver-implementation mode that spinning spacecraft such as MSL could accomplish without executing a turn. A no-turn vector mode maneuver reduces operational risk by eliminating the estimation and control of a new attitude with potentially unknown characteristics. Additionally, the existing attitude was part of the nominal

Table 8-5. Trajectory correction maneuvers planned during Cruise.

TCM	Time	OD Data Cutoff*	Description
TCM-1	L + 15 days	L + 10 days	Corrected injection errors; removed part of injection bias for planetary protection; partial retargeting to entry aimpoint for desired landing site; aimpoint biased for planetary protection. This TCM was postponed to Jan. 11, 2012 [8] TCM-1 took 3 hours. It first made a 19-minute velocity change in the direction of the axis of rotation, then more than 200 five-second timed bursts to achieve a velocity change perpendicular to that axis.
TCM-2	L + 120 days	L + 115 days	Corrected TCM-1 errors; remove part of injection bias for planetary protection; partial retargeting to entry aimpoint for desired landing site; aimpoint biased for planetary protection; vector-mode maneuver. Executed March 26, 2012. TCM-7 was 1/7 as large as TCM-1, with 3-minutes of thrust in the direction of the axis of rotation, followed by more than 60 five-second timed bursts for a velocity change perpendicular to that axis.
TCM-3	E – 60 days	E – 65 days	Corrected TCM-2 errors; target to entry aimpoint for desired landing site; vector-mode maneuver executed June 26, 2012. Four thruster firings totaling 40 seconds, to move the atmospheric entry point by 125 mi (201 km) and to advance the time of entry by about 70 seconds.
TCM-4	E – 8 days	E – 8.5 days	Corrected TCM-3 errors; vector-mode maneuver, planned prior to launch and executed on July 29, 2012. Two thruster firings totaling 6 seconds, to move the atmospheric entry point by 13 mi (21 km).
TCM-5	E – 2 days	E – 2.5 days	To correct TCM-4 errors; final entry targeting maneuver required to achieve EFPA delivery accuracy requirement, vector-mode maneuver. Cancelled, not required [8].
TCM-5X	E – 1 days	E – 1.5 days	Contingency maneuver for failure to execute TCM-5; vector mode maneuver. Because TCM-5 was cancelled, TCM-5X was (also) cancelled.
TCM-6	E – 9 hours	E – 14 hours	Contingency maneuver; final opportunity to target entry aimpoint; vector-mode maneuver. Cancelled, not required.

*Time measured from launch (L) or entry (E); OD = orbit determination; EFPA = entry flight path angle.

plan and well characterized; it provided adequate spacecraft power, and supported ground communication. The downside to vector mode maneuvers, though not a problem for MSL, is mainly higher propellant costs, especially for large "delta V" corrections [10].

8.1.5 Approach Phase

The Approach phase was defined to begin 60 days prior to entry into the Martian atmosphere and to end when the spacecraft reached the atmospheric entry interface point, defined as a radius of 3522.2 km from the center of Mars. The principal activities during the Approach phase included the acquisition and processing of navigation data needed to make decisions on the need for the final three TCMs (and to support their development if any of them had been required) and the spacecraft activities leading up to the separation from the cruise stage and start of EDL.

From a Telecom point of view, Approach was considered just late cruise in terms of range, antennas, link performance, etc. Cruise stage configurations and station configurations continued in use.

From the project's press releases and status reports [8], the following were highlights of the cruise and approach phases of the mission.

March 26, 2012. Halfway to Mars, the spacecraft adjusted its flight path for delivery of Curiosity to the surface of Mars in August. The spacecraft ignited thrusters for TCM-2 for nearly nine minutes, nudging the spacecraft one-seventh as much as TCM-1. Spacecraft data and Doppler-effect changes in radio signal from the craft indicate the TCM-2 maneuver succeeded.

June 26. TCM-3 executed, with four thruster firings totaling just 40 seconds. The maneuver served both to correct errors in the flight path that remained after earlier correction maneuvers and to carry out a decision that month to shift the landing target about 4 mi (7 km) closer to the mountain, informally named Mount Sharp. Shifting the landing target shaved months off the time needed for driving from the touchdown location to selected destinations.

TCM-3 altered the spacecraft's velocity by about one-tenth of a mile per hour (mph) (50 millimeters per second [mm/s]). The flight's first and second trajectory correction maneuvers produced velocity changes about 150 times larger on Jan. 11 and about 20 times larger on March 26, respectively.

July 11. The spacecraft completed an attitude control turn (not a TCM), adjusting its orientation for keeping its medium-gain antenna pointed toward Earth for communications. This was the third-to-last attitude control turn planned before landing day.

July 28. TCM-4 altered the flight path less than any of the spacecraft's three previous trajectory correction maneuvers on the way from Earth to Mars. Without this maneuver, the spacecraft would have hit a point at the top of the Martian atmosphere about 13 mi (21 km) east of the target entry point. The thruster firings altered the spacecraft's velocity by about one-fortieth of 1 mph (1 centimeter per second [cm/s]).

8.1.6 EDL Phase

Following a 5-day final approach, Entry Descent and Landing (EDL) was divided into three stages, lasting a total of 21 minutes. Figures 8-16 through 8-20 are pictorials of the various stages of EDL.

- First stage, Fig. 8-16 (15 minutes ending at entry), included:
  - EDL start, ending with cruise stage separation.
  - Exo-atmospheric, ending with switch to the tilted low-gain antenna (TLGA), bypassing the descent stage diplexer[9], and reaching the reference entry interface.
- Second stage, Fig. 8-17 (5 minutes beginning at entry), included:
  - Entry, including a period of potential UHF blackout due to plasma generation.
  - Parachute descent, including heat shield separation and using the landing radar, formally called the terminal descent sensor (TDS) and described in Section 8.2.
- Third stage, Fig. 8-18 (less than 1 minute starting at entry plus 309 s and ending at touchdown), included:
  - Powered descent phase, including backshell separation with switch to X-band descent low gain antenna (DLGA) and the descent UHF (DUHF) antennas.
  - Sky crane phase (during which the descent stage acted as a sky crane to lower the rover shortly before it landed), including touchdown and electrical bridle cut.
  - Flyaway of the sky crane.

Figures 8-16 and 8-17 show, respectively, the spacecraft events connected with the first two stages. Figure 8-19 shows the final stage, and Fig. 8-20 provides an overview of the entire process.

---

[9] The EDL sequence bypassed the descent stage diplexer to prevent coronal discharge within the diplexer during the repressurization by the atmosphere during entry. Since receive capability was no longer required during entry, diplexer bypass had no effect on meeting telecommunications requirements.

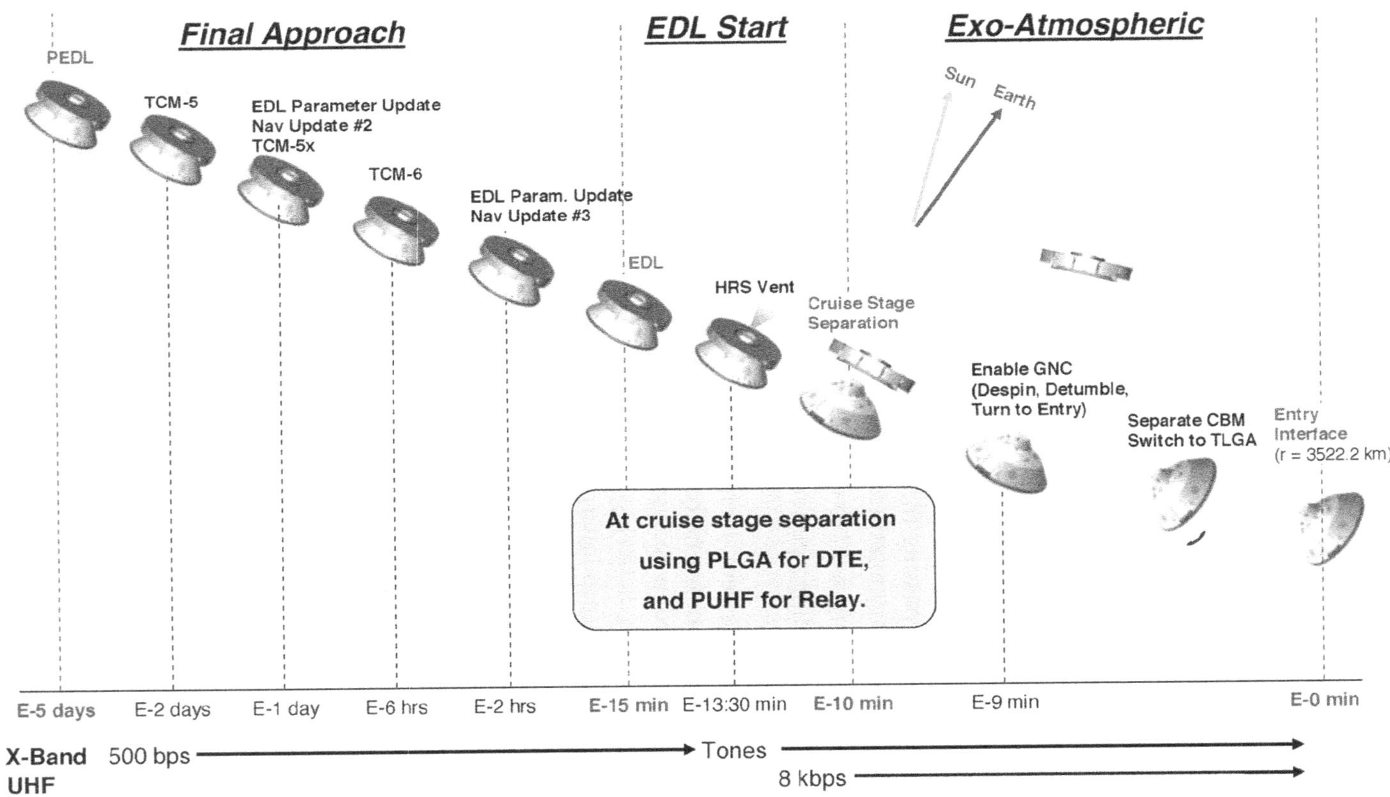

Fig. 8-16. Timeline: Cruise Stage separation to entry interface (CBM = cruise balance mass; GC = guidance, navigation, and control; HRS = heat rejection system).

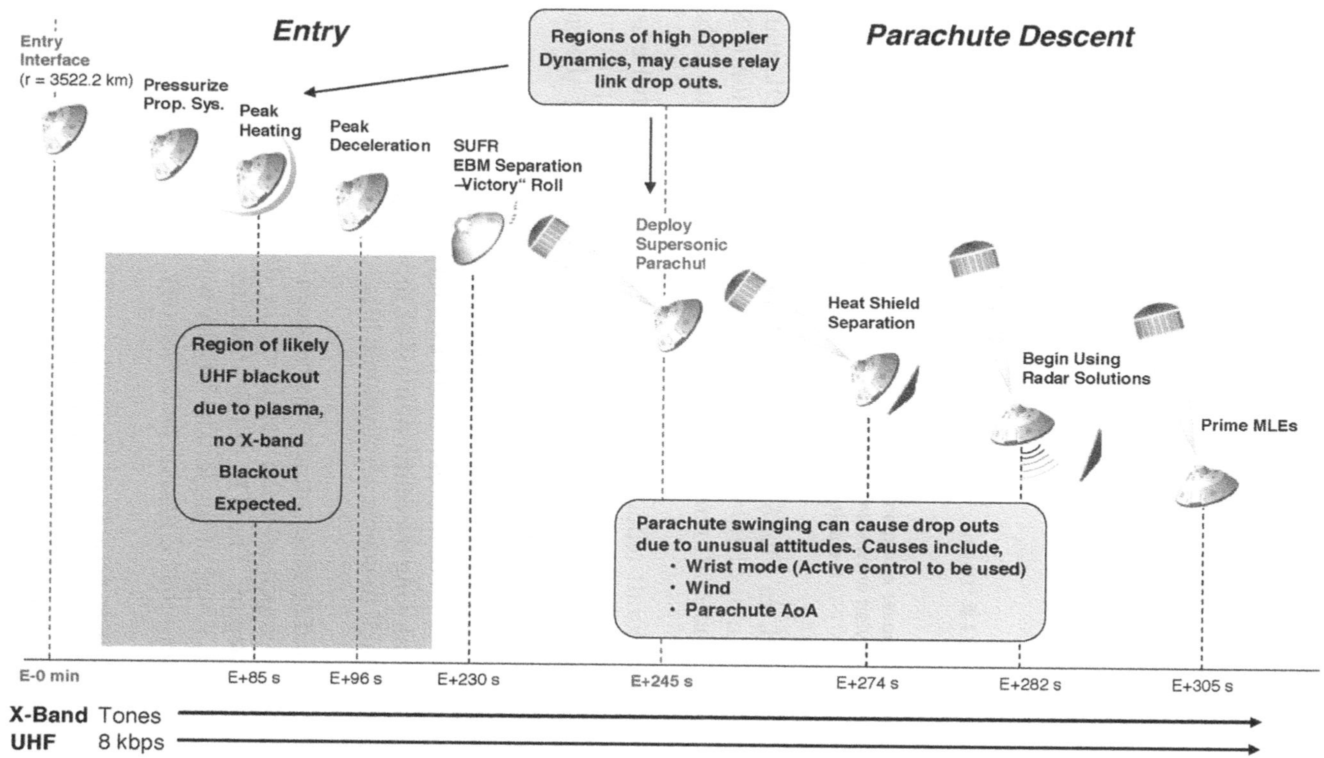

Fig. 8-17. EDL timeline: entry interface to backshell separation (EBM = entry balance mass, SUFR = straighten up and fly right).

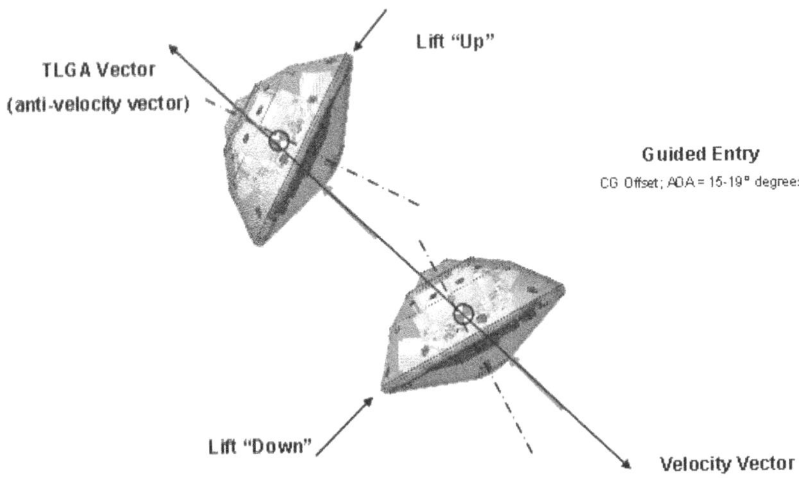

Fig. 8-18. Banking maneuver along anti-velocity vector (AOA = angle of attack, CG = center of gravity). Please visit www.wiley.com/go/deepspacecommunications to access the color version of this figure.

The pre-EDL (PEDL in Fig. 8-16) spacecraft maneuvers for EDL started approximately 10 minutes before Entry, with cruise stage separation (CSS). Unlike MER, the spacecraft remained pointed toward Earth during CSS.

After CSS, the spacecraft reduced its spin rate from the 2 rpm that existed throughout cruise. After approximately 1 minute, the spacecraft performed the turn to entry (TTE) maneuver. At this point, the attitude of the entry body was no longer optimal for Earth communications.

Prior to the spacecraft reaching the entry interface point, the EDL sequence separated the cruise balance masses (CBM), as shown in the second from last sketch in Fig. 8-16, moving the center of mass of the entry body to induce the proper angle of attack to enable aerodynamic lift.

The nominal planned angle of attack was between 15 and 19 deg. The angle of attack varied as much as ± 2.5 deg from the design value until shortly before parachute deploy. The TLGA's beamwidth and mounting direction accommodated this range of angle. The TLGA was mounted with its boresight at 17.5 deg from the –Z axis of the spacecraft.

After entry, the vehicle performed banking maneuvers to achieve a smaller landing ellipse (guided entry). These banking maneuvers used repeated short (about 20 millisecond) firings of the reaction control system (RCS) thrusters. As shown in Fig. 8-18, where Mars is at the bottom, a "lift up" maneuver was included to tilt the heat shield slightly above the velocity vector and a "lift down" maneuver to tilt it slightly below. During this time, the angle of attack (AoA) remained in the range of 15 to 19 deg. The MSL TLGA was

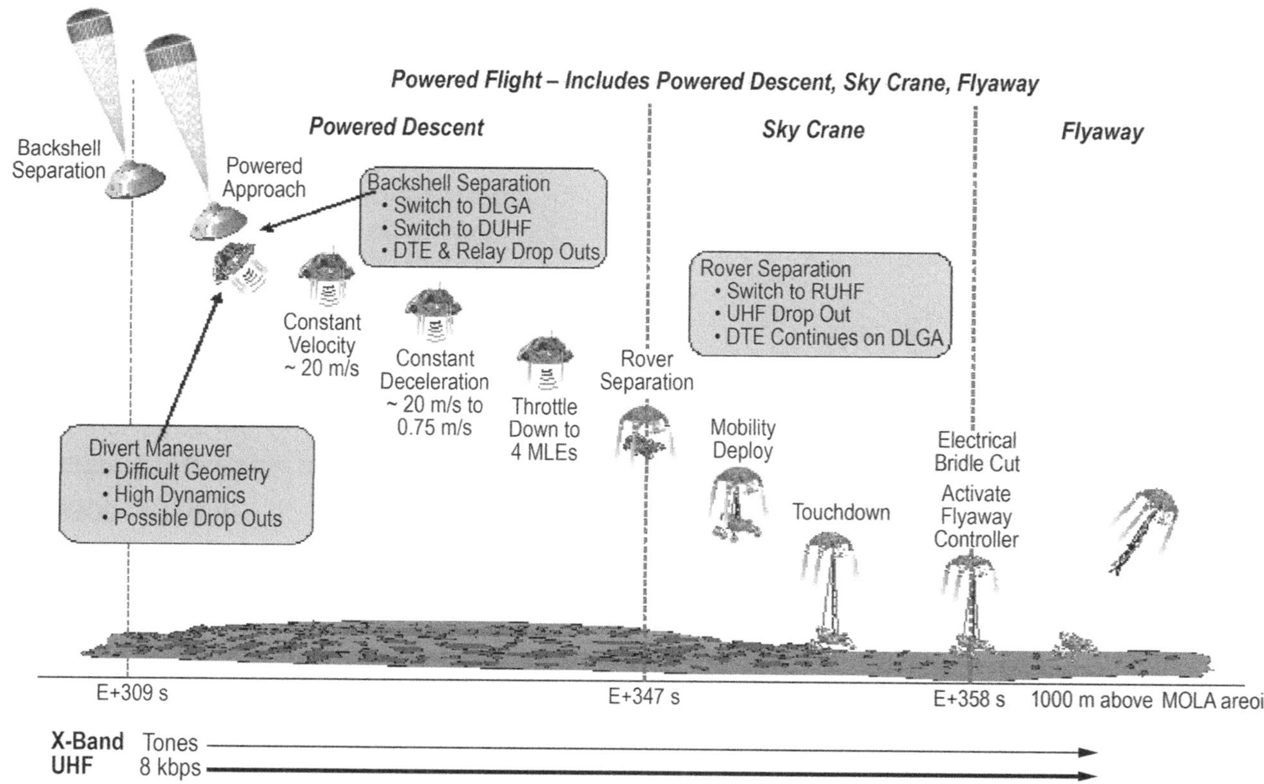

Fig. 8-19. EDL timeline: backshell separation to fly-away (MOLA = Mars Observer Laser Altimeter).

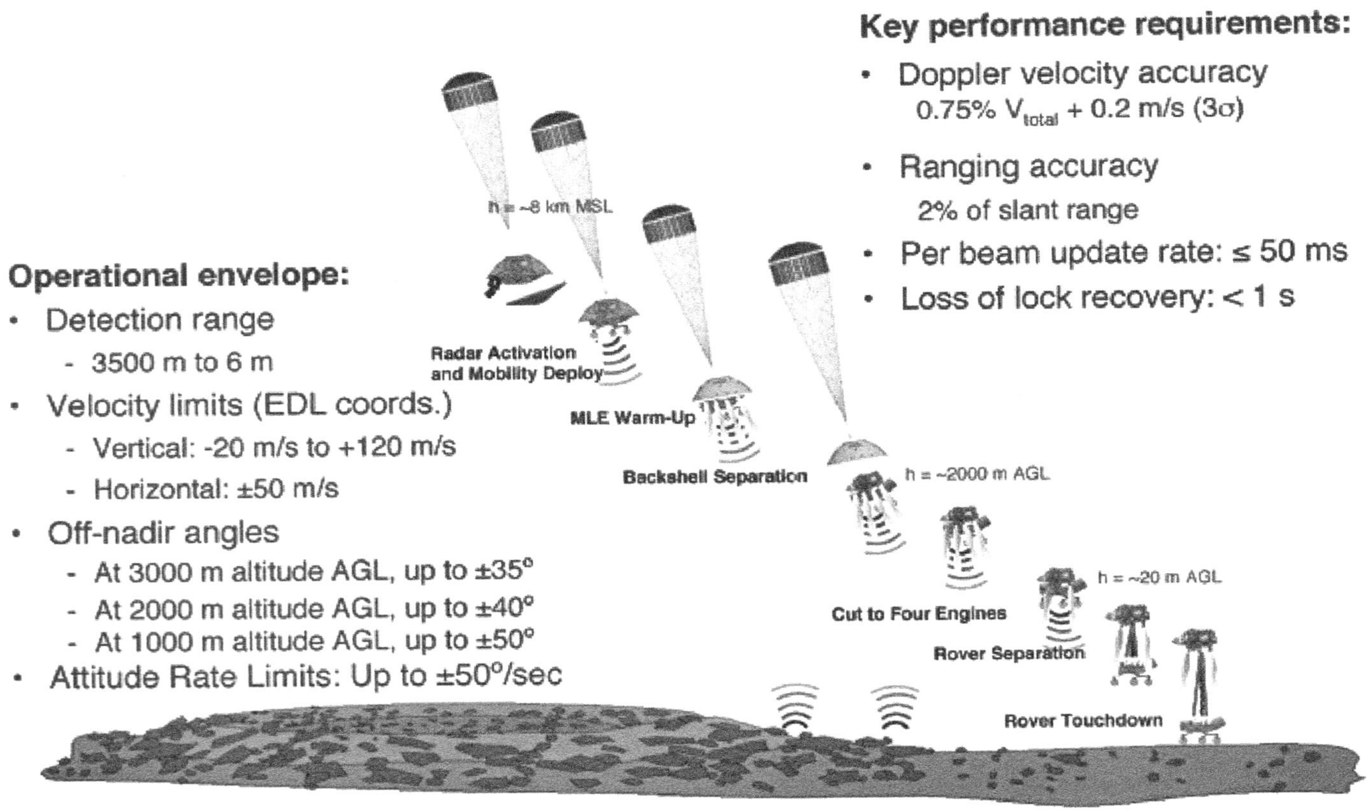

Fig. 8-20. Overview of TDS requirements during EDL (AGL = above ground level).

nominally aligned with the anti-velocity vector to minimize variations of off-boresight angles during banking. This alignment resulted in variations in off-boresight angle to be roughly equal on either side of the antenna's boresight, rather than skewed to one side of boresight.

During the period of peak heating (Fig. 8-17, centered at ~ E +85 s), it was also expected that the UHF signal could experience a dropout due to plasma shielding (see Section 8.4). The actual blackout went from about entry plus 40 s until entry plus 80 s, which was similar to MER. Next came the period of peak deceleration that caused the mission's most challenging Doppler dynamics for signal acquisition and tracking. From final approach and EDL status reports [8]:

August 1, 2012. Curiosity began flying under the control of the autonomous EDL timeline. "Those seven minutes are the most challenging part of this entire mission," said Pete Theisinger, the mission's project manager at JPL. "For the landing to succeed, hundreds of events will need to go right, many with split-second timing and all controlled autonomously by the spacecraft. We've done all we can think of to succeed. We expect to get Curiosity safely onto the ground, but there is no guarantee. The risks are real."

The flight team cancelled the build and test of the contingency version of Trajectory Correction Maneuver 5. This contingency maneuver, had it been needed, would have been used in the event an emergency prevented the team from executing the nominal scheduled TCM-5 maneuver, which was planned for Aug. 3, if needed. The project also canceled a corresponding update to parameters for the autonomous software controlling events during entry, descent and landing.

August 3. The project decided that the planned Trajectory Correction Maneuver 5 (TCM-5) and its corresponding update to parameters for the autonomous software controlling events during EDL would not be necessary. As of 19:35 UTC, the spacecraft was approximately 468,000 mi (753,200 km) from Mars, or a little less than twice the distance from Earth to the Moon. It was traveling at about 8,000 mph (3,576 meters per second (m/s). It would gradually increase in speed to about 13,200 mph (5900 m/s) by the time it reached the top of the Martian atmosphere.

August 4. The flight team uplinked and confirmed commands to make minor corrections to the spacecraft's navigation reference point parameters. As part of the onboard sequence of autonomous activities leading to the landing, catalyst bed heaters were turned on to prepare the eight Mars Lander Engines (MLEs) that were part of MSL's descent propulsion system. Also, parameters on a motion tracker were adjusted for fine-tuning determination of the spacecraft's orientation during its descent.

August 5. Flight controllers decided to forgo the sixth and final opportunity on the mission calendar for a course-correction maneuver. The spacecraft was headed for its target entry point at the top of Mars' atmosphere precisely enough that the maneuver was deemed unnecessary. Mission controllers also determined that no further updates would be necessary to the onboard information the spacecraft would use during EDL.

August 6. Touchdown in Mars' Gale Crater was confirmed to be 05:32 UTC (August 5 at 10:32 p.m. at the mission control center at JPL.) The time of day at the landing site was mid-afternoon—about 3 p.m. local Mars time at Gale Crater. Initial information received via the UHF relay links with Odyssey and MRO was that Curiosity landed facing east-southeast within Gale Crater, with a heading of 112.7 deg (±5 deg), and a few degrees of tilt. A first look at some color images taken just before landing by MSL's Mars Descent Imager provided this information on the rover's precise location. A Sol 1 overpass a few hours after EDL by Mars Odyssey provided additional information on Curiosity's position and additional imagery.

Figure 8-19 shows the last portion of EDL, starting with parachute deployment and followed by heat shield separation, radar activation, and backshell separation (BSS). At this point, the vehicle started its powered descent phase, during which the descent stage, acting as a sky-crane, lowered the rover using a bridle approximately 7 m long. After rover touchdown, the descent stage executed a fly-away to set down sufficiently far from the rover landing site to avoid the possibility of damage to the rover or contamination of the area around the rover landing site. Because the descent stage's "landing" was uncontrolled in attitude and velocity – destroying the stage, there was no expectation of further DTE X-band radiation being received from the DSDST and the TWTA.

While slowing on the parachute, the system prepared the propulsion subsystem, separated the heat shield, and began using the TDS to acquire a landing solution. (Note: A change from the 2009 mission plan was that the mobility deployment of the wheels was sequenced to occur during the sky crane maneuver in the final seconds before touchdown).

To achieve a soft controlled landing, the EDL system had to be able to accurately measure altitude and three-axis velocity (that is, horizontal and vertical velocity) beginning at several kilometers and continuing all the way down to a few meters above the surface. The TDS was designed to provide these measurements starting at heat shield jettison all the way to rover touchdown.

8.1.6.1 DTE Prime to Entry; Both DTE and Relay via MRO after Entry. Planning for communications during EDL had to be complete before landing site selection or knowledge of the specific launch date. Communications design and use had to be sufficient to allow a reasonable chance to determine what happened (using data reconstruction) in the event of an EDL failure. These mandates had been in place by Headquarters at the National Aeronautics and Space Administration (NASA) since the Mars Polar Lander failure of 1998.

Due to limitations on the range and view angles to the Mars orbiters ("relay assets"), relay during MSL EDL cannot cover the entire time from CSS to landing + 1 minute. Since the majority of important events would occur after entry, the preparation for EDL included phasing of the orbiters in their orbits to optimize a relay link from entry to landing + 1 minute. The mission design baseline was that X-band DTE should provide the coverage during the period from CSS to entry, a time when, by comparison to the after-entry period, relatively few events occur. Nevertheless DTE coverage from CSS to landing + 1 minute is desirable. The period of maximum heating was expected to create a plasma envelope sufficient to produce a link dropout at UHF relay frequencies (but not at X-band), making DTE via a plasma-penetrating frequency highly desirable during this period.

In addition to the plasma blackout period, there were several short intervals during EDL when link dropouts for both relay and DTE were accounted for in the planning. Table 8-6 lists the most important dropout periods, the cause for each dropout, and the dropout's expected and actual duration.

Table 8-6. Periods of UHF or DTE dropouts between CSS and landing [11].

Event	RF link	Cause	LOS start relative to Entry (E), actual	LOS duration, used in EDL planning	Actual duration seen in EDL
CSS	X-band	Switch to PLGA and blockage by CS	E – 600 s, sequenced	1 s	7 s (including PLGA blockage)
Turn to Entry	X-band	Switch to TLGA	E – 20 s, sequenced	1 s	4 s
Plasma Blackout	UHF	Surrounding plasma envelope	E + 40 s	25 s to 100 s	40 s (carrier), 58 s (data)
Parachute Deploy	X-band	TLGA blockage due to Sabot	E + 260 s	75 s, worst case	No dropout, but high Doppler
Backshell Deploy	Both	Change to DUHF and DLGA, blockage	E + 375 s	1 s to 6 s	6 s (UHF Txr off)
Rover Separation	UHF	Change to RUHF, blockage	E + 410 s	1 s to 6 s	6 s (UHF Txr off)

The "parachute deploy" row of the table mentions a temporary blockage of the X-band DTE from the parachute deployment sabot.[10] Figure 8-21 shows the MSL sabot in action during deployment of a parachute mass model from its canister during a ground test. The sabot is the device about a third of the way from the bottom of the picture. It is partially wrapped in a capture bag, a loose web of straps designed to keep it contained. The parachute mass model is at the top of the picture. Testing verified the movement of the sabot during parachute deployment to be quite violent, and there was a concern that when the sabot fell back to the top of the parachute canister it could impact and damage the PLGA. At backshell deployment, a steep-angled and very dynamic maneuver was performed by the powered descent vehicle to "divert" the spacecraft away from the parachute and backshell. This maneuver briefly created high Doppler rates and unusual attitudes that could cause a dropout in addition to the blockage and UHF antenna switch shown in the table.

8.1.6.2 Full-DTE EDL Communications. As discussed previously, full-DTE EDL communications is considered desirable as backup to the relay link after entry until landing + 1 minute. This so-called full-DTE coverage could not be guaranteed for both trajectory types and all landing sites. While the type II trajectory did offer full-DTE coverage for the four primary landing sites, only the northern landing site (Mawrth Vallis at 24 deg N) had full-DTE coverage for the type I trajectory.

The calculated end of DTE for the potential landing sites and cruise trajectory types and launch dates varied considerably, as early as a couple of minutes after entry to nearly as late as landing + 1 minute. For expected (nominal) communications performance and the actual type I launch on November 26, 2011, with a landing at Gale Crater on August 6, 2012, the X-band DTE loss of signal (LOS) was planned to be 2 minutes before touchdown, MRO UHF relay LOS to be 6 minutes after touchdown, and Odyssey LOS to be one minute after that.

The project relied on delay-free "bent pipe" UHF relay via the Mars Odyssey orbiter to provide immediate confirmation of a successful landing. Odyssey executed a turn to point in the right direction beforehand to listen to Curiosity during the landing. Without this Odyssey relay, a successful landing could not have been confirmed until more than 2 hours later from playback of MRO-recorded relay data from MSL.

---

[10] The term sabot referred originally to a device used in a firearm to hurl a projectile, such as a bullet, that is smaller than the bore diameter, or requires the projectile to be held in a precise position.

Fig. 8-21. Sabot, shown during ground parachute deployment test.

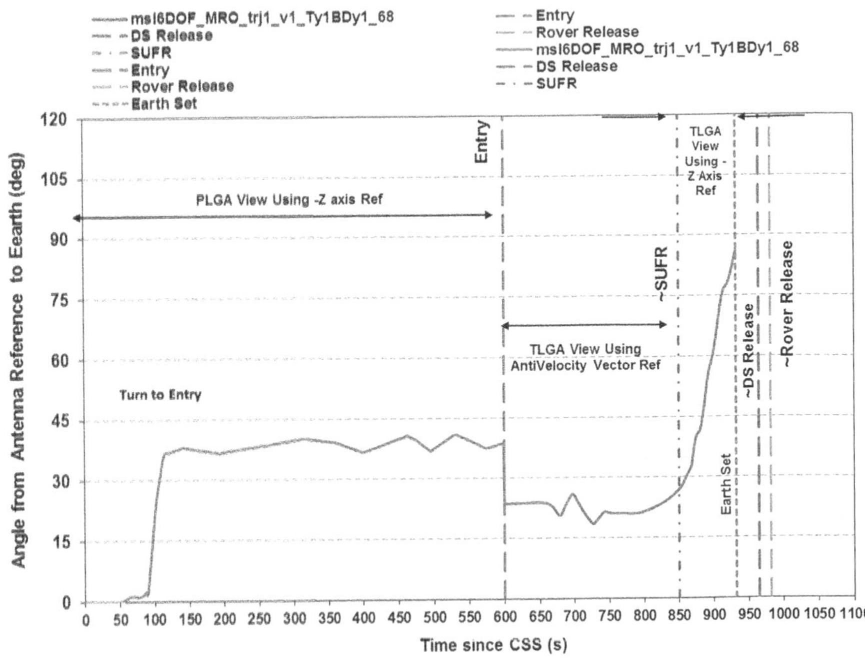

Fig. 8-22. Example DTE view angle (combination of anti-velocity and Z axis) during EDL (2011 launch data, Gale site). Please visit www.wiley.com/go/deepspacecommunications to access the color version of this figure.

The baseline sequence for using the X-band EDL antennas was to start with the PLGA (which points along the spacecraft –Z axis from CSS to entry), and then the TLGA (which points along the average tilt angle and, therefore, close to the anti-velocity vector after turn to entry) at entry. Figure 8-22 illustrates the view angles for the antennas during EDL to the Gale Crater landing site (at 4.5 deg S), for the 2011 launch. A plot of the –Z axis views would show considerably greater variation after entry due to the guided entry maneuvers (banking turns); however, using the TLGA minimized the effects of these variations in angle.

8.1.6.3 Surface Phase. The minimum surface operations duration requirement for a successful MSL mission is one Martian year (669 Martian sols or 687 24-hr Earth days). The Earth geometry during the surface mission at Gale Crater is shown in Fig. 8-23. The Mars-Earth range (solid curve, left axis) varies between 0.6 and 2.4 astronomical units (AU) (0.9 and 3.6×10^8 km, respectively). MSL landed at an Earth–Mars range of 1.7 AU (2.5×10^8 km).

Figures 8-24, 8-25, and 8-26 were developed to show the surface operational environment as a function of landing site latitude in 15-degree increments from 30 deg N to 30 deg S. Gale Crater is 4.5 deg S, thus well approximated by the solid line labeled Equator. Fig. 8-24 shows how the Earth rise time

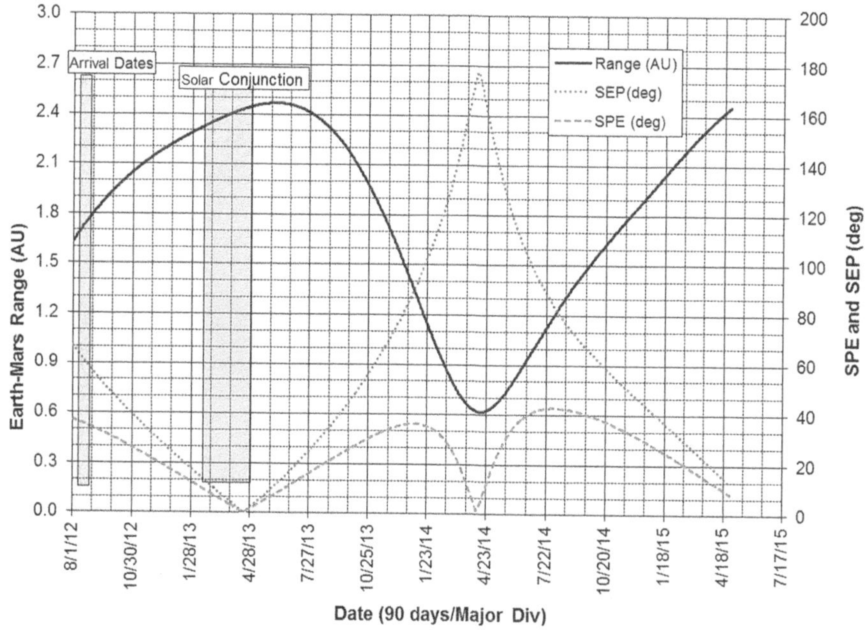

Fig. 8-23. MSL geometry during surface operations.

varies by approximately 5 hours during 1 Martian year. Fig. 8-25 shows that the Earth set-time has a similar variation. The rise and set times are in Mars Local True Solar Time (LTST) [12].[11]

The duration per day that the Earth is visible (assuming a 10-deg land mask and zero rover tilt) is shown in Fig. 8-26. This duration (8 to 13 hours) is for the Earth, not for any one DSN site on the rotating Earth. Disregarding any constraints in DSN scheduling and uplink/downlink operations, the figure shows the maximum possible time per day available to uplink commands (via a direct from Earth [DFE]) link and to get telemetry data (via a DTE link).

---

[11] Because the orbit of Mars around the Sun is not perfectly circular and the planet does not rotate about an axis perpendicular to its orbit plane, there is a seasonally variable discrepancy between the even advance of an artificially defined mean solar time and of the true solar time corresponding to the actual planet-centered position of the Sun in its sky. By analogy with the 24 time zones on Earth, Mars Mean Solar time [12] is defined on the Mars prime meridian as Mars Time Coordinated, or MTC, by analogy to the terrestrial UTC (Universal Time Coordinated). Again by analogy with the Earth, local solar time at the selected location is defined in terms of similarly constructed "Mars time zones". These Mars time zones are exactly 15 deg wide and centered on successive 15-deg multiples of longitude, at 0 deg, 15 deg, 30 deg, etc.

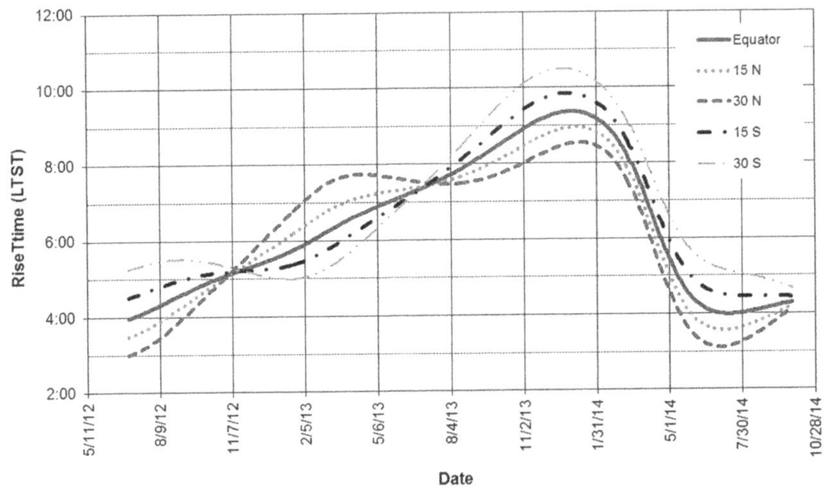

Fig. 8-24. Earth rise vs. date, for different MSL landing sites.

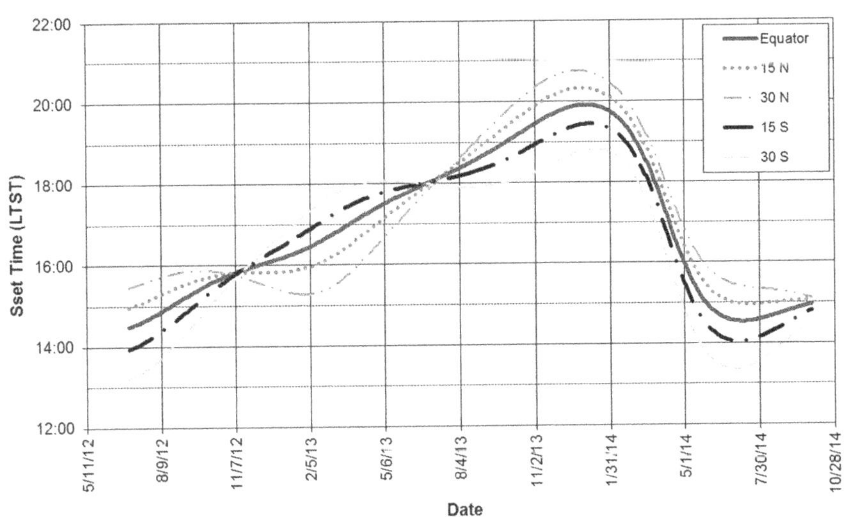

Fig. 8-25. Earth rise time for different MSL landing latitudes.

MSL relays UHF signals with the Mars Reconnaissance Orbiter (MRO) and Odyssey. Odyssey (launched in 2001) and MRO (launched in 2005) are referred to as the Mars Relay assets available during the MSL surface mission. Table 8-7 summarizes the classical orbital elements for the relay orbiters[12].

[12] See Wikipedia: Orbital elements, for definitions of Orbital Elements as used in celestial mechanics. Six parameters (Keplerian elements) are necessary to unambiguously define an arbitrary and unperturbed orbit. In Table 8-7, the epoch defines a reference time when the sixth parameter, the mean anomaly (the position of the orbiter along its ellipse) is equal to zero.

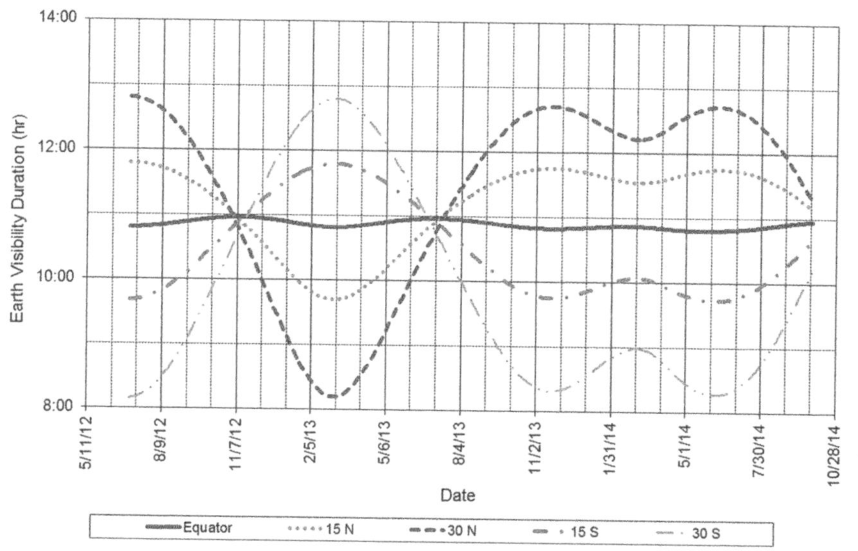

Fig. 8-26. Earth visibility duration vs. date (MSL surface mission, assumes 10-deg land mask and no lander tilt).

Table 8-7. Orbital elements for relay orbiters.

Parameter	Odyssey	MRO
Semi-Major Axis (km)	3788.1479	3648.606
Eccentricity	0.0108616	0.012176
Inclination (deg)	92.894	92.655
Longitude of Ascending Node	235.4908	−10.695
Argument of Periapsis (deg)	267.5309	−90.003
Epoch	01-Jan-2006 00:01:00	07-Dec-2006 01:00:00

Figure 8-27 shows the amount of time per sol that MRO is above an elevation of 10 deg vs. the latitude of the rover's landing site. With MSL near the Martian Equator as it is for Gale Crater, the time MRO is above 10-deg elevation is approximately 13 minutes total, divided between two passes per sol.

Figure 8-28 gives the breakdown of visibility time for each 10-deg elevation step. For a landing site between 45 deg S and 45 deg N (which includes all four candidate sites), the percentage of time is approximately independent of latitude. It is seen that more than half the time is spent between 10 deg and 20 deg elevation and that the time spent progressively decreases until it is just one quarter of one percent for elevations between 80 and 90 deg. This implies that, on average, a broader antenna pattern, which provides significant coverage at low elevation angles, is advantageous, while high gain in the zenith direction is generally not useful.

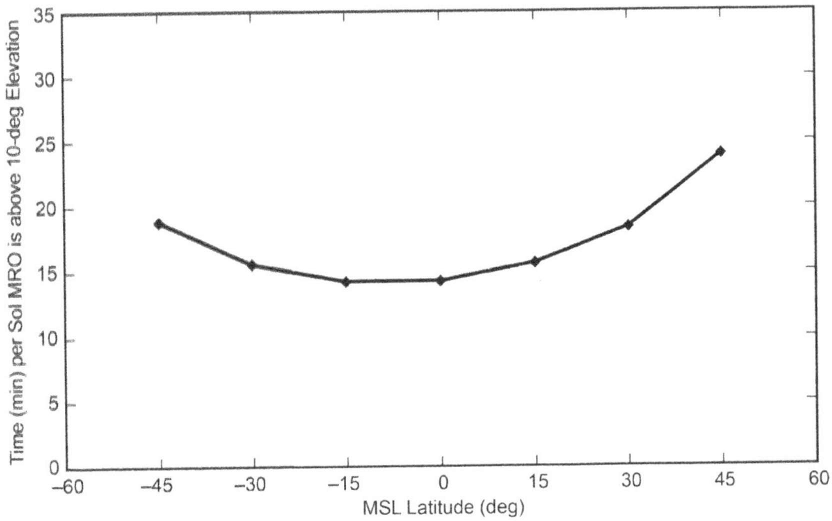

Fig. 8-27. Time (min) per sol MRO is above 10 deg elevation vs. MSL latitude.

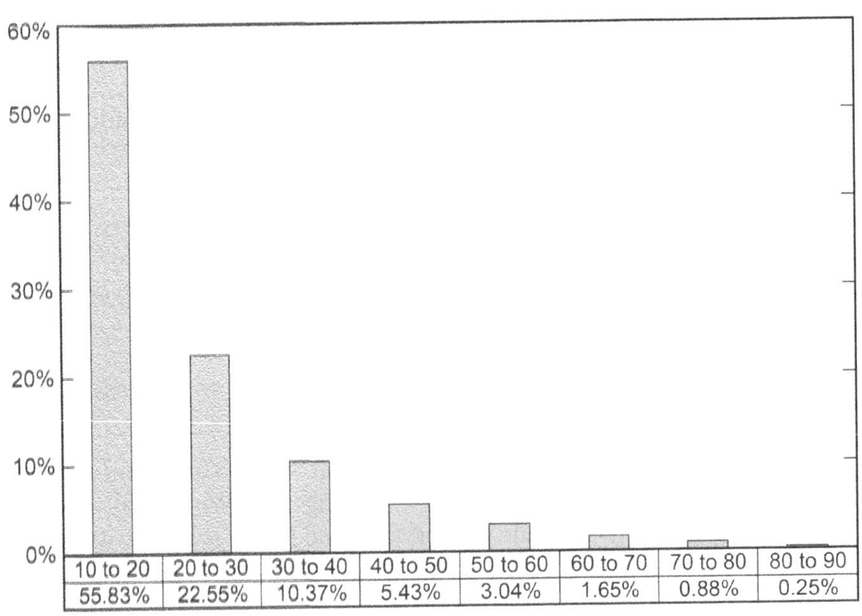

Fig. 8-28. Percentage of time spent by MRO between 10-deg steps in elevation.

8.1.7 Flight System Description

The flight system (Fig. 8-3) consists of an Earth–Mars cruise spacecraft, entry, descent, and-landing (EDL) system, and a mobile science rover with an integrated instrument package. In the figure, the MSL EDL instrumentation (MEDLI) [13] is an instrumentation suite installed in the heat shield of the

Mars Science Laboratory's (MSL) Entry Vehicle to gather data on the atmosphere and on aerothermal, Thermal Protection System (TPS), and aerodynamic characteristics of the MSL Entry Vehicle during entry and descent providing engineering data for future Mars missions.

During Cruise, the spacecraft was spin-stabilized (2 rpm); with the spin about the Z axis. In Fig. 8-3, the –Z axis is in the plane of the figure, extending at the top center of the figure.

The EDL system consisted of a mid-1970s Viking-derived aeroshell structure and propulsion system for a precision guided entry and soft landing. The soft landing was new for MSL and contrasted with the airbag designs used by the mid-1990s Mars Pathfinder (MPF) mission [14][13] and the early 2000s MER mission. Figure 8-29 is a view of the MSL rover in its fully deployed configuration.

Table 8-8 provides a top-level comparison of MSL and MER. MER telecommunication, including telecom subsystem mass and power draw, is described in Ref. 9. The MSL telecom subsystem, including mass and power draw, is described in Section 8.2. The SA-SPaH Bit Box contains drill bits.

8.1.7.1 Engineering Subsystems and Functions. The following description of the engineering subsystems related to telecom is from the MSL Mission Plan [15].

Chassis. The chassis is the central part of the rover, accommodating the flight system elements. In addition to the structural integration of components, the structural system serves as a significant part of the thermal control of the vehicle, with insulation, thermal coupling of the payload mounting panel and the Multi-Mission Radioisotope Thermoelectric Generator (MMRTG) heat exchangers. The chassis core forms the shell of the warm electronics box (WEB); panels internal to this are used to mount the flight system avionics and payload avionics. Similar to MER, the active telecom components (SDST, SSPA, and Electra UHF radios) are in the WEB.

Mobility. The rover is a scaled-up version of the six-wheel drive, four-wheel steering system from MER, utilizing the rocker-bogie configuration. Based on the center of mass, the vehicle is required to withstand a tilt of at least 50 deg in any direction without overturning. During descent, this number

---

[13] Mars Pathfinder, launched in December 1996 and landed July 1997, consisted of a lander and the Sojourner rover. MPF was originally designed as a technology demonstration of a way to deliver an instrumented lander and a free-ranging robotic rover to the surface. It accomplished this goal. Both vehicles outlived their design lives on Mars, the Pathfinder lander by nearly three times and the rover by 12 times.

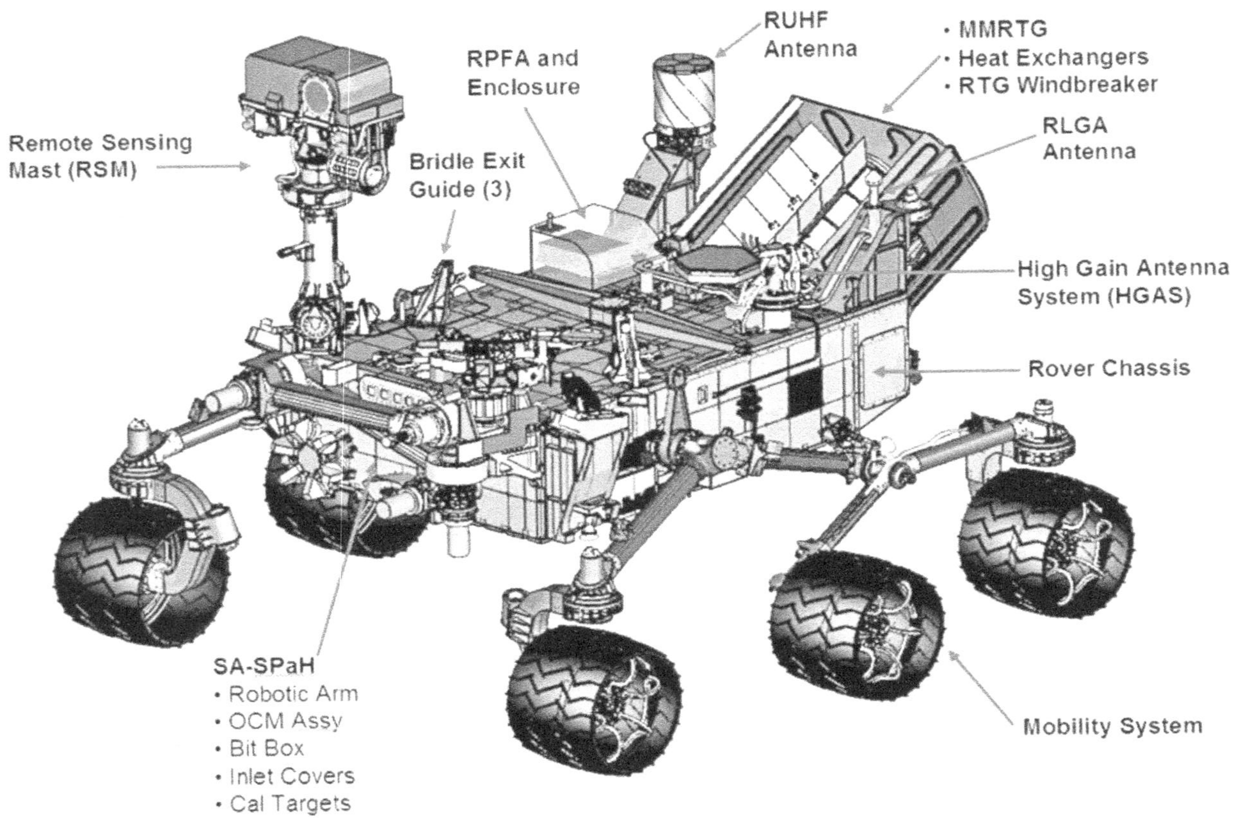

Fig. 8-29. External view of the MSL rover (OCM = organic check material, SA-SPaH = sample acquisition-sample preparation and handling). Please visit www.wiley.com/go/deepspacecommunications to access the color version of this figure.

Table 8-8. Comparison between MSL and MER.

	MSL	MER
LV/Launch Mass	Atlas V/4000 kg	Delta II/1050 kg
Design Mission Life	1 yr maximum cruise (actual cruise took 9 months) /1 Martian year surface	7 month cruise/3 month surface
Telecom Redundancy	Significant telecom redundancy/Single Mission	Limited telecom redundancy/Dual Mission
	One rover. Telecom includes a small deep space transponder (SDST) and a TWTA on the Descent Stage and an SDST and a solid-state power amplifier (SSPA) on the rover (for X-Band). The rover also has two UHF radios.	2 rovers. Each rover has 1 SDST and 2 SSPAs (X-Band), and one UHF radio
Payload	10 instruments (84 kg)	5 instrument (~9 kg)
Sample Acquisition	Arm + rock abrasion tool (RAT) + Powdering Corer + Scoop	Arm + RAT
EDL System	Guided Entry/Sky Crane	MPF Heritage/Airbags
Heat-shield Diameter	4.5 m	2.65 m
EDL Comm	Partial UHF + partial DTE or full DTE	DTE + partial UHF
Rover Mass	850 kg (allocation)	170 kg (actual)
Rover Range	>20 km designed	1 km (design) and > 37 km (actual so far for Opportunity) and 7.7 km (for Spirit)
Surface Power	MM RTG/2500 Whr/sol	Solar/<900 Whr/sol
Surface Comm	X-Band DTE + UHF	X-Band DTE + UHF

was slightly higher, as the mast had not yet deployed. Fault protection limits the rover from exceeding 30-deg tilts. The design of the rocker-bogie allows traverse over (meaning the wheel moves over) objects approximately as large as the wheel diameter. Each wheel has cleats and is independently actuated and geared, providing for climbing in soft sand and scrambling over rocks. Each front and rear wheel can be independently steered, allowing the vehicle to turn in place as well as execute arcing turns.

Power (surface operations). Rover power is primarily provided by the MMRTG, which is required to generate a constant 110 W at the start of the prime mission (though the current best estimate predicted by the Department of Energy (DOE) was approximately 110 W at the start of surface phase), decaying to approximately 104 W at the end of the 1-Mars-year mission.

Peak power from the rover activities easily exceeds the MMRTG capability, however, and the rover has two 42 amp-hour (A-hr) batteries to allow for all activities. The 2009 launch requirements for performance of the battery included that it should provide as much as 555 W-hr per sol for 670 cycles, given a starting condition of 100 percent state of charge (SOC). Such a deep discharge (555 W-hr) was not intended to occur more than once per sol. In light of the 2009 to 2011 launch slip, and after review of the surface power situation, the 555 W-hr figure was deemed not sufficient. Consequently, among other Rover changes, was a larger battery capacity than the original 20 amp-hour batteries.

Guidance, Navigation, and Control (GNC). During cruise, EDL, and surface operations, the telecom antenna orientation relative to Earth depended on GNC. The MSL Cruise Guidance and Control Subsystem made extensive use of heritage hardware and flight software algorithms from the MER and MPF missions. Three-axis inertial attitude and spin rate were determined onboard in real time using an internally redundant star scanner and one of two 4-head Sun sensors. During cruise, the flight system was spin-stabilized about the spacecraft Z axis at 2 rpm. Eight thrusters arranged in two clusters were used as actuators to control spin rate, "turn" the spacecraft (by precessing the spacecraft spin axis), and perform axial or lateral trajectory correction maneuvers (TCMs).

From the perspective of navigation, the onset of EDL and the events leading up to it were of critical importance to mission success. The EDL system differed from those used for other mission functions in that it did not require an interactive, ground-generated mission plan. On MSL, the landing radar is book kept as part of GNC. During the entire phase, the vehicle acted autonomously, based on pre-loaded software and parameters. Late parameter updates to be commanded from Earth were in the plan but each update was cancelled when the most recent navigation data and telemetry determined it was not necessary.

Thermal Control (surface operations). The rover was to be capable of landing between 45 deg N and 45 deg S and, accordingly, the thermal control system was designed to accommodate a wide variety of climates and temperatures. Across these latitudes, Mars surface temperatures can reach as high as +40 deg C and as low as –127 deg C (the freezing point of carbon dioxide [CO_2] at Martian atmospheric pressures), with daily thermal cycles as great as 145 deg C. At any latitude, the thermal system could achieve a minimum of 6 hours per sol at which the rover avionics mounting plate (RAMP) and, specifically, the Payload Mounting Module (PMP) would stay at 20 deg C or higher. In extreme cases (for example, at latitudes beyond 30 deg, such as winter at 45 deg S), the RAMP temperature must be maintained above –40 deg C throughout the sol. In addition, the thermal design also limits the daily thermal cycle to ± 30 deg C. Over the majority of the Martian year, at any latitude, the thermal system will be warming the rover. The thermal system achieves this in several ways: passively, through the dissipation of heat from internal components; by electrical heaters strategically placed on key components; and by using the rover heat rejection system (HRS). The HRS is a set of redundant integrated pump assemblies and a fluid loop that runs throughout the WEB that serves to minimize thermal gradients across the rover. The fluid loop actually serves the additional purpose of rejecting heat when the rover has become too warm, but it also can gather waste heat from the MMRTG, by pumping fluid through two heat exchangers mounted alongside the MMRTG. (Because MSL has a surface cooling loop that was not present on MER, the MSL hardware temperatures vary less than they do on MER.)

Rover Avionics. The avionics have been responsible for the command, data handling, power regulation, power distribution, and pyro functions for all mission phases (including cruise and EDL). Rover avionics have served these functions during the surface phase. At the heart of the avionics are the rover compute elements (RCEs), redundant computers which have operated one at a time, with the spare held in cold backup (except during EDL where the redundant computer acted as a hot backup). Each RCE contains a central processor (a radiation-hardened PowerPC 750 architecture system) communicating with peripheral devices using other cards connected on a compact peripheral component interconnect (cPCI) backplane interface and providing central memory storage for mission data and telemetry of 32 gigabits via a non-volatile memory/camera (NVMCAM) card. In addition to the RCEs, power switching and analog input/output are provided by the redundant rover power and analog modules (RPAMs) connected to the RCEs via MIL-STD-1553 [16] data bus connection[14,15].

Flight Software. The software in the main computer of the rover executes a control loop that monitors the status of the flight system during all phases, checks for the presence of commands to execute, maintains a buffer of telemetry for transmission, performs communication functions (manages "comm. windows" through the communications behavior manager), and

---

[14] One hardware element of the avionics that must be powered continuously on the surface is the MSL remote engineering unit (MREU) in the RPAM. The MREU in each RPAM is redundantly connected to the two Electra Lite transponders (ELTs). An ELT may receive a "hail" from an Orbiter at any time requesting the rover to wake up.

[15] MIL-STD 1553 is a standard published by the United States Department of Defense that defines the mechanical, electrical and functional characteristics of a serial data bus.

checks the overall health of the spacecraft. Central control of the entire flight system is under control of the flight software running in the RCE (the same architecture as was used for the MER mission). On the surface, activities such as imaging, driving, or instrument operations are performed under commands transmitted in a command sequence to the rover from the flight team. The rover generates constant engineering, housekeeping, and analysis (EH&A) telemetry and episodic event reports (EVRs) that are stored for eventual transmission.

Prelaunch testing showed that it was possible that other activities could generate or be affected by radio frequency interference while the rover is communicating. Though such interference with the Electra UHF radio has been seen since EDL on the MRO orbiter, none has been seen on the rover. Should any mutual interference occur during later surface operations, the flight software will ensure that incompatible activities do not run during communication windows.

Communications Behavior. The rover telecom subsystem is used to send and receive command sequences, data, telemetry, and flight-software updates. The behavior of the telecom subsystem is controlled by the interaction of flight software, ground sequences, and a set of parameter tables that define the state of telecom hardware and that control the settings and timing of communication windows. Communication windows ("comm. windows") will be sequenced whenever it is desired to communicate with the rover. Windows must fit within scheduled DSN availability and planned relay orbiter passes. A window is defined as an interval of time that contains all the activities directly associated with preparation and execution of a communication session. During surface operations, the ground system will coordinate with the DSN and the MRO project to determine a set of desired communication windows.

Communication window information is stored in two tables on the rover: the primary table and the high-priority table. The primary table can hold as many as 256 communication windows and is used for standard operations. A high-priority table can be used in the event of rover anomaly resolution or for other purposes. When communication windows are loaded into the high-priority table, they take precedence over any windows defined in the primary table. This allows new communication windows to effectively replace selected onboard windows, should the need arise, without affecting the entire set of previously planned communication events. One routine use of high priority windows is to start two varieties of carrier-only "beep" DTEs. The flight software starts a nominal beep if the daily X-band DFE has been successful and the new sequence begins execution normally; otherwise the existing sequence continues execution and starts a run-out beep. Either of these beeps is a high priority window that is started at a known time.

8.1.7.2 Payload (Science Instruments). The rover carries the largest science payload suite landed on Mars to date, with instruments sponsored by NASA and others contributed by international partners. The following, also from the Mission Plan [15], is a summary/listing of the MSL science instrumentation.

The instruments are roughly divided into four categories:

1) Remote Sensing (2):
   - Mastcam: Multi-spectral, stereo imaging, as well as video
   - ChemCam: (Chemistry and Mineralogy) Remote spectroscopy of rocks and soils from laser ablation; remote microscopic imagery
2) In-Situ (2):
   - Mars Hand Lens Imager (MAHLI): Color microscopic imager
   - Alpha-Particle X-ray Spectrometer (APXS): spectroscopy of soil and rocks using X-ray fluorescence and particle-induced X-ray emission
3) Analytical (2):
   - CheMin: Mineralogical analysis of acquired samples of rock and soil using X-ray diffraction
   - Sample Analysis at Mars (SAM): Chemical and isotopic analysis of acquired samples of rock, soil, or atmosphere (including organics) using a mass spectrometer, gas chromatographs, and a tunable laser spectrometer
4) Environmental (4):
   - Radiation Assessment Detector (RAD): Detect and measure natural high-energy radiation
   - Mars Descent Imager (MARDI): High-resolution color video of descent
   - Dynamic Albedo of Neutrons (DAN): Detect and analyze hydrogen in the near-subsurface of Mars
   - Rover Environmental Monitoring Station (REMS): To monitor the meteorology and ultraviolet (UV) environment near the rover

8.2 Telecom Subsystem Overview

The X-band subsystem (with DSDST, RSDST, TWTA, and SSPA as the active elements) was primary for cruise through EDL and is also used for DFEs, DTEs, and beeps during surface communications.

Figure 8-30 is a block diagram of the X-band portion of the telecom subsystem, and Figure 8-31 is a block diagram of the UHF portion. Table 8-9 and Table 8-10, respectively, define X-band and UHF terms used in these two figures.

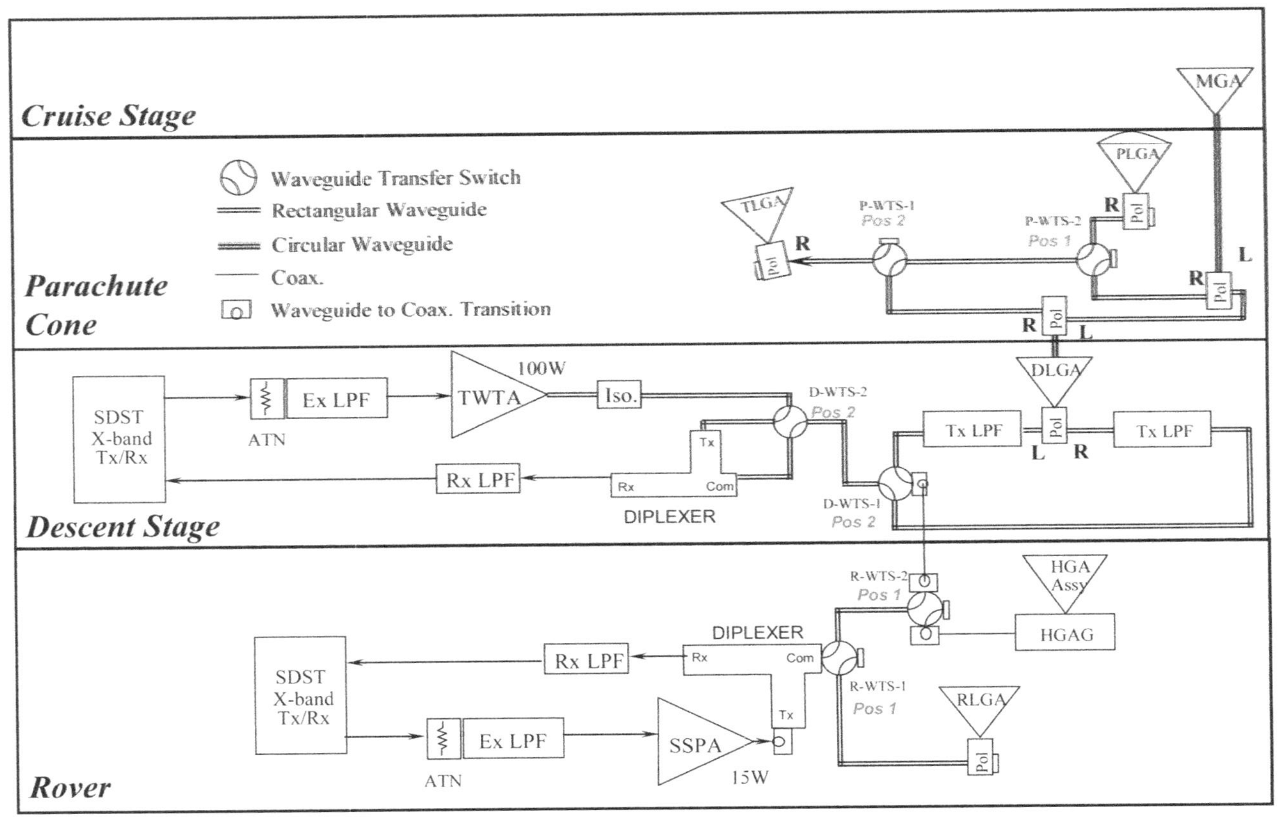

Fig. 8-30. MSL X-band subsystem launch configuration block diagram.

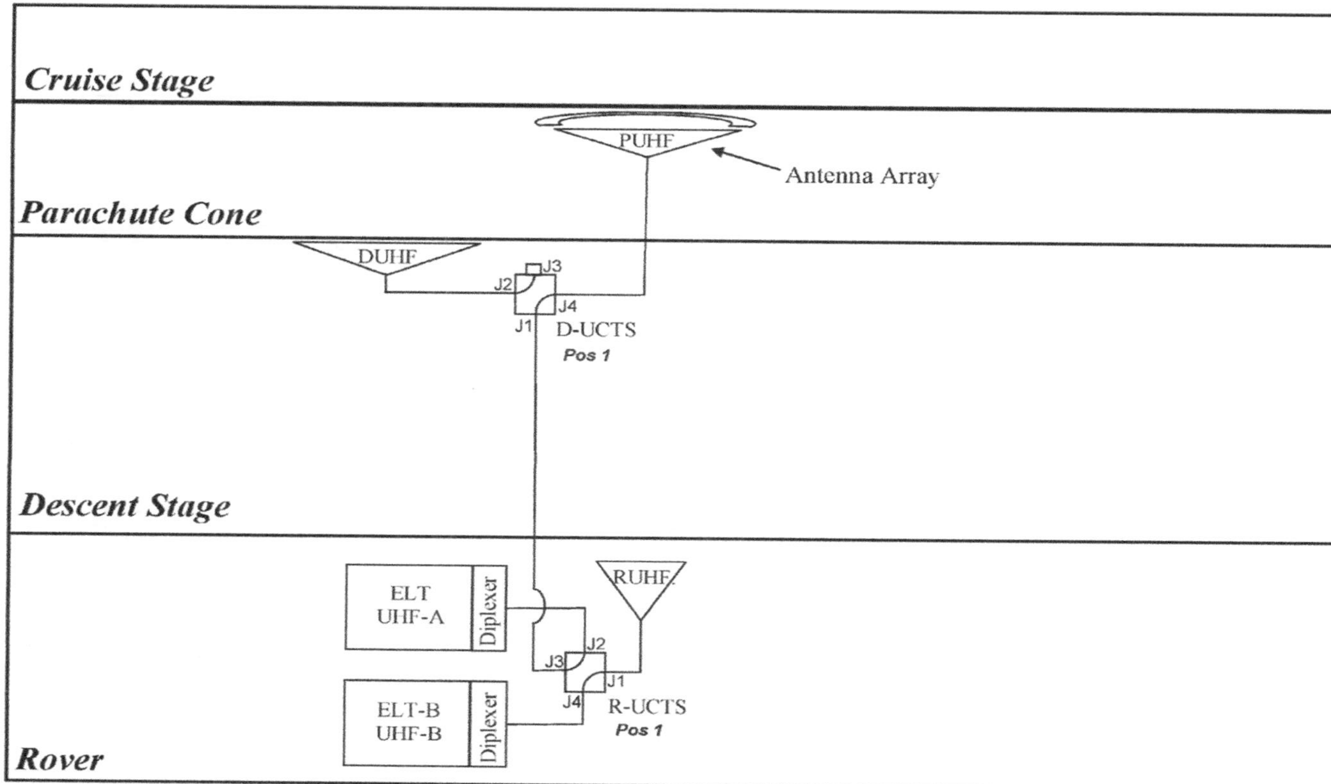

Fig. 8-31. UHF subsystem launch configuration block diagram.

Table 8-9. Acronyms and abbreviations in X-band telecom block diagram.

Term	Definition	Term	Definition	Term	Definition
Assy	Assembly	L	Left circular polarization	RLGA	Rover low gain antenna
ATN	Attenuator	LPF	Low pass filter	SDST	Small deep space transponder
Com	Common	MGA	Medium gain antenna	SSPA	Solid state power amplifier
D-	Descent	Pol	Polarizer	TLGA	Tilted low gain antenna
Ex	Exciter	P-	Parachute	TWTA	Traveling wave tube amplifier
HGA	High gain antenna	R-	Rover	Tx	Transmit
HGAG	High gain antenna gimbal	R	Right circular polarization	W	Watt
Iso	Isolator	Rx	Receive	WTS	Waveguide transfer switch

Table 8-10. Acronyms and abbreviations in UHF telecom block diagram.

Term	Definition	Term	Definition	Term	Definition
DUHF	Descent UHF antenna	J-	Jack (connector)	R-	Rover
D-	Descent	Pos	Position	RUHF	Rover UHF antenna
ELT	Electra Lite transponder	PUHF	Parachute UHF antenna	UCTS	UHF coaxial transfer switch

The four layers in Fig. 8-30 and Fig. 8-31 are called slices or stages. Each stage has an X-band antenna or two, with active X-band telecom equipment on two stages (the descent stage and the rover). Each stage except Cruise has a UHF antenna, but the UHF active elements are in the rover. The terms "Pos1" and "Pos2" in the figures define the switch positions at launch.

This complex series of EDL telecom events (involving multiple successive configurations of both X-band and UHF) is shown in Fig. 8-32. Refer back to Figs. 8-16, 8-17, and 8-19 for pictorials of the EDL events at the spacecraft level.

The DSDST and the RSDST are, respectively, the SDSTs on the descent stage and the rover.

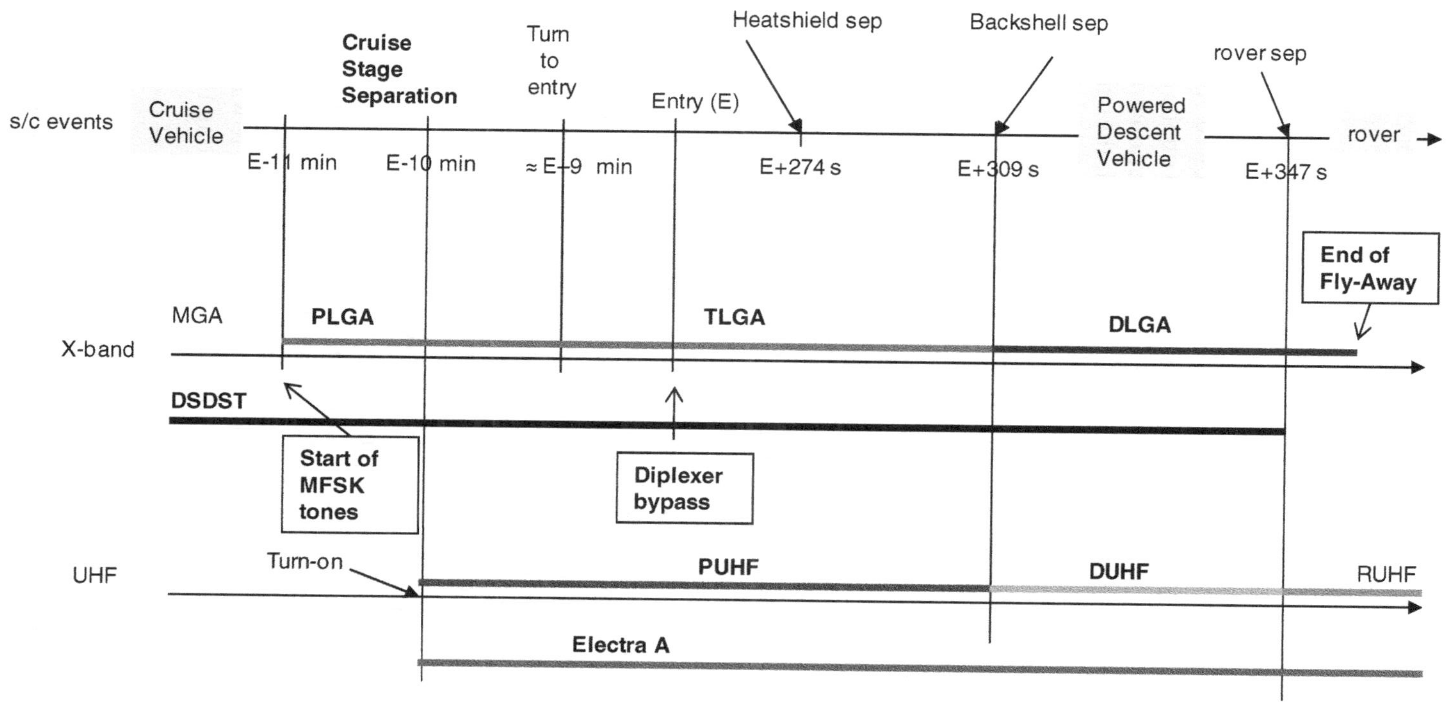

Fig. 8-32. X-band and UHF sequence during EDL. Please visit www.wiley.com/go/deepspacecommunications to access the color version of this figure.

8.2.1 Telecom for Launch, Cruise, and into EDL

The descent stage and the rover each has a small deep space transponder (SDST) [17] and a transmitter (a 100-W output TWTA on the descent stage and a 15 W output SSPA on the rover).

The nominal cruise configuration used the X-band radio (DSDST and TWTA) on the descent stage due to its lower line losses and higher output power. A backup switching arrangement would have allowed X-band to be routed to and from the X-band radio (RSDST and SSPA) on the rover. Being able to use either the TWTA or the SSPA provided functional redundancy during cruise. This redundancy proved not to be required, though the SDST and the SSPA on the rover were verified "alive" during cruise by two power on/power off test sequences of about a half hour duration each. The primary cruise SDST and TWTA continued to provide the operational RF signals during the aliveness tests.

The X-band MGA and PLGA antennas used during cruise and the first part of EDL are shown in the top two slices of Fig. 8-30 The MGA provided greater gain and a smaller beamwidth; the PLGA provided a larger beamwidth, but less gain. At any given time, one antenna was selected; that antenna simultaneously received uplink from the DSN and transmitted downlink to the DSN.

When EDL started, the EDL-main sequence sequenced CSS to occur; this ejected the top slice (the cruise stage) along with the MGA. The sequence continued, sequentially selecting the PLGA, the TLGA, and the DLGA with only downlink communications required. At the atmospheric entry interface, the sequence bypassed the diplexer in the descent stage to avoid critical-pressure high-power coronal discharge. Activating this bypass had the incidental advantage of decreasing transmit line loss somewhat.

During the banking maneuvers (Fig. 8-18), the TLGA provided the best downlink to the Earth. The original plan was to switch back to the PLGA just before parachute deployment, after the so-called "Straighten Up and Fly Right" (SUFR) maneuver, when the –Z axis and the anti-velocity vector were more co-aligned. However, because there was a good chance the parachute sabot could impact the PLGA at parachute deployment, it was decided to stay on the TLGA to maximize the likelihood of continuing the DTE link during parachute descent.

When backshell separation occurred, the second slice (parachute cone) broke away, taking the TLGA with it. For the remainder of the powered descent, the DLGA carried the downlink. No switching was needed to be on the DLGA. Because the DLGA was part of the waveguide run up to the parachute cone hardware, the DLGA began radiating upon the ejection of the backshell.

The powered descent vehicle phase concluded with rover separation and sky-crane operations. Upon touchdown, the bridle was cut and the descent stage flew away. Because they were on the descent stage, we lost effective use of the DSDST, the TWTA, and the DLGA. By now out of sight of the Earth, the SDST and TWTA continued to transmit through the DLGA until the descent stage crashed on the surface at the end of the "fly-away phase." The rover X-band system (the RSDST and SSPA) was not scheduled to operate for the first time until after the end of EDL.

The UHF block diagram is shown in Fig. 8-31. Table 8-10 defines the terms used in the figure. The two active elements are functionally redundant Electra Lite Transponders (ELT), with either ELT-A or ELT-B selected for use.

The MSL ELT is derived from the Electra UHF transponder (EUT) used on the MRO. Except for two non-radiating "aliveness" checkouts during cruise, the first use of the UHF was in the EDL phase, after cruise stage separation.

During EDL, all three UHF antennas were used: the PUHF from CSS until backshell deployment, the DUHF during powered descent, and the RUHF during and after the sky-crane stage (for the period from post landing + 1 minute through end of the surface mission).

The descent UHF coaxial transfer switch (D-UCTS) selected between the DUHF and the PUHF antennas. The rover UHF coaxial transfer switch (R-UCTS) selected between the rover UHF antenna (RUHF) and either of the descent antennas.

During surface operations, UHF is the primary mode of returning large volumes of data to the Earth (via orbiter relay). Only the RUHF is used during surface operations.

8.2.2 Surface Operations

The surface telecom system uses three antennas: two for X-band DTE/DFE and a UHF antenna for relay to an orbiting asset. Figure 8-29 shows the X-band and UHF antennas mounted on the rover deck.

The X-band antennas are the rover low-gain antenna (RLGA) and the high-gain antenna (HGA). The HGA is used for DFE commanding, DTE telemetry, and HGA beeps. The RLGA is used for low-rate (contingency) DFE commanding and RLGA beeps. The downlink signal level achievable using the RLGA is too low for DTE telemetry.

The HGA sits on a two-degree-of-freedom gimbal and is 0.28 m in diameter. The pointing accuracy requirement for the HGA is 5 deg, including rover attitude knowledge. Relative to the peak gain at boresight, the downlink gain is about 4 dB lower and the uplink gain about 3 dB lower, at 5 deg off boresight.

8.2.2.1 X-Band

8.2.2.1.1 Downlink (DTE). As an RF amplifier, the SSPA receives its X-band RF input signal from the RSDST and generates 15 W RF output. The basic telecom requirement for surface operations on the HGA is to provide a downlink capability of at least 176 bits per second (bps) to a 34-m station.

8.2.2.1.2 Uplink (DFE). The HGA is typically used for a DFE command session each sol. Depending on Earth-Mars distance, uplink rates are 1 kbps or 2 kbps. A command session takes 15 minutes, including margin for possible station transmitter delays and for packaging of commands into groups.

In safe mode, commands from the Earth are received via the RLGA, initially at the fault rate of 7.8125 bps. The RLGA is fixed-mounted to the rover and has a broad pattern. The RLGA provides link capability to command at the fault rate or higher depending on Earth-rover distance.

8.2.2.2 UHF. The primary data path for surface operations is via the UHF relay system, using the Mars orbiting assets (MRO or Odyssey[16]). The primary relay communications is via MRO, with two passes a day to return science and playback engineering data from the surface. Communications through Odyssey provides for additional data return or if MRO is unavailable, as long as there is DSN time for data return from Odyssey and sufficient energy to support UHF operations with Odyssey during the sol.

The UHF subsystem includes a pair of Electra-Lite radios. The Electra-Lite radio is a smaller version of the Electra radio flown on MRO. The MSL Electra-Lite/MRO Electra link can function using an adaptive data rate (ADR) scheme. In the ADR mode, the return link data rate will adjust to variations in signal strength due to antenna patterns, angles, and proximity between MRO and MSL throughout the overflight. In ADR, it is the MRO radio that controls the return data rate, based on its own receiver power telemetry, commanding the lander radio via the forward link to change its return link rate on the fly. The forward rate from MRO remains constant during any relay pass. Section 8.2.4.1 includes UHF frame and coding options.

A single quad-helix antenna designed especially for MSL, the RUHF, is mounted to the rover deck and used with either redundant radio.

---

[16] The Mars Explorer (MEX) orbiter is also compatible with the Electra lite radio, and it is available as an additional relay path. MEX was launched in 2003 by the European Space Agency.

The MSL/MRO relay link can also be used in a safe mode. The Electra-Lite radio can communicate a wake-up signal (via low-voltage differential signaling [LVDS]) to the rover avionics upon hearing a "hail" forward link[17] from MRO. This is most useful in a contingency mode, where the rover is effectively asleep with its ELT waiting for a communications possibility. The MSL mission intends to rely on this function as a fault response mode only. For example, a fault response would occur if the spacecraft lost its clock timing and did not know when relay passes were to occur.

The UHF functionality of the Odyssey and MEX orbiters is similar to that of the MRO UHF. However, the Odyssey and MEX radios are not Electras and do not have the adaptive data rate capability.

In some cases, an X-band DTE/DFE link of sufficient duration is not available, most likely because of scheduling contentions for the tracking stations or because the command load is larger than usual. A forward UHF link could be used in these cases to command the rover. Commanding via Odyssey and MRO has been demonstrated with the MER and Phoenix missions and more recently during MSL surface operations.

8.2.3 X-Band Flight Subsystem Description

8.2.3.1 X-Band Interfaces with MSL Control and Data Systems. The data transfer functions from the avionics subsystem to both SDST radios, include:

- Turbo codes that are the baseline for the DTE downlink. The codes have rates of 1/2, 1/3, and 1/6, and they have frame sizes of small (1784 bits) and large (8920 bits).
- Reed Solomon encoding (interleave depth 1 and 5) that can be used for the DTE downlink (in concatenation with convolutional [7, 1/2] coding[18] that is performed by the SDST).

---

[17] Forward and return links are used for communications between an orbiter and a lander. The forward link is from an orbiter to a lander. A return link is from a lander to an orbiter. Completing the end-to-end lander-Earth communications paths are the DSN X-band uplink and downlink between the Earth and the orbiter.

[18] In telecommunication, a convolutional code is a type of error-correcting code in which (a) each m-bit information symbol (each m-bit string) to be encoded is transformed into an n-bit symbol, where m/n is the code rate ($n \geq m$), and (b) the transformation is a function of the last k information symbols, where k is the constraint length of the code (from Wikipedia: Convolutional code). For MSL, the code parameter values are $k = 7$, $m = 1$, and $n = 2$; the resulting code is abbreviated (7, 1/2).

8.2.3.2 X-Band Key Hardware Components. The telecom component descriptions in the following paragraphs are organized by the stages in the Fig. 8-3 graphic. The X-band telecom block diagram (Fig. 8-30) shows the four spacecraft stages ("slices") that contain the X-band telecom subsystem elements. These four are Cruise, Backshell or Parachute Cone, Descent, and Rover. The heat shield stage had no telecom components.

8.2.3.2.1 Cruise Stage X-Band Telecom Components

8.2.3.2.1.1 Medium-Gain Antenna: The MGA, called out in Fig. 8-2, was used for mid- to late-cruise communications. The MGA was a build-to-print of the MER MGA, fed by a septum polarizer for circular polarization (CP) operation. The MSL MGA could operate either right-hand or left-hand circularly polarization (RCP or LCP), depending on which side of the polarizer was connected to the receiver or transmitter[19]. RCP was used in the mission.

The MGA, which was attached on the cruise stage, separated from the rest of the X-Band telecom subsystem at Cruise Stage Separation. In the top left drawing of Fig. 8-2, the light blue surface on top of the cruise stage is the annulus-shaped solar array with the MGA at its center.

Table 8-11 states some of the RF characteristics of the MGA.

Table 8-11. MGA RF characteristics.

Parameter	Value
Receive frequency, MHz	7150.8 (DSN channel 4)
Transmit frequency, MHz	8401.4 (DSN channel 4)
Gain, boresight, dB	18.1 ± 0.4 receive 19.2 ± 0.4 transmit
Polarization	RCP or LCP
3 dB-beamwidth, deg	±10.3 receive ± 9.3 transmit
Axial Ratio, on boresight, dB	1.01 receive; 0.27 transmit
Axial Ratio, 20 deg off boresight, dB	6.29 receive; 7.53 transmit
Design	RF conical

[19] Right-circular polarization (RCP) refers to an electromagnetic wave that propagates such that the tip of the electric field appears from the source to describe a circle in the clockwise direction. Left-circular polarization is the opposite; the tip of the electric field is seen from the source as describing a circle in the counterclockwise direction. A polarizer converts the RF beam traveling through a waveguide to an electromagnetic wave of a specific polarization. In this case the septum polarizer converts a linearly polarized wave in the waveguide run to a circularly polarized wave to be transmitted by the antenna.

Figures 8-33 and 8-34 show uplink and downlink patterns of gain as a function of angle from MGA boresight. These are based on measurements on a MER mock-up. The gain is down 3 dB from its peak at about 10 deg from boresight, as compared with about 5 deg from boresight for the rover HGA. "Beamwidth," commonly defined in terms of a total angular range on both sides of boresight, is double the above numbers.

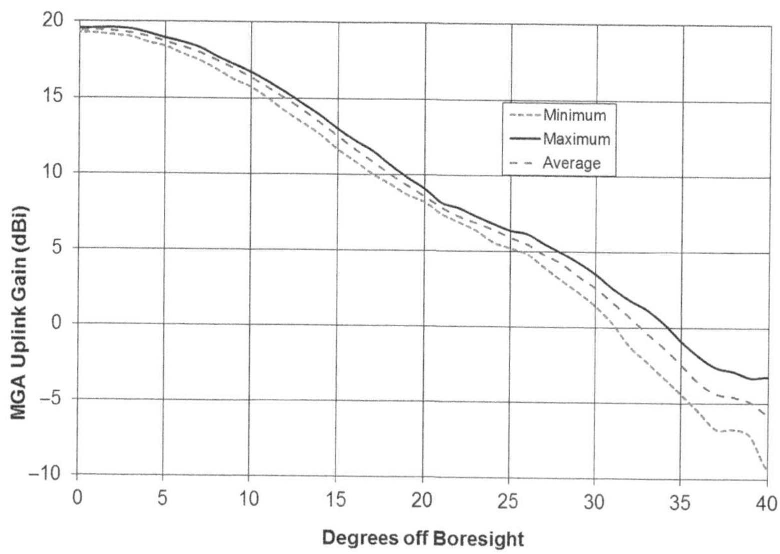

Fig. 8-33. MGA uplink gain.

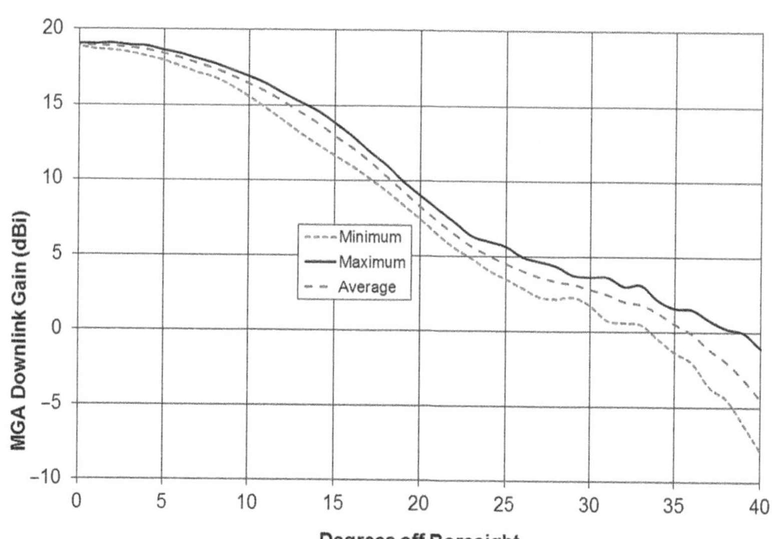

Fig. 8-34. MGA downlink gain.

8.2.3.2.2 Parachute Cone X-Band Telecom Components. The MSL X-band telecom block diagram (Fig. 8-30) shows the telecom components on the Parachute Cone, as well as on three other stages. Note that names beginning with P refer to parachute, those with D to descent, and those with R to rover.

8.2.3.2.2.1 Waveguide Transfer Switches (P-WTS-1 and P-WTS-2): All switching between X-band transponders, power amplifiers, and antennas is with waveguide transfer switches (WTS). The switches are used to connect transmit and receive functions to the proper antennas. On the parachute cone, P-WTS-1 selected the TLGA, and P-WTS-2 selected between PLGA and MGA.

8.2.3.2.2.2 Parachute Low-Gain Antenna: Figure 8-35 shows where the PLGA and the TLGA are installed. The PLGA was used from launch through the first three months of cruise and was also the default antenna for cruise safe-mode. The PLGA supported MFSK tone transmission over a wide range of pointing angles during EDL communications. The PLGA boresight was aligned along the –Z axis of the spacecraft, as shown in Fig. 8-3.

The design of all MSL X-band low-gain antennas (except for the DLGA) is the same: the PLGA, TLGA, and RLGA were each an open-ended waveguide with chokes and parasitic drooping dipoles. However, the proximity effects of spacecraft components near each LGA resulted in their individual patterns being quite different from one another.

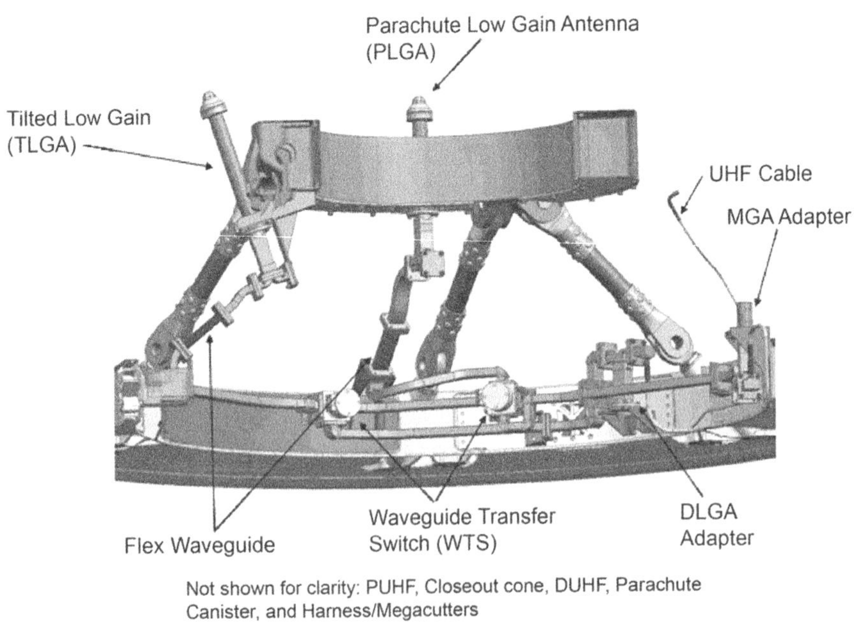

Fig. 8-35. Locations of the low-gain antennas. Please visit www.wiley.com/go/deepspace communications to access the color version of this figure.

The parasitic dipoles have the effect of broadening the pattern, as compared to the MER design. Figures 8-36 and 8-37 show, respectively, the uplink and downlink patterns of the PLGA (measured on a spacecraft mock-up). Both the maximum gain over all roll angles (red curve) and the minimum gain (blue) are shown. As the spacecraft spun at the 2 rpm rate during cruise, the peak-to-peak link performance varied by as much as several decibels.

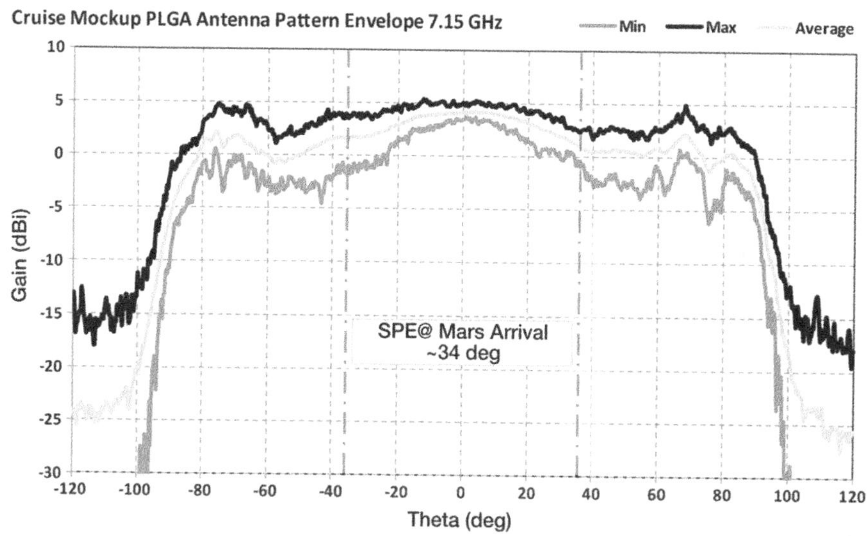

Fig. 8-36. PLGA X-Band uplink gain with spacecraft mock-up, RCP, 4/30/09. Please visit www.wiley.com/go/deepspacecommunications to access the color version of this figure.

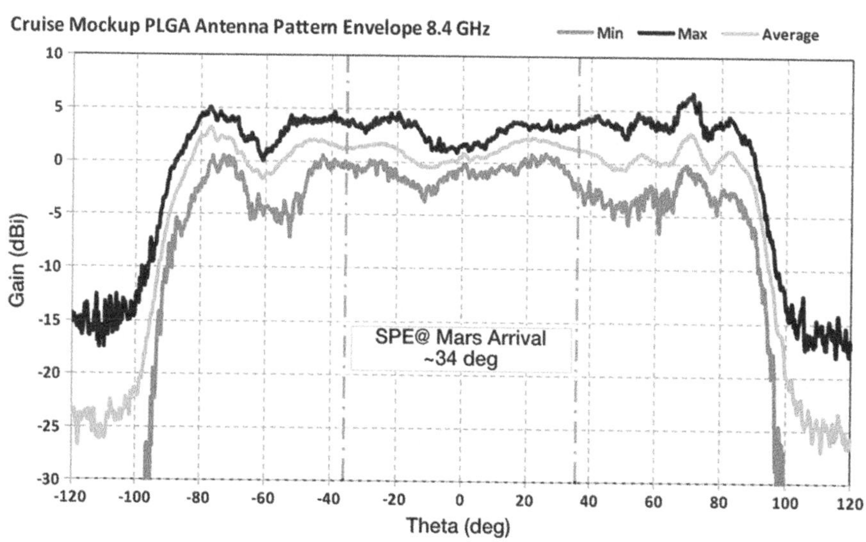

Fig. 8-37. PLGA X-Band downlink gain with spacecraft mock-up, RCP, 4/17/09. Please visit www.wiley.com/go/deepspacecommunications to access the color version of this figure.

For the very early launch dates with a type II trajectory, antenna angles could have been as large as 120 deg. Until the Type Ib trajectory was chosen for the 2011 launch, the type II cases had to be included in the telecom design and mission design tradeoffs. Communications would have been possible only for the first few hours after trans-Mars injection, while the range loss was not yet too high.

8.2.3.2.2.3 Tilted Low-Gain Antenna: The TLGA had the same design as the PLGA. One difference, however was that the TLGA's boresight was 'tilted' with respect to the spacecraft –Z axis by 17.5 deg, which was chosen to bring it close to the average anti-velocity vector direction during post-entry banking maneuvers. This minimized the span of Earth-to-boresight angles and, therefore, the link-signal level variation during the critical hypersonic and banking phases. Figure 8-18 illustrates the geometry involved for the EDL maneuvers when the TLGA is in use.

8.2.3.2.3 Descent Stage X-Band Telecom Components. The MSL X-band telecom block diagram (Fig. 8-30) shows the telecom components on the Descent stage, as well as on three other stages. Note that names beginning with D- refer to "descent."

The descent stage (DS) contained two active telecom components, the DSDST and the TWTA, as well as the DLGA. In addition, there were several components involved in routing the high-powered TWTA RF output and the much weaker RF input destined for the SDST receiver.

Figure 8-38 shows the overall layout of the DS telecom components. Most were on a telecom plate, shown in a contrasting color and detailed in Fig. 8-39.

Both the TWTA (on the far side of the plate) and the TWTA's electronic power conditioner (EPC) were powered on throughout cruise. These items dissipated relatively large amounts of spacecraft power as heat to be carried away by the cruise stage thermal control system (heat rejection system [HRS]).

8.2.3.2.3.1 Descent Low-Gain Antenna: The descent low-gain antenna (DLGA) (Fig. 8-40) is an open-ended waveguide with chokes. The relatively broad pattern of the DLGA was expected to suffer significant distortion from interaction with the surrounding structure of the descent stage. The DLGA gain as a function of angle from boresight and its variation was modeled[20] using the General Reflector Antenna Scatter Program (GRASP) antenna scattering software. Figure 8-41 shows the GRASP model used to generate the pattern.

---

[20] "DLGA on the Descent Stage: 8.400 GHz," Dan Hoppe, April 21, 2008 (internal MSL project document).

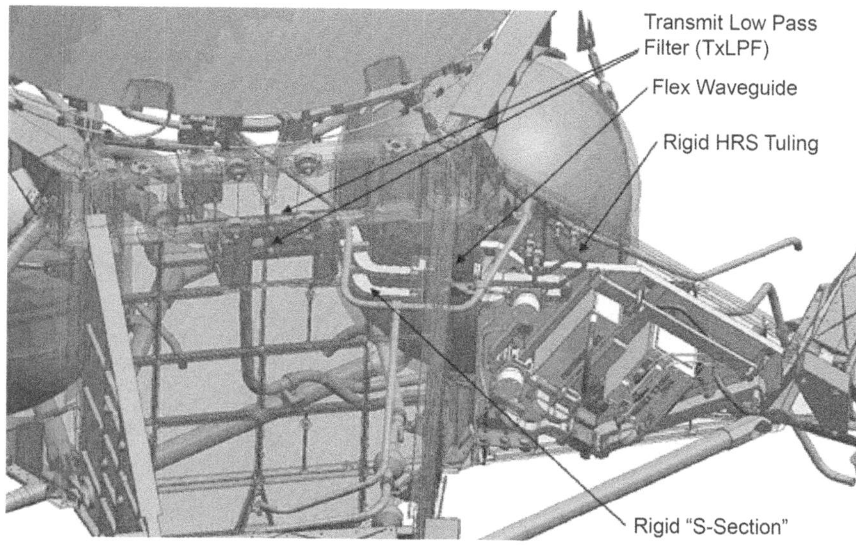

Fig. 8-38. Descent Stage X-band layout. Please visit www.wiley.com/go/deepspace communications to access the color version of this figure.

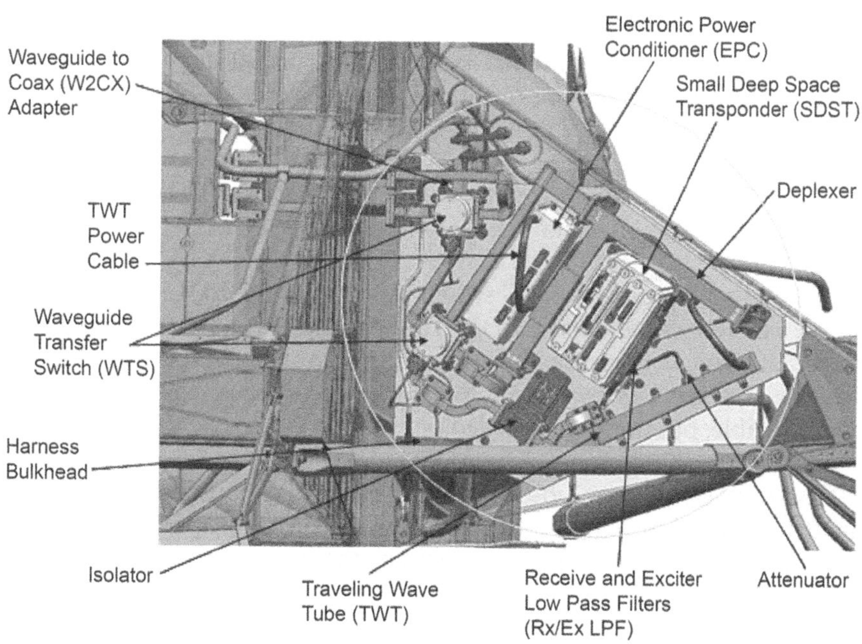

Fig. 8-39. Telecom plate assembly. Please visit www.wiley.com/go/deepspace communications to access the color version of this figure.

The actual view angles for DTE during the powered descent phase could vary widely, depending on the landing site chosen and descent geometry (such as the large tilt during the divert maneuver). As the landing date and site were not selected until well after the spacecraft design was complete, the antenna

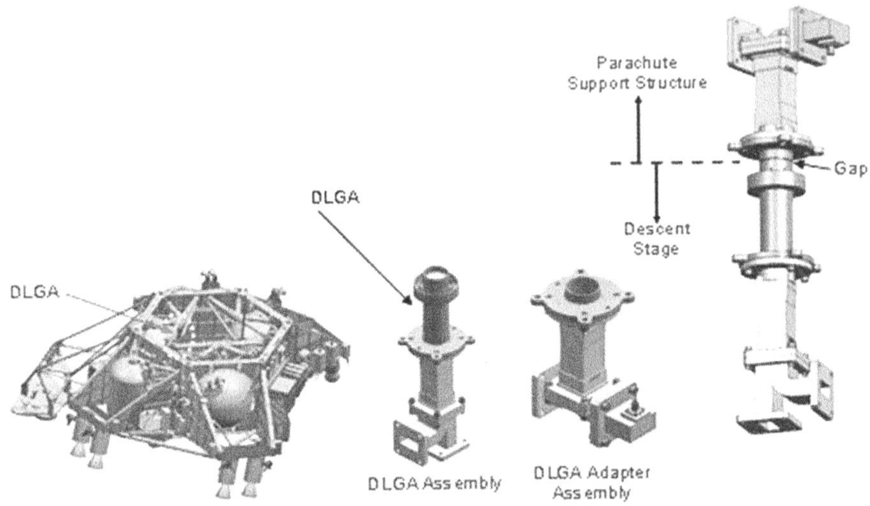

Fig. 8-40. DLGA and DLGA adapter overview. Please visit www.wiley.com/go/deepspace communications to access the color version of this figure.

design tradeoff included the possible range of boresight view angles. For some trajectories, the Earth would be quite close to the horizon at touchdown and, therefore, the antenna pattern as far as 90 deg from boresight needed to be considered. The resultant pattern in Fig. 8-42 shows that the variation is worst near the 75-deg off-boresight angle.

8.2.3.2.3.2 Descent Stage Waveguide Transfer Switches: D-WTS-1 selected one of two polarization signal paths: RCP and LCP.

D-WTS-2 is a "diplexer-bypass" switch, to be used for EDL. Bypassing the diplexer was sequenced during EDL. This avoided coronal breakdown in the diplexer at critical pressure during passage through Mars' atmosphere[21]. Once the diplexer was bypassed, X-band uplink-receive capability was no longer possible, but none was needed.

8.2.3.2.3.3 Traveling Wave Tube Amplifier: The TWTA consisted of two components: the traveling wave tube (TWT) and the electronic power conditioner (EPC) that provided the voltages required by the tube. The TWTA is of MRO heritage.

The TWT was specified to provide at least 100 W of RF output to support X-band communications and radiometric requirements during cruise and

[21] The power handling capability of the diplexer at critical pressure was found in test to be insufficient for use with the TWTA during EDL. Breakdown occurred at 85 W in test, lower than the 100-W nominal output. The bypass design avoided this problem.

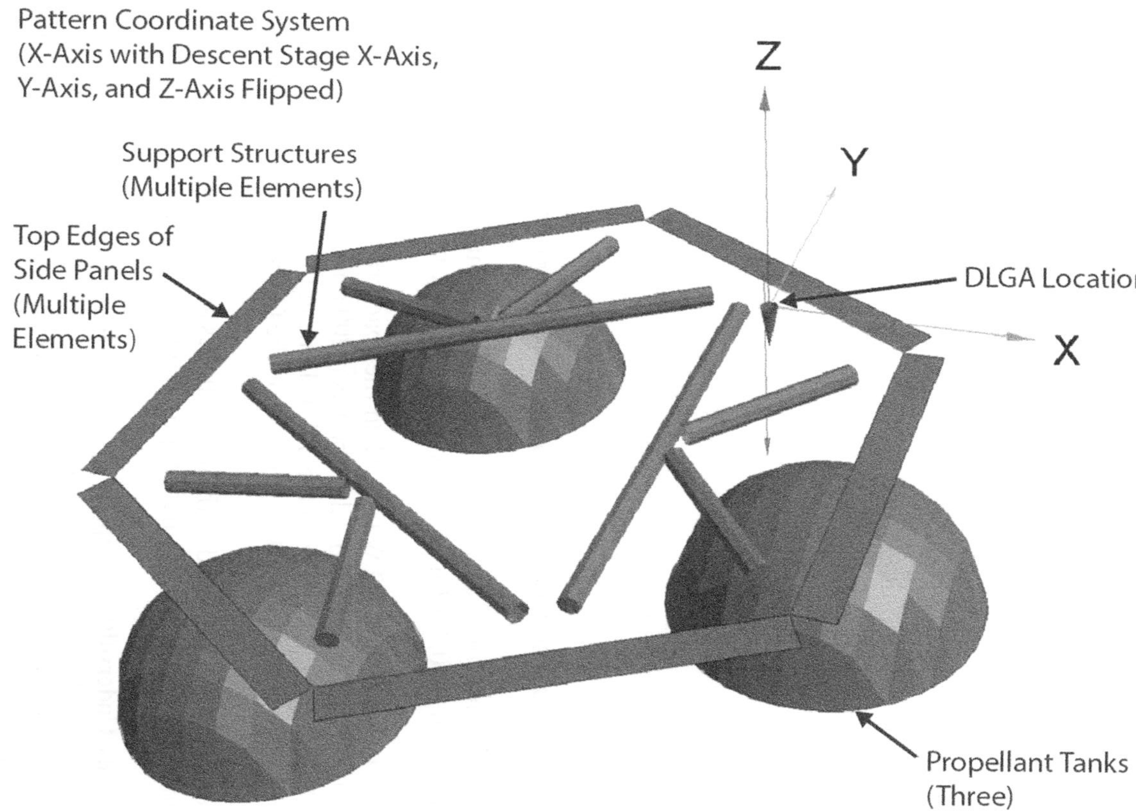

Fig. 8-41. GRASP model for DLGA scattering study.

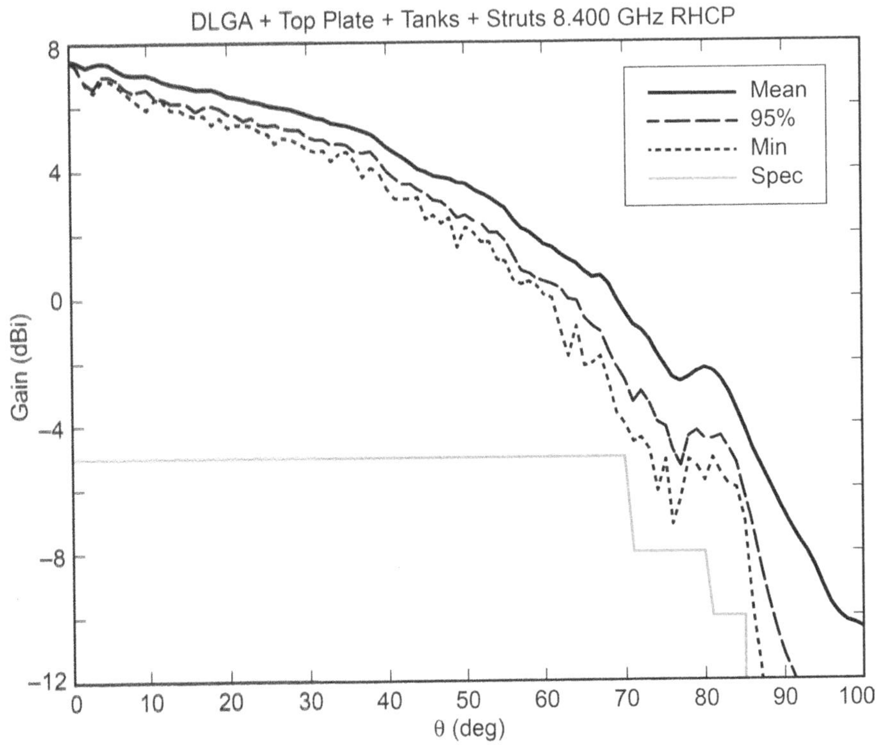

Fig. 8-42. MSL DLGA downlink pattern modeled from GRASP model.

during EDL until rover separation. The MSL flight unit had RF output of 104.7 W, which is 50.2 dBm (decibels referenced to 1 milliwatt [mW]).

As with other deep-space TWTAs, when the spacecraft bus voltage was first input to the EPC, a warm-up delay (between 200 and 240 s) occurred before the EPC applied high voltage to the TWT. During the delay, the TWTA was prevented from generating RF regardless of the On/Standby Mode control.

Because the TWTA operated with high voltages and high power levels, it was designed with three kinds of internal protection:

1) **Bus Undervoltage Trip:** The undervoltage trip would happen if the spacecraft bus voltage at the EPC input went below 20.5 V ±0.5 V. During the shutdown, period the TWTA indicated status as Under Voltage Trip. The TWTA initiated a start-up sequence when the bus input voltage rose above 21.5 V.

2) **Converter Overcurrent Trip:** The converter overcurrent trip would occur if the high-voltage converter exceeded a safe current value. The TWTA initiated an automatic restart function (ARF). The ARF

turned off the electron beam in the TWT. Within 50 ms, the electron beam came back on, and the TWTA returned to nominal operations. If a second trip occurred within 180 seconds, the TWTA would go to the start-up sequence described above.[22]

3) **Helix Current Trip:** The helix overcurrent trip was designed to occur if the TWT body (helix) current exceeds safe values (set by the TWT manufacturer); however, this trip was intentionally disabled for MSL.

8.2.3.2.3.4 X-Band Diplexer: The design of the diplexers in the descent stage and the rover is the same. The diplexers provide for the separation of the receive frequency from the antenna and the transmit frequency to the antenna.

8.2.3.2.3.5 Transmit Low-Pass Filter: Two transmit low pass filters (Tx LPF), branch out of D-WTS-1, one for each polarization.

The Tx LPF is a waveguide filter and had two purposes:

1) For near-Earth operations, the filter reduced out-of-band emissions from the TWTA. This function is similar to MRO, which also has a 100-W TWTA of the same design.

2) During EDL, the filter reduced TWTA emissions into the landing radar, especially in three frequency bands of 16.7 to 17 GHz, 25.2 to 25.5 GHz, and 33.4 to 34 GHz. The radar center frequency was 35.75 GHz. Tests with the LPF early in 2009 verified that radar operation would not be degraded by TWTA emissions into its sensitive frequency bands.

Table 8-12 documents the RF characteristics of the transmit LPF.

8.2.3.2.3.6 Exciter Low-Pass Filter: This filter attenuated the DSDST exciter broadband spurious emissions. This filter and the Tx LPF worked together to attenuate the overall out-of-band emissions sufficiently at the input to the landing radar.

---

[22] Telemetry played back via UHF after landing indicates that the TWTA tripped off late in EDL. The cessation of RF output was not seen in real time. The X-band link was not prime at the time. Assessment of the playback telemetry was that a converter overcurrent trip occurred, but the data sampling rate was insufficient to prove this. Dynamics (vibration, shock) was the most likely cause of the trip.

Table 8-12. Transmit low-pass filter RF characteristics.

Parameter	Value
Receive passband* insertion loss	0.2 dB at 7.17.2 GHz
Transmit passband insertion Loss	0.2 dB at 8.35 to 8.5 GHz
Transmit attenuation of the second harmonic (16.7 to 17 GHz)	> 50 dB
Transmit attenuation of the third harmonic (25.0 to 25.4 GHz)	> 35 dB
Transmit attenuation of the fourth harmonic (33.4 to 34 GHz)	> 30 dB
Group delay variation, over 1 MHz in receive (7.1–7.2 GHz) and transmit (8.354–8.5 GHz) passbands (*)	1 nanosecond (ns)

*A passband is the portion of the spectrum, between limiting frequencies. This portion is sent through with minimum relative loss or maximum relative gain by a filtering device.

Table 8-13. Receive low-pass filter RF characteristics.

Parameter	Value
Insertion loss	< 0.2 dB at 7.1 GHz
Transmit attenuation	> 70 dB

8.2.3.2.3.7 Receiver Low-Pass Filter: The receiver low-pass filter (Rx LPF, Table 8-13) rejected TWTA power reflected from the diplexer ('ring-around' noise) so the DSDST could detect very weak uplink signals. The SDST threshold is –155 dBm, as contrasted with the TWTA's +50 dBm RF output.

By design, the diplexer passed any signal in the receive band to the receiver while attenuating TWTA output at other frequencies. To complete the job, the Rx LPF attenuated the TWTA output at frequencies lower than the transmit band.

8.2.3.2.3.8 Waveguide: The waveguide between the TWTA isolator output and D-WTS-2 was redesigned to cut off TWTA emissions in the receive band that could "sneak back" into the SDST. The redesign was necessitated by the addition of the diplexer bypass switch D-WTS-2. Addition of the switch had introduced a new sneak path that allowed TWTA noise at the receive frequency band into the diplexer receive arm.

8.2.3.2.3.9 Descent Stage SDST: The SDSTs in the descent stage and the rover were both of the same "group buy III" design. The DSDST and RSDST transponders are discussed together in the next section.

8.2.3.2.4 Rover X-Band Telecom Components. The X-band telecom block diagram (Fig. 8-30) shows the telecom components on the rover stage as used during surface operations, as well as on three other stages. Note that names beginning with R refer to components on the rover.

The figure shows that the rover has two active components (the RSDST and the SSPA), two antennas (RLGA and HGA), the gimbal to point the HGA, and several microwave components (filters, etc.). Figure 8-43 shows the overall placement of the rover's X-band and UHF components, and Fig. 8-44 provides detail regarding the components. The UCXS in Fig. 8-43 is a UHF coaxial transfer switch.

8.2.3.2.4.1 Rover Waveguide Transfer Switches: The rover waveguide transfer switches have the designators R-WTS-1 and R-WTS-2. In the X-Band telecom block diagram (Fig. 8-30), position 1 of R-WTS-1 selects the RLGA.

R-WTS-2 selects between the HGA and the "through-path" that connected the RSDST with antennas in the cruise stage or the parachute cone. This WTS was first switched to select the HGA in Sol 1 and will not be switched again.

8.2.3.2.4.2 Rover Low-Gain Antenna: The RLGA is of the same design as the PLGA and TLGA. Fault protection selects the RLGA for safe-mode communications on the surface. Depending on Earth-Mars distance, the RLGA can also support a low-rate uplink transmitted at 18 kW from a 34-m or a 70-m antenna in case the HGA is not functional or its view of the Earth during a DFE would be obstructed by terrain or occluded by objects on the rover.

A simplified analysis [18] was performed prior to the critical design review (CDR) to evaluate the RLGA pattern using the WIPL-D commercial high-frequency electromagnetic modeling software package [19] (WI = wires, PL = plates, D = dielectrics). Only a few key components were included in the model since the structure is large compared to the wavelength (see Fig. 8-45). The pre-CDR analysis, with the pattern shown in Fig. 8-46, includes ground-plane effects. The patterns are relatively smooth; however, the worst-case variations are quite high, on the order of 10 dB over a very small percentage of the coverage region. Figure 8-47 compares the pre-CDR gain pattern for the MSL RLGA with the design and with the minimum gain value for the MER RLGA.

8.2.3.2.4.3 High-Gain Antenna: The HGA is mounted on a two-axis gimbal (Fig. 8-48) located on top of the rover deck. The 48-element microstrip patch HGA radiating element is the six-sided flat structure to the left. The antenna was provided by the European Aeronautic Defense and Space Company (EADS CASA ESPACIO).

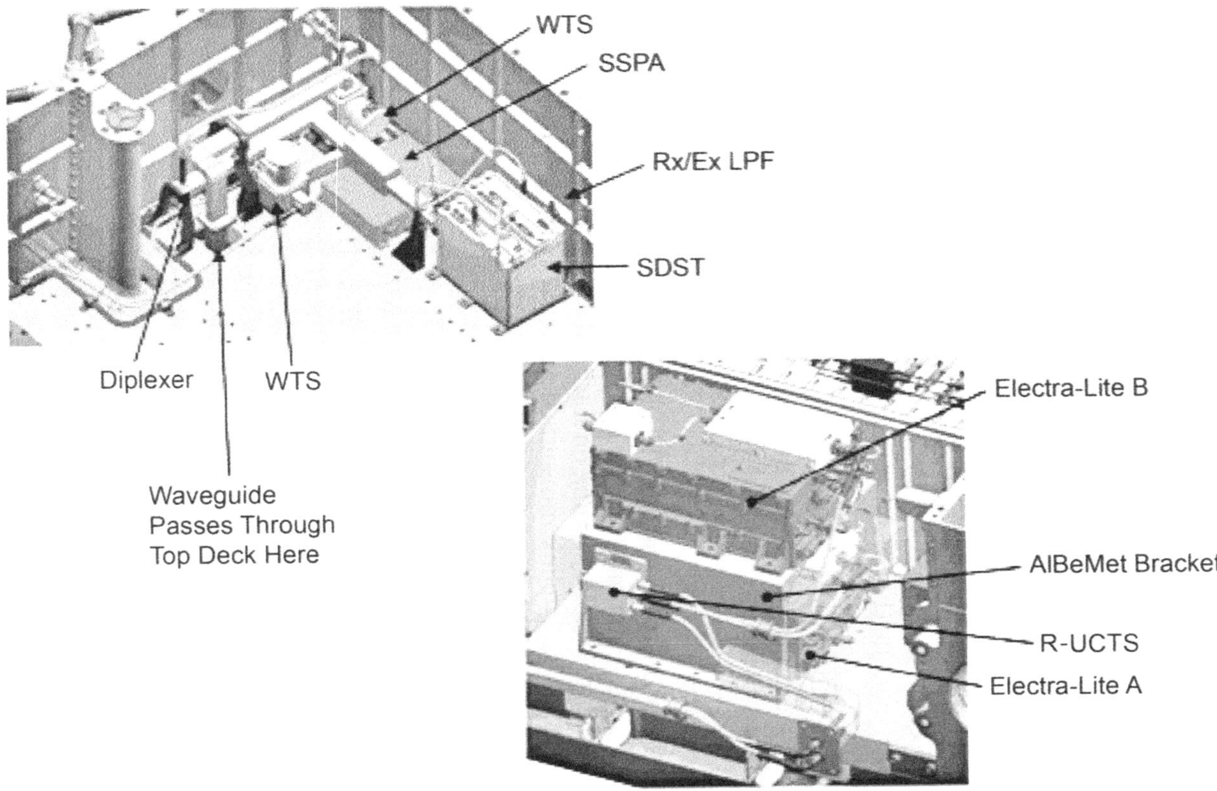

Fig. 8-43. Rover telecom internal layout showing overall placement of the X-band and UHF components. Please visit www.wiley.com/go/deepspacecommunications to access the color version of this figure.

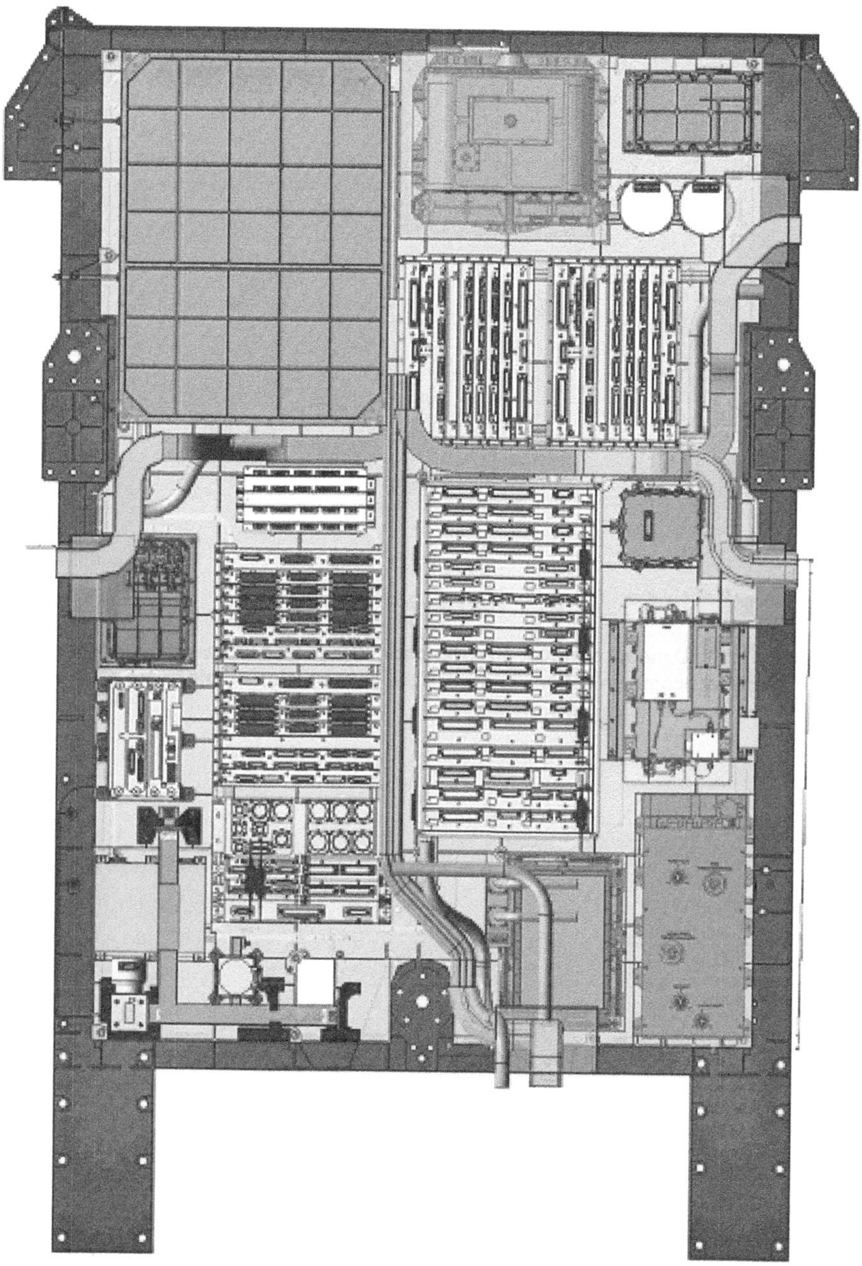

Fig. 8-44. Rover detailed internal layout. Please visit www.wiley.com/go/deepspace communications to access the color version of this figure.

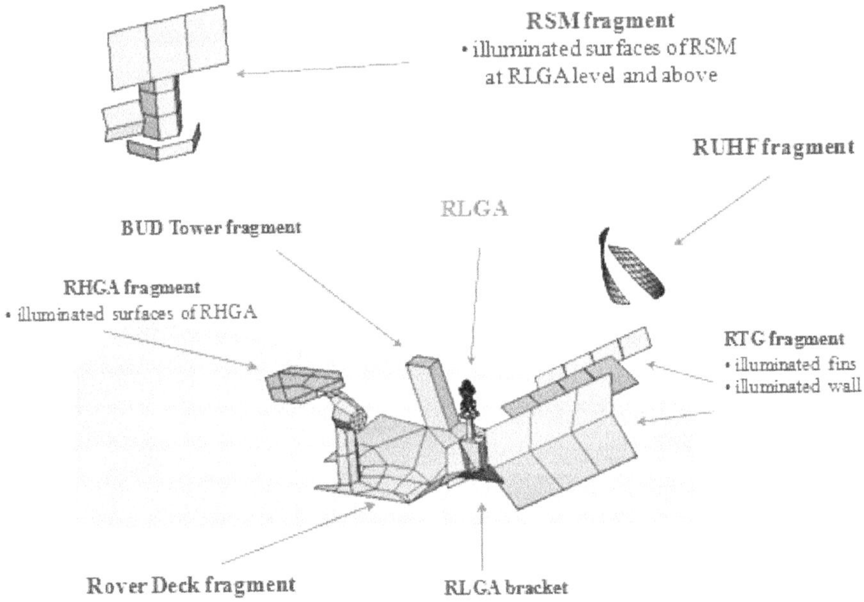

Fig. 8-45. WIPL-D model of RLGA on rover deck. Please visit www.wiley.com/go/deepspacecommunications to access the color version of this figure.

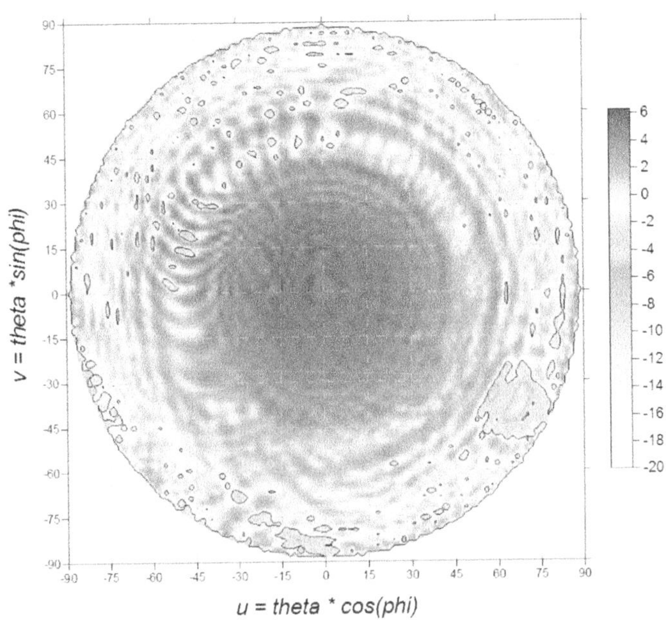

(a) Rx RLGA with groundplane effects, RCP 7.15 GHz

Fig. 8-46. RLGA patterns modeled by WIPL-D (includes ground-plane effects) for (a) 7.15 GHz and (b) 8.4 GHz. Please visit www.wiley.com/go/deepspacecommunications to access the color version of this figure.

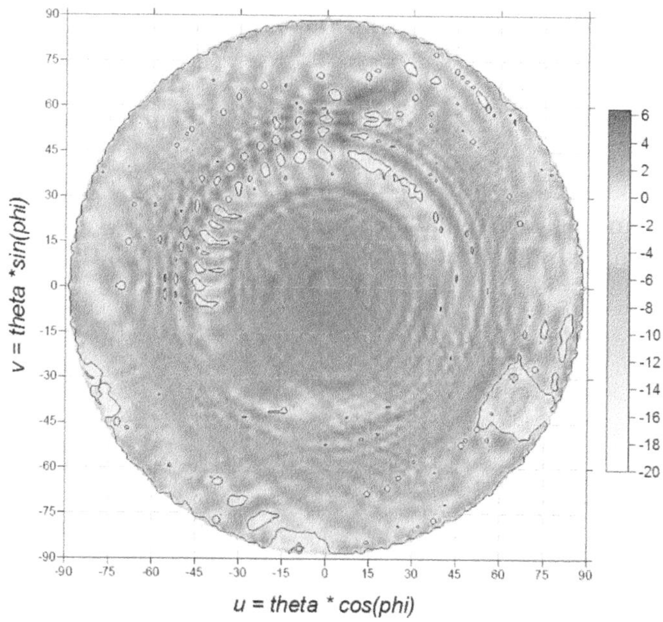

(b) Tx RLGA with groundplane effects, RCP 8.4 GHz

Fig. 8-46. (Continued)

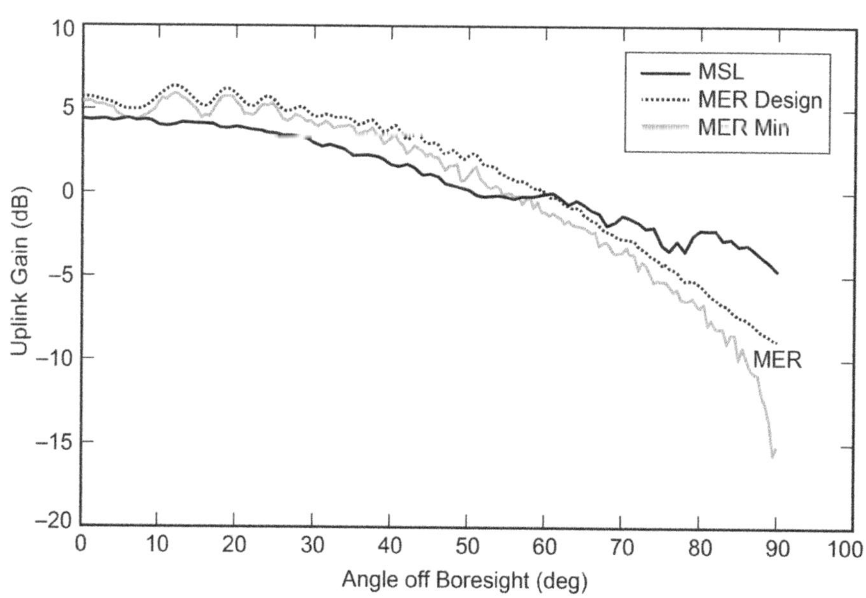

Fig. 8-47. Free-space uplink gain comparison between MER RLGA and MSL RLGA.

The HGA was deployed after the rover landed. Table 8-14 provides the rover HGA RF characteristics.

Note that Table 8-14 separates out the "circuit loss" in the gimbals from the antenna gain.

Figures 8-49 and 8-50 show the measured uplink directivity (gain relative to the gain at boresight) at a frequency of 7145 MHz. Similarly, Figs. 8-51 and

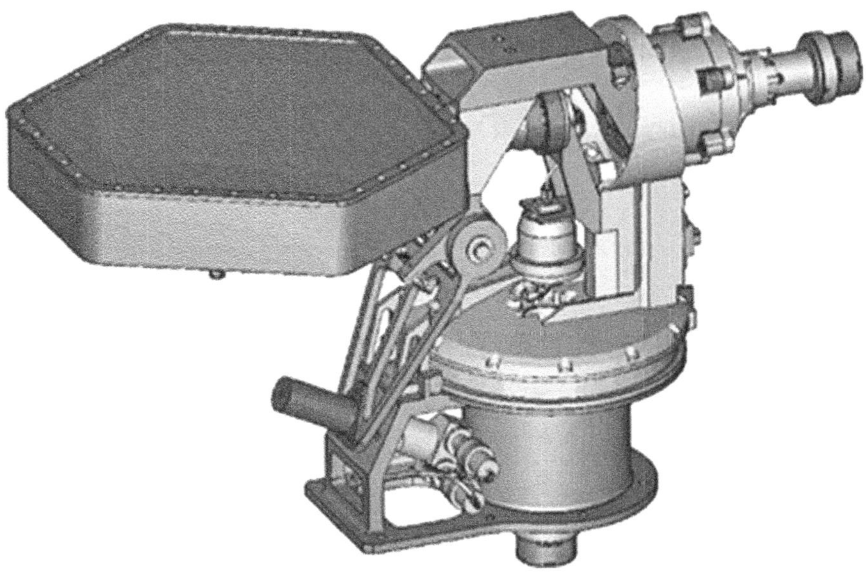

Fig. 8-48. HGA and gimbal assembly. Please visit www.wiley.com/go/deepspace communications to access the color version of this figure.

Table 8-14. Rover HGA RF characteristics.

Parameter		Value	Conditions
Dimensions	cm	25.5 by 29.4	
Transmit gain	dBi	25.5	0 deg off boresight
		24.1	2 deg off boresight
		20.4	5 deg off boresight
Receive gain	dBi	20.2	0 deg off boresight
		19.7	2 deg off boresight
		17.3	5 deg off boresight
Loss in gimbals	dB	1.2 dB	
Polarization		RCP	
Transmit axial ratio		3.0 dB	Within 5 deg from boresight
Receive axial ratio		2.4 dB	Within 5 deg from boresight

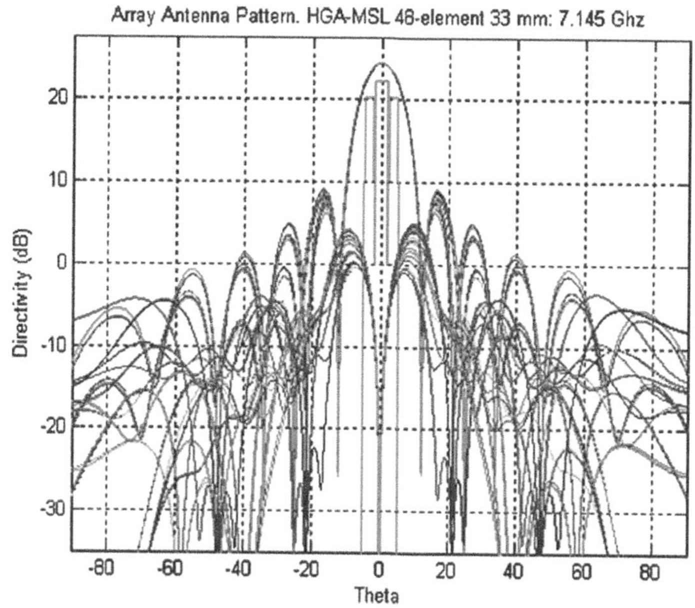

Fig. 8-49. HGA uplink directivity showing first several sidelobes.

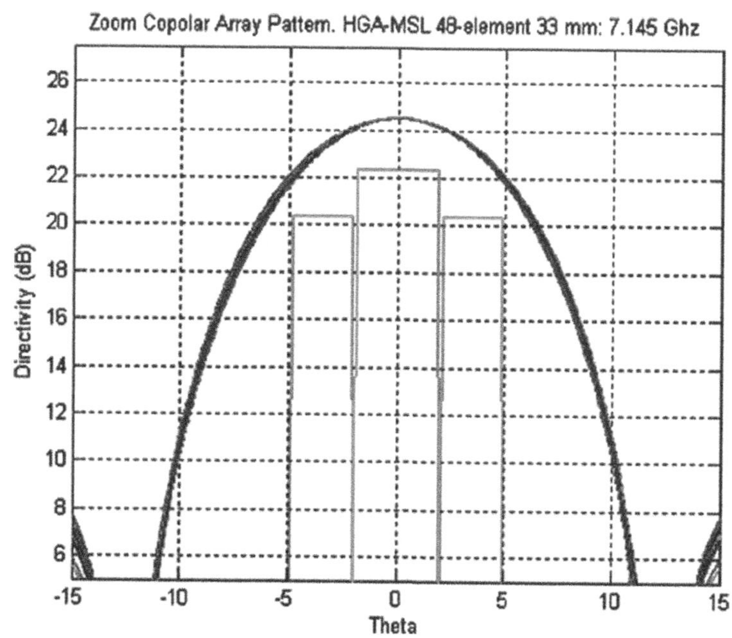

Fig. 8-50. Detail of the uplink main lobe HGA directivity.

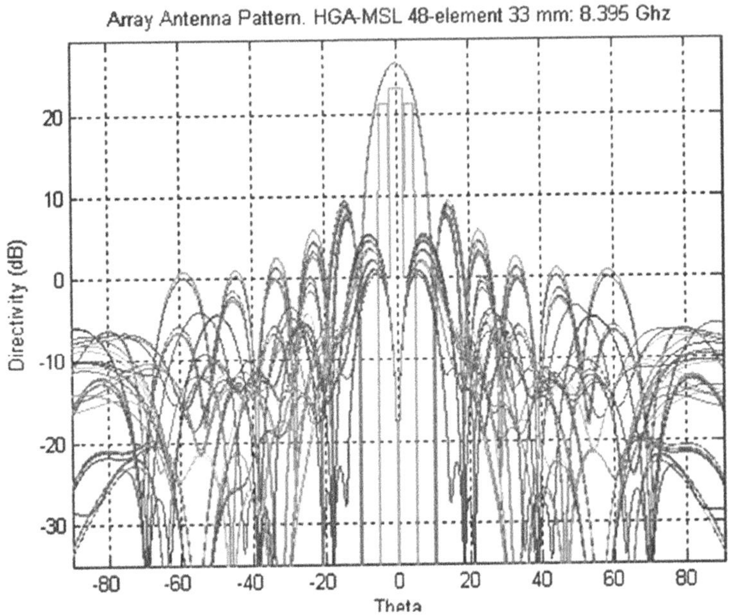

Fig. 8-51. HGA downlink directivity showing first several sidelobes.

8-52 show the measured downlink directivity at a frequency of (8395 MHz). The measured frequencies are representative of the MSL X-band frequencies.

In the patterns, theta is the angle from boresight, with the boresight planned to be Earthpointed. The multiple curves apparent in the sidelobes of the patterns represent cuts at 0, 45, 90, and 135 deg around phi, the axis orthogonal to theta.

The expanded Fig. 8-50 and Fig. 8-52 show primarily the main lobe, with the main lobe's pattern similar at each phi. The figures show that the HGA has good main lobe symmetry over the full range of phi and that the main lobe meets the required gain (shown as the rectangles beneath the main lobe).

8.2.3.2.4.4 Rover Diplexer: The design of the descent-state diplexer and the rover diplexer is the same. A diplexer provides for the frequency separation of the receive signal coming from and the transmit signal going to the antenna.

8.2.3.2.4.5 Rover Small Deep Space Transponder: This paragraph describes the rover SDST (shown in some figures as RSDST) and the identical descent SDST (DSDST). During cruise, DSDST provided the X-band links, and RSDST was powered off except during two brief "aliveness" tests. In surface operations, RSDST provides the X-band links.

To distinguish from earlier SDST designs, the MSL SDST is in the "Group Buy III." Relative to earlier designs used on MER and MRO, Group III

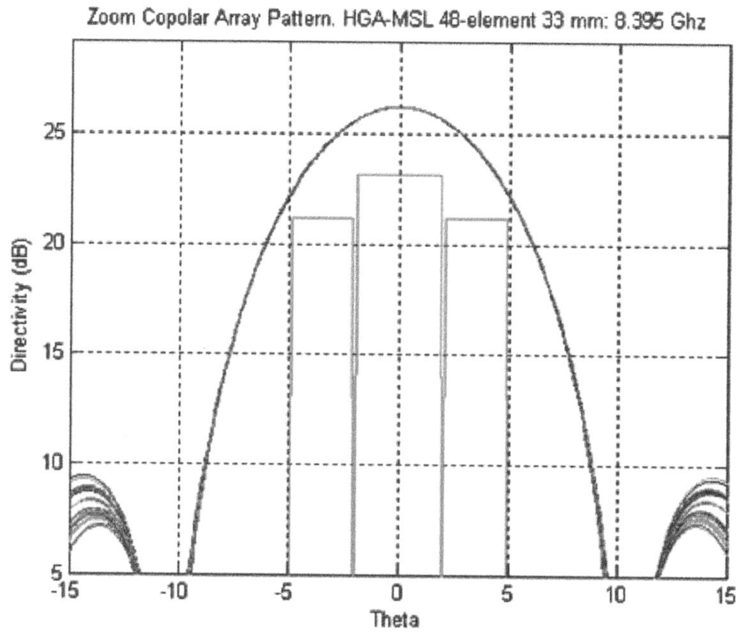

Fig. 8-52. Detail of HGA main lobe downlink directivity.

transponders have two major improvements. First, a problem has been fixed with the digital-to-analog converter in the receiver tracking loop (the "DAC glitch")[23]. This makes receiver acquisition of a swept uplink carrier frequency at any temperature easier than on previous projects. Second, Group III transponders have much less coherent leakage compared to the one used on MER-A (Spirit). As a result, the receiver static phase error (SPE) does not drift when the receiver is not locked. This also makes uplink carrier sweep frequencies easier to plan compared to MER.

Figure 8-53 shows a Group III SDST.

The SDST is composed of four different modules: the digital processing module (DPM), the downconverter module[24], the power module, and the exciter module.

---

[23] When the digital representation of the receive frequency changes from a mixture of 1s and 0s to nearly all 0s, this can cause a voltage spike at the analog output in earlier designs. This spike, more prominent at cold temperatures, can knock an already acquired SDST receiver out of lock on the uplink, particularly with sweeps in the positive direction.

[24] In a receiver, a downconverter is used to transform the signal from the passband back to the baseband for further processing. Baseband refers to the original frequency spectrum of the signal before modulation or up-conversion.

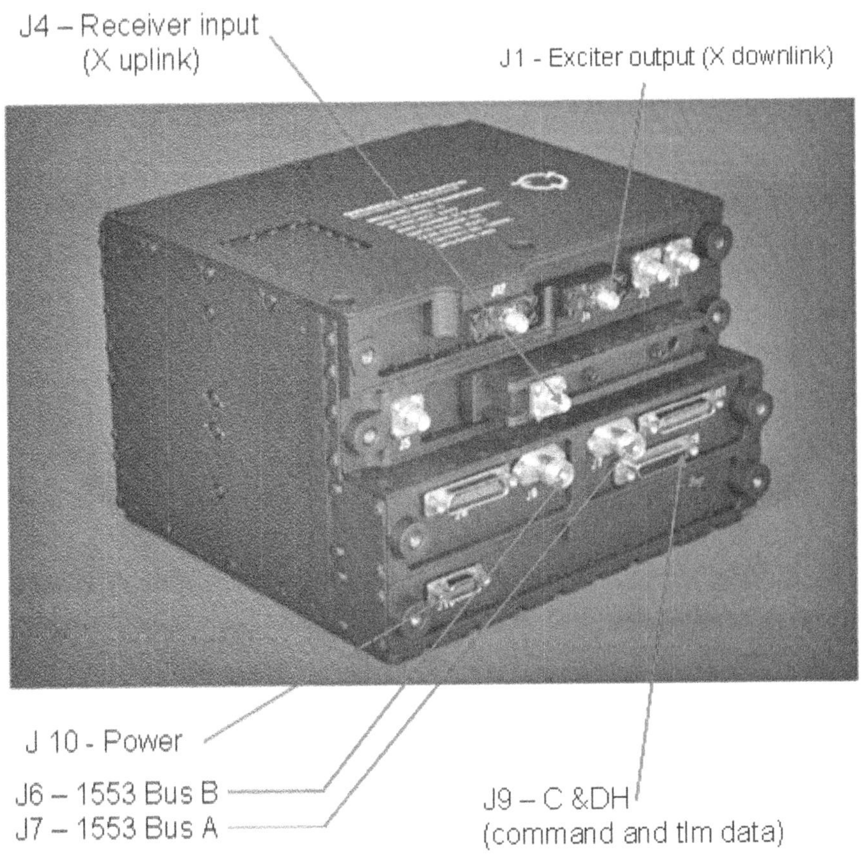

Fig. 8-53. Group III small deep space transponder. Please visit www.wiley.com/go/deepspacecommunications to access the color version of this figure.

The DPM has three main functions:

1) Convolutionally encode the data (if "coding" is enabled by 1553 control).

2) Provide X-band baseband telemetry and ranging signals to the exciter module.

3) Convert the analog output of the downconverter module into binary data.

Each SDST has two oscillators that can drive the downlink: a voltage-controlled crystal oscillator (VCXO) whose frequency is controlled by the loop's error voltage and is, therefore, related to the uplink frequency transmitted to the rover; and an auxiliary oscillator (aux osc) for which the frequency is generated on board and, therefore, varies with temperature (and to a lesser extent, atmospheric pressure).

The power converter module provides a set of steady voltages to the other SDST modules.

The downconverter module takes the 7.150-GHz received uplink signal and converts it to an intermediate frequency (IF) signal at 4/3 F1 [20][25]. The uplink signal, which may be modulated with command and ranging waveforms, gets sampled by an analog-to-digital converter (ADC) at the input of the digital processor module. These samples are provided to three "channels" to use the old analog terminology:

1) The carrier channel (for uplink carrier tracking).

2) The command channel (for demodulating the command signal).

3) The ranging channel.

The command channel has a ± 2 kHz bandpass filter centered around 2 kHz.

The ranging samples of the baseband uplink are put through a DAC to produce an analog signal. The resulting analog signal is a "turn-around" ranging waveform that modulates the downlink carrier.

Table 8-15 lists some of the SDST requirements relevant to the uplink (receive) telecom link performance. The SDST functional specification [17] provides a more complete listing.

The exciter's RF power output to the SSPA or the TWTA can be an unmodulated or modulated carrier. In surface operations, the RSDST downlink is either unmodulated (beep) or modulated by telemetry only (DTE). During cruise, the DSDST exciter module phase modulated the downlink carrier with telemetry and (if selected) either of two waveforms used for navigation.

1) Telemetry (from the DPM; this is a binary phase shift key (BPSK)-modulated square-wave subcarrier).

2) Turn-around ranging (analog, from the DPM, after its D/A converter).

---

[25] In SDST nomenclature, F1 is the fundamental frequency from which the uplink and downlink frequencies are derived. For example, the X-band downlink is 880 times F1, and the X-band uplink is 749 times F1. For MSL, operating on X-band channel 4, F1 will be approximately 9.59 MHz. The VCXO output is at two times F1.

Table 8-15. SDST receive functional characteristics.

SDST Receive Parameter	Value
Receive signal maximum power	−70 dBm (maximum to meet other performance specs) +10 dBm (maximum to cause no damage)
Carrier loop threshold bandwidth	Two bandwidth (BW) settings: 20 ± 2 Hz at receiver threshold (varies with carrier loop signal to noise ratio (SNR); max bandwidth is ~120 Hz at strong signal, 100 dB SNR) 50 Hz ± 5 Hz at receiver threshold (used in DSDST in launch phase, not planned for RSDST surface operations)
Noise Figure	< 3.2 dB over temperature, aging, and radiation, 2.1 dB typical at beginning of life (BoL), room temp
Carrier Tracking Threshold at BLF and 0-dB loop signal-to-noise ratio	–157.7 dBm typical –155.0 dBm worst case
Data Rates	7.8125 to 4000 bps

3) Differential one-way ranging (DOR) (analog, a 2 F1 [~19 MHz] sine-wave continuous wave (CW) signal, generated in the exciter module). In this case, CW refers to an analog signal as opposed to a discrete-time signal.

Table 8-16 lists some of the SDST requirements relevant to the downlink (transmit) telecom link performance; see Ref. 17 for more.

8.2.3.2.4.6 Rover Receiver Low-Pass Filter: The rover Rx LPF is of the same design as the descent stage Rx LPF. The filter rejects SSPA 'ring-around' noise (power reflected from the diplexer) so that the RSDST can detect very weak uplink signals.

8.2.3.2.4.7 Solid State Power Amplifier: The MSL SSPA is of the same design as the MER units. Figure 8-54 is an SSPA block diagram (© 2005 IEEE). The X-band SSPA consists of a solid-state RF amplifier, an electronic power converter (EPC), mode control and telemetry circuitry, and input and output isolators. Table 8-17 shows key characteristics. More detail, diagrams, and photographs of the MER SSPA are in Ref. 21.

8.2.3.3 Functional Redundancy (Rover as Backup to Descent Stage). The MSL RF switch complement allowed the RSDST to act as a back-up to the DSDST during cruise. The downside is that the rover has a weaker transmitter (15-W SSPA rather than the 100-W TWTA) and more circuit losses.

Table 8-16. SDST exciter characteristics at 880f1.

X-Band 880f1 Transmit Parameter	Value
Output power level of X-band exciter	13.0 + 3/−2 dBm over temperature, tolerance, end of life, radiation
Phase noise	<−20 dBc/Hz @ 1 Hz (aux osc mode)
Aux Osc short term frequency stability	0.06 ppm at any constant temp from 10°C to 40°C (1 sec integration measured at 5 minute intervals over 30-min span)
NCO subcarrier tone short term stability	1 ppm (as of July 2009)
NCO subcarrier tone long-term stability	50 ppm (as of July 2009)
Harmonics	<−50 dBc
In-band and out-of-band spurious	<−50 dBc
Minimum symbol rate	0 symbols per second (sps) for subcarrier; 2000 sps for direct modulation
Maximum symbol rate	Filtered mode: 4.4 million symbols per second (Msps) Wideband (unfiltered) mode: > 4.4 Msps
Modulation Index accuracy	±10%
Ranging modulation indices (peak)	4.375, 8.75, 17.5, 35, 70 deg
Ranging modulation Index accuracy	±10%
Ranging modulation index stability (over temp., radiation, and EOL)	< ±20%
Ranging delay variation over flight acceptance (FA) temperature range	< 20 ns typical
DOR modulation index (peak)	70 deg nominal
DOR modulation index accuracy	± 10%
DOR modulation Index stability (over temp., radiation, and EOL)	< ±25%
Telemetry Modulation Index	0 to 138 degrees (128 steps, approximately 1.1 degrees per step)

The difference between rover and descent stage downlink capability is about 9 dB in effective isotropic radiated power (EIRP). Except for the power difference all downlink functions were available through either path. Note that there is no provision to bypass the diplexer in the rover. The diplexer could have handled the SSPA RF power output at critical pressure during EDL and does handle it on the surface.

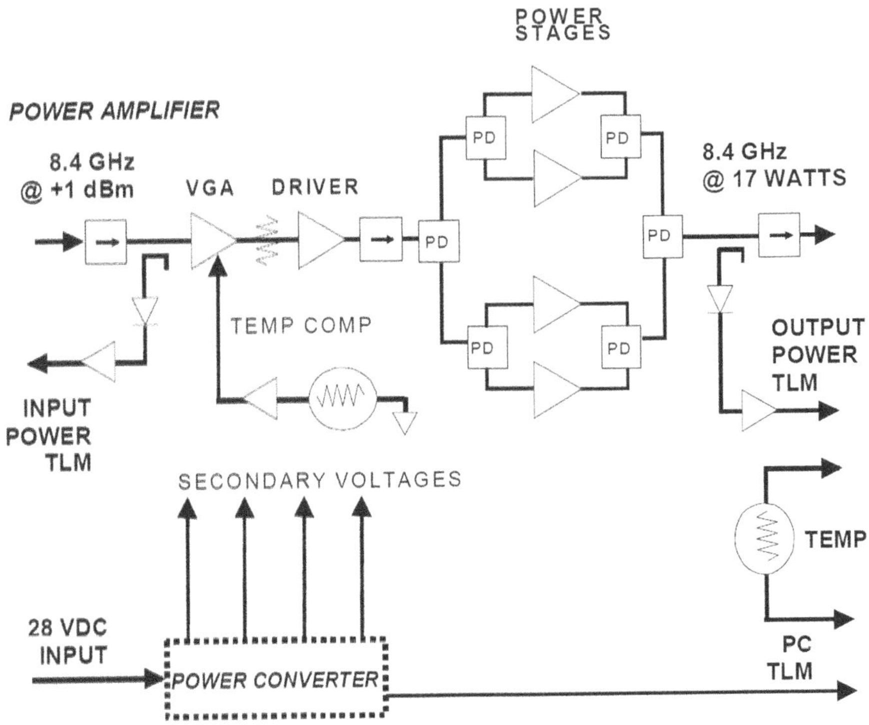

Fig. 8-54. MSL SSPA block diagram (PD = passive device in power amplifier, VGA =variable gain amplifier). From W. Boger et al., "X-band, 17 Watt, solid-state power amplifier for space applications", Microwave Symposium Digest, 2005 IEEE MTT-S International (© 2005 IEEE).

Table 8-17. SSPA key characteristics.

Parameter	Value
Frequency Range	8.395–8.455 GHz
Output Power (RF)	15 W
DC Input Power	55 W nominal, 64 W max
Output voltage standing wave ratio (VSWR)	1.5:1 max

8.2.3.4 Encoding Modes/Frame Sizes. The X-band downlink has three Turbo codes (1/2, 1/3, and 1/6); the data rates range from 10 bps to 62,500 bps. RF spectrum management bandwidth limitations to prevent interference between missions preclude the use of the combination of 62,500 bps bit rate with turbo 1/6 coding. That combination would produce too high a symbol rate; for Mars missions, the symbol rate is limited to 300,000 sps.

An effort was made to reduce the telemetry frame size (1784 bits) at low rates while keeping some coding efficiency. MSL can use an interleave depth of 1, whereas MER had only an interleave depth of 5. From these values, the effective information rate (the ratio of information bits in a frame to total bits) on MER was 1760/3040 = 0.58; on MSL it is 1784/2072 = 0.86, an improvement of 1.7 dB. The improved efficiency means MSL has significantly less frame overhead than MER at 10 and 40 bps.

8.2.4 UHF Flight Subsystem Description

8.2.4.1 UHF Interfaces with MSL Control and Data Systems. The Electra-Lite (ELT, used for UHF relay) differs from the SDST (used for X-band DTE/DFE) in that the transmitted data rate is controlled by the radio, not the multi mission system architectural platform (MSAP) telecommunications interface board (MTIF). The MTIF clocks data into the ELT internal buffer. When the buffer-fill threshold (a settable parameter) is passed, the ELT forces the data flow control line high and the MTIF stops clocking in data until the line drops again. MSL selected three ELT buffer fill rates as baseline:

- 8250 Hz used for low transmit data rates to minimize latency (primarily for use during EDL), with a telemetry frame size of 1784 bits.
- 33,000 Hz used for transmit data rates between 2 kbps to 32 kbps (primarily for safe-mode low-latency applications), with a telemetry frame size of 1784.
- 2,062,500 Hz for normal operations, with a telemetry frame size of 8920 bits. This fill rate can keep up with even the highest transmit rates of 2 Mbps.

Two kinds of data encoding are possible for the UHF links:

1) Reed Solomon encoding (interleave depth 1 and 5) for unreliable (bit-stream or non-Proximity-1 protocol) MSL-to-Orbiter communications.

2) Checksum-frame for nominal reliable (Proximity-1) mode UHF return link (EDL and surface operations).

The ELT has redundant uplink (command) and downlink (telemetry) LVDS interfaces that are cross-strapped to the MTIF card (each MTIF has four command/telemetry ports). The active downlink port on the telecom side must be selected via a 1553 command. Both uplink ports in ELT are always active.

The rover bus controller (BC) on the MTIF controls the primary 1553 bus. ELTs are connected via the remote terminal (RT). The MTIF resides in the rover computer element (RCE).

8.2.4.2 UHF Key Hardware Components. Refer to Fig. 8-31 for the UHF components on the parachute cone, the descent stage, and the rover, as discussed in the next three sections.

8.2.4.2.1 Parachute Cone UHF Telecom Components

8.2.4.2.1.1 Parachute UHF Antenna (PUHF): The PUHF was used only from the CSS to the backshell deployment portion of EDL. Refer to Fig. 8-16 for this part of EDL. The antenna provided communication with relay orbiters over a wide range of view angles. If a major spacecraft failure such as an event resulting in a tumbling attitude had occurred, the antenna would have permitted reconstruction data to be received in all but the most extreme attitudes.

The PUHF (see Fig. 8-55) was a wrap-around antenna of the type used previously on launch vehicles and during the Phoenix lander EDL. It was designed and manufactured by Haigh-Faar of New Hampshire, in close cooperation with JPL.

The antenna consisted of four segments individually mounted on the parachute cone, and connected via a one- to-four power divider. Each segment had two radiating patch antenna elements, making a total of eight radiating patch antennas in a conical array.

Excitation of the antenna was via a coaxial cable between the descent stage switch D-UCTS and the one- to-four power divider mounted on the inside of the parachute cone. The pattern was semi-omnidirectional, roughly azimuthally symmetric, with a null aligned with the spacecraft –Z axis. Pattern cuts for a full 0 to 360 degrees in theta (red labels), for different phi angles (rotation about the –Z axis) overlaid, are shown in Fig. 8-56.

The figure is a polar plot with angle from the reference direction around the plot's circumference and the gain increasing radially outward. The reference

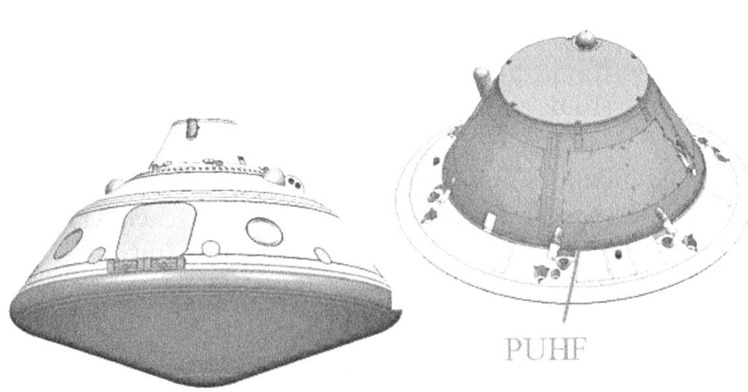

Fig. 8-55. PUHF antenna mounted on parachute cone. Please visit www.wiley.com/go/deepspacecommunications to access the color version of this figure.

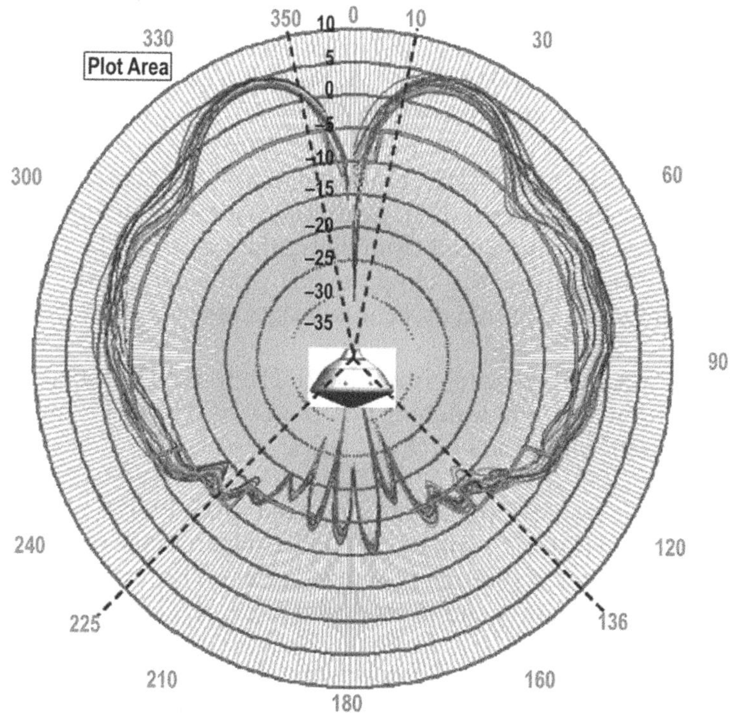

Fig. 8-56. PUHF antenna required and measured radiation pattern (angles and curves as defined in text). Please visit www.wiley.com/go/deepspacecommunications to access the color version of this figure.

direction is the –Z axis (theta = 0 deg) at the top of the figure. Theta angles are labeled in red 0 to 360 degrees clockwise. Gain is labeled from −35 dBi near the center of the plot to +10 dBi at the outer edge.

The several colored curves in this figure plot gain vs. theta. They are analogous to the curves for the HGA in Figs. 8-49 through 8-52. However, the HGA figures are rectangular plots with theta on the X-axis and gain on the Y-axis.

- Required gain: indicated by the thicker red curve.
- PUHF measured gain: indicated by the remaining six curves. Each curve shows the gain vs. theta at a fixed phi (0, 30, 60, 90, 120, or 150 deg).

The figure shows that the PUHF gain pattern generally met the requirement (measured gain greater than required gain) for the range of theta angles specified. Over most theta angles, the six phi curves do not differ from each other by more than 6 dB. The antenna had a narrow but deep null along the –Z axis, and its performance within about 20 deg of the +Z axis was highly variable with small changes in pointing angle.

8.2.4.2.2 Descent Stage UHF Telecom Components. The UHF components on the descent stage are the D-UTS switch and the DUHF antenna (plus connecting cables). The DUHF (Fig. 8-57) is a sleeve dipole design that provides an azimuthally symmetric pattern.

As shown in Fig. 8-58, the DUHF transmit pattern was significantly affected by the descent stage hardware [22]. Scattering from the descent stage distorted and moved the dipole null from along the −Z axis. The pattern coverage, however, was sufficient to close the link to the relay assets during the critical few minutes of powered descent and sky crane activity.

The D-UCTS was used to switch from the PUHF to the DUHF at backshell deployment. To avoid "hot switching" of the D-UCTS, UHF transmission was stopped briefly while the switch actuated.

The antenna cable from the parachute cone (where the PUHF was mounted) to the descent stage (DUHF) and the antenna cable from the descent stage to the rover (RUHF) passed through the mega-cutters. The two cables were severed at backshell separation and rover deployment, respectively.

8.2.4.2.3 Rover UHF Telecom Components. The Electra-Lite radios are mounted inside the rover WEB, as shown in Fig. 8-59. ELT-A and ELT-B are redundant radios. ELT-A has been in use since EDL, and this use will

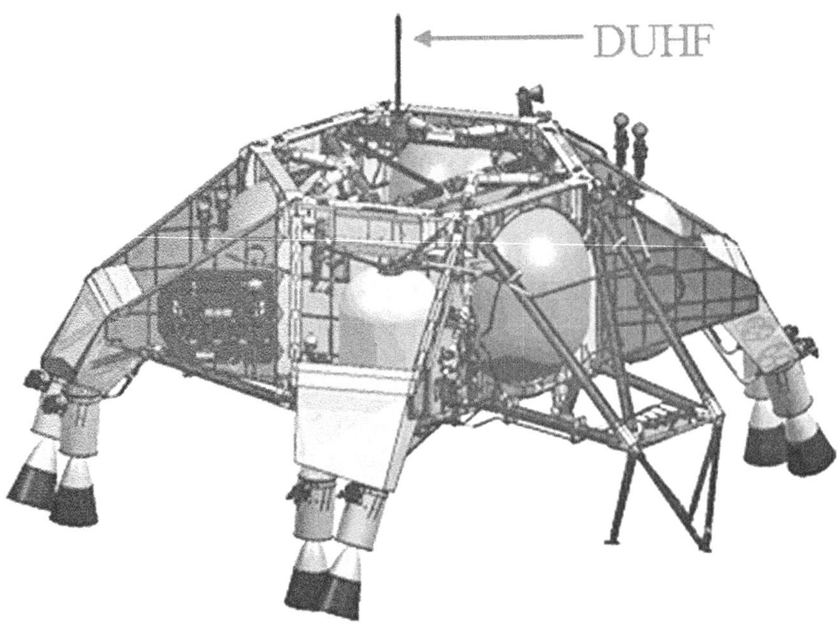

Fig. 8-57. DUHF antenna mounted on descent stage. Please visit www.wiley.com/go/deepspacecommunications to access the color version of this figure.

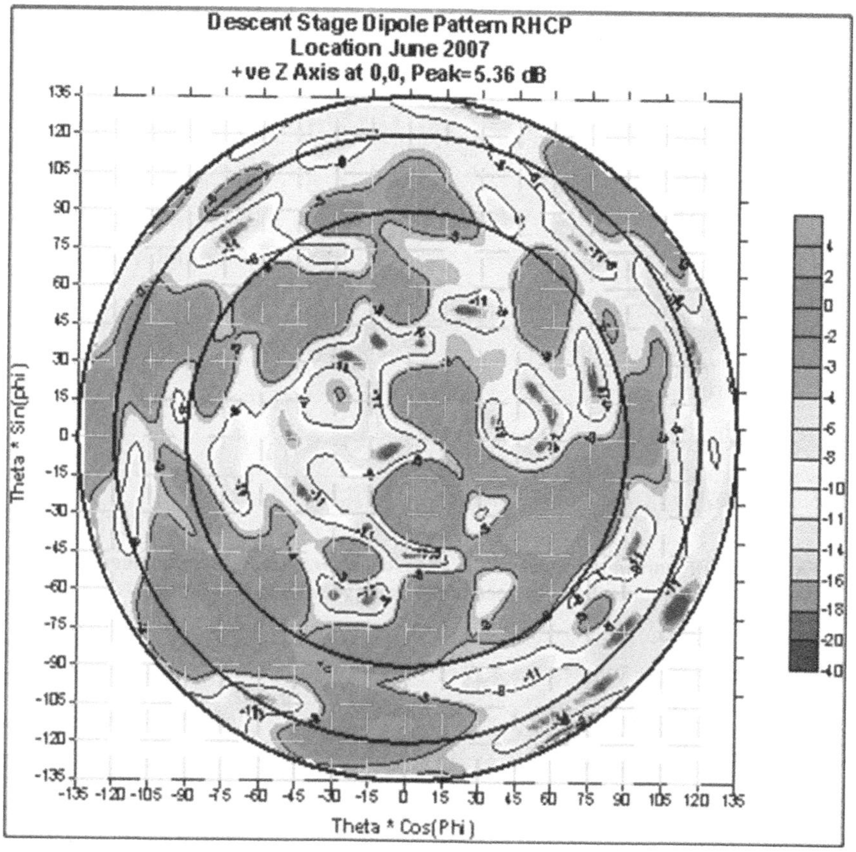

Fig. 8-58. Transmit pattern for DUHF antenna mounted on descent stage. Please visit www.wiley.com/go/deepspacecommunications to access the color version of this figure.

continue unless there is a problem with it. The ELT-B radio is mounted on a bracket (made from AlBeMet[26]) above the ELT-A radio. The R-UCTS mounts on the side of the bracket. Because there is extra thermal isolation of the ELT-B due to the bracket, ELT-B would run warmer than ELT-A during return-link transmission. ELT-B thermal control would only be an issue if ELT-B was required for a long overflight, and the WEB thermal environment was severe.

The RUHF is a quadrafilar helix antenna specially designed for the MSL mission. It is used for all of the surface activities in the mission and was used

[26] AlBeMet is the trade name held by the Brush Wellman Company for a beryllium and aluminum composite material derived by a powder metallurgy process. AlBeMet is formed by heating fine beryllium and aluminum powder under high pressure to form a uniform material. These alloys are significantly less dense than aluminum. (http://en.wikipedia.org/wiki/AlBeMet)

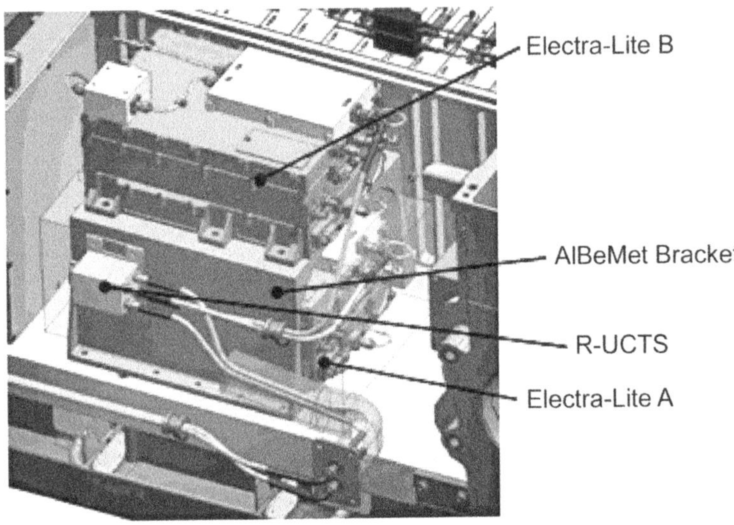

Fig. 8-59. UHF hardware in rover WEB. Please visit www.wiley.com/go/deepspace communications to access the color version of this figure.

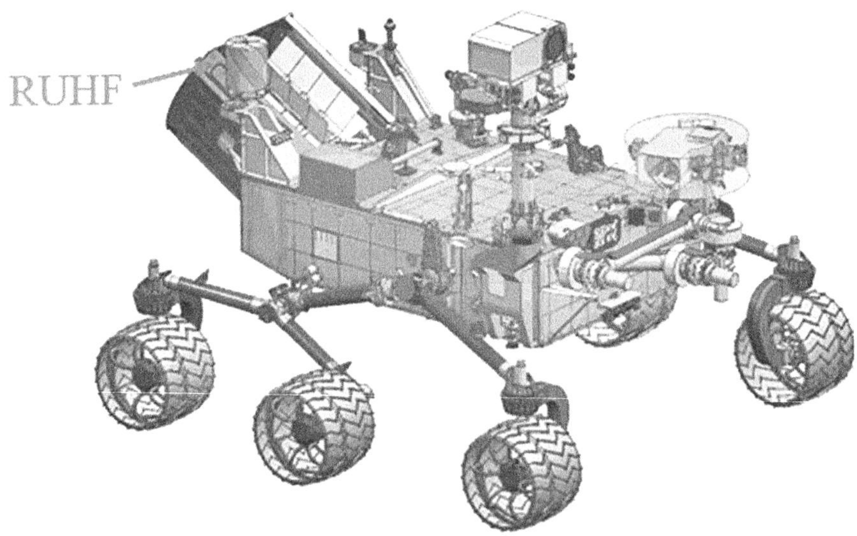

Fig. 8-60. RUHF antenna mounted on rover. Please visit www.wiley.com/go/deepspace communications to access the color version of this figure.

during the last portion of EDL. Figure 8-60 shows the RUHF mounted on the rover. Figure 8-61 shows a detailed view of the antenna itself.

The EDL sequence commanded the R-UCTS to switch UHF from the DUHF to the RUHF for the sky crane activity (Fig. 8-62). During this activity, the RUHF was shadowed partially and to a varying degree by the descent stage above it. In this sketch, the term BUD refers to the bridle, umbilical, descent rate limiter device.

Fig. 8-61. MSL RUHF Quadrafilar helix antenna. Please visit www.wiley.com/go/deepspacecommunications to access the color version of this figure.

To study the effects on the pattern, we performed an analysis using WIPL-D at three representative heights (distances) below the descent stage (as illustrated in Fig. 8-62). The results at the 401 MHz return link frequency are shown in the three parts of Fig. 8-63. It is evident from the top part that, as expected, the RHUF pattern shows much distortion initially (when the rover is still close to the descent stage (1 m away). Near the end of the sky crane deployment (7.5 m away from the descent stage), the bottom part of the figure shows a pattern similar to the surface pattern. (The surface pattern at 401 MHz is in Fig. 8-66.)

After the landing, the broad pattern of the RUHF has provided coverage at RCP over most of the sky to very low on the horizon.

A WIPL-D analysis [19] was performed early in the telecom development to assess the pattern distortion in surface operations due to the rover deck, the RTG, and other objects in close proximity to the antenna. At the UHF frequencies, most of the deck and its payload can be considered close to the antenna. Fig. 8-64 shows the WIPL-D model.

Knowledge of the RUHF performance on the Martian surface is critical to the mission. The RHUF 437 MHz forward link pattern and the 401 MHz return link pattern were measured using an MSL mockup in a fashion similar

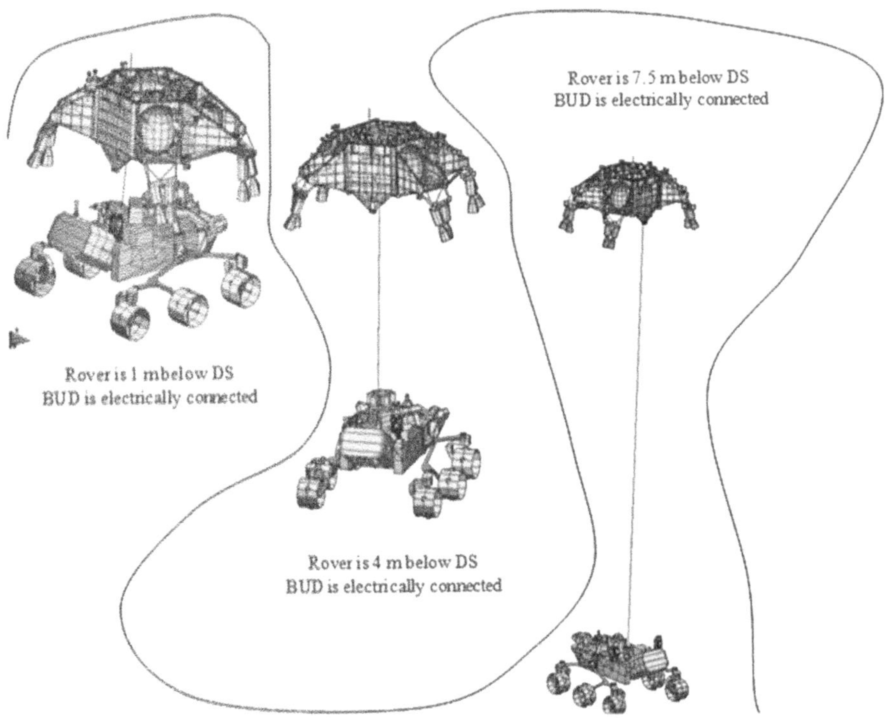

Fig. 8-62. RUHF pattern study for sky crane. Please visit www.wiley.com/go/deepspace communications to access the color version of this figure.

to the measurements made for the Phoenix lander mission [23]. These measurement patterns are included in the data volume prediction tool used by the flight team for planning sol-to-sol activities.

Figures 8-65 and 8-66 show the analysis results, including ground effects [24], for the return and forward links, respectively. The patterns are significantly better than the corresponding MER UHF monopole patterns, with no deep nulls except at the low elevation angles near the horizon.

8.2.4.2.3.1 Electra-Lite Radios: The dual-redundant Electra-Lite radios (ELT-A and ELT-B) implement the functions for relay communications with the Mars Odyssey and MRO orbiters (and other compatible orbiters such as MEX). ELT-A has provided the rover UHF communications for EDL and all surface operations so far. ELT-B was tested for "aliveness" during cruise, and is standing by during surface operations.

The ELT is a variant of the JPL Electra style software defined radio (SDR) that is intended for use in landers. The name Electra-Lite refers mainly to a reduced weight relative to the standard Electra and also to a lower power consumption. The only function not present in the ELT is a half-duplex mode signal path. The ELT provides full capability in all the other radio functions

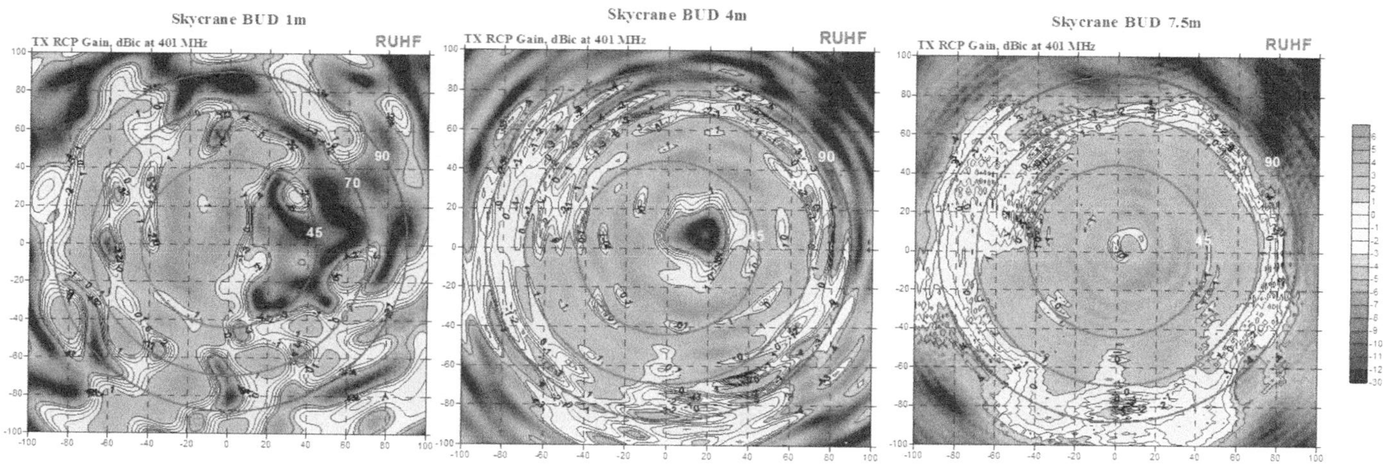

Fig. 8-63. RUHF pattern analysis results for sky crane BUD at 1 m, 4 m, and 7.5 m. Please visit www.wiley.com/go/deepspace communications to access the color version of this figure.

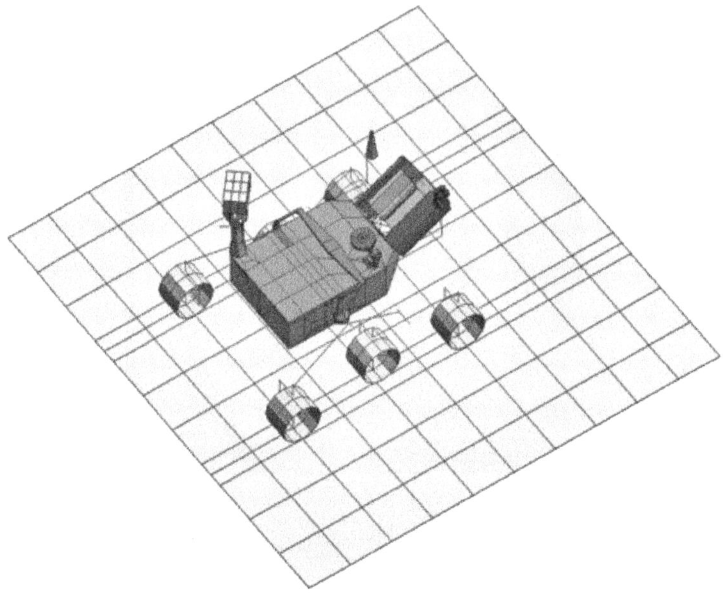

Fig. 8-64. WIPL-D model for RUHF pattern analysis. Please visit www.wiley.com/go/deepspacecommunications to access the color version of this figure.

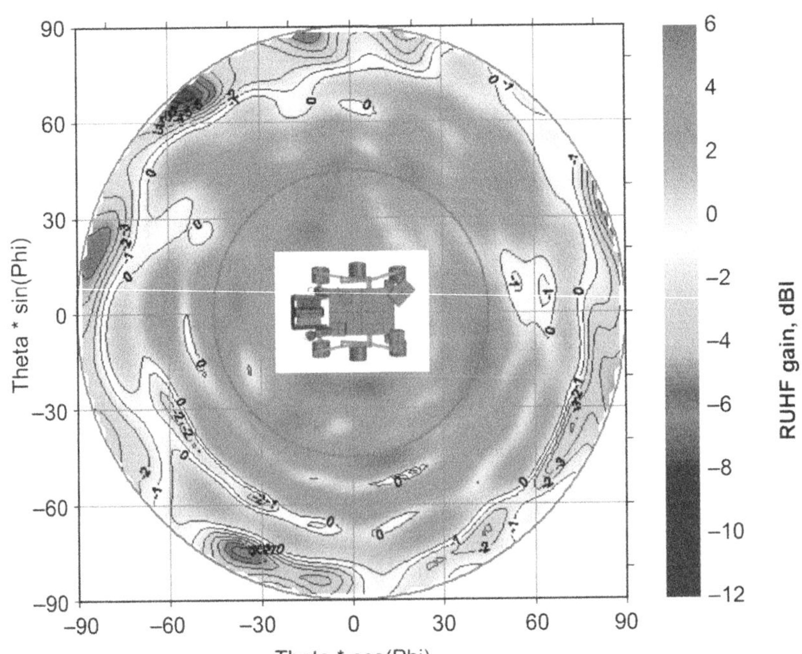

Fig. 8-65. Surface RUHF antenna pattern, Rx. Please visit www.wiley.com/go/deepspacecommunications to access the color version of this figure.

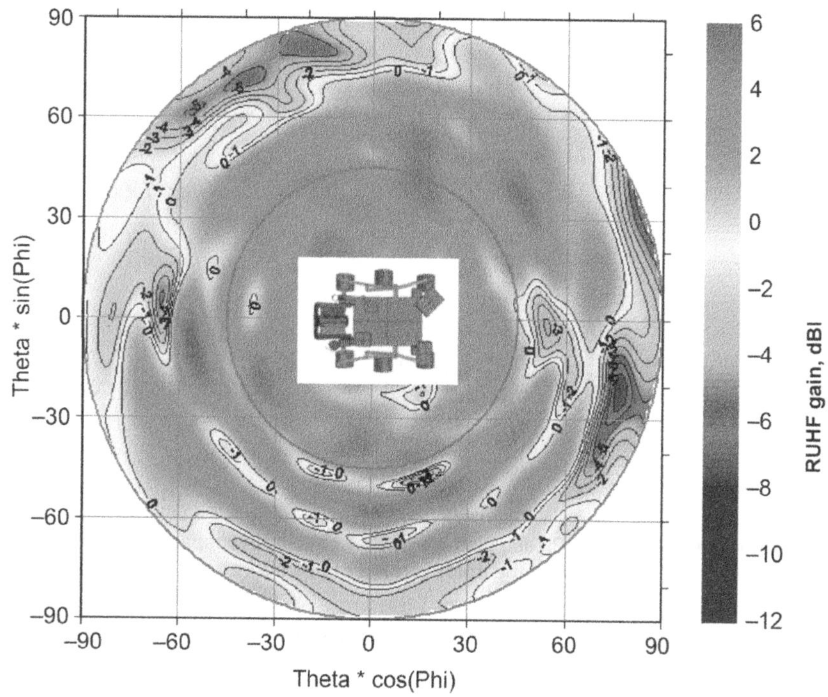

Fig. 8-66. Surface RUHF antenna pattern, Tx. Please visit www.wiley.com/go/deepspace communications to access the color version of this figure.

of the standard Electra, and Ref. 25 provides more detail on the Electra and Electra-Lite radios.

Unlike the MSL X-band subsystem with its transponders and its separate power amplifiers (TWTA or SSPA) and diplexers, each ELT has an integrated transponder, power amplifier, and diplexer. The UHF RF power delivered to the antenna from the diplexer is greater than 8.5 W.

The flight radio FM-002 (ELT-A) is shown in Fig. 8-67.

The downlink and uplink operational parameters are summarized in the next two subsections, followed by information on the Prox-1 parameters, buffer data management, and other ELT functions.

8.2.4.2.3.1.1 ELT Downlink: Downlink is the link from MSL to the orbiter, also referred as the return link.

- Downlink rates from 2 kbps to 2048 kbps are available. The CE505 radio onboard Odyssey can only support rates of 8, 32, 128, and 256 kbps. The nominal return link with MRO uses suppressed carrier modulation with adaptive data rates (ADR) during a relay pass (overflight).

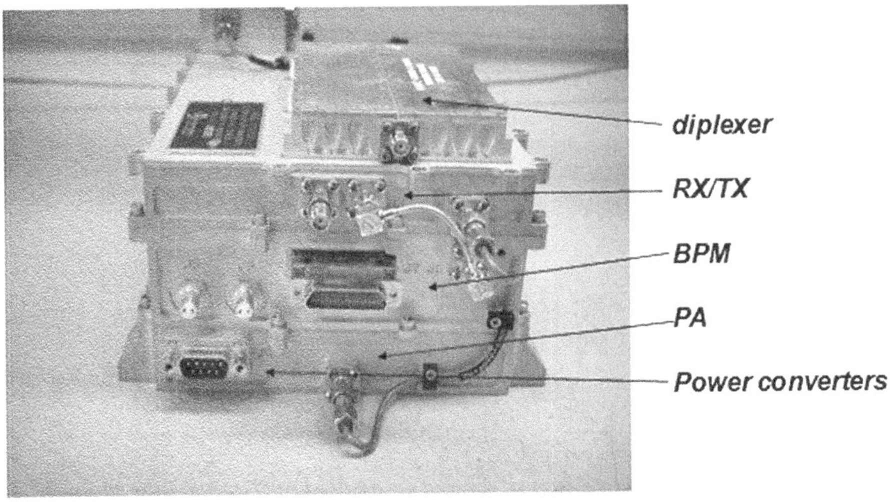

Fig. 8-67. Electra-Lite FM-002. Please visit www.wiley.com/go/deepspacecommunications to access the color version of this figure.

Odyssey links must use residual carrier with a single data rate for each relay pass. Which data rate or the ADR mode (allows for multiple data rates) is used is determined by the "hail" by the orbiter.

- Throughput efficiencies at either the MRO or MSL end of the link, or both, limit the effective maximum rate to approximately 1.35 Mbps, depending on whether there is significant forward data being sent by MRO.
- Bypass and (7,1/2) convolutional coding.
- As defined by the Consultative Committee for Space Data Systems (CCSDS) Proximity-1 standard [26], three frequency channels are available: Channel 0: 401.585625 MHz; Channel 1: 404.4 MHz; or Channel 2: 397.5 MHz. Which channel the ELT uses is determined by the hailing from the orbiter.
  - Channel 0 was the baseline at the time of landing and is the only one compatible with Odyssey (so is used with Odyssey).
  - With MRO, however, relay operations during the early sols confirmed the presence of return link electromagnetic interference/electromagnetic compatibility (EMI/EMC) issues with the preferred MRO science instrument operating modes. From in-flight link characterization and performance tests between MSL and MRO, we found the best return link frequency to MRO was 391 MHz. This assessment also included antenna patterns and gains.
- Modulation: Residual Carrier with fixed modulation index (with biphase-L baseband modulation), and suppressed carrier (with non-return to zero, NRZ).
- Coherency enabled/disabled.

8.2.4.2.3.1.2 ELT Uplink: Uplink is the link to MSL from the orbiter, also referred as the forward link. The planned nominal forward link with MRO will use residual carrier modulation at 32 kbps. Odyssey links must use residual carrier.

- Uplink rates are from 2 to 256 kbps.
  - The following UHF uplink rates can be used by MSL: 2, 8, 16, 32, 64, 128, 256 kbps.
  - Only 8 and 32 kbps is supported by Odyssey.
- Bypass and (7,1/2) convolutional coding.
- Three CCSDS standard [26] frequency channels: 0: 437.1 MHz, 1: 435.6 MHz, 2: 439.2 MHz.
  - Channel 0 is the baseline and used for the forward link from both Odyssey and MRO.
  - There are no EMI/EMC issues identified with MRO with the use of Channel 0 for the forward link.
- Modulation: Residual carrier with fixed modulation index (with byphase-L), and suppressed carrier (with NRZ).

8.2.4.2.3.1.3 Proximity-1 Parameters (Forward and Return Links)

- Sequence controlled (reliable) link is the nominal protocol.
- Bit stream (unreliable) mode that bypasses the Proximity-1 protocol was used for EDL and is available for off-nominal conditions in surface operations.
- The adaptive data rate mode (ADR), avaliable when relaying between MRO and MSL, takes advantage of the ability of the Proximity-1 protocol to command different data rates on the fly. (ADR is available only with the MRO radio, not with Odyssey. Also with MSL only the return link data rates are being changed this way; the baseline forward rate between MRO and MSL is fixed at 32 kbps.)
- The baseline hailing interaction data rate for MSL is 8 kbps.

8.2.4.2.3.1.4 Buffer Data Management (Forward and Return Links)

- ELT provides a transmit (downlink) flow control signal to the MTIF using an additional LVDS line ("ready for data") so that the ELT transmit buffer does not overflow. Flight software (FSW) can also set various buffer parameters (buffer depth and two watermarks) to control the latency in the transmit link.
- On the MSL receive (forward link) side, there is no flow control. The content of the Proximity-1 transfer frame data field is sent out of the LVDS line as soon as the frame is validated.

8.2.4.2.3.1.5 Other ELT Functions

- Collection of Proximity-1 time-packets via the 1553B interface is nominally planned for every overflight.
- Collection of radiometric data via 1553B is nominally done only in troubleshooting scenarios.
- Wake signaling is baselined for use only in a fault-protection situation. Send "spacecraft wake-up" signal to LVDS upon receipt of Proximity-1 hail from an orbiter.

8.2.4.2.3.1.6 UHF Coaxial Transfer Switches (D-UCTS and R-UCTS): The coaxial transfer switches (in the Fig. 8-31 UHF block diagram) were used to switch between the antennas during EDL. The switch in the rover stage selected the radio to either the descent stage switch or to the RUHF antenna. During surface operations this switch selects between ELT-A and ELT-B.

These coaxial switches are double-port double-throw (position 1 and position 2) switches, manufactured by Sector Microwave. The switches have a port-to-port isolation of > 60 dB and a maximum insertion loss of 0.2 dB. Their switching time is < 50 ms. They are rated to handle as much as 15 W of RF power, with a maximum return loss of –20 dB.

8.2.5 Terminal Descent Sensor (Landing Radar) Description

The MSL terminal descent sensor (TDS), used only during the last part of the EDL activity, was a six-beam Ka-band pulse-Doppler radar designed to measure the three-axis velocity and altitude of the spacecraft from about 100 m/s and 4 km altitude until rover touchdown [27]. Specifically, the radar provided line-of-sight velocity and range for each radar beam to the spacecraft's navigation filter, to compute the three-axis velocity and altitude of the descent stage.

Figure 8-68 is a system overview block diagram of the TDS. Table 8-18 details the TDS high-level performance and physical characteristics. Chapin (2011) [27] provides the high-level sensor specifications and physical characteristics of the TDS.

Although the TDS was not book kept during development as part of the telecom subsystem, it transmitted and received RF power. The MSL X-band telecom subsystem as configured for EDL included an exciter low-pass filter and transmitter low-pass filters to ensure that the radar's performance was not degraded by the simultaneous operation of the X-band SDST and TWTA during the powered descent and sky-crane portions of the descent.

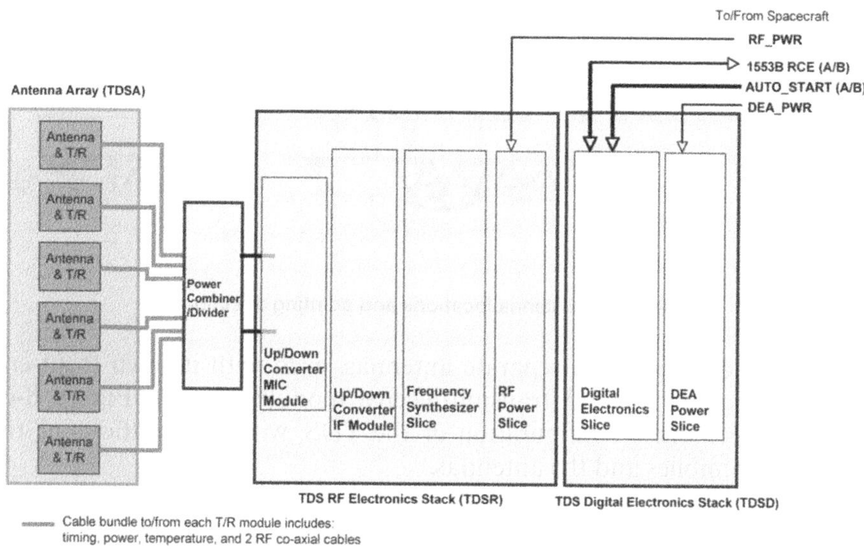

Fig. 8-68. System overview block diagram of the TDS (T/R = transmit/receive, TRM = transmit/receive module). Please visit www.wiley.com/go/deepspace communications to access the color version of this figure.

Table 8-18. TDS high level characteristics.

Parameter or condition	Value
Center frequency	36 GHz
Antenna beamwidth	3 deg
Transmit power (per beam)	2 W
Pulse width	4 to 16 ns
Altitude above Mars	6 to 3500 m
Velocity of descent stage	200 m/s maximum
Spacecraft bus power	30 W
Transmitting	120 W
Mass	25 kg
Dimensions	1.3 × 0.5 × 0.4 m

The TDS hardware consists of the TDS Digital Stack (TDSD), the TDS RF Stack (TDSR), the RF power combiner and divider, and the antenna (TDSA). The block diagram shows the electrical interfaces among the TDSD, TDSR, and the TDSA.

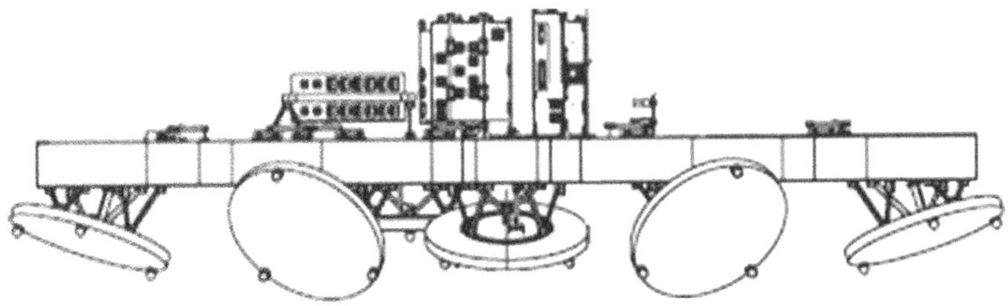

Fig. 8-69. TDS antenna locations and pointing directions.

The TDSA consists of six separate antennas, each with its own front-end filter assembly (FFA) and transmit/receive module (TRM). Figure 8-69 shows the mechanical configuration of the TDS, with the locations of the electronic assemblies and the antennas.

The TDSD is a stack of two "slices" of electronic assemblies: the digital electronics assembly (DEA) slice and the digital power distribution unit (DPDU) slice.

The TDSR is a stack of four "slices" of electronic assemblies: the RF power distribution unit (RPDU) slice, the frequency synthesizer slice, the up/down converter intermediate frequency (IF) module (UDIM) slice, and the up/down converter microwave integrated circuit (MIC) module (UDMM) slice.

The power slices supplied conditioned direct current (DC) voltages to the electronic assemblies. The frequency synthesizer slice consists mainly of a voltage-controlled oscillator and supporting circuitry that generated the reference frequencies for the DEA slice and the RF electronics. The frequency upconverter generated the Ka-band pulse for the transmitter, and the frequency downconverter module converted the received signal down to video frequency for digitization.

The DEA served as the controller and digital signal processor of the TDS. The DEA consists of a 1553 transceiver chip supporting 1553B command and telemetry transfers and a large (~1 mega gates) reprogrammable field-programmable gate array (FPGA), referred to as the radar processor (RP), for generating timing signals for the TDS and handling all the digital signal processing of radar data. A 12-bit ADC was utilized to digitize radar video signal with a bandwidth as high as 240 MHz. A radiation-hardened scalable processor architecture (SPARC) processor served as the radar controller (RC); this item handled spacecraft commands, time tagging, post-processing, packaging of telemetry messages, and other functions.

The TDS cycled through the antenna beams, making measurements with the six antenna beams one at a time at 20 Hz (50-ms intervals). A limited number of beam sequences were pre-stored in the TDS memory to be recalled by using a beam sequence table ID command. The use of the beam sequence allowed the exclusion of an anomalous beam or a blocked beam.

In Normal mode, the TDS was set up to make line-of-sight velocity and range measurements at 20-Hz rate by cycling through the radar beams according to the specified beam-cycling pattern. The Normal mode of TDS operation included Acquire, Dop1, and Dop2 dwells. A dwell consisted of a group of radar pulses with identical radar parameters. The Dop1 and Dop2 dwells were executed only when a target was detected (valid range) in the Acquire dwell. In Dop1, the TDS determined the Doppler velocity by using the pulse-pair Doppler technique. In Dop2, the TDS used a different inter-pulse period (IPP) to resolve the Doppler velocity ambiguity. We also determined the slant range in Dop2.

8.2.6 MSL Telecom Hardware Mass and Power Summary

8.2.6.1 X-Band Mass. Table 8-19 lists the masses for the major X-band telecom assemblies, as defined in the MSL Mass Equipment List (MEL) [28]. The table is organized by the stages previously defined: cruise, EDL (combined parachute cone and descent), and rover.

8.2.6.2 X-Band Spacecraft Power Consumption. Table 8-20 summarizes the power consumption of the active subsystem elements (SDSTs and power amplifiers) at a nominal bus voltage of 28 V.

The SDST values are indicated for receiver-only operation, receiver and exciter in the non-coherent mode, and receiver and exciter in the coherent mode. The exciter is required for a downlink, and the non-coherent mode also requires the auxiliary oscillator to be powered. The DSDST was powered during cruise and EDL, and the RSDST is powered for X-band activities in surface operations.

When the TWTA is on, it has two modes: standby (the unit producing no RF output) and operating. When the TWTA is off, it draws no power.

Throughout cruise and EDL, the DSDST (receiver and exciter) and the TWTA remained powered. For the most part, the TWTA remained in the operating mode, being sequenced to the standby mode only briefly during antenna switches.

During surface operations, only the RSDST receiver is powered on in the background mode or when a DFE is active. The RSDST exciter and the SSPA are also powered on when a DTE or a beep is active.

Table 8-19. X-band telecom mass summary.

Spacecraft Stage	X-Band Subsystem Element	Quantity	Unit Mass (kg)	Total Mass (kg)
Cruise	MGA and adapter	1		0.65
EDL backshell	PLGA	1		0.4
	TLGA	1		0.4
	WTS	2	0.45	0.9
	Microwave components			2.8
EDL descent	DLGA	1		0.35
	Microwave components			2.5
EDL descent telecom plate	TWTA	1		2.5
	WTS	2	0.45	0.9
	SDST	1		3.0
	Telecom plate	1		6.6
	Microwave components			2.5
Rover X-band	SDST	1		3.0
	SSPA	1		1.4
	WTS	2	0.45	0.9
	RLGA	1		0.4
	HGA	1		1.4
	HGA gimbal	1		6.6
	Microwave components			3.7
TOTAL				**40.9**

8.2.6.3 UHF Mass and Power Consumption. The TDS had four power states[27]:

- **Off** – Both power switches from the 28-V spacecraft bus were open (no power to the TDS).
- **Low** – The power switch to the digital power distribution unit (DPDU) was closed while the power switch to the RF power distribution unit

[27] The TDS was powered off during cruise, except for self-checks when other power modes were active for brief periods. The TDS modes were controlled by an onboard sequence during EDL.

Table 8-20. X-band telecom spacecraft power consumption.

Subassembly	Descent Stage Telecom Subsystem Mode	Nominal Bus Voltage (28.0 V)
DSDST	RX	11.4
	RX + noncoherent TX (TLM only: SAFE_MODE)	14.7
	RX + Coherent TX (TLM Only: CRUISE MODE)	14.4
TWTA	Standby mode (no RF drive) (exciter OFF)	62.4
	Operating (with RF drive) (exciter ON)	175.2
RSDST	RX	11.3
	RX + noncoherent TX (TLM only: SAFE_MODE)	14.8
	RX + coherent TX (TLM only: CRUISE_MODE)	14.2
SSPA	Exciter OFF (no RF drive) Exciter ON	45.1 62.9

(RPDU) remained opened. At this lower power state, only the DEA was powered on, and the RP (Xilinx[28] field programmable gate array [FPGA]) was not running because there was no 1-GHz clock to the FPGA. The power draw in this state was about 30 W.

- **Quiescent** – Both power switches were closed, essentially powering up both the DEA and the RF electronics. The TDS was in standby mode in which the high-power amplifiers in the TRMs were properly biased but not transmitting.
- **Full** – Radar timing was running in the RP and pulsing the TRMs to transmit RF power. The power usage is about 120 W.

8.3 Ground Systems EDL Operations: EDL Data Analysis (EDA)

EDL was a critical and the most anticipated communications mission phase. Obviously, for purposes of redundancy and signal diversity, all possible communications paths were to be used to their fullest extent. Both X-band and UHF were on when they could provide signals to be received. The DSN operated with multiple redundant antennas. MSL transmitted a UHF return link

[28] Xilinx, Inc. is the inventor of the field programmable gate array (FPGA). An FPGA is a semiconductor device that can be configured by the customer or designer after manufacturing—hence the name "field-programmable." FPGAs contain programmable logic components called "logic blocks," and a hierarchy of reconfigurable interconnects that allow the blocks to be "wired together"—somewhat like a one-chip programmable breadboard. (Wikipedia:Xilinx and FPGA).

to MRO and Odyssey (and MEX as backup). Also the Parkes Observatory in Australia tracked the UHF carrier.[29]

In addition to the standard closed loop receivers, the DSN antennas were also connected to a special EDL Analysis (EDA[30]) system that performed fast Fourier transform (FFT) signal processing on the signal captured by the open-loop radio science recorder (RSR) receivers.

A NASA Tech Brief [29] documents the MER EDA, which is described as a system of signal-processing software and computer hardware for acquiring status data conveyed by M-FSK tone signals transmitted by a spacecraft during descent to the surface of a remote planet.

MSL undertook an EDA rebuild to use modern computer hardware and the Linux operating system.

8.4 Telecom Subsystem Link Performance

8.4.1 X-Band

8.4.1.1 Cruise Link Performance. Telecom performance prediction and analysis played a major role during cruise and approach. In particular, many of the checkout and subsystem maintenance activities scheduled during these mission phases required relatively high telemetry data rates with the consequent low link margin.

During cruise there were five main X-band functions: downlink carrier tracking, uplink command, downlink telemetry, differential one-way ranging (delta-DOR), and turnaround ranging. Because the turnaround ranging signal levels during late cruise were low relative to threshold (especially on the downlink), ranging was the most difficult of the functions to achieve.

For surface operations, the primary X-band functions are uplink command, downlink carrier-only beeps, and downlink telemetry. Because the landing

---

[29] On schedule at 05:16 UTC on August 6, the signal was detected. It was slightly stronger than expected. Parkes tracked the descent of the rover until just after the parachute deployment and heat-shield separation. Less than two minutes before scheduled landing, MSL dropped below the Martian horizon, out of sight of Earth. See http://www.parkes.atnf.csiro.au/people/sar049/msl_tracks/edl/ for details.

[30] The EDA was first developed for MER and is described in Chapter 7. The EDA, which has been updated for MSL, is a system for carrier and tone detection and tracking the high Doppler dynamics and low SNR of EDL.

site is well characterized in location, the turnaround ranging and delta-DOR navigation data are not required at the surface, and the two-way Doppler is rarely used.

Sequential ranging is a two-way measurement. Ranging is degraded by three sources of thermal noise:

1) Noise on the uplink (from the finite SNR on the uplink).

2) Noise in the transponder ranging channel.

3) Noise in the station receiver.

Of these, the noise in the ranging channel is the largest contributor because of the channel's bandwidth. The double-sided bandwidth is 3 MHz, adequate to pass a ranging waveform that has a clock-component fundamental frequency of 1 MHz. The receiver noise results from a relatively low spacecraft transmit power and consequently low SNR.

Figure 8-70 shows the ranging P_r/N_0 during cruise for several configurations. There are curves for early-cruise at 34-m stations on the PLGA using either the TWTA or the SSPA, for later cruise at 34-m stations on the MGA using

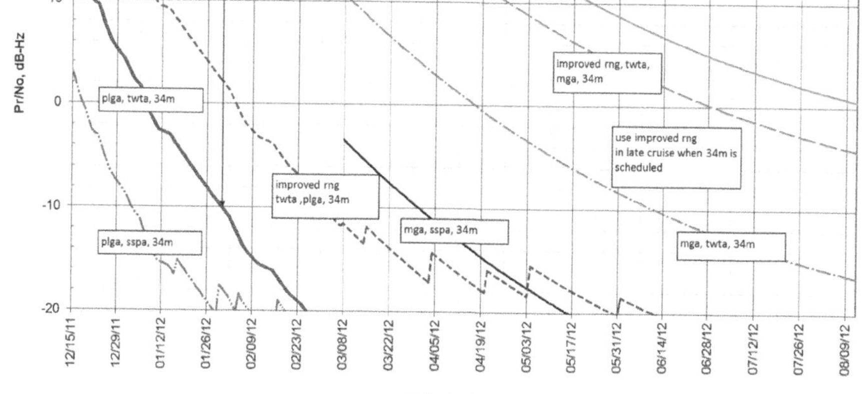

Fig. 8-70. Turnaround ranging P_r/N_0 in cruise for type Ic trajectory (rng = ranging). Please visit www.wiley.com/go/deepspacecommunications to access the color version of this figure.

either of these amplifiers, and for 70-m stations late in cruise using the MGA and the TWTA.

MSL had available the following three "improved ranging" techniques to use separately or together as necessary. "Improved ranging" is a mode that can

- Configure the SDST downlink ranging modulation index at 35 deg,
- Set the station uplink ranging mod index to 5 dB carrier suppression, the highest available, and
- Decrease the telemetry modulation index to 45 deg.

Because the X-band system performed normally, only the first was needed, and this reduced telemetry capability (bit rate) very slightly. The second would have reduced command capability (bit rate) and the third would have reduced telemetry capability (bit rate) significantly.

A more drastic form of increasing ranging performance, which fortunately was not required, is called "ranging max". "Ranging max" would have

- Set the SDST ranging modulation index to its highest value of 70 deg and turned off the telemetry modulation, and
- Turned off the station command modulation, with the station ranging mod index set to 5-dB carrier suppression.

The DSN can process ranging samples at signal levels as low as P_r/N_0 of –20 dB-Hz, albeit at the cost of integration times that increase as P_r/N_0 decreases. Figure 8-71 shows the cycle time for the collection of each ranging point as a function of P_r/N_0. This figure is based on a required minimum probability of ranging acquisition of 99% and a one-sigma ranging accuracy of 3 m. Because the cycle time becomes very large as the P_r/N_0 goes below –10 dB-Hz, –10 dB is taken as the ranging threshold. Cycle time is defined in the ranging module of the *DSN Telecommunications Link Design Handbook* [30].

$$\text{Cycle time in seconds} = T1 + 3 + (L - C)*(T2 + 1)$$

where $T1$ is the integration time of the clock component and

$T2$ is the integration time of each of the other components,
L is the component number of the last (lowest frequency) ambiguity resolving component, and
C is the component number of the range.

During the cruise mission phase, the telecom lead on the Spacecraft Team coordinated several times with the Navigation Team lead and with the DSN to define suitable ranging integration times T1 and T2 to use for the next series of tracking passes. These times are not only a function of the changing P_r/N_0 but also of factors such as the change in round-trip light time (RTLT)

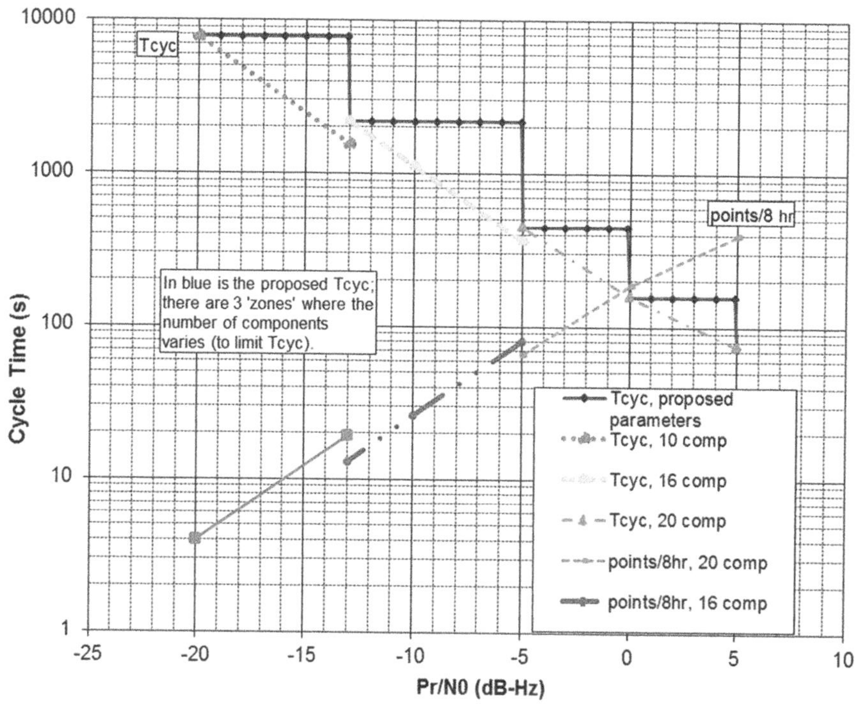

Fig. 8-71. Required ranging cycle time as a function of P_r/N_0. Please visit www.wiley.com/go/deepspacecommunications to access the color version of this figure.

during a maximum-duration station pass and the ambiguity resolution required.

The P_r/N_0 available at a given time and station/spacecraft geometry and configuration can be increased by:

- Increasing the uplink ranging modulation index
- Lowering the command modulation index by reducing uplink rate (or turning command modulation off)
- Increasing the downlink ranging modulation index
- Lowering the telemetry modulation index (or turning telemetry off)
- Increasing the ground station size from 34 m to a 70 m.

In early cruise, while the spacecraft was on the PLGA, the –Z axis had to be pointed within 80 deg of Earth to provide telecom coverage, as discussed in Section 8.1.3 and shown in Fig. 8-15. Beyond 80 deg, significant scattering off the spacecraft would make antenna pattern modeling for prediction unacceptably unreliable.

About 3 months after launch, due to the increasing Earth–MSL distance, the ranging performance drove a telecom configuration change from the PLGA to the MGA. Pointing requirements on the MGA were significantly tighter than for the MGA, as the Earth had to be within 15 deg of the –Z axis immediately after the transition from the PLGA and to 9 deg later in the cruise. The cruise pointing strategy, in fact, achieved an MGA pointing error of less than 4 deg.

The telecom subsystem capability drove much of the scheduling of station coverage during cruise, as key events required 70-m coverage to satisfy real-time data rate requirements or additional 34-m coverage to return larger data volumes.

Figure 8-72 is a prediction of uplink data rate capability from a 34-m station from launch to Mars arrival. The figure is for a Type Ic launch at window-open (Day 1). The 2-kbps maximum rate shown is the highest that the SDST can support.

Figure 8-73 shows the corresponding set of predictions for the downlink data rates. Part (a) of the figure is capability to a 34-m station, and Part (b) is capability to a 70-m station. MSL can produce telemetry rates up to a maximum of 62.5 kbps, but the highest rate used during cruise was 25 kbps.

The pointing conditions for each antenna in each figure are indicated. Given this pointing, the stair-step fall-off in capability is due to going through thresholds for each bit rate as the Earth-spacecraft distance steadily changes.

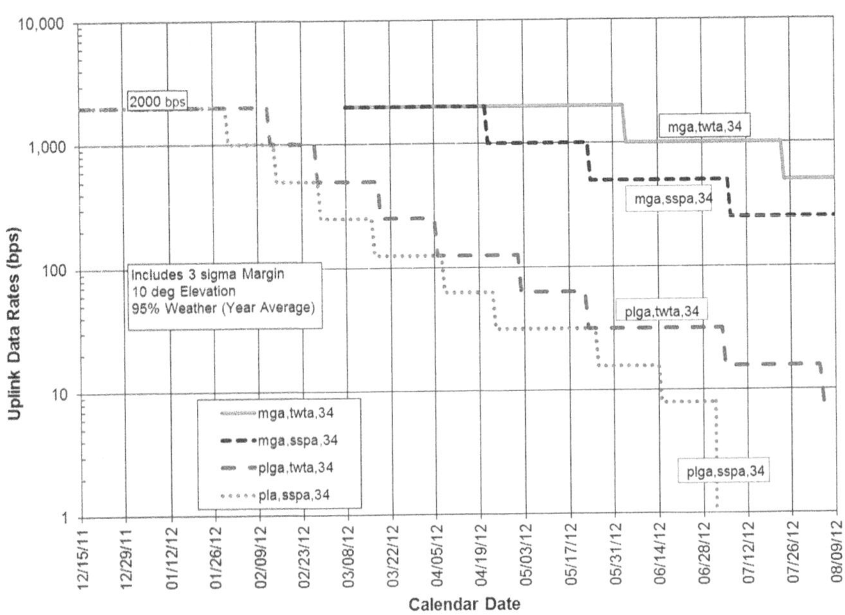

Fig. 8-72. MSL X-band uplink data rates during cruise (after TCM-1, Type Ic trajectory).

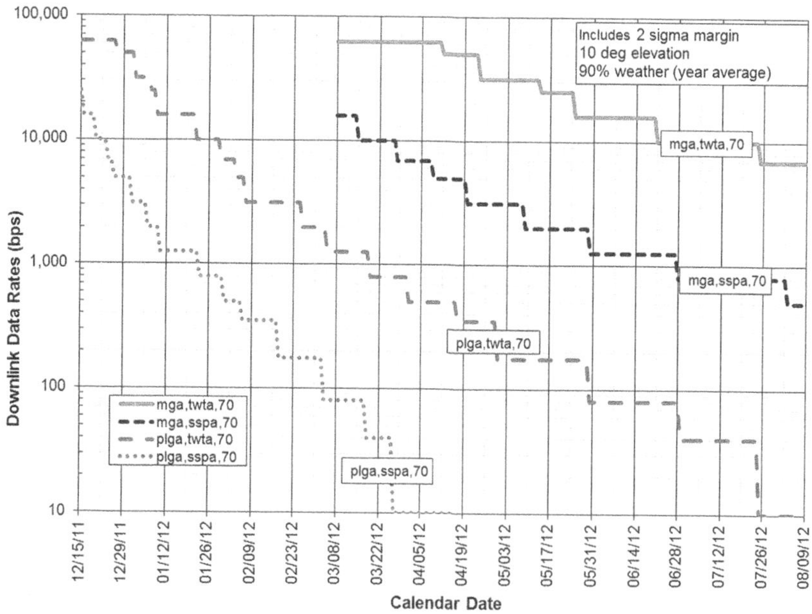

Fig. 8-73. MSL X-band downlink data rates via TWTA during cruise, type Ic trajectory day 1 (after TCM-1).

8.4.1.2 EDL (X-Band) Performance. At a planned time during EDL, the SDST telemetry mode was set to "TLM_OFF," which meant the normal engineering housekeeping, and analysis (EH&A) telemetry data stream no longer modulated the outgoing signal. After that time the X-Band subsystem transmitted information to Earth concerning the state of the spacecraft in the form of a series of discrete frequencies (referred to as MFSK tones or semaphores). Each tone was actually an unmodulated telemetry subcarrier (at a specific frequency) of the normal SDST transmitted signal. Each tone in the series indicated the occurrence of an event or condition. Each timeline segment of EDL had its unique event schedule (parachute deployment, for example), and each segment had a unique set of nominal and off nominal tones. The collection of tones, of which 256 were available, constituted the MFSK dictionary.[31]

[31] Multiple-frequency-shift keying (MFSK) is a variation of frequency-shift keying (FSK) that uses more than two frequencies. MFSK is a form of M-ary orthogonal modulation, where each symbol consists of one element from an alphabet of orthogonal waveforms. M, the size of the alphabet, is usually a power of two so that each symbol represents log2M bits. Like other M-ary orthogonal schemes, the required E_b/N_0 ratio for a given probability of error decreases as M increases without the need for multisymbol coherent detection. In fact, as M approaches infinity the required E_b/N_0 ratio decreases asymptotically to the Shannon limit of –1.6 dB. (from Wikipedia:MFSK).

The geometry of EDL was challenging for DTE links, with large view angle variations during the descent. The plot in Fig. 8-22 illustrates the type of angle variation that occurred for DTE. Because the RTLT was significantly longer than the EDL duration, EDL had to be entirely pre-programmed. Once the last command was sent, observers on the Earth were simply along for the ride, to observe the signals radioed back after all the actions had already been completed at Mars. Consequently, the real value and purpose of EDL communications (besides public outreach) was to permit reconstruction of what happened in the event of a failure.

To observe EDL as it played out over the varying Earth view angle, we relied on a series of spacecraft low-gain broadbeam antennas: the PLGA, TLGA, and DLGA. Each LGA was essentially an open-ended waveguide antenna, with the PLGA and TLGA having added parasitic dipoles to broaden the beams even more off boresight.

The ground detection system for EDL consisted of the DSN antennas and radio science receivers. They recorded and piped the raw received signal to the EDA computers. The EDA did the spectral analysis on the signal and extracted the tones from the noise. Figure 8-74 shows the EDA configuration at a station complex (for example, Goldstone): several RSRs recording data from several antenna assets and feeding data to the EDAs. As the diagram shows, the distribution of signals to the RSRs (dotted lines) was at IF via the full spectrum processing (FSP) subsystem.[32]

We did not need to co-locate the EDAs at the DSN complexes for MSL as was done for MER; instead the connections from the stations to the EDAs were via the network to JPL.

To maximize the number of semaphore tones reliably detected, we planned for as much off-nominal coverage and, therefore, far off-boresight coverage, as possible. Prior to the EDLs of the two MER rovers in January 2004 and again in the early studies for MSL, analysis had been performed to quantify the SNR levels that could be expected to produce acceptable tone detection probabilities. Figure 8-75 shows an example analysis simulation where challenging high Doppler events could cause tones to be missed. (Carrier frequency tracking results are presented using 5-Hz FFT and 1.0-s update with $T = 1$ s integration with correctly interpolated Doppler profile and the nominal rate search space of ± 40 Hz/s.) For MSL EDL, such events included the parachute deployment and the divert maneuver, when large Doppler rates and high off boresight angles occur.

[32] Distribution of telemetry and Doppler signals is via a separate IF switch to the Downlink Tracking and Telemetry (DTT) subsystem.

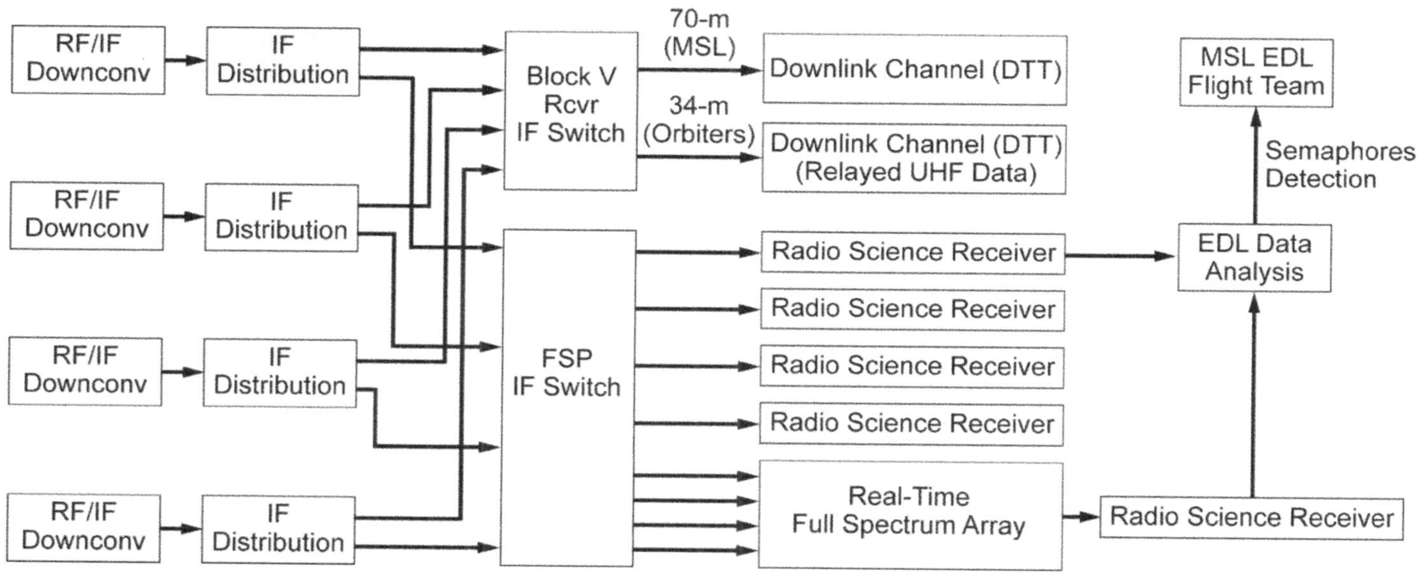

Fig. 8-74. Configuration of EDAs for EDL support.

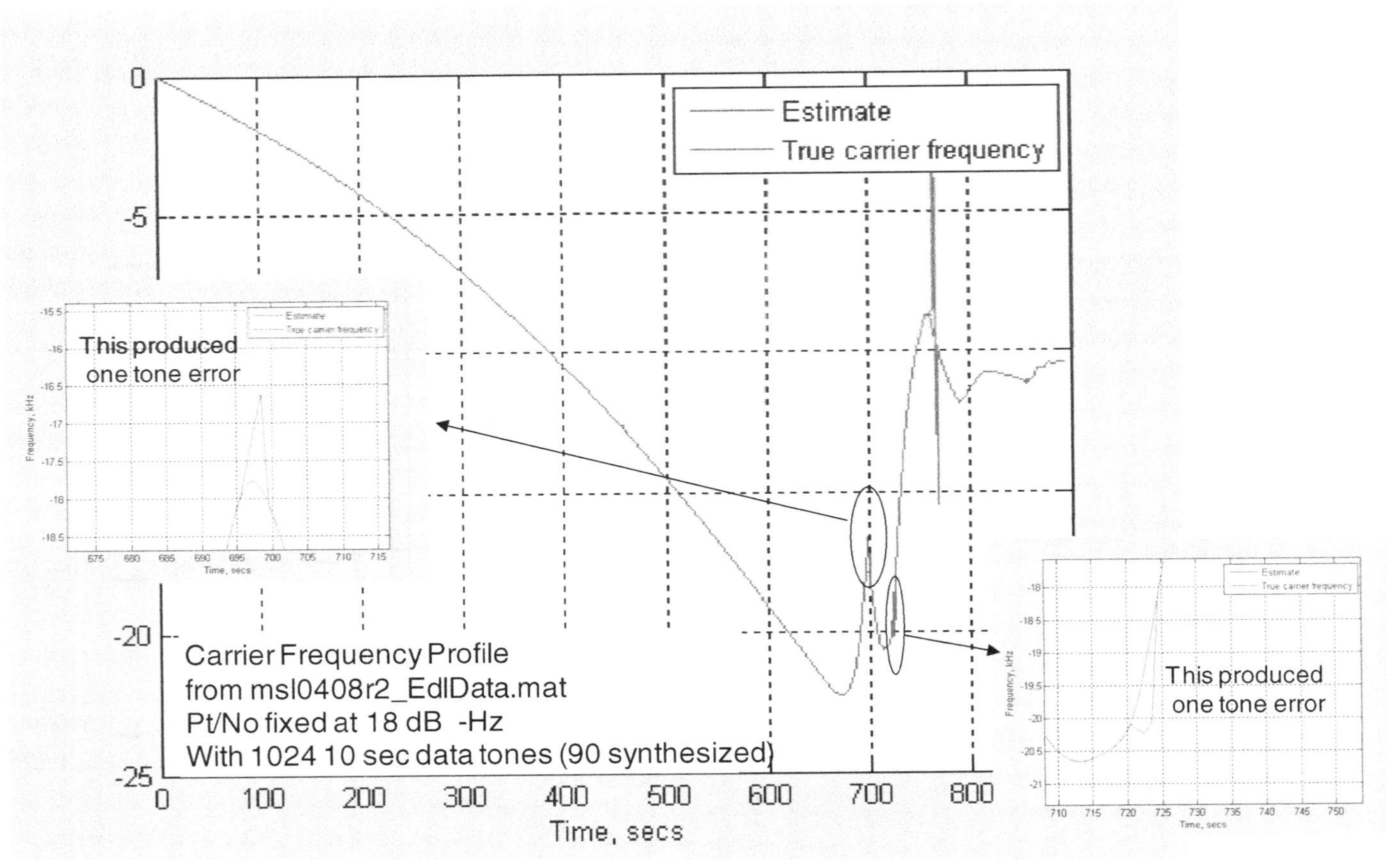

Fig. 8-75. MER EDA analysis example showing missed tone cases. Note: missed tones 70 (690–700 s) and 73 (720–730s). Please visit www.wiley.com/go/deepspacecommunications to access the color version of this figure.

When the SNR is strong, the job is easy. When the angles are far off boresight or the Doppler environment is challenging (such as when the Doppler rates and higher derivatives are large), the task becomes difficult or, in some cases, impossible. It is possible to array the DSN antennas during EDL to maximize the probability of tone detection if we expected difficulties in detection. DSN antenna arraying was studied for MSL use but determined not necessary.

Nevertheless, in some cases, and assuming nominal entry profile, the DTE signal was predicted to be reasonably strong, such as in the case shown in Fig. 8-76, for the Gale landing site.

In this example, we see the pre-entry period has a sufficiently strong DTE signal, near the 30 dB-Hz level, well above the low dynamics threshold for high-probability correct detection. This period also has very low Doppler rates, and therefore the Doppler predictions, were expected to be accurate. The left halves of Figs. 8-76, 8-77, and 8-78 show, respectively, the signal level, the Doppler frequency, and the Doppler rate during this "low dynamics" period. In this relatively benign environment, DTE was expected to be easily achievable.

In the subsequent "high-dynamics" period, shown in the right halves of these figures, the UHF return link relayed via an orbiter would be the more reliable signal.

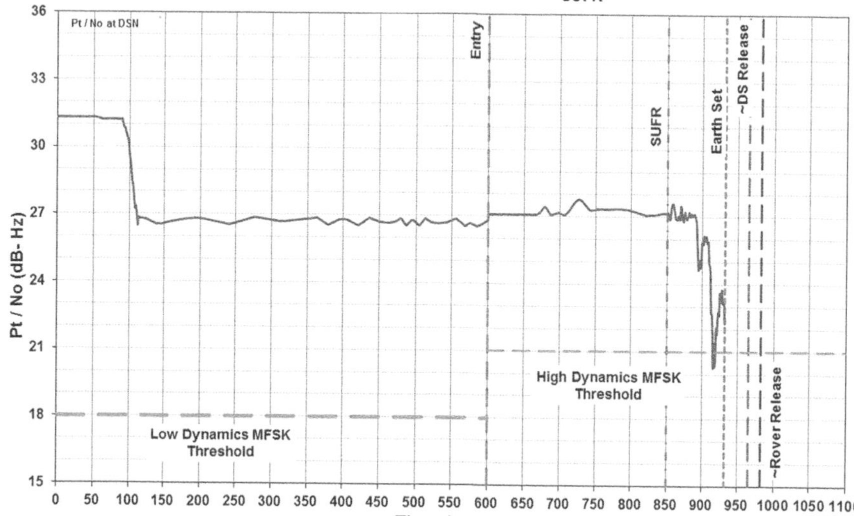

Fig. 8-76. DTE P_t/N_0 at DSN, Gale site, 2011 launch. Please visit www.wiley.com/go/deepspacecommunications to access the color version of this figure.

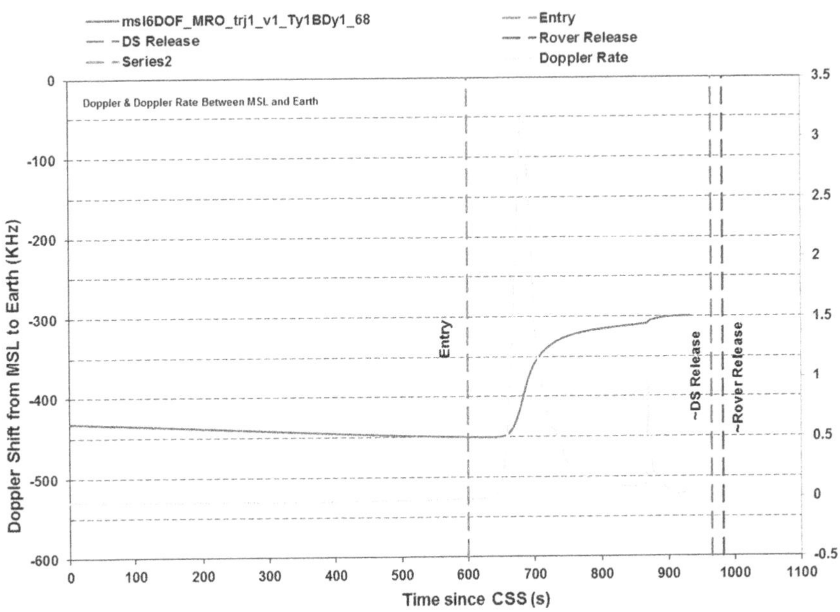

Fig. 8-77. DTE Doppler Frequency Shift and Doppler Rate at DSN, Gale site, 2011 launch. Please visit www.wiley.com/go/deepspacecommunications to access the color version of this figure.

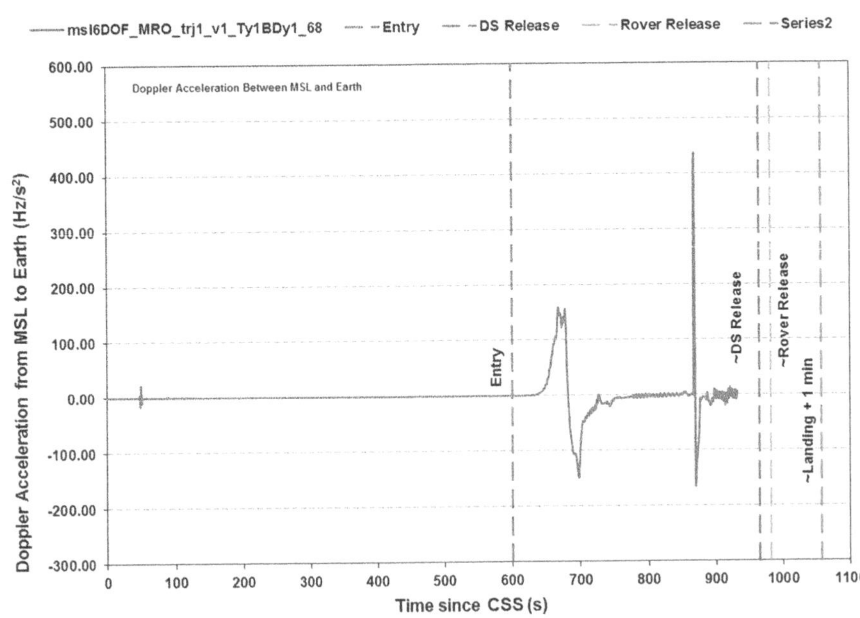

Fig. 8-78. DTE Doppler Acceleration at DSN, Gale site, 2011 launch. Please visit www.wiley.com/go/deepspacecommunications to access the color version of this figure.

After entry, the large deceleration during hypersonic entry produced large Doppler rates and acceleration (shown in Figs. 8-77 and 8-78), and the view angles also became challenging. Reliable X-band DTE provided a separate link for EDL events during the period of UHF plasma blackout, which coincided with the large deceleration period. The DTE reliability, however, depended largely on the accuracy of the profile predictions, for the receiver to be able to compensate for the Doppler changes.

Planning for real-time DTE coverage after landing was ultimately limited by the Earth setting below the horizon at the landing site, the time of which was affected by the launch date. Planned coverage was to landing plus one minute.

8.4.1.3 Surface Performance (X-Band). Throughout surface operations, the X-band requirements are:

- Data return from the HGA (at least 160 bps at max range, 34-m BWG station, 5-deg HGA pointing error).
- Command capability via the HGA (225 kbps in 20 min, equivalent to a 190 bps uplink rate, via a 34-m, with 5-deg HGA pointing error).
- Emergency command capability via the RLGA (support of safemode 7.8125 bps uplink rate via a 70-m antenna, assuming the Earth is 70 deg off the RLGA boresight).

X-band surface communications, unlike cruise, assumes that turnaround ranging is never required—we know where Mars is and have other means of determining more accurately than ranging where the rover is on Mars. Therefore, these DFE command rate capabilities assume only command modulation is on the uplink.

The Telecom Forecaster Predictor (TFP) tool, first operationally used to predict spacecraft-DSN links for the Deep Space 1 mission in 1998, has been since updated for newer missions, including MSL. The first year of surface operations included maximum range and a solar conjunction in April 2013. Excluding the solar conjunction within a Sun–Earth–probe angle of 3 degrees, the requirements above have all been met: 176 bps downlink rate and 1 kbps uplink rate on the HGA, and 7.8125 bps uplink rate on the RLGA.

The X-band DFE is used for transmitting command sequences almost every morning on Mars. X-band DFE communication sessions include allowance for a 7-minute preparation period to set up the communication window and a 10-minute rover activity keep-out afterwards for post-pass processing.

The X-band DTE downlink is used to send limited amounts of engineering data to Earth independent of the relay orbiters. Though the rover can receive DFEs or transmit DTEs independent from each other, the practice is almost always to open a single communication window per sol, for either a DFE only or a combined DFE/DTE.

X-band communications are often combined into a single pass station pass per sol, with the pass duration made long enough to accommodate verification of DFE commanding. Verification can be via the DTE telemetry that follows, accounting for RTLT. More often, verification is by the station receiving a separate carrier-only "beep" that follows the DFE window.[33] After the first 400 sols on the surface, the combined DFE/DTE was scheduled about once a week, with the sols in between DTEs having beeps only.

The next three figures are for 2011 mission surface operations, with arrival from a Type Ic cruise trajectory. Figure 8-79 shows how uplink performance varies with time for the HGA and also with off-Earth angle for the RLGA. The DFE command data rate capability for 34-m and HGA never falls below 500 bps. In contrast, the 34-m/RLGA capability dips as low as 15.625 bps at larger Earth–Mars ranges, assuming a 40-deg off-boresight angle and as low as 7.8125 bps at a 70-deg off-boresight angle. With a 70-m station scheduled, the RLGA DFE capability is always at least 31.25 bps, disregarding 70-m antenna aberration effects[34]. Note that the "40-deg" and "70-deg" curves refer to different RLGA off-boresight angles.

Figure 8-80 shows how limited the X-band DTE downlink capability is (SSPA on the surface instead of the TWTA used during cruise). Telemetry via the X-band downlink is available only via the HGA.

Figure 8-81 shows that downlink P_t/N_0 via the RLGA is too low to support even the minimum 10 bps downlink rate at high elevation angles over a

---

[33] The beep is the simplest form of providing an X-band downlink capability to send discrete messages (such as "operation normal," or "need help") at the equivalent of very low bit rates. This concept is similar to an EDL-type signaling scheme, but could have much longer integration times than the minimum 10-second interval between semaphores at EDL.

[34] Aberration is the name of an effect resulting from the Earth-Mars geometry. Station antenna pointing toward Mars is optimized for the signal being received on the downlink at a given time. The antenna pointing for an uplink that was transmitted a round trip light time earlier has a slight error when Mars' coordinates (declination and right ascension) are changing. The 70-m antenna beamwidth is narrow enough that aberration-caused uplink pointing error can cause the received uplink to be several decibels lower than it would be without pointing error. The 31.25-bps capability may become 7.8125-bps capability with aberration. MSL takes aberration into effect either by optimizing the pointing for uplink during commanding or by reducing the command rate. Aberration works on both uplink and downlink. Optimizing the pointing for the uplink will decrease the downlink capability for any mission that is transmitting to that station from Mars.

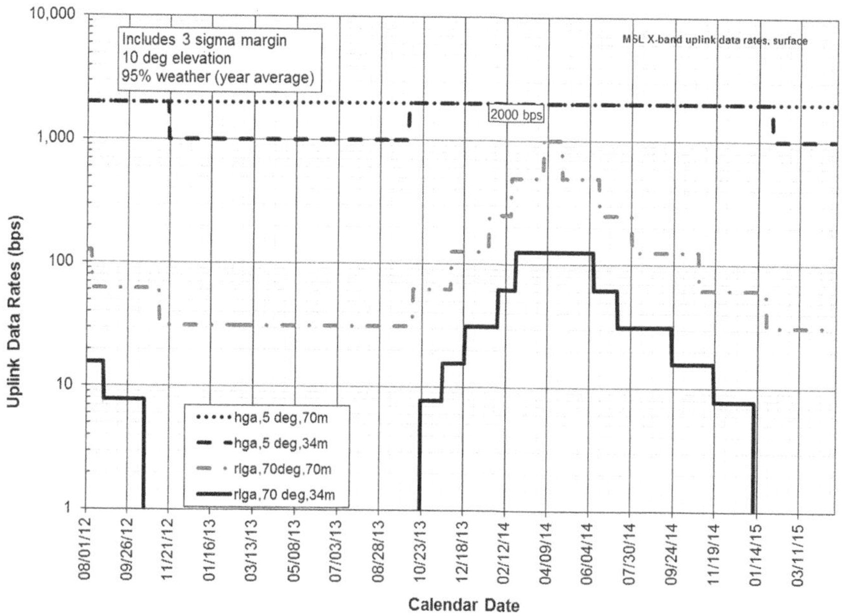

Fig. 8-79. Supportable X-band uplink data rates, Mars surface.

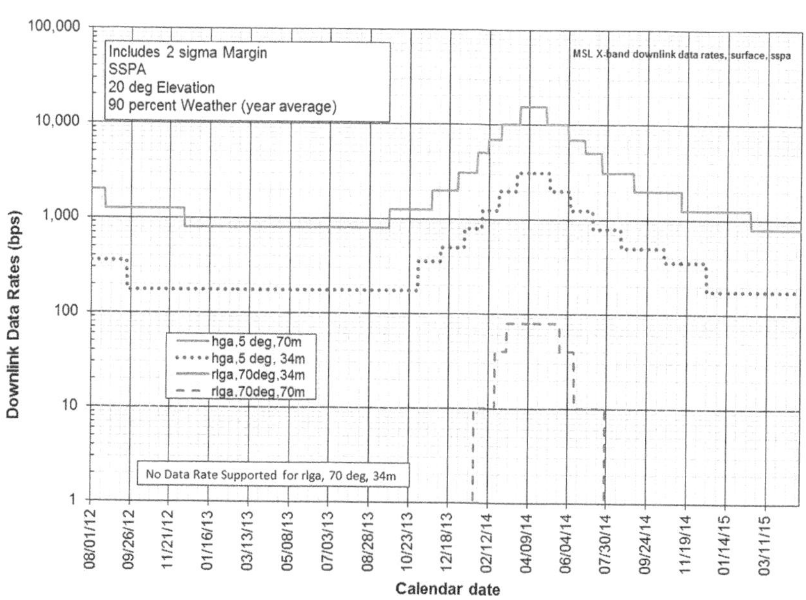

Fig. 8-80. Supportable X-band downlink rates via SSPA, Mars surface.

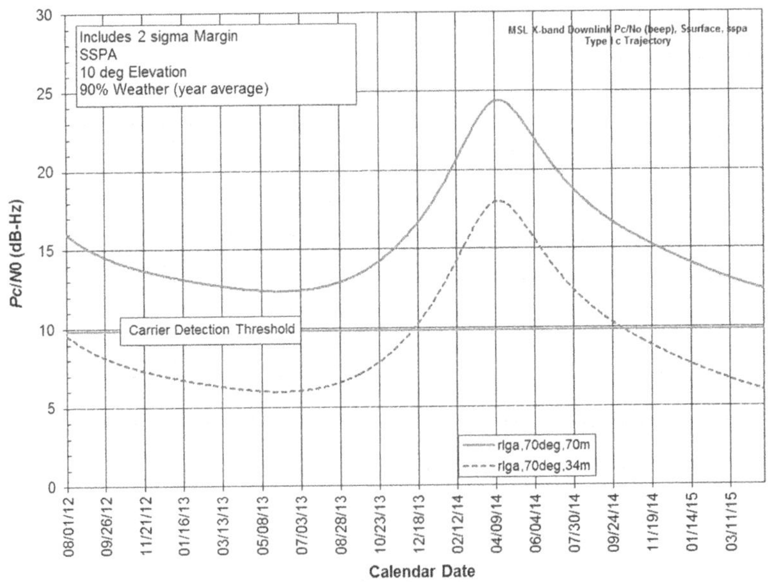

Fig. 8-81. X-band downlink P_t/N_0 for carrier-only "beep" via SSPA, Mars surface, Type Ic trajectory.

significant fraction of the Earth-Mars ranges. It is this lack of telemetry capability that has caused MSL (like MER before) to sequence carrier-only beeps. The unmodulated DTE "beep" carrier to a 34-m station is above threshold via the RLGA throughout the span of Earth-Mars distances.[35]

8.4.2 UHF

The two prime functions for the UHF subsystem have been relay support during EDL and surface science data relay.

8.4.2.1 EDL (UHF). One of the top (Level 1) requirements for deep space missions is that the spacecraft provide communication to Earth of data throughout all mission critical events at a rate sufficient to determine the state of the spacecraft in support of fault reconstruction. Previous missions landing on Mars provided data communication during the critical events of

---

[35] RLGA beeps include both the runout beep that signals if the new sol's sequence did not take over and the off-nominal beep that signals if fault protection has detected a problem requiring safe mode. To ensure receipt of potential off-nominal beeps, the project provided for radio science support going into the second year of surface operations. The RSR, used in parallel with the station's closed-loop receiver, works at P_t/N_0 several dB lower than the 10-dB closed-loop threshold. Radio science post-processing can be used to search for a suspected beep over an extended frequency range and after the fact.

EDL (Mars Pathfinder [MPF] in 1997 and MER in 2004) by making use of DTE semaphore tones to indicate the spacecraft condition. The X-band link and the semaphore tones used during EDL for the two MER spacecraft provided an increase in information content compared to that of MPF.

As compared with MER, the use of EDL guided entry and propulsive descent on MSL required subsystems whose status would change more quickly than on MER. These subsystems had more moving parts that were moving more quickly than MER's airbags. The higher degree of activity resulted in a need for a higher information rate than tones alone could provide during EDL[36].

The use on MSL of a UHF communications relay to a Mars orbiting asset during EDL greatly enhanced communications capability by providing spacecraft telemetry. MER successfully demonstrated a UHF link with the MGS orbiter for the terminal descent (post parachute deployment) portion of EDL. However, the MER mission opted to not pursue any UHF capability prior to lander/backshell separation due to the significant development risks of placing a UHF antenna on the backshell.

Similar to their support provided for the Phoenix EDL, both MRO and Odyssey orbiters were used to relay MSL lander data to the Earth.[37] After extensive study of MSL visibility by asset (and redundancies) across the full range of launch and arrival periods (both primary and contingency), it was concluded that, for latitudes in the 45 deg S to 45 deg N range, DTE coverage using MFSK tones was considered the primary planned source of information from cruise stage separation to at least entry. MRO and Odyssey UHF relay coverage were considered the primary telecom link from at least entry to rover landing. The amount of overlap—that is, the time when both links would be useful—depended on the landing site selected.

Even a UHF telemetry link would not allow all faults to be detected, particularly if the failure interfered with the spacecraft's ability to maintain the link. X-band semaphores were used to provide information on the major events and anomalies during EDL. This is an example of how X-band and UHF were complementary during EDL, and both continue in use for surface operations.

---

[36] The information rate conveyed from tones is quite limited in comparison to true spacecraft telemetry. For MSL, one tone every 10 seconds from an alphabet size of 256 provided an information rate of 8/10 = 0.8 bps. The MSL X-band telemetry data rate providing the greatest margin would have been only 10 bps. In contrast to the 0.8 bps effective X-band DTE rate, MSL UHF telemetry provided 8000 bps during EDL.

[37] The MEX orbiter also provided a secondary EDL relay opportunity in addition to either MRO or Odyssey as prime.

MRO could not provide delay-free (bent-pipe) relay, as it first recorded the return-link relay data as it was received from the descending spacecraft during the overflight and then sent the data to the DSN. Both relay of demodulated telemetry (also called unreliable return link bit-stream reception) and open-loop recording of the modulated UHF carrier (also called canister mode in CE505 radio terminology) were considered for the relay reception onboard the orbiters. Open-loop recording had been successful during Phoenix EDL. An advantage of open-loop recording is that the signal would still be recorded even if the carrier dropped below lock threshold. Thus, the possibility would exist to recover data lost in real time with non-real time analysis techniques. In contrast, with bit-stream mode, telemetry transmitted during periods of carrier unlock would be lost.

Based on predicted signal strength and variability, MSL chose the bit stream mode for EDL relay to all three orbiters.

To achieve sufficient return-link signal strength, it was necessary for the relay orbiters to turn to point their UHF antennas at MSL to the best of their capabilities. The baseline plan was that they would point to the descending MSL within 30 deg of their UHF antenna boresights for EDL.

The link to the orbiter was characterized through the EDL descent phases by a large change in range, large variations in antenna view angles, high Doppler rates, and, consequently, large changes in signal to noise ratios. The next three figures show the range (Figure 8-82), the view angles (Fig. 8-83), and the Doppler and received power (Fig. 8-84) for Gale crater and the 2011 launch. In each figure, part (a) is for the link to MRO, and part (b) for the link to Odyssey. Figures 8-84 (a) and (b) illustrate that there was sufficient post-entry margin to MRO and Odyssey to close the link most of the time.

As highlighted, UHF blackout was likely to occur in the period from 600 to 700 seconds after CSS. A reliable X-band DTE would be desirable during the UHF blackout. Discontinuities at about 965 and 982 seconds after CSS represent changes in predicted performance as the UHF link changed from the PUHF to the DUHF and then to the RUHF (Fig. 8-19).

The received-power plot analysis does not include "signal smearing" due to high Doppler rates, which can be many tens of hertz per second. Smearing degrades the signal-to-noise ratio (SNR) by effectively spreading the signal over many frequency bins and decreasing the signal-to-noise ratio by on the order of 10 dB in the worst cases (during maximum deceleration for example). Signal level and Doppler profile predictions were made for the specific launch date and landing site. After-the-fact analysis of the open-loop recorded data from MRO would have proved invaluable if a major entry failure had occurred, making the actual X-band Doppler rate profile significantly different from

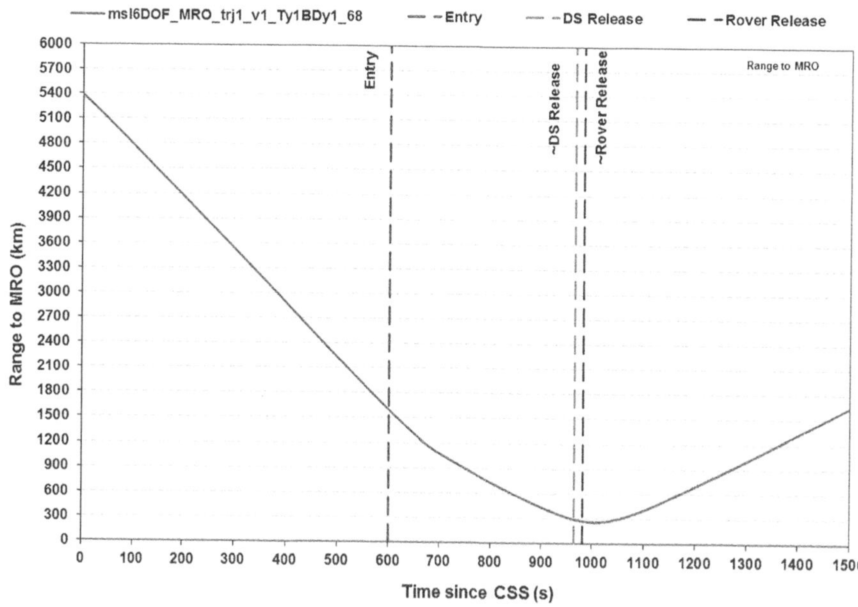

(a) Rover to MRO

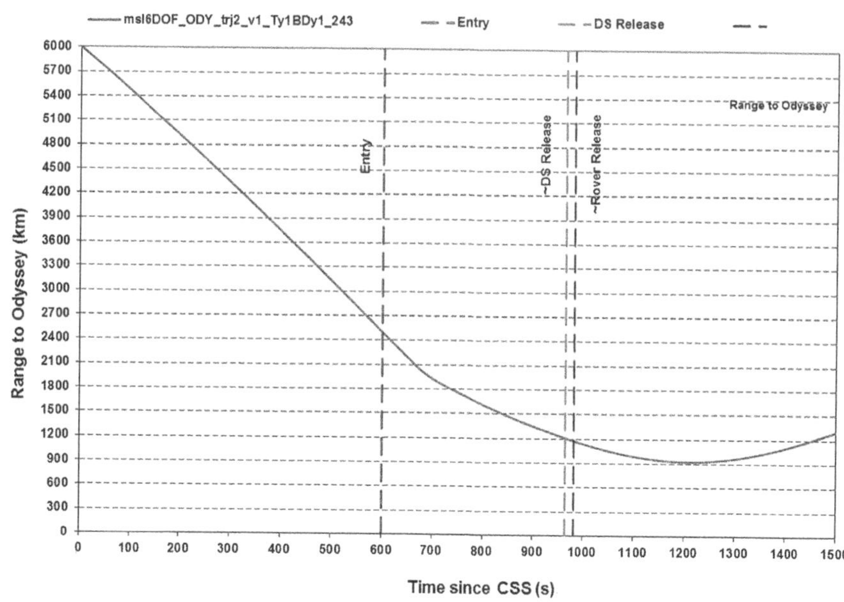

(b) Rover to Odyssey (green dash = Rover release)

Fig. 8-82. Range variation during EDL (a) rover to MRO and (b) rover to Odyssey. (a) Rover to MRO (Series13=wrap around min angle; Series1=wrap around max angle) (b) Rover to Odyssey. Please visit www.wiley.com/go/deepspacecommunications to access the color version of this figure.

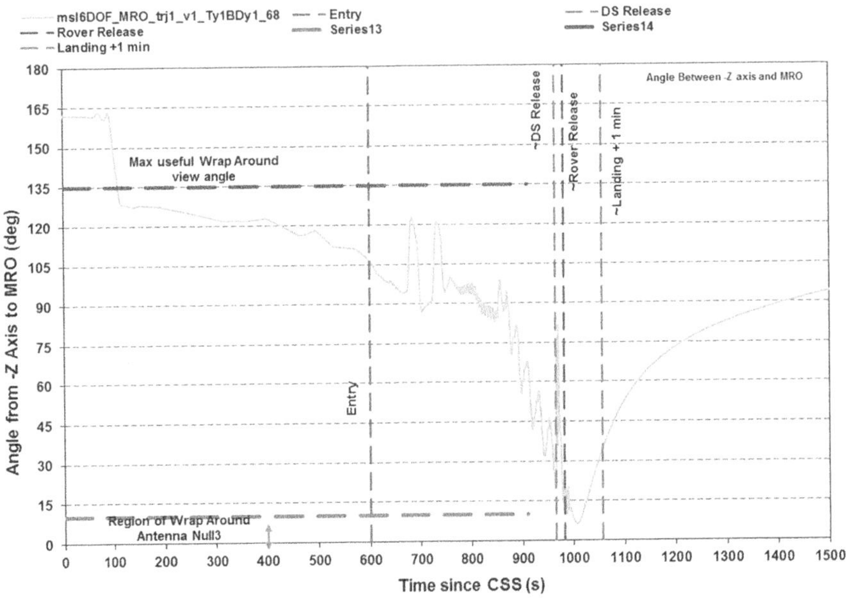

(a) Rover to MRO (Series13=wrap around min angle; Series1=wrap around max angle)

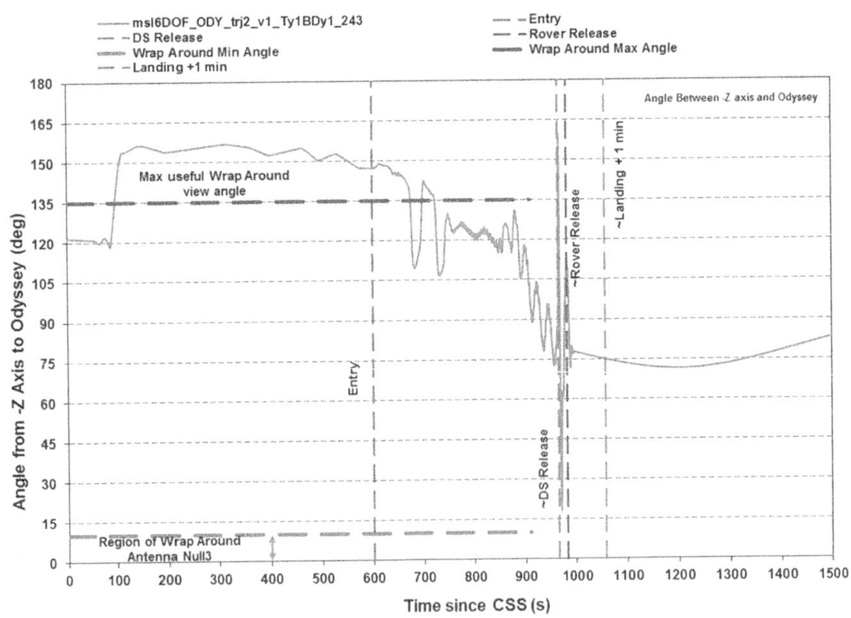

(b) Rover to Odyssey

Fig. 8-83. View angle variation during EDL (a) rover to MRO and (b) rover to Odyssey. (a) Rover to MRO (b) rover to Odyssey. Please visit www.wiley.com/go/deepspace communications to access the color version of this figure.

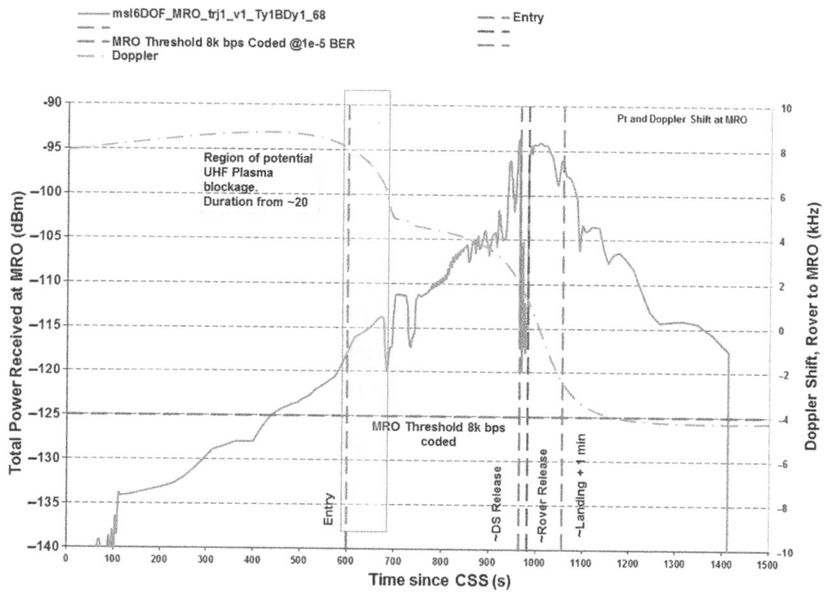

(a) Rover to MRO

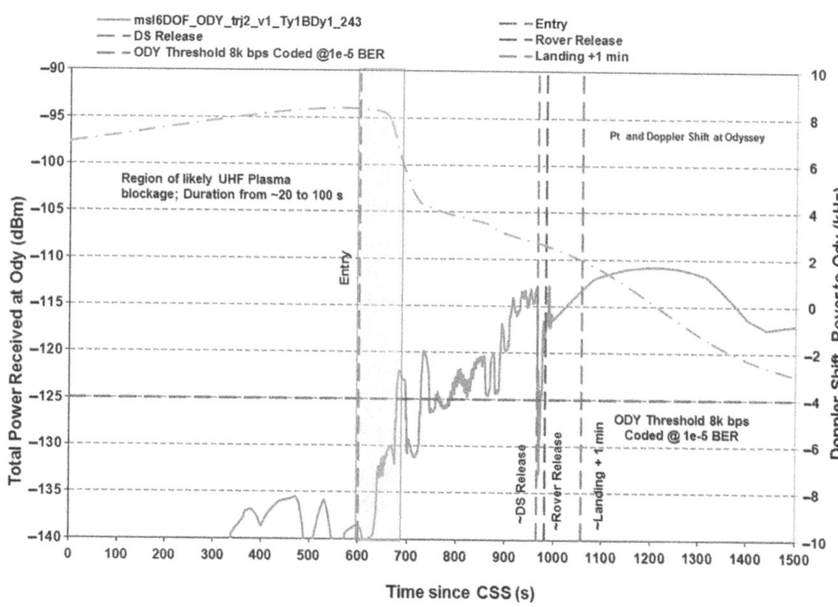

(b) Rover to Odyssey

Fig. 8-84. Doppler and return link received total power during EDL (a) rover to MRO and (b) rover to Odyssey [green = Rover release, magenta = Landing + 1 min]. Please visit www.wiley.com/go/deepspacecommunications to access the color version of this figure.

that planned. In particular, post processing would have been used to recover the telemetry if the actual Doppler profile could be reconstructed. Fortunately for MSL, none of these contingencies had to be put into use.

The rover's autonomous EDL software behavior remained in control for a short while after landing. The EDL behavior instructed the Spacecraft Mode Manager (SMM) to transition from EDL mode to surface mode.

The EDL UHF return link to MRO continued for several minutes after landing. As planned, surface-related data (e.g., Hazcam images) was prioritized and put into this initial stream. The time limitations (to send a required amount of higher-priority data at a certain bit rate) meant that lower-priority data (for example, from MEDLI or MARDI) awaited a subsequent post-landing relay pass. Data prioritization had been defined in advance by the flight team.

Engineering data gathered during EDL (estimated at ~100 Mbits) was not required to be immediately downlinked; however, this data was prioritized aboard the rover and the orbiters for playback to the DSN within 10 sols of landing and without risk of being overwritten. This prioritization enabled the EDL engineering team to complete an in-depth analysis of EDL performance to feed forward to future missions.

8.4.2.2 Surface (UHF Relay). MSL has relied on UHF-relay telecommunications passes as the primary method of returning science and engineering data to Earth during surface operations. Relay passes also are occasionally used to uplink commands and large flight software files to the rover; however, operations are nominally designed around the use of the DFE X-band link on the rover HGA as the primary uplink method.

UHF passes make use of MRO, in a 3-p.m. ascending Sun-synchronous orbit. The MRO pass pattern repeats every 17 sols and provides anywhere from 30 to 600 Mbits per pass (100 to 1150 Mbits per sol). The expected average performance per sol was well above the 250 Mbits requirement total for the two passes—and this volume has been achieved.

Relay data planning strategies include the use of adjacent passes (use of one high-volume pass instead of two low-volume passes).

MSL has also taken advantage of relay support by the Odyssey spacecraft to augment UHF passes with MRO. As long as MRO is available, the baseline MSL surface mission plan does not depend on Odyssey support.

Accurate daily relay data volume predictions are vital to the operations tactical process. Several tools of MER and Phoenix heritage have been updated with MSL specific data (RUHF antenna patterns, for example). The generalized telecom predictor (GTP), a variant of the telecom forecaster predictor (TFP), was the primary tool during operations to make UHF predictions for the first several months. These predictions account for the orientation (yaw, pitch, and roll) of the rover. Improved predict models are based on experience to date. They account for local terrain elevation "masks" and orbiter and rover antenna patterns that are not symmetric. The links use different portions of the patterns for orbiter overflights to the left or the right of the rover, and the rover port side or starboard side as seen by the orbiter.

8.5 Surface Operations (Plans)

This section comes mostly from the MSL Mission Plan [15].

8.5.1 Mission Operations System Approach

The flight team has been staffed to support intense surface operations over a 669-sol (~687 Earth days) period, some of which was conducted 7 days per week on Mars time (selected staff reported 40 minutes later each day) to minimize the end-to-end time between receipt of data from one sol to the uplinking of activities for the next sol. With experience gained, the staffing first dropped back to Earth time (but with shift start and shift end made earlier by up to 2 hours or made later by up to 4 hours to accommodate planning for a particular Mars time/Earth time alignment. Later, the staffing was further dropped back to 5 days per week, allowing weekend or holiday sols to be planned together for a single command load. The second or third such sol was called "run out."

To support a long-duration surface mission, the MSL mission operations system (MOS) uses a distributed operations concept similar to MER's (Chapter 7). This means, in particular, that both data processing and subsequent analysis and planning might be in several locations: at JPL and at the home institutions (usually universities) of the science team members. JPL is the central data distribution hub where selected data products are provided to remote science operations sites as needed. JPL is also the central hub for the uplink process, though participants are distributed at their respective home institutions.

The uplink process is dominated by a tactical uplink process. "Tactical" refers to work that is necessary to get a final set of commands up to a rover for each sol (or group of runout sols). Analysis of yesterday's downlink data is used to decide and plan where and what today's rover activities should be. The uplink communication to the spacecraft is either with X-band DFE with the DSN or UHF through MRO. Downlink, governed by data volume requirements, is UHF relay only, as shown in Fig. 8-85. Relay links are defined as

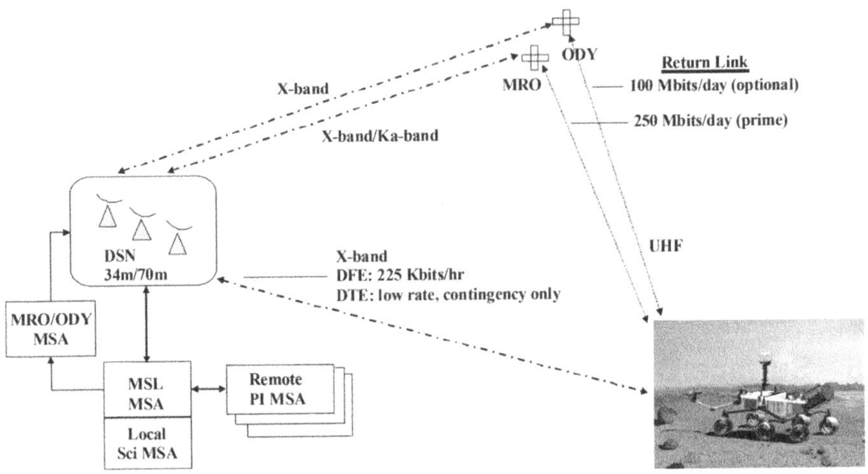

Fig. 8-85. MSL surface operations MOS overview.

"decisional" or not, depending on whether their information is essential to the tactical planning.

Figure 8-86 is a diagram identifying the elements of the operations functional architecture and the major Uplink (command) and Downlink (data) processes that they support.

To the left are science analysis and planning functions. The center has the sequencing, data collection and engineering, and navigation analysis functions. To the right are the multi-mission data-processing functions. The multi-mission ground data system (MGDS) of the Interplanetary Network Directorate (IND) at JPL interfaces with the DSN.

Each element (box) represents a set of related software and facilities, people, and local processes.

8.5.2 Initial Surface Ground Operations

For the first 60 to 90 days of operations (consisting of rover initial configuration for surface operations, rover checkout, and first surface location operations), all teams were located at JPL. Tactical operations were on Sol (Mars) time. This allowed up to 18 hours per day of planning for one-sol turnaround. Operations at JPL provided face-to-face coordination and learning. Once the rover moved into steady-state operations and the operations teams demonstrated a one-shift turnaround, the flight team transitioned to tactical operations on Earth time. Shortly thereafter, the team became distributed, with science teams operating from their home sites.

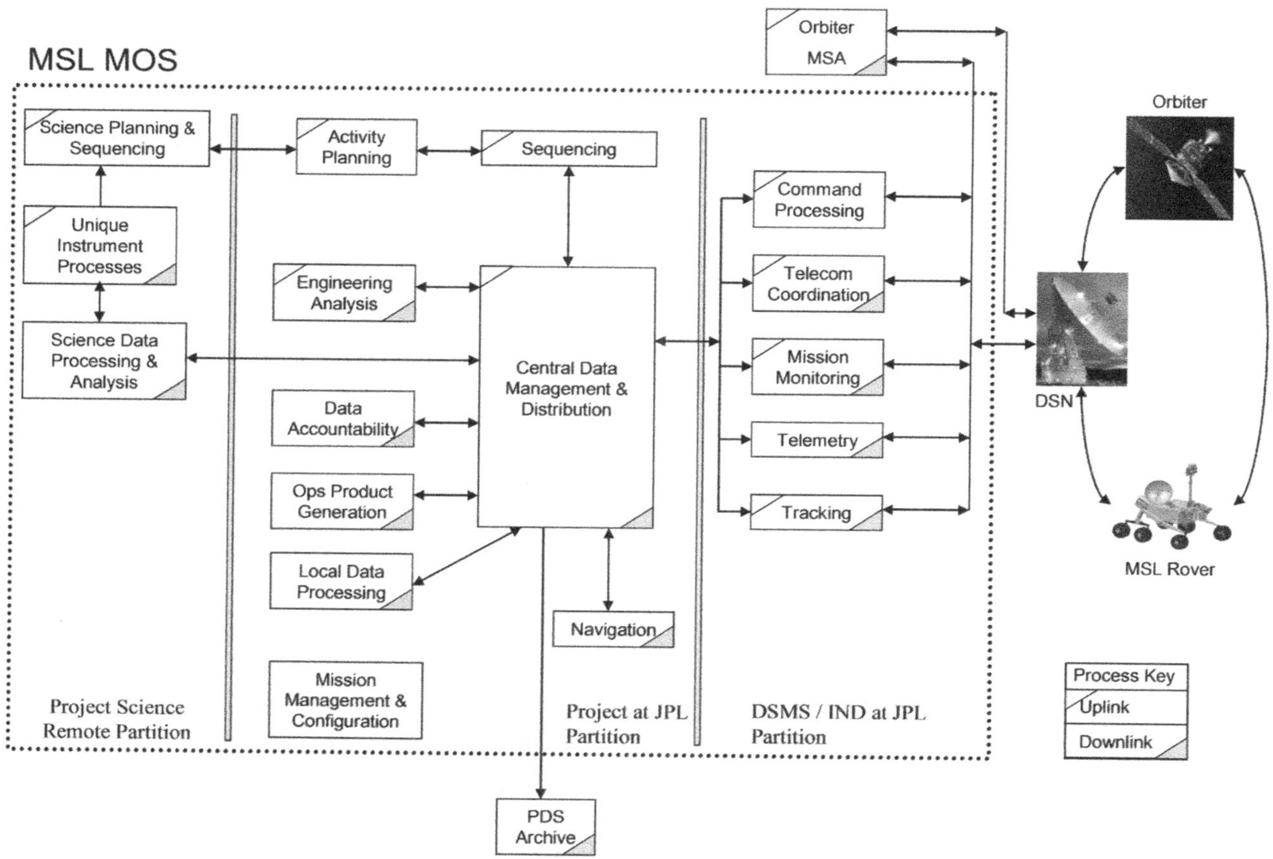

Fig. 8-86. Functional architecture of MSL MOS.

From the beginning, the tactical flight team has been a virtual team, comprising members from across the organizations such as science, spacecraft, mission planning, and the Deep Space Network (DSN). The virtual team is a focused, multi-disciplinary group that is formed from members across the MSL mission to work a particular issue (in this case, carrying out the tactical uplink process). In these particulars, the MSL Flight Team organization was based on how the MER Flight Team was organized and deployed. In addition, the organization of and allocation of activities within the MSL team was designed to increase integration and reduce the total number of separate teams and inter-team interfaces.

8.5.3 Tactical Operations after First 90 Sols

The MSL Tactical Surface Operations is tailored to support nearly daily commanding of the rover, based on today's science evaluation of yesterday's returned data.

Figure 8-87 shows the command and data flow and is a top-level timeline of the tactical process.

The steady-state tactical process was initially performed 7 days per week through the first months of the prime surface mission. Based on the evolution of the MER surface team, the core of the tactical team works four 10-hour shifts per week, with a sliding start time between 6 a.m. and 1 p.m. The driver for the sliding start time is to have a "modified" prime shift at JPL. This is more conducive to family life over the long duration and also helps maintain normal sleep cycles. The start time is made as close to 8 a.m. as possible. The normal uplink cycle is 8 hours between receiving telemetry data in the MSL MSA to uplink approval. This enables 1-day turnaround cycles. A small number of tactical planning templates were developed to facilitate the rapid turnaround. Some examples of activities have been site recognizance, target approach, sample acquisition and processing, science instrument analysis on acquired sample, and traverse to new science site. These templates were initially developed in response to pre-defined ("canned") Mission Scenarios. After landing, particularly after the first 30 sols, the process became discovery-driven. The MOS approach and GDS tools provide for re-use of Sequences and conversion of once-unique Sequences into Activities for future use.

8.5.4 UHF Telecom Constraints

Though X-band uplink continues to be prime, the relay orbiters can be used to uplink sequences, flight software, and any other data we might need to send to the rover.

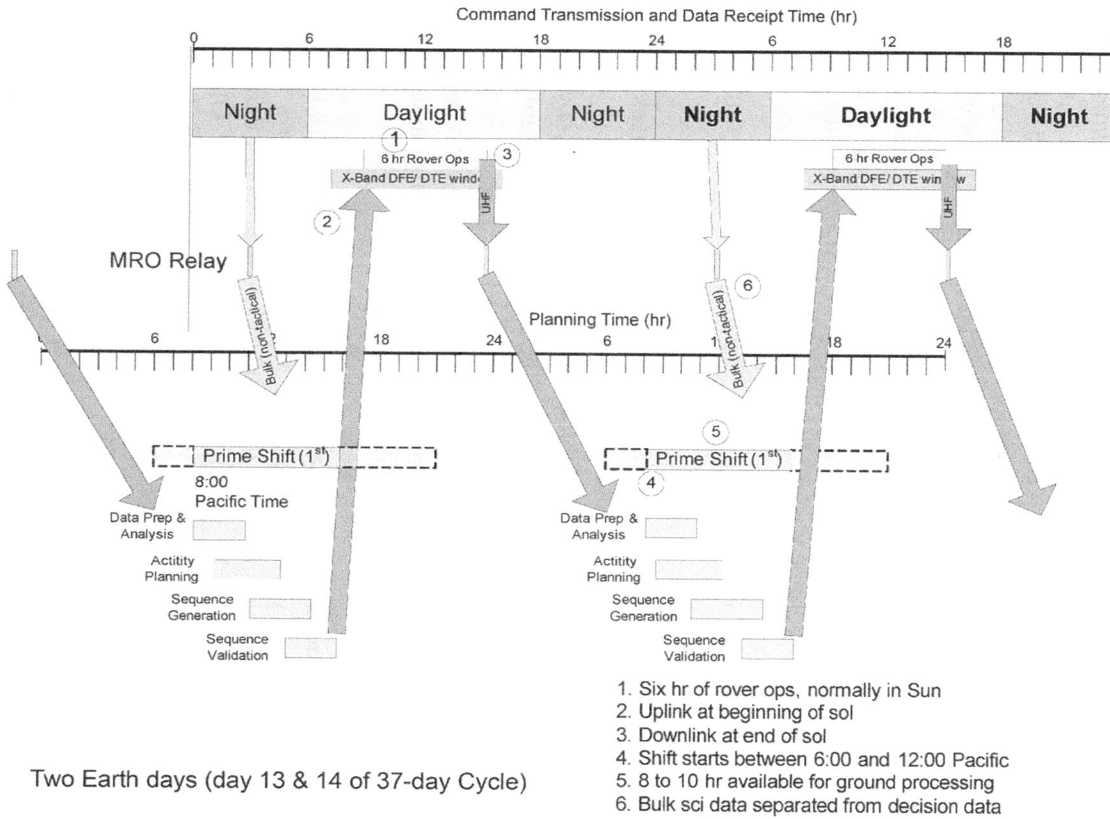

Fig. 8-87. MSL tactical process during surface ops (at top level).

With ground and surface process durations as currently defined, we would be able to use the 3 a.m. MRO pass over 50 percent of the time to uplink our sequences if necessary. Data cleanup commands, which may be large and which are usually not time critical, if not sent on X-band, would typically be sent in time for relay to the rover during the 3 a.m. pass.

MSL relies on UHF communication with relay orbiters to downlink the data generated during surface operations. For planning purposes, the baseline for the rover and the MRO and Odyssey orbiters assumed that the downlink relay bandwidth would average 125 Mbits per pass for a total of 250 Mbits per sol and that the volume of decisional data collected each sol would not exceed 100 Mbits.

Figure 8-88 shows the expected return link data volume the MRO relay orbiter. Relay capacity follows a 5 or 6 sol pattern of two low-volume followed by three or four high-volume passes.

This figure is based on the following assumptions regarding MRO performance:

- Every pass, MRO performs a roll maneuver, up to 30 degrees, to point its UHF antenna as close as possible to MSL.
- The rover is at the Equator (a reasonable assumption for Gale Crater).
- MRO must rise at least 10 deg above the horizon to communicate with MSL.
- The relay return link utilizes adaptive data rates.
- Performance is based on the best morning and evening passes.
- A 2-sigma margin for the return link is applied.

Given these assumptions, the worst-case performance is no less than 125 Mbits per sol and the average performance is 687 Mbits per sol. On days when the expected return link volume is less than 100 Mbits, decisional data will be downlinked according to the priority assigned by the Science Operations Working Group (SOWG). On days when the expected volume is large, any backlog of decisional data can be downlinked using the additional capacity over 125 Mbits.

In addition to data volume constraints, the start time of the relay pass imposes an important constraint on surface operations. In a typical sol, the command sequence is uplinked via DFE (currently starting at 9:30 local mean solar time, LMST) and any activity that produces data needed to support a decision during the next ground processing cycle must be completed prior to the afternoon MRO pass.

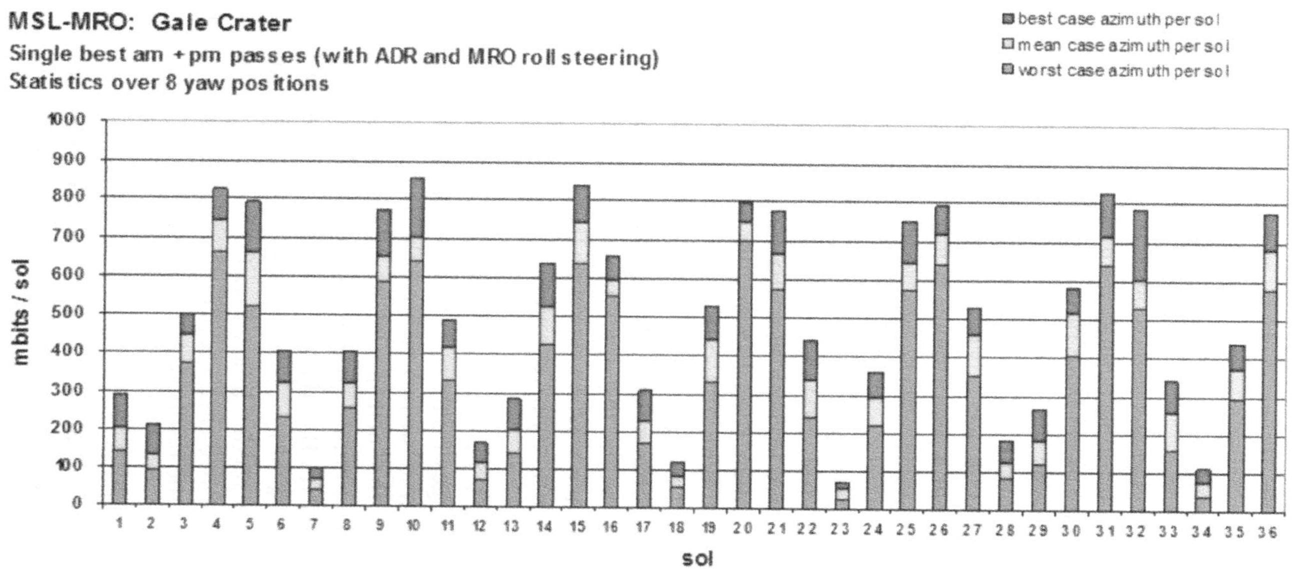

Fig. 8-88. Estimated sol-by-sol return data volume through MRO.

The actual MRO pass start times vary by about ±1 hour around 15:00 LMST for the rover at the equator. At 30 N and 30 S, MRO pass start times can range from 1:45 to 4:00 LMST for the a.m. pass and from 14:00 to 16:15 LMST for the p.m. pass.

8.6 Surface Operations (Characterized in Flight)

The MSL and MRO Electra radios and the relay link operations and performance characterizations described in this section are mainly from Ref. 31 as they were late in the MSL prime mission in 2014.

The use of new generation Electra software- and firmware-defined radios on both the MSL and MRO spacecraft has enabled new operational modes that provide three times the link performance compared to other current and past Mars relay links.

The previous generation relay radio used for NASA Mars missions is the Cincinnati Electronics 505 (CE-505) radio. As has been described in Chapters 6 and 7, this radio was used on both Mars Exploration Rovers (MER) and the 2001 Mars Odyssey orbiter. The CE-505 radio has four return-link data rates: 8, 32, 128 and 256 kbps. With a CE-505 at either end of the relay link, mission operators choose one of the four rates to use for each relay pass based on their models of what the relay geometry of the link will support. Within the data rate limitation, this radio and this operations mode have been very successful. The Odyssey relay orbiter provides this same fixed data rate per pass relay service to the MER Opportunity rover and the MSL Curiosity rover.

For the MSL/MRO link, new relay modes relative to the MSL–Odyssey link are available. A full range of data rates from 1 kbps to 2048 kbps in factor-of-two steps (1, 2, 4, 8 and so on) is possible. Perhaps the most important functional upgrade is the inclusion of an adaptive data rate (ADR). With ADR in use, MRO continuously monitors the signal-to-noise ratio (SNR) on the return link from MSL. Based on this live metric, the MRO Electra directs the MSL Electra to raise or lower its return-link data rate to match as closely as possible the instantaneous link capacity. Supplementing ADR, Electra also allows fully suppressed carrier operations providing a 1.2-dB increase in data power relative to the residual carrier mode.

The ADR mode eliminates the need for the conservatism that stems from having to choose a single rate prior to a pass. It increases transferred data volume by matching link data rate to link capacity over the course of the pass as the slant range and antenna boresight angles vary. Combined with the higher data rates that are now available with the Electras, the net effect for MSL is a three times increase in average data volume per MSL-to-MRO pass compared to an MSL-to-Odyssey pass.

8.6.1 Mitigating the Effects of Electromagnetic Interference

Two instruments on board MRO, the Compact Reconnaissance Imaging Spectrometer for Mars (CRISM) and the Mars Climate Sounder (MCS), each had electronics that integrated power, digital processing, and the scanning portion of their hardware. The scanning twist capsules of these instruments left a gap in the electronics boxes, breaking the Faraday cage model and allowing overtones of digital switching and power supplies to leak high power level electromagnetic interference (EMI) into the UHF receive band. The net effect for Electra was a 10-dB or more reduction in receiver sensitivity at the standard 401.6-MHz return-link frequency when these instruments were on during UHF relay operations.

Given the frequency agility of Electra, MRO embarked on a search for new frequencies in the nominal 390 to 405 MHz return-link frequency band that might have less EMI. A return link center frequency of 391 MHz was identified as minimally affected by EMI, particularly at the new higher data rates to be used in support of MSL. Even so, it was expected that an inter-project trade to idle some of the MRO science instruments to achieve a quieter EMI environment might be necessary to achieve the required volume of science data returned from MSL. This trade would prevent MRO's instruments from collecting their maximal science data while flying over the rover.

Before the MSL landing, the MSL and MRO telecom teams recommended moving the surface operations return-link center frequency to 391 MHz, operating in suppressed carrier mode with ADR enabled with a required EMI-based performance loss (margin) of 6 dB ± 1.0 dB included in MSL data volume return planning. This margin was later reduced to a span of 3.1 to 3.4 dB from an in-flight baseline bit-error rate (BER) versus received signal level curve for 391 MHz.

The "heritage" fixed data rate and residual carrier modes that were used by both Odyssey and MRO to support MER became the starting point for MSL-MRO landed operations. To manage the remaining uncertainties, the MSL and MRO telecom and operations teams created a post-landing relay link characterization plan. This plan included a low risk strategy to introduce and evaluate each new capability on the MSL-MRO link and to quantitatively assess the impact of EMI as a function of the MRO science instrument mode.

8.6.2 Data Volume Achieved with MRO and Odyssey Links

During the first 62 sols, and then confirmed through the first 200 sols, the in-flight EMI was characterized, and a satisfactory MRO "quiet mode" was established. During this characterization, the MRO science instruments were powered on sequentially from the smallest EMI producers to the largest, based on expectations from prior testing. Baseline was HiRISE only, as this camera had not shown any EMI issues prelaunch. CRISM was the last of the instruments to be turned on. As the second scanning instrument, it was also placed in a quiet mode; that is, parked with no data collection enabled. Even in this quiet mode, CRISM produces UHF EMI tones at the Electra antenna with more power than would come from MSL or any expected future lander UHF uplink signal. The MRO science "quiet mode" configuration, with MCS and CRISM in their quiet modes, became the baseline for nominal MSL support. With the quiet-mode, EMI degradation is estimated at 3.1 to 3.4 dB relative to the HiRISE-only mode. This suggests that at any given instant, the EMI from the instruments cuts the achievable data rate from MSL in half; that is, reduces it by ~3 dB. In this mode we are still able to return an average of 240 Mbits per pass.

Figure 8-89 shows the variation in transferred data volume per sol with MRO in relay quiet mode and ADR on from sol 40 to sol 67. The 5.2-sol modulation in per-sol data volume performance is due to a roughly 10-day cyclic variation in the overflight geometry and the interaction of maximum elevation angle with data volume. Gaps in the plot are sols where only one pass was exercised and therefore not a valid statistical value for the per-sol data volume. Based on the successful relay performance, well exceeding Curiosity's 250 Mbit/sol data return requirement, the decision was made on Sol 65 to continue relay operations with MRO in the relay quiet mode.

Besides the MSL-to-MRO relay passes, MSL continues to receive relay support from the Odyssey orbiter, returning roughly 130 Mbits per sol. For additional robustness in the Curiosity surface relay plan, MEX was prepared to provide backup relay support if Odyssey (ODY) and/or MRO became unavailable for some period of time. Over the first years of Curiosity surface operations, ten demonstration relay passes were performed between Curiosity and MEX confirming return-link performance at rates up to 128 kbps

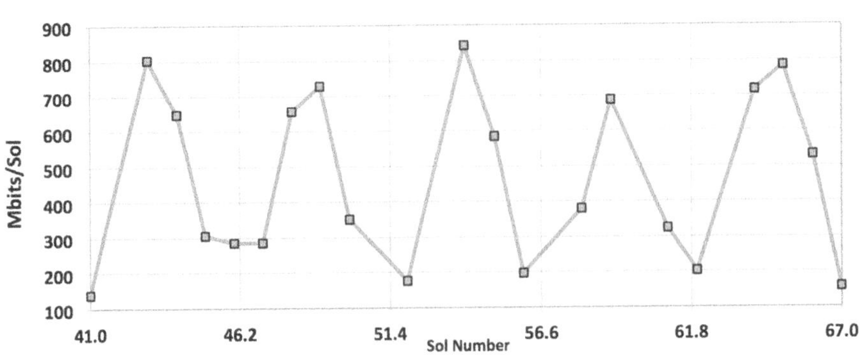

Fig. 8-89. MSL–MRO ADR data volume per sol with relay quiet mode.

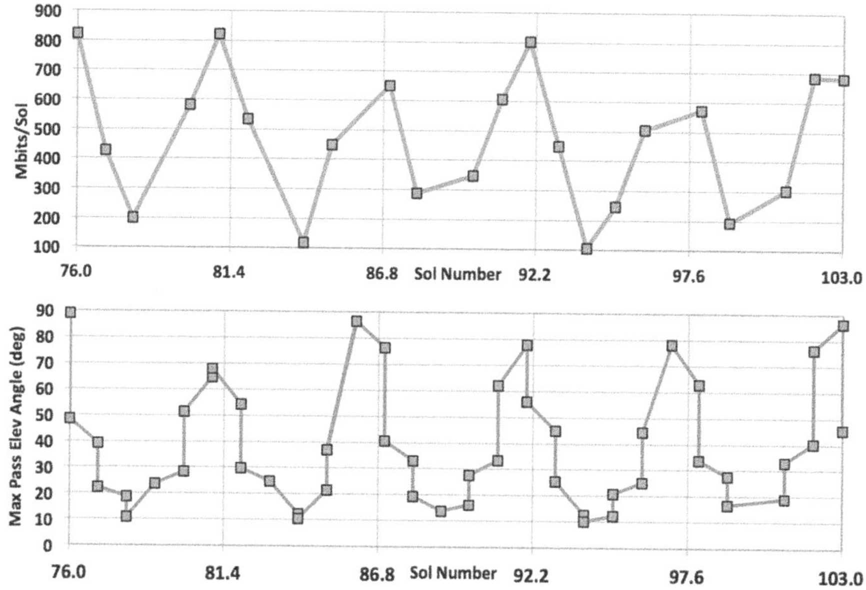

Fig. 8-90. MSL–MRO geometry repeat cycle and data volume cycle.

and the capability to deliver MSL command products on the MEX forward link.

8.6.3 Relay Link Models

MSL has completed one full Mars year on the surface (prime mission) and is in its second Mars year (extended mission). The MSL project has collected relay link performance data using the standard MRO relax quiet mode.

8.6.3.1 Elevation Angle. Comparing the maximum elevation angle during relay passes with the per-sol data returns from sol 76 through sol 103 illustrates a short-term cycle of 5.4 days. Figure 8-90 shows five of these cycles over this 27-sol period. There, the X-axis marker spacing is 5.4 sols.

MSL exercised passes with maximum elevation angles as low as 10 deg. Passes with higher maximum elevation angle will be of longer duration and have shorter slant range to the MSL lander when compared to passes that have a lower maximum elevation angle.

For operational simplicity, MRO scheduled every relay pass to begin hailing at roughly 120 seconds after MRO rose above MSL's local horizon. This approach also avoided initial pass hailing during difficult low elevation angle signal conditions. Conversely, the end of a pass was not scripted but was dictated by these difficult low-elevation-angle RF signal conditions near the

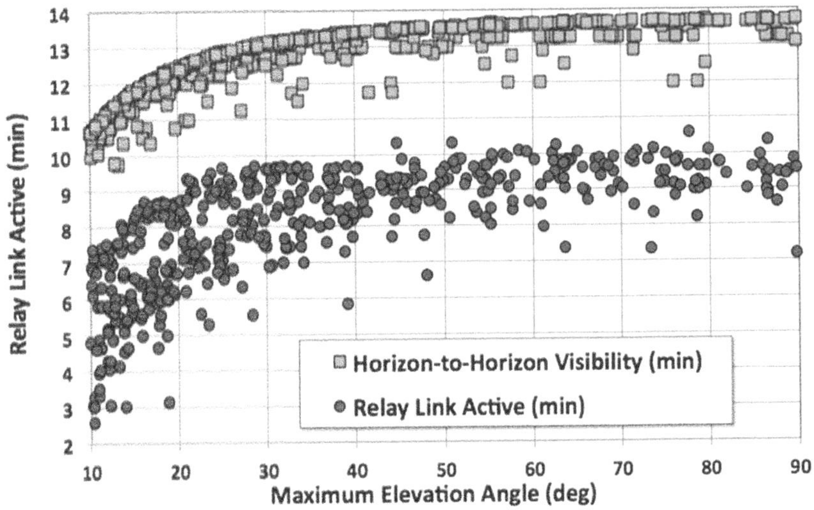

Fig. 8-91. MSL-to-MRO pass durations as a function of elevation angle.

time that MRO set below the local horizon. Terrain signal blockage at the rim of Gale Crater or multipath signal variations at low-elevation angles would terminate the link before MRO reached zero degrees in elevation angle. The actual relay session duration was typically 200 to 300 seconds less than the horizon-to-horizon geometric view period.

The widest spread in session time (defined as proximity-1 link established) is at the lower elevation angles. Figure 8-91 compares the geometric horizon-to-horizon pass duration and the (lesser) prox-1 session time for sessions with maximum elevation angles from 10 to 90 deg.

Short link dropouts would often occur during the lower elevation passes as the link struggled to keep up with faster signal fluctuations due to multipath fading. The result was a larger variation in pass duration and a higher uncertainty in the data volume actually transferred for the low elevation passes. This point became important for planning science data return.

8.6.3.2 East-versus-West. When MRO passed to the west side of MSL, there was a higher average returned data volume than when MRO passed to the east of MSL. After 120 sols of performance data was collected in the MRO relay quiet mode, improved east-versus-west pass performance predictors were generated, as shown in Fig. 8-92.

8.6.3.3 Evaluation of Other Variables. *MRO Port versus Starboard* – All p.m. passes are south-to-north on the sunlit side of Mars, and all a.m. passes

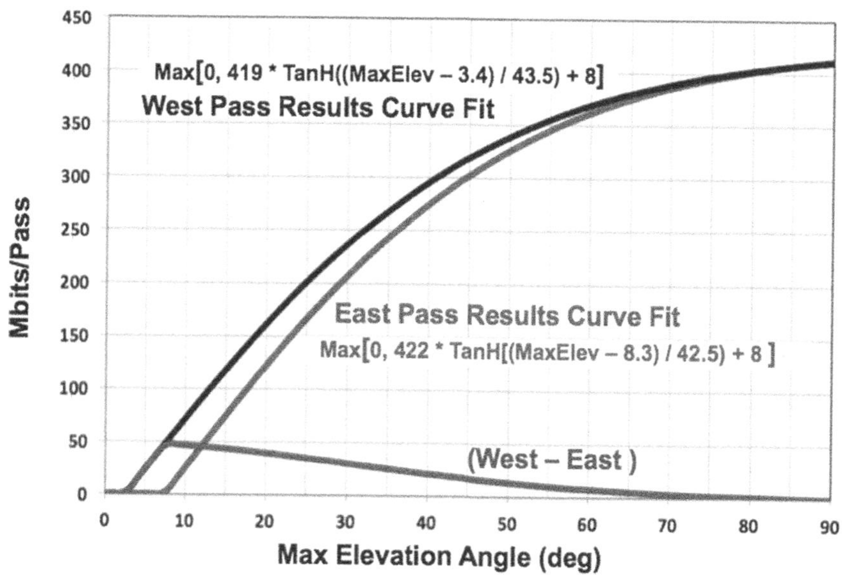

Fig. 8-92. Best-fit predictor of MRO–MSL east-versus-west performance.

are north-to-south on the Mars night side. Thus, MSL west-side passes are MRO port side in the p.m. and starboard side in the a.m. Conversely, MSL east-side passes are MRO port side in the a.m. and starboard side in the p.m. MRO port-versus-starboard comparisons yielded no statistical dependency, thus demonstrating there are no significant MRO antenna pattern asymmetries.

MSL Orientation – Early in the mission, MSL moved very little and only limited statistics were available to detect and quantify the impact of MSL's in-situ orientation. After the first year of surface operations, there were some indications of the impact of MSL yaw on data volume. We have used the yaw information to improve the data volume predictions significantly.

MRO Roll Steer – One of the operational mitigations for EMI-induced data loss is to perform MRO roll steering of up to 30 degrees toward the direction of the MSL rover. This provides a bit more antenna gain in the direction of the rover, and it was expected that this would increase data volume relative to a constant-nadir pointing strategy. After a year on the surface, the larger data set shows that roll steering yields 40 to 70 more Mbits per pass for both east- and west-side passes, with larger performance gains experienced during lower elevation passes that are further down the gain-slope of a nadir-pointed MRO antenna. This is exactly the result that was expected and it is now quantified.

References

[1] L. D'Amario, *Mars Science Laboratory Interplanetary Baseline 2011 Interplanetary Trajectory Characteristics Document*, JPL D-27210 (internal document), Jet Propulsion Laboratory, California Institute of Technology, Pasadena, California, July 15, 2009.

[2] F. Abilleira, *Mars Science Laboratory DSN Initial Acquisition Geometry Report*, Rev. A, D-27213, also MSL-377-0317 (internal document), Jet Propulsion Laboratory, California Institute of Technology, Pasadena, California, November 1, 2010.

[3] "Interplanetary Trajectories," *Basics of Space Flight*, web site, Jet Propulsion Laboratory, California Institute of Technology, Pasadena, California. http://solarsystem.nasa.gov/basics/bsf4-1.php (accessed October 30, 2014)

[4] B. Florow, K. Breitenbach, and S. Udomkesmalee, *Positioning, Phasing & Coordinate Systems (3PCS), Volume 1, MSL Coordinate Systems*, D-34642 (JPL internal document), Jet Propulsion Laboratory, California Institute of Technology, Pasadena, California, May 29, 2007.

[5] L. D'Amario, T. Martin-Mur, *Launch/Cruise/Approach Design, in Launch/Cruise/approach Technical Interface Meeting*, D-64186 (internal document), Jet Propulsion Laboratory, California Institute of Technology, Pasadena, California, May 19, 2009.

[6] A. Makovsky and M. Danos, *Mars Science Laboratory X-Band Surface and Cruise Data Rate Analysis*, D-68027 (internal document) (project number MSL-566-3346, also internal document), Jet Propulsion Laboratory, California Institute of Technology, Pasadena, California, February 23, 2011.

[7] "Lockheed Martin's Atlas V Selected to Launch Mars Science Laboratory in 2009" (press release), Lockheed Martin website, June 7, 2006. http://www.spaceref.com/news/viewpr.html?pid=20031 (accessed April 27, 2011)

[8] *Mars Science Laboratory Curiosity Rover*, public website, News Room section, Jet Propulsion Laboratory, California Institute of Technology, Pasadena, California. http://mars.jpl.nasa.gov/msl/news/newsroom/ (accessed October 16, 2014)

[9] J. Taylor, A. Makovsky, A. Barbieri, R. Tung, P. Estabrook, and A. G. Thomas, *Mars Exploration Rover Telecommunications, DESCANSO Design and Performance Summary Series*, Article 10, Jet Propulsion Laboratory, California Institute of Technology, Pasadena, California, October 2005. http://descanso.jpl.nasa.gov/DPSummary/summary.html (accessed October 30, 2014)

[10] E. Chapin, *MSL Terminal Descent Sensor User's Guide*, Ver. 4.0.3, D-37174 (internal document), Jet Propulsion Laboratory, California Institute of Technology, Pasadena, California, July 9, 2012. https://charlie-lib.jpl.nasa.gov/docushare/dsweb/View/Collection-47120 (accessed October 30, 2014)

[11] B. Schratz, M. Soriano, P Ilott, J. Shidner, A. Chen, and K. Bruvold, "Telecommunications Performance During the Entry, Descent and Landing of the Mars Science Laboratory," *Journal of Spacecraft and Rockets*, vol. 51, no. 4, pp. 1237–1250, July–August, 2014.

[12] "Mars24, Technical Notes on Mars Solar Time," Goddard Institute for Space Studies website, Goddard Space Flight Center, Greenbelt, Maryland. http://www.giss.nasa.gov/tools/mars24/help/index.html (accessed October 30, 2014)

[13] J. Herath, "Project Overview: System Design Review (SDR)," [MSL EDL instrumentation (MEDLI) website], MEDLI-SDR-0210, University of Idaho, Moscow, Idaho, May 1, 2007. http://www.mrc.uidaho.edu/~atkinson/SeniorDesign/ThermEx/MEDLI/MEDLI_SDR_Project_Overview.pdf (accessed October 30, 2014)

[14] "Mars Pathfinder," fact sheet, Jet Propulsion Laboratory, California Institute of Technology, Pasadena, California, http://www.jpl.nasa.gov/news/fact_sheets/mpf.pdf

[15] *Mars Science Laboratory Mission Plan*, JPL D-27162, Rev. B (internal document), Jet Propulsion Laboratory, California Institute of Technology, Pasadena, California, July 31, 2010.

[16] *Digital Time Division Command/Response Multiplex Data Bus*, MIL-STD-1553B, United States Department of Defense, continual updates.

[17] *Small Deep Space Transponder (SDST), Functional Specification and Interface Control Document for MSL and Juno Projects*, D-33672 (internal document), Jet Propulsion Laboratory, California Institute of Technology, Pasadena, California, May 15, 2007.

[18] N. Blyznyuk, *Mars Science Laboratory Telecom System Engineering Pre-CDR Peer Review*, Telecom Antennas Pattern Analysis," JPL D-64361 (internal document), Jet Propulsion Laboratory, California Institute of Technology, Pasadena, California, April 24, 2007.

[19] B.M. Kolundzija and A.R. Djordjevic, "Theoretical Background of WIPL-D," *Electromagnetic Modeling of Composite Metallic and Dielectric Structures*, WIPL-D d.o.o. http://www.wipl-d.com/resources.php?cont=theoretical-background (accessed October 30, 2014)

[20] P. Illot, *Mars Science Laboratory Telecommunications Functional Design Description*, Rev. B, D-34199 (internal document), Jet Propulsion Laboratory, California Institute of Technology, Pasadena, California, October 27, 2008.

[21] W. Boger, D. Burgess, R. Honda, C. Nuckolls, "X-Band, 17 Watt, Solid-State Power Amplifier for Space Applications," *IEEE MTT-S International Microwave Symposium Digest*, vol. 3, pp. 1379–1382, 2005. http://ieeexplore.ieee.org/xpls/abs_all.jsp?arnumber=1516940 (accessed October 30, 2014)

[22] N. Blyznyuk, *DUHF Accommodation Study*, JPL D-64238 (internal document), Jet Propulsion Laboratory, California Institute of Technology, Pasadena, California, June 14, 2006.

[23] P. Ilott, J. Harrel, B. Arnold, N. Bliznyuk, R. Nielsen, D. Dawson, and J. McGee, "UHF Relay Antenna Measurements on Phoenix Mars Lander Mockup," *Antenna Measurements and Techniques Association (AMTA)*, Austin, Texas, October 22–27, 2006.

[24] N. Blyznyuk, *RUHF Accommodation Study*, JPL D-64394 (internal document), *Jet Propulsion Laboratory, California Institute of Technology, Pasadena, California, June* 14, 2006.

[25] C. D. Edwards, Jr., T. C. Jedrey, E. Schwarzbaum, A. S. Devereaux, R. DePaula, M. Dapore, and T. W. Fischer, "The Electra Proximity Link Payload for Mars Relay for Telecommunications and Navigation," IAC-03-Q.3.A.06, *54th International Astronautical Congress*, Bremen, Germany, 29 September 29–October 3, 2003. http://trs-new.jpl.nasa.gov/dspace/bitstream/2014/7832/1/03-2150.pdf (accessed April 27, 2011)

[26] *Proximity-1 Space Link Protocol—Physical Layer, Recommendation for Space Data System Standards*, CCSDS 211.1-B-4 (Blue Book), Consultative Committee for Space Data Systems Secretariat, National Aeronautics and Space Administration, Washington, District of Columbia, December 2013. http://public.ccsds.org/publications/archive/211x1b4.pdf (accessed October 30, 2014)

[27] E. Chapin, *MSL Terminal Descent Sensor User's Guide*, Ver. 4.0.1, D-37174, MSL-576-1520 (internal document), Jet propulsion Laboratory, California Institute of Technology, Pasadena, California, June 6, 2011.

[28] A. Chen, *Unified Mass Equipment List*, JPL D-38142, MSL-386-1731 (internal document), Jet Propulsion Laboratory, California Institute of Technology, Pasadena, California, June 30, 2011.

[29] T. Pham, C. Chang, E. Satorius, S. Finley, L. White, P. Estabrook, and D. Fort, *Entry Descent Landing Data Analysis (EDA)*, NASA Tech Brief NPO-41220 BR, National Aeronautics and Space Administration, Washington, District of Columbia, October 13, 2004.

[30] *DSN Telecommunications Link Design Handbook*, Rev. E, 810-005 also labeled JPL D-19379, Jet Propulsion Laboratory, California Institute of Technology, Pasadena, California. http://deepspace.jpl.nasa.gov/dsndocs/810-005/ Note: Turnaround ranging is in module

203c of 810-5, at the specific link http://deepspace.jpl.nasa.gov/dsndocs/810-005/203/203C.pdf (both accessed October 30, 2014)

[31] D. J. Bell, S. Allen, N. Chamberlain, M. Danos, C. Edwards, R. Gladden, D. Herman, S. Huh, P. Ilott, T. Jedrey, T. Khanampornpan, A. Kwok, R. Mendoza, K. Peters, S. Sburlan, M. Shihabi, and R. Thomas, "MRO Relay Telecom Support of Mars Science Laboratory Surface Operations." *2014 Aerospace Conference*," Big Sky, Montana, March 1–8, 2014.

Acronyms and Abbreviations

AACS	attitude and articulation control subsystem
ABM	aerobraking maneuver
ABX	aerobraking exit
Ac	acquisition (also ACQ)
ACE	Advanced Composition Explorer (spacecraft)
ACE	call sign for project real-time mission controller
ACIS	antenna control and interface subsystem
ACK	acknowledgment
ACQ	acquisition (also Ac)
ACS	attitude control system
A/D	analog to digital
ADC	analog to digital converter
ADR	adaptive data rates
AGC	automatic gain control (received carrier power)
AGL	above ground level
Ahr	ampere hour
AIS	ancillary information subsystem
AlBeMet	trade name of aluminum beryllium composite produced by Brush Wellman Inc.
ALC	automatic level control
AM	ante meridian
AMMOS	Advanced Multimission Operations System
AOA, AoA	angle of attack
AOS	advanced orbiting systems
APID	application process identifier
APXS	alpha particle X-ray spectrometer
ARC	Ames Research Center
ARF	automatic restart function

Deep Space Communications, First Edition. Edited by Jim Taylor.

ARQ	automatic repeat queuing
ARQ	automatic repeat reQuest
ASCII	American Standard Code for Information Interchange
ASI	Italian Space Agency
ASIC	application-specific integrated circuit
Assy	assembly
ATCM	auto trajectory correction maneuver
ATDF	archived tracking and data file
ATLO	Assembly, Test, and Launch Operations
ATN	attenuator
ATS	aft transition structure
AU	astronomical unit (~1.496×10^8 km)
AutoNav	autonomous navigation
aux osc	auxiliary oscillator
AWGN	additive white Gaussian noise
b	bit (note: telecommunications uses bits [b]; data storage uses bytes [B], which are 8 bits per byte)
BBPP	base-band patch panel
BC	buss controller
BECO	booster engine cutoff
BER	bit-error rate
BIP	backshell interface plate
BLF	best-lock frequency
BLGA	backshell low gain antenna on MER
B_{LO}	one-sided loop threshold noise bandwidth
BML	backup mission load bps (bits per second (bps)
BMOX	Beacon-Monitor Operations Experiment (on DS1)
BODA	burnout detection algorithm
BoL	beginning of life
BPF	bandpass filter
BPM	UHF radio frequency module (MRO)
bps	bits per second
BPSK	binary phase-shift keying
BS	bachelor of science
BSS	backshell separation

BTD	buffered telemetry demodulator
BTU	British thermal unit (1.06 kilojoule)
BUD	bridle, umbilical, descent-rate-limiter device
BVR	Block V Receiver
BW	bandwidth
BWG	beam waveguide
C3	launch-specific energy
CAN	Canberra (Deep Space Communications Complex)
CAS	Cassini (Project, spacecraft)
CBE	current best estimate
CBM	Communications Behavior Manager
CBM	cruise balance mass
C&C	command and control
CCAFS	Cape Canaveral Air Force Station
CCAM	collision and contamination avoidance maneuver
CCB	common core booster
CCS	computer command subsystem
CCSDS	Consultative Committee for Space Data Systems
CD	cumulative distribution
C&DH	command and data handling
CDR	Central Data Recorder
CDR	Critical Design Review
CDS	command data subsystem
CDSCC	Canberra Deep Space Communications Complex (DSCC)
CDU	command detector unit
CEP	circular error probability
C.E.	convolutional encoding
CFDP	CCSDS (Consultative Committee for Space Data Systems) file delivery protocol
CG	center of gravity
CLBW	carrier-loop bandwidth
ChemCam	Chemistry and mineralogy camera (MSL)
CheMin	Chemistry and Mineralogy (makes X-ray diffraction analyses of rock and soil samples, MSL)
C-ISA	Centaur interstage adapter

CLA	carrier-lock accumulator
CLGA	cruise low-gain antenna on MER
Clk	clock
CLK GEN	clock generator
CLTU	command link transmission unit
cm	centimeter
CMA	command modulator assembly
Cmd	command
CMDRAD	command radiation
CME	coronal-mass ejection
CNR	carrier-to-noise ratio
comm	communication, common port
conscan	conical scanning
CP	coupler
cPCI	compact peripheral component interconnect
CPA	command processor assembly
CRC	cyclic redundancy check
CRISM	Compact Reconnaissance Imaging Spectrometer for Mars (instrument on MRO)
CRS	cosmic ray system (VGR)
CS	checksum
CSIRO	Commonwealth Scientific and Industrial Research Organization (Australia)
CSS	cruise stage separation
CSS	channel select synthesizer
CTS	coaxial transfer switch (also referred to as CXS)
CTT	compatibility test trailer
CTX	Context Camera
CW	continuous wave
CXS	coaxial switch (also referred to as CTS)
D-	descent
D/A	digital to analog
DAC	digital-to-analog converter
DACS	data acquisition and command subsystem
DAN	Dynamic albedo of neutrons (instrument on MSL)

dB	decibel
dBc	dB below carrier
dB-Hz	decibel-hertz
dBi	decibel with respect to isotropic gain
dBic	decibel with respect to isotropic antenna
dBm	decibels referenced to milliwatts
DC	direct current; steady-state bias
DCC	downlink channel controller
DCD	data capture and display
DCO	digital control oscillator
DCPC	Downlink Channel Processing Cabinet
DCT	design control table
DCT	discrete cosine transform
DDUT	drop-dead uplink time
DEA	digital electronics assembly
DEA	descent engine assembly
DEC	Dual-Engine Centaur
deg	degree
delta-DOR	delta differential one-way ranging
DESCANSO	Deep Space Communications and Navigation Systems Center of Excellence
DFE	direct-from Earth
DGT	Deep Space Communications Complex (DSCC) Galileo Telemetry (system)
DI	Deep Impact (Project, spacecraft)
DIXI	Deep Impact Extended Investigation
DKF	Deep Space Network keyword files
DL, D/L	downlink
DLGA	descent LGA (low-gain antenna on MSL)
DM	data mover
DMC	DSCC meteorological computer
DMD	data monitor and display
DN	data number
DOE	Department of Energy
DOFF	degrees off (boresight)

DOM	distributed object manager
DOR	differential one-way ranging
DOY	day of year
DPDU	digital power distribution unit
DPM	digital processing module (in Small Deep Space Transponder)
DPT	data priority table
DR	Discrepancy Report
DRCF	data-rate capability file
Drv	drive
DS	descent stage
DS1	Deep Space 1 (Project, spacecraft)
DSCC	Deep Space Communications Complex
DSDST	descent stage small deep space transponder
DSMS	Deep Space Mission Systems
DSN	Deep Space Network
DSS	Deep Space Station
DSS-14	70-m Deep Space Station at Goldstone, California
DSS-15	34-m Deep Space Station at Goldstone, California
DSS 19	Very Large Array (San Agustin, near Socorro, New Mexico
DSS-25	34-m Deep Space Station at Goldstone, California
DSS-26	34-m Deep Space Station at Goldstone, California
DSS-34	34-m Deep Space Station at Canberra, Australia
DSS-42	26-m Deep Space Station at Canberra, Australia
DSS-43	70-m Deep Space Station at Canberra, Australia
DSS-45	34-m Deep Space Station at Canberra, Australia
DSS-49	Parkes 64-m radio telescope (Parkes, Australia)
DSS-55	34-m Deep Space Station at Madrid, Spain
DSS-63	70-m Deep Space Station at Madrid, Spain
DSS-65	34-m Deep Space Station at Madrid, Spain
DTE	direct-to Earth
DTF	DSN Test Facility
DTR	digital tape recorder (on GLL)
DTT	Downlink Telemetry and Tracking Subsystem

DTTL	data transition tracking loop
DTV	digital television
D-UCTS	descent UHF (ultrahigh frequency) coaxial transfer switch (on MSL spacecraft)
DUHF	descent UHF (ultrahigh frequency) antenna
DVCF	data-volume capability file
DX	diplexer
EADS	Casa Espacio, European Aeronautic Defense and Space Company (MSL HGA vendor)
EAS	engineering analysis subsystem
EBM	entry balance mass
E_b	energy per command bit
E_b/N_0	bit-energy-to-noise spectral-density ratio
EAS	engineering analysis subsystem
EBM	entry balance mass
EDA	EDL data analysis
EDL	entry, descent, and landing
EFPA	entry flight path angle
EH&A	engineering health (or housekeeping) and accountability (rate)
EH&A	engineering housekeeping, and analysis
EIRP	effective isotropic radiated power
ELT	Electra Lite transponder (on MSL and on future rovers and landers)
EM	electromagnetic
EMC	electromagnetic compatibility
EMI	electromagnetic interference
Eng/HK	Engineering/housekeeping
EOL	end of life
EOM	end of mission
EPC	electronic power converter
EPOCh	Extrasolar Planet Observations and Characterization
EPOXI	(combination of two DI Project mission names) Extrasolar Planet Observations and Characterization (EPOCh), and Deep Impact Extended Investigation (DIXI)
EPS	electrical power system

ERT	Earth-received time
E_s	energy per telemetry symbol
E_s/N_0	symbol energy-to-noise spectral-density ratio
ESA	European Space Agency
EST	Eastern Standard Time
ESTEC	European Space Administration Research and Technology Centre
ETX	exciter transmitter
EU	engineering unit
EUT	Electra UHF (ultrahigh frequency) transceiver
eV	electron volt
EVR	Event Report
Ex	exciter (provides the radio frequency drive to the transmitter)
EXC	exciter
F1	fundamental frequency of the small deep space transponder (SDST)
F66	frequency input to the sequential ranging assembly
FA	flight acceptance
FA	flight allowable
FCD	feedback concatenated decoder
FDS	flight data subsystem
FET	field effect transistor
FEM	flight-engineering model
FFA	front-end filter assembly
FFT	fast Fourier transform
FPA	fault-protection algorithm
FPGA	field programmable gate array
FRO	frequency reference offset
FS	Flight System
FSC	full-spectrum combiner
FSK	frequency-shift keying
FSP	full-spectrum processing subsystem
FSR	full-spectrum recorder
FSS	Frame Synchronizer System (subsystem in Voyager)

FSS	frequency selective surface (on Voyager S-/X-band antenna (SXA)
FSU	filtering and switch unit
FSW	flight software
ft	transmitted frequency from sequential ranging assembly
FTE	full-time equivalent
FTS	frequency and timing system
FY	fiscal year
G	ground
GaAs	gallium arsenide
Gb	gigabit
GB	gigabyte
GCN	Ground Communications Network
GDS	Ground Data System
GEM	Galileo Europa Mission
GDS	Goldstone (Deep Space Communications Complex)
GDSCC	Goldstone Deep Space Communications Complex (DSCC)
GFE	Government furnished equipment
GHz	gigahertz
GLL	Galileo (Project, spacecraft)
GMM	Galileo Millennium Mission
GMSK	Gaussian-filtered minimum-shift keying
GNC	guidance, navigation, and control
GRACE	Gravity Recovery and Climate Experiment
GRASP	General Reflector Antenna Scatter Program
GSE	ground support equipment
GS&E	general science and engineering
GSFC	Goddard Space Flight Center
GSOC	German Space Operations Center
G/T	ratio of antenna gain to noise temperature
GTP	generalized telecom predictor
GUI	graphical user interface
GWU	George Washington University

H	hydrogen
Hazcam	hazard camera (on MER rover)
HCD	hardware command decoder
HDO	half-duplex overlay
HDTV	high definition television
HEF	high efficiency (antenna)
HEMT	high electron mobility transistor
HGA	high-gain antenna
HGAG	HGA (high-gain antenna) gimbal
HGAS	high-gain antenna system
hh:mm	hours:minutes
hi-fi	high fidelity
HiRISE	High Resolution Imaging Science Experiment (instrument on MRO)
HK	housekeeping telemetry
HLAN	high-speed LAN (local area network)
HP	high power
HPCW	high-priority comm window
HRS	heat rejection system (on MER and MSL spacecraft)
HSD	high-speed data
HVPS	high-voltage power supply
Hz	hertz
IBEX	Interstellar Boundary Explorer (spacecraft)
ICA	IF (interface) channel assembly
ICD	interface control document
ICT	integer cosine transform
ID	identification
IDC	image data compression
IDC	intermediate frequency to digital converter
IDD	(MER) instrument deployment device
IDS	IPS (ion-propulsion system) diagnostic sensors
IEM	integrated-electronics module
IF	intermediate frequency
IMP	Interplanetary Monitoring Platform (spacecraft)

IMU	inertial-measurement unit
IND	Interplanetary Network Directorate (now IPN-ISD)
IPN-ISD	InterPlanetary Network and Information Systems Directorate (formerly TDA, TMOD, and IND
IPP	inter pulse period
IPS	ion-propulsion system
IR	improved ranging
IR	infrared
IRIS	infrared interferometer spectrometer
IS	isolator
ISA	Incident, Surprise, Anomaly (report)
ISA	interstage adapter
ISO	isolator (at traveling wave tube amplifier output)
ISS	imaging science subsystem
ITE	impact through egress
IUS	Inertial Upper Stage
JOI	Jupiter orbit insertion
JPL	Jet Propulsion Laboratory
JSX	Jupiter Saturn Explorer (previous name for Voyager 2 spacecraft)
K	kelvin
Ka-band	frequencies in the range 26.5–40 GHz
KaPA	Ka-band power amplifier
KaTS	Ka-band Translator System
kb	kilobit (note: telecommunications uses bits [b]; data storage uses bytes [B], which are 8 bits per byte)
kbps	thousand(s) of bits per second
kg	kilogram
KHA	Ka-band horn antenna
km	kilometer
km/hr	kilometers/per hour
km^2/s^2	kilometers squared per second squared
kN	kilonewton
kW	kilowatt

LAN	local-area network
L-band	frequency range between 390 and 1550 MHz
LC	inductance capacitance filter
LCCD	level clock conversion distribution (interface)
LCP	left-circular polarization; left circularly polarized
LECP	low-energy charged particle
LEM	lander electronics module (MER lander)
LGA	low-gain antenna
LGAX	LGA (low-gain antenna) directed along the +x-axis
LGAZ	LGA (low-gain antenna) directed along the +z-axis
LGAZ–LGA	LGA (low-gain antenna) directed along the –z-axis
LH_2	liquid hydrogen
LH	left hand
LHCP	left-hand circular polarized (or polarization)
LMA	Lockheed Martin Aerospace
LO_2	liquid oxygen
$\log_{10}$	logarithm base ten
LOS	loss of signal
Lox	liquid oxygen
LP	low power
LMC	link monitor and control
LNA	low-noise amplifier
LPA	lander petal actuator (MER)
LPE	low-power electronics
LPF	low-pass filter
LPSA	lander pyro switching assembly
LST	local solar time
LTST	local true solar time
LV	launch vehicle
LVDS	low-voltage differential signaling
m	meter
MAD	Madrid DSCC (Deep Space Communications Complex)
MAG	magnetometer
MAHLI	Mars hand lens imager (on MSL)

MAQ	Magellan acquisition
MARCI	Mars Color Imager (on MRO)
MARDI	Mars descent imager (on MSL)
Mastcam	(multi-spectral, stereo imaging and video camera on MSL mast)
Mbit	megabit (note: telecommunications uses bits [b]; data storage uses bytes [B], which are 8 bits per byte)
Mbps	megabit per second
MBR	Mars Balloon Relay (protocol used on MGS)
MCD	maximum-likelihood convolutional decoder
MCD/FS	maximum-likelihood convolutional decoder frame synchronizer (subsystem)
MCIC	motor controller interface card
MCIF	motor controller interface
MCS	Mars Climate Sounder (on MRO)
MDA	metric data assembly
mdeg	millidegree
MDNS	mission design and navigation subsystem
MDS	modulation-demodulation subsystem
MDSCC	Madrid Deep Space Control Center (DSCC)
MECO	main engine cutoff
MEDLI	MSL EDL instrumentation
MEP	Mars Exploration Program
MER	Mars Exploration Rover (Project, spacecraft, or rover)
MER-A	Mars Exploration Rover "Spirit"
MER-B	Mars Exploration Rover "Opportunity"
MES	main engine start
MEX	Mars Express (Project, spacecraft)
MFS	Multi-Functional Structure (experiment)
MFSK, M-FSK	multiple frequency shift keying (tones)
MGA	medium-gain antenna
Mgate	million gates
MGDS	Multi-mission Ground Data System
MGS	Mars Global Surveyor (spacecraft)
MHz	megahertz

MIC	microwave integrated circuit
MICAS	miniature integrated-camera spectrometer
MIL-71	name of Deep Space Network station at Cape Canaveral (not an acronym)
MIL-STD	Military Standard [usually followed by a document number]
MIT	Massachusetts Institute of Technology
MLE	Mars landing engine (on MSL rover)
MM	multi mission
MOC	Mars Orbiter Camera (on MGS)
MMO	(JPL) Mission Management Office
MMRTG	multimission radioisotope thermoelectric generator
MOD	modulation
MOI	Mars orbit insertion
MOLA	Mars Observer laser altimeter (on MGS)
MON	monitor system
MOS	mission operations system
MP	modem processor
MPa	megapascal
MPCS	mission data processing and control subsystem
MPF	Mars Pathfinder (spacecraft, rover)
MPST	Mission Planning and Sequencing Team
mrad	milliradian
MREU	MSL remote electronics unit (also called REU)
MRO	Mars Reconnaissance Orbiter (Project, spacecraft)
MRO	memory readout
ms	millisecond
m/s	meter(s) per second
MS	master of science
MSA	mission-support area
MSAP	multi mission system architecture platform
MSAT	Mobile Satellite
MSL	Mars Science Laboratory (Project, spacecraft, or rover)
MSLICE	MSL InterfaCE (a surface operations visualization and planning tool)
MSPA	multiple spacecraft per aperture

MSR	Mars Sample Return (Project)
Msps	megasymbols per second
MSSS	Malin Space Science Systems
MST	monitor sample time
MTC	Mars Time coordinated
MTIF	MSAP telecommunications interface board
MUKOW	MRO uplink keep out window
mW	milliwatt
N	newton
Na	sodium
NAIF	Navigation and Ancillary Information Facility
NASA	National Aeronautics and Space Administration
NAV	navigation
NCO	numerically controlled oscillator
NEAR	Near Earth Asteroid Rendezvous (spacecraft)
NEN	Near Earth Network
net	network
NISN	NASA Integrated Services Network
NMC	Network Monitor and Control
NMP	New Millennium Program
NOCC RT	Network Operations Control Center real time (System)
NOP	network operations plan
NPO	NPO Energomash (Russian manufacturer)
NPP	Network Planning and Preparation system
nrad	nanoradians
NRAO	National Radio Astronomy Observatory
NRZ	non-return to zero
ns	nanosecond
NSP	Network Simplification Project
NVM	nonvolatile memory
NVM	nonvolatile memory/camera
NVM/CAM	nonvolatile memory/camera

OCM organic check material
OD orbit determination
ODB operational database
ODF orbit data file
ODY Odyssey (Project, spacecraft)
ONC Optical Navigation Camera Experiment (MRO)
ORSC orbiter relay state change
Osc oscillator
OWLT one-way light time

p parachute (on MER spacecraft)
P- parachute (cone) (on MSL spacecraft)
PASM power actuation and switching module
P_c carrier power
PCM pulse code modulation
P_c/N_0 carrier power to noise-spectral-density ratio
PD passive device (in solid state power amplifier, a microwave coupler or combiner)
PDF Portable Document Format
pdf probability density function
P_d/N_0 data power-to-noise spectral-density ratio
PDS Planetary Data System (label)
PDS Parachute Deceleration System (on MER aeroshell)
PDU protocol data unit
PEDL pre-entry, descent, and landing
PEPE Plasma Experiment for Planetary Exploration (on DS1)
P-file prediction file
P/FR Problem/Failure Report
PhD doctor of philosophy
PI principal investigator
PLAR Post-Launch Assessment Review
PLF payload fairing
PLGA Parachute Low-Gain Antenna (MSL, chapter 8)
PLGA Petal Low-Gain Antenna (MER, chapter 7)
PLL phase-locked loop

PLS	plasma science
PM	post meridian
PMA	Pancam Mast Assembly (on MER)
PMP	payload mounting module
PN	pseudonoise
Pol	polarizer
POR	power-on reset
PPI	Planetary Plasma Interactions (part of Planetary Data System)
ppm	parts per million
PPS	photopolarimeter system
P_r	downlink ranging power
PRA	planetary radio astronomy
P_r/N_0	ranging power-to-noise spectral-density ratio
Prox-1	Proximity-1 protocol
PR/TSA	project requirements/TMOD [Telecommunications and Mission Operations Directorate (now called IPN-ISD)] support agreement
psf	pounds per square foot
PSM	power supply module (MRO)
PSS	parachute support structure (on MSL spacecraft)
P_t/N_0	quantity downlink
P_t/N_0	total power-to-noise spectral-density ratio
P_{tone}	power in differential one-way ranging (DOR) tone
PUHF	parachute UHF (ultrahigh frequency) antenna (on MSL spacecraft)
PWS	plasma wave spectrometer (Galileo)
PWS	plasma wave system (on voyager)
QPSK	quadrature phase-shift keying
R-	rover
RA	radar altimeter (on MER lander)
RAAR	relay antenna assembly receive
RAAT	relay antenna assembly transmit
rad	radian (57.3 degrees)

RAD	radiation assessment detector (instrument on MSL rover)
RAD	Rocket-Assisted Deceleration (system in MER)
RAMP	rover avionics mounting plate
RAT	rock abrasion tool (on MER)
RAX	Remote-Agent Experiment (of DS1 Project)
RC	radar controller
RC	request command
RC	radar controller
RCC	receiver control computer
RCE	rover computer element (on MSL)
RCP	receiver channel processor
RCP	right-circular polarization; right-circularly polarized
RCS	reaction control system
RCVR	receiver
RED	Rover Equipment Deck (MER)
Reff	effective information rate
REM	rover electronics module (on MER)
REMS	Rover environmental monitoring station (instrument on MSL)
REU	remote electronics unit (on MSL rover)
RF	radio frequency
RFM	(ultrahigh frequency, UHF) radio frequency module (MRO)
RFPDU	radio frequency power distribution unit (in terminal descent sensor of MSL spacecraft)
RFS	radio frequency subsystem
RFSTLC	radio frequency subsystem tracking-loop capacitor
RH	right hand
RHCP	right-hand circular polarized (or polarization)
RID	radio-frequency to intermediate-frequency downconverter
$\mathbf{R_J}$	Jupiter radii
RLAN	receiver local area network (LAN)
RLGA	Rover low-gain antenna (on MER and on MSL rover)
RM	ranging maximum
rng	ranging
RNS	Reliable Network Service

RP	radar processor (on MSL spacecraft)
RP	rocket propellant or refined petroleum (kerosene)
RPAM	rover power and analog module
RPDU	RF (radio frequency) power distribution unit (in TDS)
RPFA	rover pyro fire assembly (on MSL)
rpm	revolutions per minute
RRA	relay receiving [or radio] antenna (on GLL orbiter)
RRH	relay radio hardware (on GLL orbiter or probe)
RRH	relay receiving hardware (on GLL orbiter)
RRP	receiver and ranging processor
RS	Reed-Solomon (code)
RSDL	RS (Reed-Solomon) downlink
RSDS	Raw Science Data Server
RSDST	Rover Small Deep Space Transponder (on MSL)
RSM	remote sensing mast (on MSL rover)
RSR	radio science receiver
RT	remote terminal
R/T	receive/transmit
RTG	radioisotope thermoelectric generator
RTLT	round-trip light time
RTN	return
ru	range unit
R-UCTS	rover UHF (ultrahigh frequency) coax transfer switch (on MSL)
RUHF	rover UHF (ultrahigh frequency) antenna (on MSL)
R-WTS	rover waveguide transfer switch (on MSL)
Rx	receive
SA	sample acquisition
SAF	station allocation file
SAIC	Science Applications International Corporation
S&L	standards and limits
SAM	sample analysis at Mars (instrument on MSL rovers)
SAMPEX	Solar Anomalous and Magnetospheric Particle EXplorer (spacecraft)

SA-SPaH	sample acquisition-sample preparation and handling
S-band	radio frequencies between 2 and 4 GHz (2–2.5 in deep space communications)
S/C	spacecraft
SCaN	Space Communications and Navigation (Office)
SCARLET	Solar Concentrator Array Using Refractive Linear Element Technology (Deep Space 1)
SCET	spacecraft event time
SCID	spacecraft identifier
SCMF	spacecraft command message file
SDR	software defined radio
SDST	Small Deep-Space Transponder
SEC	Single-Engine Centaur
SEC	Sun–Earth–Craft (angle)
S–E–C	Sun–Earth–Craft (angle)
SEP	separation
SEP	solar-electric propulsion (technology)
S–E–P	Sun-Earth-probe (angle)
SeqGen	sequence generation (spacecraft activity planning software)
S-EXC	S-band exciter
SFOC	Space Flight Operations Center
SFOS	spaceflight operations schedule
SHARAD	Shallow (Subsurface) Radar (MRO)
sigma	spelled out form of Greek character, σ (for 1 standard deviation)
SIM	Space Interferometry Mission (proposed space radio telescope; the program was cancelled in 2010)
SIT	select in test
SLC-17A	Space Launch Complex 17A
SLE	Space Link Extension
SMAP	Soil Moisture Active & Passive Mapping (satellite)
SMM	spacecraft mode manager
SNR	signal-to-noise ratio
SNT	system-noise temperature
SOAS	science operations analysis subsystem

SOC	state of charge
SOE	sequence of events
sol	Martian day
SOWG	Science Operations Working Group
SP	service package
SPARC	scalable processor architecture
SPC	Signal Processing Center
SPD	S-band polarization diversity
SPE	static phase error
SPE	Sun–Probe–Earth (angle)
SPIE	Society of Photo-Optical Instrumentation Engineers
sps	symbols per second
SR	sweep range
SRA	sequential ranging assembly
SRB	solid rocket booster
SRBJ	solid rocket booster jettison
S-RCVR	S-band receiver
SRM	solid rocket motor
SRU	stellar-reference unit (on DS1)
SSA	solid state amplifier (on VGR)
SSA	Sun-sensor assembly (on DS1)
SSD	symbol stream distribution
SSI	solid state imaging (on GLL)
SSK	soft symbol controller
SSNR	symbol signal-to-noise ratio
SSPA	solid state power amplifier
SSR	solid-state recorder (on GLL)
STDN	NASA Spaceflight Tracking and Data Network
STE	system test equipment
ST/N_0	ratio to product of power in data sidebands (S) and bit duration (T) to noise spectral density; equivalent to E_b/N_0
STS	Space Transportation System
STS-34	Space Transpiration System 34 (Shuttle Atlantis)
S-TWTA	S-band traveling-wave tube amplifier (the S-band traveling wave tube and its power supply)

SUFR	straighten up and fly right (MSL entry, descent and landing phase)
SUPAERO	École nationale supérieure de l'aéronautique et de l'espace
SXA	S- and X-band antenna (on VGR)
sync	synchronization
TBOT	telecom at beginning of tracks
TC	transmit command
TCA	telemetry channel assembly (Voyager)
TCM	trajectory correction maneuver
T_{cyc}	cycle time
TDA	Telecommunications and Data Acquisition (previous name for Deep Space Network)
TDDS	Tracking Data Delivery Subsystem
TDL	telecom development lab
TDM	time-division multiplexing
TDRS	Tracking and Data Relay Satellite
TDRSS	Tracking and Data Relay Satellite System
TDS	terminal descent sensor (landing radar on MSL spacecraft)
TDSA	TDS (terminal descent sensor) antenna array
TDSD	TDS (terminal descent sensor) digital stack
TDSR	TDS (terminal descent sensor) radio frequency stack
TEMIC	TEMIC Telefunken (decoder vendor)
TEOT	telecom at end of tracks
TFP	telecom forecaster predictor
TFREQ	transmit frequency (Deep Space Network)
TIP	target interface point
TIRS	Transverse Impulse Rocket System (in MER)
TLGA	tilted low-gain antenna (on MSL spacecraft)
tlm	telemetry
TLM	telemetry system
TLMGEN	telemetry-rate generator
TLP	telemetry processor
TMOD	Telecommunications and Mission Operations Directorate (now called IPN-ISD)
TMU	telemetry modulation unit

TOAST	Telecom Orbit Analysis and Simulation Tool
TPAP	telecom prediction and analysis program
T/R	transmit/receive; transmitter/receiver
TRK	tracking system
TRM	transmit/receive module
TSB	telecom support board (on MER)
TTE	turn to entry
TTS	test and telemetry system
TW	truth window
T/W	ratio of thrust and weight on pad
TWNC	two-way noncoherent (mode, TWNC-on and TWNC-off)
TWT	traveling-wave tube
TWTA	traveling-wave tube amplifier (the TWT plus its amplifier)
Tx	transmit (or transmitter)
UA	University of Arizona
UCTS	UHF (ultrahigh frequency) coax transfer switch (on MSL, rover and spacecraft)
UDIM	up/down IF (intermediate frequency) module
UDMM	up down MIC [microwave integrated circuit] module (on terminal descent sensor of MSL)
UHF	ultra-high frequency; radio frequencies between 0.3 and 3 GHz (about 0.4 GHz for Mars relay communications)
UL, U/L	uplink
ULDL	uplink–downlink
UPA	uplink processor assembly
UPL	Uplink Subsystem
URA	uplink ranging assembly
USA	United States of America
USN	Universal Space Network
USO	ultrastable oscillator
UTC	Universal Time Coordinated (also known as Greenwich Mean Time)
UV	ultraviolet
UVS	ultraviolet spectrometer (on VGR)

VCO	voltage-controlled oscillator
VCXO	voltage-controlled crystal oscillator (list only one)
VDC	volts direct current
VEEGA	Venus-Earth-Earth gravity assist
VGA	variable gain amplifier
VGR	Voyager
VIM	Voyager Interstellar Mission
VLA	Very Large Array (San Agustin, near Socorro, New Mexico)
VLBI	very long baseline interferometry
VSR	very long baseline interferometry (VLBI) science receiver
VSWR	voltage standing wave ratio
W	watt
W2CX	waveguide to coaxial
WEB	warm electronics box (MER rover)
WG	waveguide
Whr	watt-hour
WIND	full name (not an acronym) of a spacecraft studying near-Earth solar wind
WIPL-D	Wires, Plates, Dielectrics (a commercial high frequency electromagnetic modeling software package)
WTS	waveguide transfer switch
X-band	radio frequencies between 7 and 11 GHz (7–9 GHz in deep space communications)
X-EXC	X-band exciter
XFMR	transformer
XPA	X-band power amplifier
XSDC	X- to S-band downconverter (GLL)
X-TWTA	X-band traveling-wave tube amplifier (the X-band traveling-wave tube and its power supply)
XTR	X-band transmit receive

About the Companion Website

This book is accompanied by a companion website:

www.wiley.com/go/deepspacecommunications

The website includes color images and higher resolution versions of the following figures:

Chapter 2
Figure 2-6
Figure 2-7

Chapter 3
Figure 3-6

Chapter 4
Figure 4-1
Figure 4-2
Figure 4-5
Figure 4-6
Figure 4-7

Chapter 5
Figure 5-1
Figure 5-3
Figure 5-4
Figure 5-5
Figure 5-8
Figure 5-14
Figure 5-15
Figure 5-16

Chapter 6
Figure 6-2
Figure 6-7
Figure 6-13
Figure 6-14
Figure 6-15
Figure 6-19
Figure 6-20
Figure 6-22
Figure 6-23

Chapter 7
Figure 7-2
Figure 7-3
Figure 7-4
Figure 7-4
Figure 7-6
Figure 7-7
Figure 7-8
Figure 7-9
Figure 7-10
Figure 7-11
Figure 7-13
Figure 7-16
Figure 7-19
Figure 7-23
Figure 7-24
Figure 7-25
Figure 7-45
Figure 7-48
Figure 7-49
Figure 7-50
Figure 7-51
Figure 7-52

Chapter 8
Figure 8-3

Deep Space Communications, First Edition. Edited by Jim Taylor.

Figure 8-4
Figure 8-5
Figure 8-6
Figure 8-8
Figure 8-9
Figure 8-10
Figure 8-11
Figure 8-12
Figure 8-16
Figure 8-17
Figure 8-18
Figure 8-20
Figure 8-29
Figure 8-31
Figure 8-40
Figure 8-41
Figure 8-45
Figure 8-49
Figure 8-50
Figure 8-51
Figure 8-52
Figure 8-53
Figure 8-55
Figure 8-56
Figure 8-57
Figure 8-58
Figure 8-59
Figure 8-60
Figure 8-62
Figure 8-63
Figure 8-67
Figure 8-69
Figure 8-74

Index

Page numbers in *italics* refer to figures and tables.
Page numbers followed by *n* indicate that the word is in a footnote.

Printed and bound by CPI Group (UK) Ltd, Croydon, CR0 4YY
21/09/2026
14965814-0002